BIOLOGY OF FIBROBLAST

ORGANIZATION COMMITTEE

Sven Gardell   K. I. Kivirikko
E. Kulonen   T. C. Laurent
         J. Pikkarainen

BIOLOGY OF FIBROBLAST

Fourth Sigrid Jusélius Foundation Symposium
Turku, Finland, August 1972

Edited by

E. KULONEN

Department of Medical Chemistry
University of Turku, Finland

and

J. PIKKARAINEN

Central Public Health Laboratory,
Helsinki, Finland

ACADEMIC PRESS . LONDON AND NEW YORK
A Subsidiary of Harcourt Brace Jovanovich, Publishers

ACADEMIC PRESS INC. (LONDON) LTD
24–28 Oval Road
London NW1

U.S. Edition published by
ACADEMIC PRESS INC.
111 Fifth Avenue
New York, New York 10003

Copyright © 1973 by ACADEMIC PRESS INC. (LONDON) LTD

All Rights Reserved
No part of this book may be reproduced in any form by photostat, microfilm, or any other means, without written permission from the publishers

Library of Congress Catalog Card Number: 73–9467
ISBN: 0–12–428950–9

Text set in 11/12 pt. Monotype Baskerville, printed by letterpress, and bound in Great Britain at The Pitman Press, Bath

LIST OF CONTRIBUTORS

(Full addresses are included in the first page of each paper)

AHONEN, J., Helsinki, Finland
ALLEN, R., San Francisco, Calif., U.S.A.
AULA, P., Helsinki, Finland
AUTIO, S., Helsinki, Finland
BAILEY, A. J., Langford, England
BALAZS, E. A., Boston, Mass., U.S.A.
BARD, J., Edinburgh, Scotland
BARKER, E., Tucson, Ariz., U.S.A.
BATES, C. J., Cambridge, England
BAZIN, S., Garches, France
BERG, R. A., New Brunswick, N.J., U.S.A.
BJÖRKERUD, S., Göteborg, Sweden
BLENNOW, G., Lund, Sweden
BLOUGH, H. A., Philadelphia, Pa., U.S.A.
BORNSTEIN, P., Seattle, Wash., U.S.A.
BRUNK, U., Uppsala, Sweden
BUCK, C. A., Manhattan, Kan., U.S.A.
BYERS, P. H., Bethesda, Md. U.S.A.
CARLSSON, S. A., Stockholm, Sweden
CASTOR, C. W., Ann Arbor, Mich., U.S.A.
CIFONELLI, J. A., Chicago, Ju. U.S.A.
DARZYNKIEWICZ, Z., Boston, Mass., U.S.A.
DEHM, P., New Brunswick, N.J., U.S.A.
DELAUNAY, A., Garches, France
DORFMAN, A., Chicago, Ill., U.S.A.
DORLING, J., Maidenhead, Berks, England
EGE, T., Stockholm, Sweden
EHRLICH, H. P., Seattle, Wash., U.S.A.
ELSDALE, T., Edinburgh, Scotland
FRANSSON, L. A., Uppsala, Sweden
FUHRER, J. P., Philadelphia, Pa, U.S.A.
GABBIANI, G., Geneva, Switzerland
GLINOS, A. D., Washington, D.C., U.S.A.

GRAHAM, J. M., London, England
GRANT, M. E., Manchester, England
GRASEDYCK, K., Hamburg, Germany
HASTINGS, J. C., Tucson, Ariz., U.S.A.
HEPPLESTON, A. G., Newcastle-upon-Tyne, England
HOBZA, P., Prague, Czechoslovakia
HOLTZER, H., Philadelphia, U.S.A.
HÖÖK, M., Uppsala, Sweden
HOVI, T., Helsinki, Finland
HUNT, T., San Francisco, Calif., U.S.A.
HURYCH, J., Prague, Czechoslovakia
IVASKA, K., Turku, Finland
JACKSON, D. S., Manchester, England
JUNQUA, S., Créteil, France
KENT, P. W., Durham, England
KITTRIDGE, D., San Francisco, Calif., U.S.A.
KIVIRIKKO, K. I., Oulu, Finland
KIVISAARI, J., Turku, Finland
KULONEN, E., Turku, Finland
LAIPIO, M. L., Helsinki, Finland
LANGNESS, U., Kiel, Germany
LAPIÈRE, C. M., Liège, Belgium
LENAERS, A., Liège, Belgium
LEVENE, C. I., Cambridge, England
LEVITT, D., Chicago, Ill., U.S.A.
LINDAHL, U., Uppsala, Sweden
LINDNER, J., Hamburg, Germany
LORENZEN, I., Copenhagen, Denmark
MACEK, M., Prague, Czechoslovakia
McGOODWIN, E. B., Bethesda, Md., U.S.A.
MÄENPÄÄ, P. K., Helsinki, Finland
MAJNO, G., Geneva, Switzerland
MALMSTRÖM, A., Uppsala, Sweden
MARTIN, G. R., Bethesda, Md., U.S.A.
MATALON, R., Chicago, Ju., U.S.A.
MAYNE, R., Philadelphia, Pa., U.S.A.

MOCZAR, M., Créteil, France
MORA, P. T., Bethesda, Md., U.S.A.
MUIR, H., London, England
MÜLLER, P. K., Munich, Germany
NÄNTÖ, V., Helsinki, Finland
NICOLETIS, C., Garches, France
NIINIKOSKI, J., Turku, Finland
NORDLING, S., Helsinki, Finland
OGSTON, A. G., Oxford, England
OLSEN, B. R., New Brunswick, N.J., U.S.A.
PARTRIDGE, S. M., Langford, England
PASTERNAK, C. A., Oxford, England
PERTOFT, H., Uppsala, Sweden
PIERARD, G., Liège, Belgium
PIKKARAINEN, J., Helsinki, Finland
POHJANPELTO, P., Helsinki, Finland
PONTÉN, J., Uppsala, Sweden
PRESTON, B. N., Clayton, Victoria, Australia
PROCKOP, D. J., New Brunswick, N.J., U.S.A.
RAEKALLIO, J., Turku, Finland
RENCOVÁ, J., Prague, Czechoslovakia
RINGERTZ, N. R., Stockholm, Sweden
RITCHIE, J. C., Ann Arbor, Mich., U.S.A.
ROBERT, A. M., Créteil, France
ROBERT, B., Créteil, France
ROBERT, L., Créteil, France
ROBINS, S. P., Langford, England
ROSS, R., Seattle, Wash., U.S.A.
RUOSLAHTI, E., Helsinki, Finland
RYAN, G. B., Geneva, Switzerland
SCHILTZ, J. R., Philadelphia, Pa., U.S.A.
SCOTT, J. E., Maidenhead, Berks. England
SCOTT, M. E., Ann Arbor, Mich., U.S.A.
SILVER, I. A., Bristol, England
SJÖBERG, I., Lund, Sweden
SMETANA, K., Prague, Czechoslovakia
SMITH, B. D., Bethesda, Md, U.S.A.
SMITH, S. F., Ann Arbor, Mich., U.S.A.
SNOWDEN, J. M., Clayton, Victoria, Australia
SPECTOR, W. G., London, England
STRAŽIŠČAR, Š., Ljubljana, Yugoslavia
STRUDEL, G., Nogent-sur-Marne, France
THOMPSON, J. N., Chicago, JU., U.S.A.
TURK, V., Ljubljana, Yugoslavia
UDENFRIEND, S., Nutley, N.J., U.S.A.
UITTO, J., New Brunswick, N. J., U.S.A.
VAHERI, A., Helsinki, Finland
VAINIO, K., Heinola, Finland
VASTAMÄKI, M., Turku, Finland
WARREN, L., Philadelphia, U.S.A.
WASTESON, A., Uppsala, Sweden
WEINSTEIN, D. W., Philadelphia, Pa., U.S.A.
WESTERMARK, B., Uppsala, Sweden
WIEBKIN, O. W., London, England
WINKLE JR., W. VAN, Tucson, Ariz., U.S.A.
WINTER, G. D., Stanmore, Middx, England
WOHLAND, W., Tucson, Ariz., U.S.A.
ZAHRADNIK, R., Prague, Czechoslovakia

THE SIGRID JUSÉLIUS FOUNDATION

Aleksanterinkatu 48B, Alexandersgatan
Helsinki 10, Helsingfors

BOARD OF TRUSTEES
Ordinary Members

C-E. Olin
Ph.D., Chairman

L. E. Taxell
Professor, Vice-Chairman

Bertel von Bonsdorff
Professor

Hugo E. Pipping
Professor

Filip Pettersson
Banker

Bo Palmgren
Professor

Jorma Pätiälä
Professor

Deputy Members

Nils Oker-Blom
Professor

Lars Sjöblom
Professor

Harald Brunou
Former Member of the Supreme Court

Niilo Hallman
Professor

James Ek
Physician-in-Chief, Pori

Henrik Nybergh
Banker

Eero Schroderus
Banker

ADVISORY MEDICAL COMMITTEE

Bertel von Bonsdorff
Professor, Chairman

James Ek
Physician-in-Chief, Pori

Osmo Järvi
Professor

Niilo Hallman
Professor

Anders Langenskiöld
Professor

Nils Oker-Blom
Professor, Secretary

DIRECTOR
Marcus Nykopp
ML

FOREWORD

The Sigrid Jusélius Foundation was created by the late Mr. Fritz Arthur Jusélius in memory of his daughter, Sigrid, who died in 1889, when only eleven years old. F. A. Jusélius, who was born in Pori in 1855 and died there at the age of seventy-five, was a successful businessman and a colourful personality. From a comparatively modest beginning—a small family shop, selling groceries and household goods—he developed a flourishing timber export business and built up a fortune which, however modest by international standards, ultimately enabled him to establish the biggest trust for public welfare in Finland.

Soon after the death of his beloved daughter Sigrid he had begun to contemplate a lasting monument to her memory. Having thought of many different possibilities the idea was finally realised in his last will and testament dated May 3rd 1927, although it had existed in draft form since 1916. In this document he surrendered his entire fortune "for the purpose of supporting and promoting medical research—independently of language and nationality —to fight those diseases which are particularly harmful to mankind". To this effect he created the Sigrid Jusélius Foundation.

The Sigrid Jusélius Foundation is administered by a Board of Trustees, composed of seven ordinary members and the same number of deputy members, all of them appointed for periods of five years.

The Board of Trustees is assisted by an Advisory Medical Committee. Grants from the Foundation are finally allocated after consultation with the two medical societies of Finland, Suomalainen Lääkäriseura Duodecim and Finska Läkaresällskapet.

In 1948 the Foundation was able to start functioning for the purpose set out in the testament and it has since then awarded grants for medical research to a total of Finnish Marks 17,967,694, equivalent to US $4.3 million. The total number of recipients to date is 1,892. It may thus be said that most of the post war medical research in Finland has been supported by the Sigrid Jusélius Foundation.

The desire and hope of Fritz Arthur Jusélius, however, was to promote international scientific contacts. The Foundation has attempted to realize these hopes in two ways: recently some grants have, on application from Finnish grantees, been allocated to foreign scientists to enable them to work in Finnish institutions and hospitals. Experience has so far been restricted but if this form of activity turns out to be fruitful in improving international scientific contacts it may, in the future, be increased and continued, probably in collaboration with some corresponding foreign Foundations.

The second and more important effort so far has, however, been the arrangement of international symposia. The present symposium 'on the biology of the fibroblast' is the fourth in the series. The first Symposium. was held in 1965 on the "Control of Cellular Growth in Adult Organism". The theme of the second symposium in 1967 was the "Regulatory Functions of Biological Membranes" and that of the third Symposium in 1970, "Cell Interactions and Receptor Antibodies in Immune Response".

The intention has been to concentrate as much as possible on restricted and well defined problems which would benefit most from a multi-disciplinary approach. The present symposium on the biology of the fibroblast is a good example of this and it is hoped that this volume containing the work presented at the symposium will be of interest for those working with fibroblasts either on the cellular level or on the level of the organism.

Nils Oker-Blom

CONTENTS

List of Speakers vii
Foreword ix

I. General Papers

Reactivity of the connective tissue. *By Eino Kulonen* 3
The role of the extracellular space. *By A. G. Ogston* 9
Elastin biosynthesis and the influence of some extracellular factors of macromolecular organization. *By S. M. Partridge* 13
The generation and maintenance of parallel arrays in cultures of diploid fibroblasts. *By Tom Elsdale* 41

II. Cytological Aspects and Differentiation

Some overt and covert properties of chondrogenic cells. *By Richard Mayne, John R. Schiltz and Howard Holtzer* 61
The differentiation of cartilage. *By Daniel Levitt and Albert Dorfman* . 69
Relationship between the chick periaxial metachromatic extracellular material and vertebral chondrogenesis. *By Georges Strudel* . . 93
Studies, using sponge implants, on the mechanism of osteogenesis. *By George D. Winter* 103
The collagen protein synthesis in long-term cultures of human foetal peripheral blood. *By M. Macek, J. Hurych and K. Smetana* . . 127
The fibroblast as a contractile cell: The myofibroblast. *By Giulio Gabbiani, Guido Majno and Graeme B. Ryan* 139
Density dependent regulation of growth and differentiated function in suspension cultures of mouse fibroblasts. *By André D. Glinos* . . 155
Phases of experimental granulation tissue formation and nucleic acid metabolism. *By J. Ahonen, M. Vastamäki and Pekka H. Mäenpää* . 169
Mechanism for cellular ageing in long-term culture. *By Jan Pontén, Bengt Westermark and Ulf Brunk* 183

Nucleolus specific antigens in human fibroblasts and hybrid cells studied with patient autoantibodies. *By Nils R. Ringertz, Thorfinn Ege and Sten-Anders Carlsson* 189

Putrescine as a growth factor. *By Pirkko Pohjanpelto* 195

Determination of proteolytically most active cell population in experimental granuloma and purification of its cathepsins. *By Š. Stražiščar and V. Turk* 199

Isolation of cells from experimental granulation tissue. *By K. Ivaska and H. Pertoft* 205

The three-dimensional culturing of fibroblasts in collagen. *By Jonathan Bard and Tom Elsdale* 209

III. Extracellular Space and Cell Surface

Diffusion properties of model extracellular systems. *By B. N. Preston and J. McK. Snowden* 215

Factors affecting the biosynthesis of sulphated glycosaminoglycans by chondrocytes in short-term maintenance culture isolated from adult tissue. *By O. W. Wiebkin and Helen Muir* 231

The effect of hyaluronic acid on fibroblasts, mononuclear phagocytes and lymphocytes. *By E. A. Balazs and Z. Darzynkiewicz* . . . 237

Aggregation of feline lymphoma cells by hyaluronic acid. *By Bengt Westermark and Åke Wasteson* 253

The assembly of fibroblast plasma membranes. *By C. A. Pasternak and J. M. Graham* 261

Cell surface and initiation of proliferation. *By Antti Vaheri, Erkki Ruoslahti, Tapani Hovi and Stig Nordling* 267

The nature of the surface of normal and virus-transformed tissue culture cells. *By L. Warren, G. P. Fuhrer and C. A. Buck* . . . 273

Modification of glycoprotein biosynthesis in transformed mouse cell lines. *By Paul W. Kent and Peter T. Mora* 287

Effect of influenza virus infection on lipid metabolism of chick embryo fibroblasts. *By H. A. Blough and D. B. Weinstein* 303

IV. Specific Synthetic Functions

Recent studies on the biosynthesis of collagen. *By Darwin J. Prockop, Peter Dehm, Bjørn R. Olsen, Richard A. Berg, Michael E. Grant, Jouni Uitto and Kari I. Kivirikko* 311

CONTENTS

The intracellular translocation and secretion of collagen. *By Paul Bornstein and H. Paul Ehrlich* 321

On the nature of the polypeptide precursors of collagen. *By George R. Martin, Peter H. Byers and Barbara D. Smith* 339

Intracellular forms of collagen—underhydroxylated collagen. *By Peter K. Müller, Ermona B. McGoodwin and George R. Martin* . . . 349

Approach to the hydroxylation of collagenous proline. *By Josef Hurych, Pavel Hobza, Jiřina Rencová and Rudolf Zahradnik* 365

Collagen proline hydroxylase activity and anaerobic metabolism. *By U. Langness and S. Udenfriend* 373

Procollagen and procollagen peptidase in skin as a functional system. *By Ch. M. Lapière, A. Lenaers and G. Pierard* 379

The role of hydroxylysine in the stabilization of the collagen fibre. *By Allen J. Bailey and Simon P. Robins* 385

Ascorbic acid and collagen synthesis. *By C. I. Levene and C. J. Bates* . 397

Biosynthesis of collagen in the axial organs of young chick embryos. *By Suzanne Bazin and Georges Strudel* 411

Collagen degradation in *vivo* and *in vitro*. *By D. S. Jackson* . . 417

Intracellular localisation of connective tissue polyanions. *By J. E. Scott and J. Dorling* 421

Biosynthesis of the L-iduronic acid unit of heparin. *By Ulf Lindahl, Magnus Höök, Anders Malmström and Lars-Åke Fransson* . . . 431

Biosynthesis of dermatan sulfate in fibroblasts. *By Lars-Åke Fransson, Anders Malmström, Ulf Lindahl and Magnus Höök* 439

Genetic defects in the degradation of glycosaminoglycans: the mucopolysaccharidoses. *By Albert Dorfman, Reuben Matalon, Jerry N. Thompson and J. Anthony Cifonelli* 449

Chemistry of excretory products in the Hunter syndrome during plasma infusion. *By Lars-Åke Fransson, Ingrid Sjöberg and Gösta Blennow* 463

Aspartylglycosaminuria *in vitro;* electronmicroscopic and enzymatic studies on cultured fibroblasts. *By Pertti Aula, Seppo Autio, Veikko Näntö and Marja-Leena Laipio* 473

V. Inflammation, Repair and Fibrosis

Connective tissue activation: mechanism and significance. *By C. William Castor, Susan F. Smith, James C. Ritchie and Mary E. Scott* . 483

Activation of connective tissue enzymes in early wound healing. *By J. Raekallio* 501

Local and systemic factors which affect the proliferation of fibroblasts. *By I. A. Silver* 507

Probes for the measurement of the microenvironment. *By I. A. Silver* . 521

The fibroblast and inflammation. *By W. G. Spector* 525

The biological response to silica. *By A. G. Heppleston* . . . 529

Experimental fibrosis in liver and other organs. *By J. Lindner and K. Grasedyck* 539

Role of the fibroblast in controlling rate and extent of repair in wounds of various tissues. *By Walton Van Winkle, Jr., J. Christopher Hastings, Ellen Barker and Waltraud Wohland* 559

Intercellular matrix of hypertrophic scars and keloids. *By Suzanne Bazin, Claude Nicoletis and Albert Delaunay* 571

Factors influencing wound healing in rheumatoid arthritis. *By Kauko Vainio* 579

Respiratory patterns of fibroblasts in reparative tissue. *By T. Hunt, R. Allen, D. Kittridge, J. Niinikoski and P. Ehrlich* 585

Oxygen and wound healing: a new technique for determining respiratory gas tensions in human wounds. *By Juha Niinikoski and Jaakko Kivisaari* 591

Injury and repair in vascular connective tissue. *By Ib Lorenzen* . . 601

Connective tissue, smooth muscle and leukocytes in repair processes of the arterial wall. *By S. Björkerud* 615

The smooth muscle cell in connective tissue metabolism and atherosclerosis. *By Russell Ross* 627

In vitro incorporation of labelled precursors in the macromolecules of the polymeric stroma of normal and pathological arterial wall. *By L. Robert, S. Junqua, A. M. Robert, M. Moczar and B. Robert* . . 637

VI. Summarizing Remarks

The fibroblast, a retrospective view of the meeting *By Russell Ross* . 653

Postscript and acknowledgements. *By J. Pikkarainen and E. Kulonen* . 661

Author Index 665

Subject Index 683

I
General Papers

Reactivity of the Connective Tissue

Eino Kulonen

Department of Medical Chemistry, University of Turku
Turku 52, Finland

Since the Second World War I have been impressed by the different ways in which researchers have approached connective tissue. I feel that the time is now ripe for an attempt to integrate the connective tissue research from a biological and medical point of view. Originally the research was motivated by the technical applications on skin and bone for the leather and gelatine industries, respectively, which prompted further approaches with organic-chemical methods resulting in advanced knowledge of the primary structures, amplified by macromolecular and colloid sciences to give three-dimensional images of the materials. The medical work has been concentrated mainly on rheumatoid disease, synovial fluid, wound healing and genetic defects, but the really important and extensive medical applications have not emerged as yet.

While the technical and strictly structural elaborations seem to have passed the peak of interest, large resources are invested at present on the metabolic work and on the dynamic aspects of connective tissue, with the final aim to be able to regulate its amount. The experimental pathologists and the cytologists use the fibroblasts from connective tissue as model cells. However, it is not correct to consider the fibroblast as a ubiquitous and vigorous cell only, like a mammalian equivalent of *E. coli*. On the contrary, it exerts most specific synthetic functions which determine the internal environment for all the cells in the body and these functions have been a condition for the evolution of multicellular organisms with differentiated tissues and regulated growth. The fibroblasts and related connective tissue cells should be treated with the same esteem as the blood cells and the lymphocytes.

The chemical and metabolic characterization of the subcellular structure of the fibroblasts in their various functional states has not advanced very far although much work has been motivated by the effects of the viral transformations and genetic defects. No knowledge is available on the biochemical evolution of the fibroblast, only on the development of collagen and acid mucopolysaccharides.

The reactivity is an essential attribute of all the living cells but in the fibroblasts this capacity is well developed. In Fig. 1 I have sketched the course of the fibroblast reaction, which results in the intercellular matrix and fibres. The parallel between this scheme and the production of antibodies in the lymphoid system is obvious. The so-called adjuvants are good inducers of granuloma. The first phase is a cellular proliferation, checked by the contact inhibition. The granulation tissue so

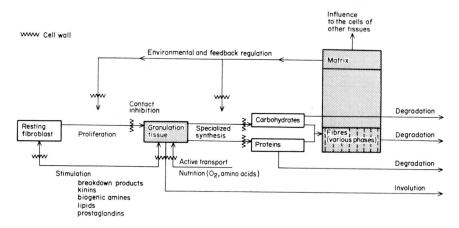

Fig. 1. Cycle of connective tissue activation.

formed synthesizes acid mucopolysaccharides and glycoproteins, which in part remain in contact with the cell surface and can be considered as extensions of the cell wall. Another specific synthesis leads to the formation of collagenous fibres, which happens in multicellular organisms only. The insoluble fibres give the cell aggregates a mechanical stability.

The main actions of the fibroblasts are extracellular and, furthermore, their plasma membrane contains the receptors of the extracellular influences. The cell walls take the key role in the cell-to-cell, cell-to-environment and environment-to-cell interactions at the molecular level. The surface of the cell wall is acidic, mainly due to the neuraminic acid moieties. The actual proton density is influenced by the adjacent acid macromolecules, but also by the actions of neuraminidase and synthetic transferases. The contact inhibition, cellular adhesion, certain immunological properties, secretion of proteins, active transport and the binding of serotonin have been claimed to depend on neuraminic acid (for review see Winzler, 1970). The proton density may affect the cell functions (Rubin, 1971). The eventual changes in the conformation of the cell surface may spread to the contiguous structures inside the cell. The study of the membranes of the fibroblasts and their neuraminic acid-containing fragments should yield clues for understanding the reactivity of the fibroblast, and also of other cells. The activation of connective tissue can be studied experimentally with various models which exhibit the proliferative and synthetic phases (Kulonen, 1970; Lampiaho and Kulonen, 1967; Ahonen, 1968).

When the stimulus of the connective tissue reaction disappears, an involution takes place. However, at the formation of permanent connective tissues during the growth, a further differentiation to cartilage and bone may occur. The role and significance of connective tissue in morphogenesis still remain obscure. During ageing of the individual the mutual relations of the different functions and of the structural fractions are altered in the connective tissue also.

The medical interest in the reactivity of the connective tissue embraces such a wide variety of conditions such as the reparation of tissue injury (Table I), reactions to ill-defined factors resulting in fibroses, scleroses or rheumatoid disease and even the disturbances of the growth.

Table I
Medical applications of the reactivity of the connective tissue

Reparation of tissue injury (e.g. wound healing)
Rheumatoid disease in synovial tissue and elsewhere
Hypertrophic scars (keloids, adhesions)
Fibroses (liver cirrhosis, silicosis, sequelae of O_2-intoxication)
Scleroses (atherosclerosis, otosclerosis, nephrosclerosis)
Degenerative diseases (e.g. paradentosis, osteoarthrosis)
Morphological defects and disturbances of growth
Weakness of connective tissue (e.g. hernia)

Although the details of the synthetic functions in the connective tissue have been under intensive investigation with emphasis on their regulation, less attention has been paid to the initial stimulation of the fibroblast or its precursor cell. The production of connective tissue depends on the function of each cell but also on the number of the cells. The possibilities to feedback control are numerous but poorly understood. The reactions of fibroblast, smooth muscle, and nerve cells have much in common. Granulation tissue responds with the synthesis of collagen to substances (bradykinin, serotonin, histamin, prostaglandin) better known for their effects on the smooth muscle or the nervous tissue (Aalto and Kulonen, 1972; Boucek and Alvarez, 1970; Majno *et al.*, 1971; Blumenkrantz and Søndergaard, 1972). The mechanism of the stimulation of different cells may be similar although the response is variable: cellular proliferation, synthetic functions, muscular contraction or nerve impulse. This confirms the uniformity of Nature at the molecular level, simultaneously providing a link between somatic and mental responses. Whether the stimulation of the fibroblasts both to divide and to synthesize proteins and carbohydrates can be induced by the same agents, remains to be explored.

The factors stimulating the fibroblasts are heterogeneous. The most common cause for the reaction of the connective tissue is a tissue injury when ill-defined breakdown products are liberated. Many substances have been suggested as the 'mediators of inflammation' and belong to the stimuli of the fibroblasts (Marx, 1972). In many cases the nature of the stimulus is unknown, with regard to rheumatoid disease, for example. General irritation of the fibroblasts in the body may be the mechanism of the so frequent but ill-defined musculoskeletal disorders. There are certain more specified forms of tissue reaction, which are interesting in their own right. I have in mind the silicosis and the complications of oxygen intoxication which both cause a fibrosis in the lungs.

Lipids are generally implicated in the etiology of arteriosclerosis and of liver cirrhosis. The lipoproteins form complexes with acid mucopolysaccharides but it is not known whether they are bound with the cell coat carbohydrates also. The catabolism of saturated fats entails the release of hydrogen and the redox-balance is shifted accordingly to favour an enhanced synthesis of both collagen and acid mucopolysaccharides through the increased formation of proline and galactosamine, respectively. There are several fragments of evidence to show that the ingestion of abundant fat increases the synthesis of proteins, including collagen, in liver (Häkkinen and Kulonen, 1972) and in granulation tissue (T.-T. Pelliniemi, personal communication). Without doubt, the arteriosclerotic process involves a connective tissue reaction, either because of a mild injury or perhaps against lipids (Hauss *et al.*, 1965). Considering the great public investments in arteriosclerosis research and the heated discussions on the mass media on the merits of various dietary lipids it is appalling how the molecular side of the problem has been neglected.

Manifestations of the reactivity of the connective tissue such as the diseases of the joints, especially of a rheumatoid nature, and those of the vascular wall are very prevalent in Finland. They are acknowledged as important causes of human suffering and of large economic losses both to the community and to the individual. The Finnish Research Council for Medical Sciences has selected the cardiovascular and connective tissue diseases among its three main targets of medical research in Finland. No government can afford not to fight rheumatism. However, there has been much discussion on the practical applications of this decision. The fundamental research is in the defensive, although it costs only a small fraction of the expense of extended surveys and field trials. Therefore they are no alternatives. The experience of tuberculosis or rickets indicates that we need a penetrating etiological knowledge before diseases can be effectively suppressed and prevented by medical policies.

ACKNOWLEDGEMENTS

The original research in our laboratory has been supported by the Sigrid Jusélius Foundation and the Finnish Research Council for Medical Sciences.

REFERENCES

Aalto, M. and Kulonen, E. (1972). *Biochem. Pharmacol.* **21**, 2835–2840.
Ahonen, J. (1968). *Acta Physiol. Scand.* suppl. **315**, 1–73.
Blumenkrantz, N. and Søndergaard, J. (1972). *Nature [New Biol.]* **239**, 246.
Boucek, R. J. and Alvarez, T. R. (1970). *Science* **167**, 898–899.
Häkkinen, H. –M., Franssila, K. and Kulonen, E. (1972). *Scand. J. Clin. Lab. Invest.* suppl. **122**, 29.
Hauss, W. H., Junge-Hülsing, G., Matthes, J. K. and Wirth, W. (1965). *J. Atheroscler. Res.* **5**, 451–465.
Kulonen, E. (1970). *In* "Chemistry and Molecular Biology of the Intercellular Matrix" (E. A. Balazs, ed.) Vol. 3, pp. 1811–1820. Academic Press, London and New York.
Lampiaho, K. and Kulonen, E. (1967). *Biochem. J.* **105**, 333–341.
Majno, G., Gabbiani, G., Hirschel, B. J., Ryan, G. B. and Statkov, P. R. (1971). *Science* **173**, 548–550.
Marx, J. L. (1972). *Science* **177**, 780–781.
Rubin, H. (1971). *In* "Growth Control in Cell Cultures" (G. S. W. Wolstenholme and J. Knight, eds.) pp. 127–149. Churchill Livingstone, Edinburgh and London.
Winzler, R. J. (1970). *Int. Rev. Cytol.* **29**, 77–125.

The Role of the Extracellular Space

A. G. Ogston

Trinity College and the Department of Biochemistry
Oxford, England

I am tempted to quote the schoolchild's essay on the cow, which said: 'The head is for holding the horns and so that the mouth can be somewhere'. There has to be a space for the cell to be in; it is where the cell is. Yet the word 'is' implies at least the cell's capacity to survive, and this sets limits on what the space may or may not contain; and in multicellular organisms, each cell may require a particular location in which and from which it can, not merely survive, but also perform those chemical and physical exchanges that constitute its part in the economy of the whole.

Neither the cell, nor its environment, can be profitably considered in isolation. Each is determined by a whole series of relationships between them. This has been illustrated in several papers; for example, by effects of cell concentration. Perhaps we can sharpen the question 'What is the role of the extracellular space?' by asking instead 'What sort of relationships exist between a cell and its environment?' Problems of definition, referred to beforehand then tend to disappear.

The cell's first need is to survive. Viewed physicochemically, it is surprising that it can do so. In relation to its environment it is a region of instability, of lowered entropy, both with respect to the substances that form it and to their distribution in space. Since the processes that lead inevitably towards uniformity and increase of entropy cannot be entirely stopped, the low entropic state of the cell can be maintained only by a continuous diversion of external sources of energy or lowered entropy so as to keep the cell, not in equilibrium, but in a steady state of disequilibrium. This is what the internal metabolism of the cell is all about.

Obviously there is a very close dependence of the cell upon its environment, not only for sources of energy, but also for the exchange (supply and removal) of the chemical substances needed for the maintenance of its structure and function. If this were all, we might not be justified in thinking of the cell and its environment as being separable; rather we might think of them as a single integrated system. But then, the very instability which makes the cell so closely dependent on its environment gives it also a certain degree of independence. It is able, for short periods at least, to react chemically or physically in ways that are more than passive responses to environmental influences; it can respond to relatively small signals; it can survive unfavourable changes.

We should consider also the reciprocal ability of the cell to change the environment. This capacity reaches its highest development in the homoiostatic mechanisms of the mammal, aimed at maintaining the constancy of the *milieu interne*. Even beyond this, multicellular organisms have developed complex systems of humoral control which involve interactions in both directions between particular cell types and the *milieu interne*, to which (in imitation of Sherrington) we might well give the title of '*the integrative action of the metabolism*'. Finally man has developed means of controlling widely, though often unintentionally and unfavourably, his own external environment.

But we are concerned here with a particular type of cell in its immediate environment. What kinds of relationship should we look for? Because of its small size and, consequently, large surface/volume ratio, a cell is likely to be strongly influenced by the *intensive* properties of its environment; by these we mean those properties, such as temperature, pressure, composition or pH which do not depend upon the volume or quantity of environment to which the cell has access. For example, a cell can never differ much in temperature from that of its immediate surroundings nor, except for cells with strong rigid walls, much in pressure.

It can, however, maintain considerable differences of concentration (or of activity or chemical potential) of individual chemical substances, in either direction; that is, an intracellular concentration can be maintained many times higher or lower than that in the environment. Two devices are used to achieve this: first, a relative impermeability of the cell envelope so that in general the universal tendency for the substance to diffuse to equality of chemical potential is hindered; secondly, specific, asymmetric transport processes that couple energy available from metabolism to supply that needed to set up and to maintain the disequilibrium. Since the coupling is stoichiometric, there will be a thermodynamically determined limit to the concentration ratio that can be maintained in any given case; for example, with perfect coupling and minimal back-diffusion, a coupled process yielding 10,000 cal/equivalent could maintain a concentration ratio up to 10^7.

Beyond these remarks, I do not know of any generalization that I can usefully make. There is so wide a variety of substances and of their interacting actions on a variety of cells—whether as metabolites, modulators or structural units, useful or antagonistic, coming randomly from the environment at large or (as in the internal medium) from the activities of other cells—that each must become a problem for special investigation, and discussion. We have had some of this and shall have more.

It has been suggested that I should say something about *osmotic pressure* as a factor in the relationship between cell and environment. In doing this, I am merely choosing a particular case of the partition of one chemical component, namely water, between the cell and its environment; it is exceptional only because it is normally present in both in large molar excess over all other components. Because of this excess, the lowering of its activity brought about by solutes is small compared with the activity of pure water; it is for this reason that we use *osmotic pressure* to express the relatively small differences of activity which are, nevertheless, important in determining its distribution. Moreover, osmotic pressure is, by its definition, the exact analogue, in the way in which it describes change of activity, of hydrostatic pressure.

Osmotic pressure differences can arise only from the effects of solutes which themselves are not in equilibrium distribution. In particular, the characteristic contents of a cell (proteins etc.) lower the activity of water within it. Consequently, if the solutes characteristic of the environment (salts etc.) were distributed at equilibrium, there would always be a positive difference of osmotic pressure between the inside and the outside of the cell. Water would tend to pass inwards, diluting its contents and causing it to swell. This would continue until the cell contents are uniformly distributed through the available environment, unless it is prevented in one of three ways:

(*i*) if the cell possesses a rigid or elastic envelope, the inward passage of water will build up an internal hydrostatic pressure which will eventually balance the osmotic pressure and prevent further swelling; many bacterial cells appear to do this;

(*ii*) by an active, specific means of extruding water from the cell; this mechanism may be present in the vacuoles of protozoa;

(*iii*) by an active, specific means of extruding or excluding from the cell some solute component of the environment, so as to maintain an opposite, compensating difference of osmotic pressure.

Mammalian cells appear to use the last method. There is evidence that the sodium ion (with accompanying anions) is the principal balancing agent, and it seems possible that this is the original significance of the 'sodium pump', adapted for more special purposes in excitable cells.

The presence in the environment of solutes which, because of their large molecular size (e.g. glycosaminoglycans) cannot equilibrate with the cell contents, will contribute an effective external osmotic pressure which will help to balance, or may even exceed, that of the cell contents. We can conceive that the remaining difference (either way) may be compensated by maintained disequilibrium of ions such as sodium. There will remain, in both compartments, a lowering of the activity of the water which one might expect to affect processes into which water enters as a reactant. However, as has been pointed out, this effect is likely to be negligibly small; for example, in a solution whose osmotic pressure is 1 atmosphere (10^6 dyn/cm^2) the activity of water is lowered by only 0·1%.

The situation is not very different where, as in cartilage, the extracellular polymer is entrapped as though in a semipermeable sac. The spaces within the collagen network, being in equilibrium with external tissue fluid, will have a raised hydrostatic pressure corresponding to the osmotic pressure of the contents of those spaces; an included cell will be exposed to the same hydrostatic pressure as that of its immediate environment. Its problem of osmotic adjustment is therefore merely that of operating at a somewhat increased (perhaps doubled) atmospheric pressure. The increase of pressure will largely or wholly compensate for the lowering of the activity of water due to the osmotic pressure of the extracellular solute.

Elastin Biosynthesis and the Influence of some Extracellular Factors of Macromolecular Organization

S. M. PARTRIDGE

*Meat Research Institute, Agricultural Research Council
Langford, Bristol, England*

I would like to run rapidly through some of the salient features of what has been learnt about elastin since about 1955 and then to try to spend rather more time on the discussion of some interesting physical and chemical properties which now seem to be coming to light. I should say that I am a protein chemist, not a cell biologist or a biophysicist: and my main object in dealing with the physical structure and possible routes to elastogenesis is to try to engage interest in something that I feel to be important in the medical and nutritional field: and in the hope that others more qualified than I will be attracted to the solution of these problems.

It is really only in relatively recent years that serious attempts have been made to study the macromolecular structure of elastin and to try to relate this to the function of elastica in such tissues as skin; the yellow ligaments or the walls of the blood vessels and perhaps, just as importantly, the lung, pleural membranes and mammary tissue.

It seems worth mentioning at once, that elastin only shows long-range extensibility and its characteristic property of rubber-like recoil when it is fully swollen with water. If the water is removed from elastin fibres the dry protein becomes horny and inextensible. However, elastin is not the only protein to show this property. About ten years ago another elastic protein named 'Resilin' was discovered in the wing hinges of locusts and grass-hoppers by Weis-Fogh (1961) while he was still working in Copenhagen. Resilin has a very different amino acid composition from elastin and contains many more hydrophilic side chains than the mammalian elastomeric protein but nevertheless samples of elastin from many different birds and mammals all seem to have a closely similar pattern of amino acid composition. This seems to indicate that the amino acid sequence is a rather critically important factor for the manifestation of elasticity: but also that elastomeric properties may be based on different principles in vertebrates than in the 'Resilin' structures of flying insects.

However this may be, in the yellow ligaments of mammals or the walls of arteries, long range extensibility is the main requirement. Elastin is a structural protein and the successful performance of this function must depend, just as in synthetic elastomers, on its chemical composition and physical constitution. For these reasons it seems to be important to learn what we can about the macromolecular and submicroscopic structure of elastin and to investigate the way in which its parts are put together during biosynthesis.

SOME BROAD INDICATIONS ABOUT THE COURSE OF ELASTIN BIOSYNTHESIS

Unfortunately elastin has none of the beautiful regularity of conformation found in muscle or collagen and X-ray diffraction studies have yielded little more than a pair of amorphous rings. Quite a different story, one might say, from the enzyme 'elastase'. Pancreatic elastase was first discovered by Balo and Banga in 1949 but it was exciting to be able to read two years ago of the final determination of the complete amino acid sequence of porcine pancreatic elastase by Shotton and Hartley (1970) and the complete mapping of the three-dimensional structure by Shotton and H. W. Watson (1970) of Bristol University.

However, some progress has been made with the fibres of elastin, in spite of the difficulty that there are at present no general methods available for the solution of the problems of the structure of an insoluble protein that is also a three-dimensional continuously crosslinked gel. Indeed the basic difference between the protein elastomer and synthetic rubbers appears to be that the former contains up to two-thirds of its volume as water while all man-made elastomers so far produced have been single-phase systems containing random hydrocarbon chains (Partridge, 1962).

A great deal of work has been spent in the last ten years in a number of laboratories in the attempt to determine the detailed chemistry of the crosslinkages in the elastin structure. I do not propose to discuss crosslink biosynthesis in detail here; however, in order to remind those who are not closely concerned with work in this field my next few figures summarize in broad outline the type of structures that are generated as crosslinks from the side-chains of lysine after some of them have been oxidized to aldehydes.

SUMMARY OF THE INFORMATION FROM BIOSYNTHETIC STUDIES

Basically all the lysine-derived crosslinks formed by elastin, as is now known with collagen, are produced by variations of the same simple theme. Some of the side-chain lysine or hydroxylysine residues of the soluble precursor proteins become oxidized to aldehydes. These, as a first step, may either combine together to form a more or less labile

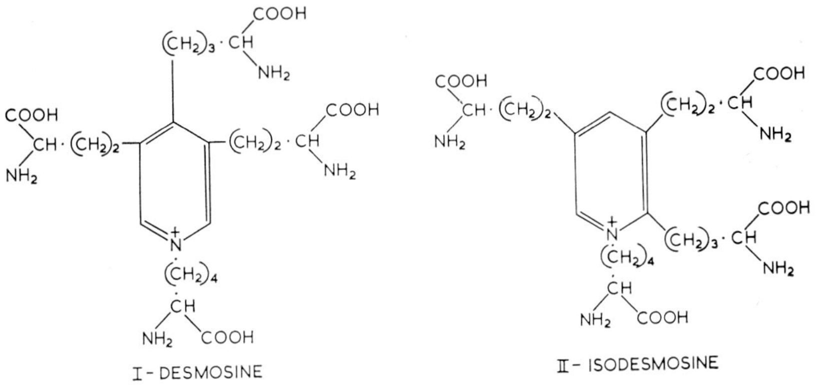

Fig. 1. Structural formulae of the desmosine isomers.

aldol or an aldehyde residue may combine with an unchanged lysine residue to form a Schiff base or aldimine. In the case of collagen, many of the labile aldimines derive from preformed hydroxylysine residues (Bailey et al., 1969; Mechanic et al., 1971; Davis and Bailey, 1971) and indeed, this may also be true of some lysine-derived compounds found in minor amounts in elastin. However, mature elastin from all tissues examined invariably contains as major crosslinks very substantial amounts of two biologically stable isomeric substances known as desmosine and isodesmosine. These survive acid hydrolysis and can be readily identified by their u.v. absorption spectra. Fig. 1 shows the structural formulae of the desmosine isomers as isolated by Dr. J. Thomas and myself about ten years ago (Partridge et al., 1963; Thomas et al., 1963). It will be seen that these substances have highly condensed structures containing positively charged pyridinium rings. The ultraviolet absorption spectra of the desmosine isomers is compared in Fig. 2 to the equivalent tetramethylpyridinium chlorides and this comparison

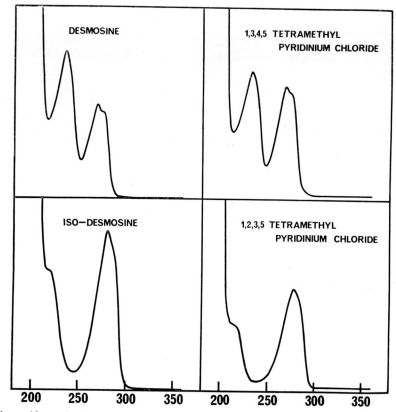

Fig. 2. Absorption in the ultraviolet region of desmosine and isodesmosine compared with the 1, 3, 4, 5- and 1, 2, 3, 5-tetramethyl pyridinium chlorides. The molecular extinctions measured in 0·05 N hydrochloric acid at the maxima were: Desmosine, λ 268 mμ, ε 4900; 1, 3, 4, 5-tetramethyl pyridinium, λ 266, ε 4730; isodesmosine λ 278 mμ, ε 7850; 1, 2, 3, 5-tetramethyl pyridinium λ 276, ε 6940. (Reprinted from Thomas *et al.*, 1963.)

is the main evidence on which the position of the side-chains in the desmosines is assigned.

The desmosines are C_{24} compounds and this suggests that the route of biosynthesis could be from four lysine residues (Partridge, 1965). Fig. 3 shows the result of incorporating ^{14}C-lysine into a cultured slice of duckling aorta (Partridge *et al.*, 1964). The elastin was isolated in a pure condition by alkali treatment and then hydrolysed. The lower chromatogram shows the radioactive peaks obtained from the hydrolysis products of purified elastin that had been reduced with borohydride before purification. The chromatogram above is of the same material but

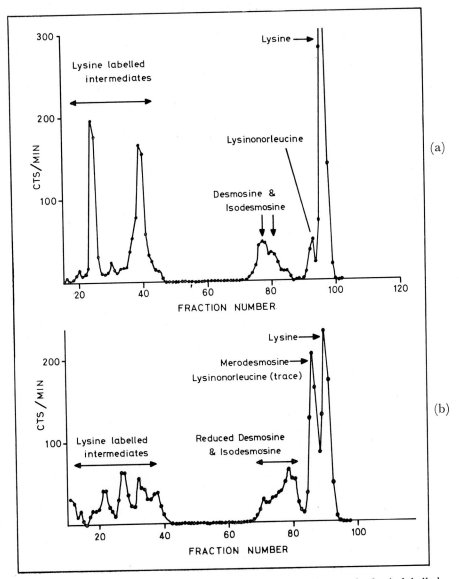

Fig. 3. Chromatograms of the hydrolysis products of duckling aortic elastin labelled with ^{14}C-lysine in tissue culture: (a) Control elastin; (b) Elastin isolated after borohydride reduction. (From Starcher et al., 1967. Courtesy of The American Chemical Society.)

the elastin was isolated and hydrolysed without prior reduction. The chromatograms show the formation of labelled desmosine, isodesmosine, lysinonorleucine and merodesmosine together with a number of other substances of less basic character (Starcher et al., 1967).

Fig. 4 shows in summary, some of the suggested biosynthetic pathways leading to desmosine and isodesmosine and some other permanent crosslinks found in smaller amounts in elastin. Compound IV, the Schiff base between lysine and the aldol is considered to be a very labile intermediate but it is now known that some at least of the reduced form exists naturally as merodesmosine in mature elastin. Merodesmosine was isolated by Dr. Barry Starcher and myself (Starcher et al., 1967) and clearly arises from combination of a free lysine residue with the aldol formed from condensation of two further allysine residues. As indicated above, a small amount of this substance occurs naturally in a reduced form, but with most tissue culture preparations the greater part of the merodesmosine is in the form of the labile Schiff base. Reduction with borohydride appears to influence the equilibrium in favour of merodesmosine and quite large quantities may be isolated after reducing the whole aortic tissue with borohydride prior to isolating the elastin component with hot alkali.

To summarize this story so far we appear to have shown that for a short period of time before the formation of permanent irreversible crosslinks a dynamic equilibrium of aldols and Schiff bases exists. This dynamic equilibrium of temporary chemical linkages may represent a mechanism to allow remodelling during growth of the tissue.

The information gained so far about biosynthesis of the crosslinks allows us to form some views about the process of elastogenesis and at this point it may be worthwhile to say at once, briefly, how we visualize the process at the present time. The difficulty of course is that in general any process which involves such a large number of separate reactions at specific sites would normally be regarded as very slow and inefficient and it seems certain that a biological property of the proteins concerned must be employed in order to ensure that the reactants find themselves spatially oriented and in a position to form the desmosine ring. It does seem necessary to make the assumption that the specific surface pattern of the tertiary structure of a globular protein precursor must be employed for this purpose.

Fig. 5 shows how globular protein molecules could bring this about if their properties were such that they could fit into some sort of paracrystalline array. Of course the figure shows a two-dimensional drawing. In three dimensions the proportion of free space would look rather different. In Fig. 5 the basic assumption made was that the fibres and

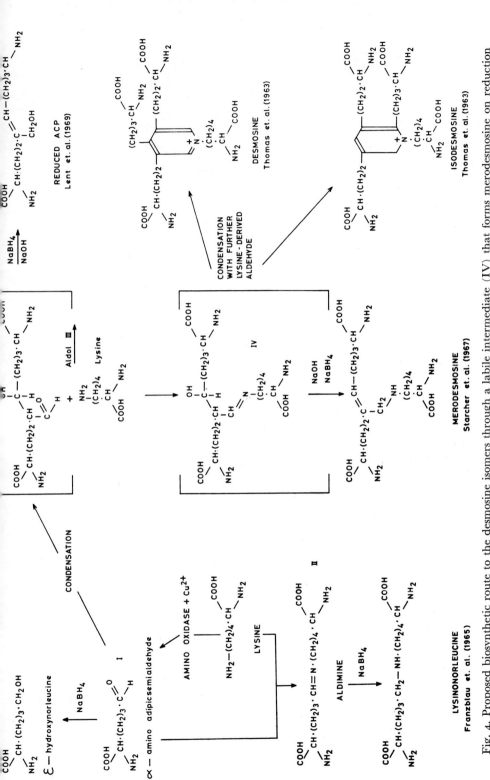

Fig. 4. Proposed biosynthetic route to the desmosine isomers through a labile intermediate (IV) that forms merodesmosine on reduction of elastin with borohydride (ACP, Aldol condensation product). (From Starcher *et al.*, 1967. Courtesy of The American Chemical Society.)

Fig. 5. Schematic representation of the formation of a crosslinked, three-dimensional isotropic elastomer from a globular protein monomer. The specific tertiary structure of the protein precursor is important in bringing the reacting groups into apposition and is an important factor determining the type of crosslinkage to be formed. In aqueous solution at physiological temperatures it is probable that the hydrophobic centres of the original protein molecules are in large part retained. (From Partridge, 1966. Courtesy of the American Chemical Society.)

membranes of elastin arise from the polymerization of globular protein molecules and that the three-dimensional continuous structure of the biopolymer is automatically generated simply because the repeating unit has a minimum of four similar 'valency zones' at its surface, directed outwards from the centre.

I should say that up to quite recently the existence of a soluble protein regarded as the elastin precursor had been an inference as nobody had been able to isolate such a protein. However, in 1968, a definitive paper appeared from Professor Carnes' group in Salt Lake City reporting the isolation of a soluble protein with the composition of elastin from the aortas of copper deficient pigs (Smith *et al.*, 1968). I will say no more about the precursor protein and the process of elastogenesis for the moment, but I would like to return to this fascinating problem later; after we have considered some of the physical properties of elastin and the results obtained from a number of biophysical techniques.

BIOPHYSICAL TECHNIQUES

Unfortunately the range of experimental techniques that are available for the study of an insoluble non-crystalline protein polymer is very small, but at least with elastin from bovine *ligamentum nuchae*, we can make use of the fairly uniform size and shape of the fibres resulting from purification by autoclaving. These purified fibres of elastin are about six or seven microns in diameter and about a quarter of a millimeter in length. They thus make an excellent column packing material and when fully swelled in dilute acetic acid they contain about 65% of solvent on a volume basis. Columns packed with highly purified elastin fibres can be used as gel filtration columns, and with some types of solute the separations are as good or better than commonly obtained with crosslinked dextrans or crosslinked agarose.

Fig. 6 shows the separation of a synthetic mixture of sugars, glycols and alcohols on a column packed with elastin fibres. This separation was carried out in order to show the degree of resolution obtainable. In Fig. 7 the distribution coefficients calculated from the column data of a mixture of aliphatic alcohols, glycols and sugars are plotted against molecular weight of the solutes. In fact the cube root of the molecular weight has been used but this was done simply to get all the data on the same graph. It will be seen that the sugars and polyglycols up to a molecular weight of about 1000 Daltons are retarded in *inverse proportion* to molecular weight and are obviously taking part in *excluded volume phenomena*. Aliphatic alcohols on the other hand are retarded by a greater amount and the retardation is in *direct* proportion to molecular weight. They are therefore absorbed on internal surfaces (Partridge, 1967a, b).

While we are considering the various kinds of phenomenon that may lead to absorption by water-swollen elastin fibres it may be appropriate to say a few words in discussion of the possibility of using this sort of technique for the study of the absorption by elastin fibres of such surface-active substances as sterols, prostaglandins and fat-decomposition products. It is proposed to carry out further work of this kind but in our first trials, complications were

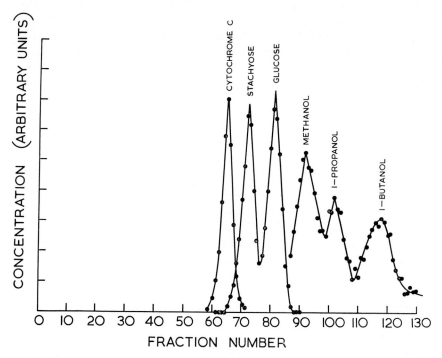

Fig. 6. The separation of a mixture of sugars and alcohols using a column packed with water-swollen elastin fibres. (From Partridge, 1967a.)

met and it became rather clear that further details of the structure and internal organization of elastin as a biopolymer are needed if valid interpretations of the experimental results with these more complex substances are to be made.

Unfortunately, quantitative theories relating molecular exclusion phenomena to the structure of the gel and the Stokes radius of the solute are still not available except for the simplest systems. Thus the equations of Gelotte (1960) relate to a system of uniform water-filled channels of circular cross-section while the plots of Laurent (1964) and of Siegel and Monty (1966) refer to a system randomly distributed uniform rods.

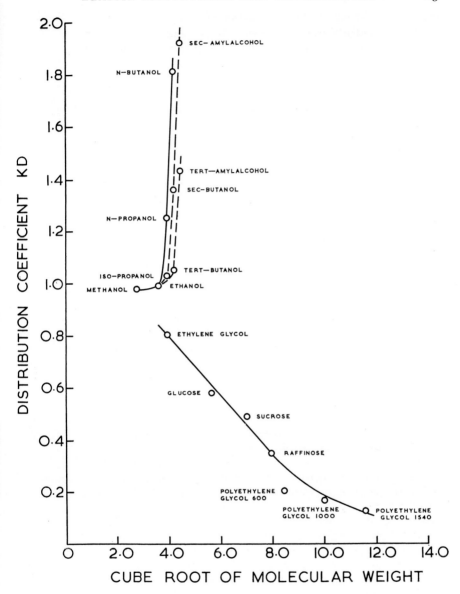

Fig. 7. The distribution ion (K_D) between the water spaces in the elastin gel and the external water for a number of sugars, glycols and alcohols. Values for K_D greater than unity indicate absorption on internal surfaces of the gel. (From Partridge, 1967a.)

Table I
Calculated values of the effective radius of the pores in elastin (r) for solutes of different molecular size

	Molecular Weight	Stokes Radius, α (Å)	Elution Volume V_e (ml)	Distribution Coefficient K_D	Radius of Pore, r (Å)
Ethylene glycol	62	1·61	87·72	0·81	16·1
Glucose	180	3·63	81·18	0·58	15·1
Sucrose	342	4·88	78·63	0·49	16·3
Raffinose	504	6·20	74·62	0·35	15·1
PEG 600[a]	600	8·00	70·35	0·20	14·5
PEG 1000[a]	1000	9·93	69·22	0·17	16·6

[a] Polyethylene glycols, commercial preparations.

Table I shows the results calculated on the basis of a system of channels with circular cross-sections according to Gelotte (1960). The calculated pore radius is 15–16 Å. This may be regarded as the radius of the equivalent cylindrical pore and as can be seen from the table it is independent of the size of the solute used in the experiment for the whole molecular weight range up to about 1000 Daltons. I should say that soon after obtaining the results in Table I, calculated according to Gelotte, I sent the data to Professor Torvard Laurent to ask his opinion since he was working with similar systems at the same time. Almost immediately he returned a straight line graph in which exactly the same data were plotted according to Siegel and Monty (1966). Extrapolation of the best straight line gave 8 Å as the radius of the parallel rods assumed (in Siegel and Monty's mathematical treatment) to form the structural elements of the gel.

Thus, with non-adsorbing solutes up to about 1000 Daltons, a reasonably good fit is obtained with both theories: indicating the obvious fact that neither of the simple structures examined can represent more than a broad approximation of the native structure of the swollen elastin gel. Indeed it would be as difficult to visualise the structure of elastin as a system of exactly cylindrical water channels occupying two-thirds of the volume as it would be to imagine it as a system of straight parallel rods, randomly distributed. However, the measurements give a fairly reasonable indication of the size range of the assumed macromolecular repeating units (which make up the three-dimensional isotropic structure) whatever shape these repeating units adopt in the swollen gel.

CORPUSCULAR STRUCTURES

Basically the structure we are now thinking about is one in which there is a corpuscular arrangement, the centres of the particulate elements being relatively hydrophobic, with the intercorpuscular spaces filled with water. This kind of structure was also suggested by Robert and Poullain (1963) starting from different premises and from an independent study of a different experimental situation.

As will be seen later there is now much evidence in favour of a structure of condensed hydrophobic regions surrounded by some kind of interface with the water of swelling and clearly this is a situation which must affect the physical properties of the protein fibre. Of course structures such as this are capable of experimental investigation by quite a large number of independent methods.

A structure of this type would come about if the essential configuration of the soluble globular protein precursor of elastin survives the process of crosslink formation via the lysine-derived aldehyde residues. This has already been demonstrated, for instance in the crosslinking of protein crystals by double-ended aldehydes such as glutaraldehyde. Thus F. M. Richards and his colleagues at Oxford have shown that *some* crystalline enzymes can be crosslinked into a three-dimensional network without disordering the crystal structure. In fact with some protein crystals the structure is so little disorganized by the formation of crosslinks that very similar single-crystal X-ray diffraction patterns result from the treated or the untreated proteins (Richards and Knowles, 1968).

Of course, both purified elastin fibres and the soluble α-elastin prepared from them have suffered a variety of high temperature treatments in the course of preparation and purification, and these treatments would normally be expected to bring about almost complete randomization of the original native configuration. Under these conditions it is probable that any refolding which occurs spontaneously in the fibres or the soluble protein, is rather poor and incomplete and does not give rise to a single unique configuration. The lack of such a unique configuration is indicated by the very diffuse X-ray diffraction rings given by the fibre preparations of elastin.

SOLUBLE ELASTINS PRODUCED BY PARTIAL HYDROLYSIS OF THE MATURE FIBROUS PROTEIN

As far as myself and my colleagues were concerned the first real clue to the crosslinked structure of elastin came as long ago as 1955 from analysis of the soluble protein which we found could be prepared by careful partial hydrolysis of purified fibrous elastin with hot oxalic acid. This soluble protein, the so-called α-elastin, was polydisperse, but a purified fraction had an osmotic molecular weight of 60–80,000 Daltons and by N-group analyses it contained 15–20 separate peptide chains. A variety of physical and chemical techniques all led to the same conclusion: that the independent peptide chains were cross-linked together by firm covalent bonds.

I should say that the soluble α-elastin has rather remarkable physical properties, since on raising the temperature of a solution in dilute buffer near the isoelectric point a dense coacervate phase separates with a water

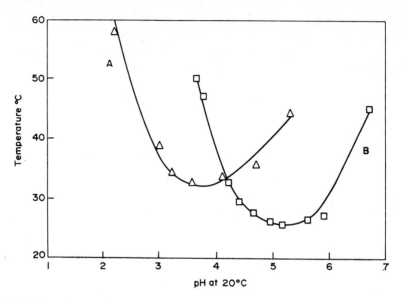

Fig. 8. Temperature of first appearance of coacervate. The curve on the right is for soluble α-elastin produced from fibrous bovine elastin by partial hydrolysis with oxalic acid: that on the left is for α-elastin after the loss of peptide material by controlled treatment with an elastolytic enzyme isolated from a mould. Acetate or phosphate buffers adjusted to ionic strength 0·01. The coacervate appeared on raising and disappeared on lowering the temperature.

content of about 66% v/v; not much different from the water content of the native elastin fibre. Coacervate formation is reversible and if a tube containing the freshly precipitated coacervate is dipped in iced water or simply held under the cold tap the droplets of coacervate rapidly redissolve and a clear solution results (Partridge et al., 1955; Partridge and Davis, 1955).

Curve B in Fig. 8 is a very old one dating back to 1955. The U-shaped curve shows the temperature at which the two-phase coacervate system forms as a function of the pH of the solution. It will be seen that the curve is very similar to the U-shaped curve given by phenol-water systems and obviously the lowest point of the U indicates the isoionic point of the protein in the buffer used.

The U-shaped curve was obtained in buffered solutions with a sodium ion concentration of about one hundredth molar. For some time the mechanism of this change in structure remained something of a mystery but it was felt that the reversal of the usual situation whereby the system shows reduced solubility as the temperature is raised may well indicate the intervention of hydrophobic interactions and entropic changes due to changes in the water structure at internal surfaces.

As has already been observed the clear α-elastin solution separates sharply into two phases as the temperature of an isoelectric solution is raised from room temperature to about 30°C. It became of interest then to learn if the formation of the coacervate was brought about by an underlying configurational change. That this was true was strongly suggested by the observations of Mammi et al. (1968) who examined the circular dichroism of α-elastin in water-ethanol mixture and reported a continuous increase in the content of an apparently ordered helical component as the ethanol concentration of the solvent increased.

Fig. 9 shows the circular dichroism curves for the clear solution (–·–·– curve) and the coacervate phase prepared from it (– – – –) as measured in collaboration with D. W. Urry at the American Medical Association Laboratories in Chicago (Urry et al., 1969). The coacervate phase was observed as a film deposited on the surface of a fused silica window and the dotted curve represents the film spectrum after applying a correction for adsorption flattening and dispersion distortions. Obviously the correction for absorption flattening due to the particulate system is very large but the change in the position of the bands clearly indicates a marked configurational change accompanying coacervate formation. In the coacervate phase, configurational asymmetry makes a high contribution to the total optical rotation, while the spectrum of the molecularly dispersed elastin was typical of disordered or denatured proteins.

HYDROPHOBIC BONDING

While these indications of a basically corpuscular structure continued to be revealed by a variety of different experimental approaches, some more deliberate attempts were made to test for hydrophobic interactions particularly by L. Robert and his colleagues in Paris and by L. Gotte and the Biophysical Group at Padua. The researches of Robert on the effect of a range of aliphatic alcohols on the alkaline hydrolysis of peptide bonds in elastin can be regarded as direct evidence for extensive regions of hydrophobic bonding. At the same time the concerted use of the widest possible range of histological and biophysical techniques, on highly purified samples of elastin by Professor Gotte and his colleagues has led to the solution of some previously puzzling phenomena and has established further evidence for the essential hydrophobic nature of the interchain reactions of elastin.

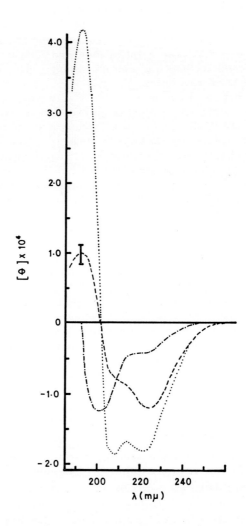

Fig. 9. Circular dichroism curves for the solution and coacervate phases of α-elastin produced by partial hydrolysis with dilute oxalic acid. The mean residue ellipticity (θ) was calculated using a mean residue molecular weight of 80·7. —·—·—, solution spectrum; ———, coacervate film spectrum; ·····, coacervate film spectrum after making corrections for absorption flattening and dispersion distortions. (From Urry et al., 1969.)

The discovery of what amounts to two microphases in α-elastin is of considerable importance if only because it illuminates applicability of the classical thermodynamic theory of rubber-like elasticity to the water-swollen elastin fibre. This question was studied again recently by D. P. Mukherjee (1969) at the Massachusetts Institute of Technology. Mukherjee concluded that when fully swelled in distilled water the peptide chains of elastin exist in an associated form due to hydrophobic interactions as is also indicated by much recent work on soluble protein derivatives of elastin; but in good swelling solvents, such as formamide-water mixtures the structure assumes a dissociated random coil conformation. Swelling in aqueous formamide is isotropic and the elastic behaviour of the formamide-swollen elastin approaches that predicted from the classical kinetic theory as developed for anhydrous cross-linked rubbers. In 44% v/v formamide the calculated average molecular weight of the peptide chain between crosslinks (M_c) reaches a maximum of 7000–8000. In formamide of this concentration it is believed that the value of M_c is due solely to covalent linkages and the value agrees fairly well with a model containing some lysinonorleucine crosslinks and in which desmosines are assumed to be the major intermolecular crosslink.

Returning to the structure of elastin in a more physiological condition that is—in equilibrium with dilute salt solutions near physiological pH—it seems apparent that under these conditions discrete centres of hydrophobic interactions must be present. If these are large enough they will approach a compact globular shape to reduce the interfacial area and to ensure that the (Gibbs) free energy of the system as a whole is minimal.

It was clear from the work of Mukherjee that we needed a new theory of long range elasticity to cover the situation of protein elastomers with a high water content. Fortunately, from their calorimetric studies, Weis-Fogh and Anderson (1970a, b) were able to provide a substantial advance in this direction and J. M. Gosline, working in the same laboratory, has now confirmed and extended their theory of 'oil droplet elastomers'. In recent work, J. M. Gosline shows that elastic energy is stored during stretching as a change in water structure, and he suggests the term 'Solvent Active Elastomer' as a name for elastic mechanisms of this type. Gosline's results provide further confirmation that elastin contains a system of compact hydrophobic globules in which the peptide chains have reduced conformational freedom.

There is one further point that may be of general biological interest and that arises from attempts to understand the behaviour of protein elastomers. As Weis-Fogh and Anderson (1970a, b) have already pointed out, in some biological situations a liquid drop elastomer could have the great advantage of being able to produce fibrils of much smaller diameter than is possible with rubber-like elastomers. A single linear row of linked globular molecules could constitute a fibril thick enough to show the required elastic behaviour.

ELECTRON MICROSCOPY

We are thus led to the conclusion that at physiological temperatures the structure of water-swollen elastin fibres includes hydrophobic globular centres, with fluid properties and little resistance to shape deformation arising either from interchain hydrogen bonding or coulumbic interactions.

Of course it would be valuable to have a visual demonstration of the structure of purified elastin through the electron microscope at high magnification; but there are difficulties about this. Elastin, because of its non-polar character, does not easily take up heavy metal electron stains. Also, elastin fibres consist of a very closely textured crosslinked gel containing about 2/3 by volume of water. As this gel shrinks its structure collapses.

However, Fig. 10 shows the result of negative staining the edge of a very small particle of elastin purified by treatment with alkali. This structure is very similar to that shown in the electron micrographs of Cox and O'Dell obtained as early as 1966, and to those of other workers since, including some of the micrographs obtained by Dr. Anna Kadar (Kadar *et al.*, 1971a, b).

The structure seems to show only an isotropic pattern of short, bent and tangled basket-work-like elements of about 11–14 Å diameter but mainly at about 30–40 Å centres. The pattern has obviously been compressed and distorted by volume collapse during drying.

Unfortunately there has been much lack of understanding between different authors who have attempted to interpret the different appearances of electron micrographs of elastica prepared in different ways. It may therefore be worthwhile to define what the word 'elastin' means to the chemist, and to consider how the pure fibrous protein 'elastin' may differ from the 'elastica' of tissues and also from various degraded elastins.

The structures that histologists call elastic fibres, elastic membranes 'elastica' or simply 'elastic' are of course multicomponent systems, only one component of which is the macroscopically amorphous isotropic three-dimensional protein polymer known as 'elastin'. This matter has been clarified recently by the careful researches of Ross

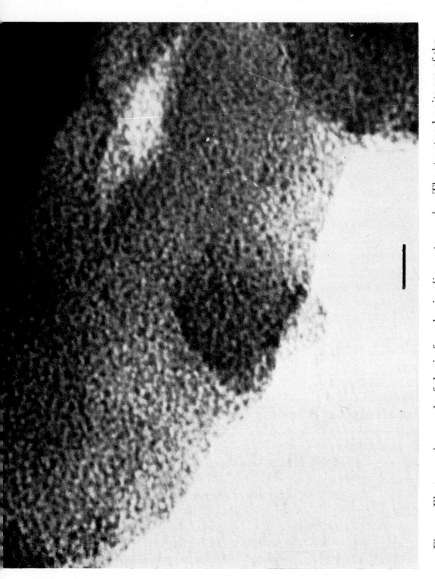

Fig. 10. Electron micrograph of elastin from bovine *ligamentum nuchae*. The structural units are of the order of 12–15 Å in diameter and appear to be arranged as a three-dimensional network. Preparation mordanted with pyrogallol and stained with uranyl acetate. Bar = 100 Å. (Reprinted from Partridge, 1970).

and Bornstein (1970). Electron microscopic observations by many authors in the last few years have demonstrated that elastic fibres in arteries and elastic ligaments of embryonic or mature animals contain two morphologically different components in intimate association (cf. Greenlee et al., 1966). One of these components consists of thin, straight, easily stainable microfibrils (110–120 Å in diameter). The second component is amorphous and normally has little affinity for lead or uranyl acetate. In the embryonic development of elastic tissue the microfibrils appear first and the amorphous component, secreted by the cells, is deposited between and around the microfibrils. The microfibrils appear to remain at about 110–120 Å diameter throughout the process to maturity, whereas the elastin fibre as a whole may reach a diameter of several microns. Thus, in contrast to embryonic elastic fibres, the mature fibre contains only a small content, by weight, of the microfibrillar protein.

Ross and Bornstein (1970) demonstrated that trypsin, chymotrypsin and pepsin selectively digested the microfibrillar component leaving the amorphous component intact and provided convincing proof that the two components consisted of entirely different proteins. The microfibrillar protein had a very different amino acid composition from alkali-purified elastin. Thus it contained significant amounts of sugars and appeared to be crosslinked by -SS- linkage through cystine rather than through the desmosine or other lysine-derived crosslinks characteristic of elastin or collagen. Ross and Bornstein provided ready means of separating the two components by gentle procedures involving reduction of the microfibrillar component by dithiothreitol in the presence of 6 M guanidine followed by alkylation of the cysteine with ethylenimine.

From its amino acid analysis and sugar content the microfibrillar substance appears to be similar to the glycoprotein extracted from bovine aortae by Barnes and Partridge (1968). This glycoprotein had been studied separately with no knowledge of its electron microscopic appearance. However, both the microfibrils and the extracted glycoprotein can be dissolved into true solution and reprecipitated by a variety of methods; there is no evidence that the glycoprotein microfibrils are capable of displaying elastomeric properties (i.e. long range extensibility) under any conditions either as reprecipitated material *in vitro* or in the mature or the immature elastin fibre. The function of the microfibril thus appears to be very little related to the elastomeric properties of the tissue but seems to be important embryologically and may play a role in determining the shape and character of the elastin fibre.

CHEMICAL DAMAGE TO THE ELASTIN COMPONENT DURING PURIFICATION

Fortunately elastin is relatively very resistant to hydrolytic procedures and purification can be regarded as complete when the residue of elastin reaches constant composition. This of course is advice of perfection and it is never quite possible, at least with aorta elastin, to obtain complete purity without causing a small amount of hydrolytic damage. Of course the acquisition of a pure undegraded sample is the corner-stone of all chemical work; but additionally both the presence of foreign proteins and hydrolytic damage may be detected by some histological stains: and both are usually readily detectable by the heavy metal stains used for electron microscopy. It was observed as long ago as 1955 that carefully purified elastin from *ligamentum nuchae* of cattle contains no more than about 0·03 residues of terminal amino and terminal carboxyl groups per 1000 amino acid residues (Partridge *et al.*, 1955a, b). Because of this general paucity of positive or negative charges in crosslinked elastin, ultra thin sections are almost transparent in an electron beam even after staining with lead or uranyl salts.

However, if sections of fibrous elastin are subjected to partial hydrolysis with elastolytic enzymes the material in the surface layer may contain up to 4–5% of positively or negatively charged terminal residues before the section passes into solution. (The terminal amino residues can readily be detected by reaction with dansyl chloride followed by complete hydrolysis.) With very thin sections such partial hydrolysis can produce considerable changes in *cation binding* and general staining properties.

Changes of a similar sort are sometimes expectable when glutaraldehyde is used for fixing. Glutaraldehyde crosslinking can reduce very considerably the net positive charge due to N-terminal or side chain amino residues and leave what amounts to a crosslinked polyanion. Thus some of these processes, which seem innocent enough, can in fact convert very thin sections of elastin into a macromolecular sheet of amphoteric or polyanionic material which in fact can bind heavy metal electron stains extremely efficiently.

SPECULATIONS ABOUT THE PROCESS OF ELASTOGENESIS

Having ranged, in a somewhat untidy way, through the chemistry of the crosslinks and such chemical and physical data concerning the whole polymer or its breakdown products as are presently possible to obtain; perhaps this is the time to pause to think a little about the physical processes that occur during elastogenesis. In making these speculations I have to apologize to the cell biologists. Of course I am well aware that, as has indeed recently been stressed by Pikkarainen and Kulonen (1972), the processes of elastogenesis—and particularly the morphogenesis of such beautiful and efficient structures as the concentric fenestrated membranes of the aorta wall—follows a pathway that has been beaten out since the beginning of biological time. It seems rather impertinent to have any views except those based on actual observation: nevertheless, each individual step in the process of elastogenesis is a chemical reaction or a physical process; and these reactions and processes must obey the ordinary rules of chemistry and physics wherever they occur.

I have already briefly mentioned the work that Carnes, Weissman, Smith, Sandberg and their colleagues carried out at Salt Lake City from about 1968. These authors established the basis of methods of extracting and purifying the soluble protein precursor of elastin from the aortae of copper deficient pigs (Smith *et al.*, 1968; Sandberg *et al.*, 1969; Sandberg *et al.*, 1971).

Now there is still some question of doubt as to the molecular weight of the soluble, apparently globular, protein that is the precursor of fibrous elastin. This is in part due to doubt as to whether the protein extracted from copper deficient aorta contains some extension of its peptide chain indicative of a transport form of the protein. On the one hand, the apparently over-high content of polar amino acids found in preparations of 'tropoelastin' isolated by making use of its solubility in acidified organic solvents may be due to impurity; on the other hand, it may indicate covalently bound 'extra peptide', perhaps with a quite special biological significance.

However, quite recently (1972) Smith *et al.* reported the isolation of a neutral salt soluble precursor protein from copper deficient pig aorta by a process which avoided the use of organic solvents acidified with formic acid. This 'neutral salt soluble elastin' was of an unambiguous purity since it has the amino acid composition expected for a highly purified porcine elastin except for the expected additional lysine residues.

The analysis of 'neutral salt soluble elastin' is given in Table II. As might be expectable the soluble elastin has no detectable content of methionine or histidine: both long suspected as being present as impurities in porcine fibrous elastin. Estimates for the molecular weight of this protein, when measured in several different ways, approached the single minimum value of 74,000 Daltons; a little higher than the 68,000 Daltons previously given by Sandberg *et al.* (1969) for formic acid/solvent extracted 'tropoelastin'.

It will be seen that the assumed precursor protein contains no desmosine or other lysine-derived amino acids associated with inter- or intramolecular crosslinks. However, the lysine content at 38 residues per mole of 74,000 Daltons is considerably more than the minimum of about 20 residues required to account for the known permanent intermolecular crosslinks.

The question arises as to what happens to the remaining 18 or

Table II
Amino acid composition of salt-soluble elastin from copper deficient porcine aorta

Amino acid	Residues per minimal subunit weight 74,000	
	Salt-soluble elastin	Insoluble porcine elastin
Hydroxyproline	9	10
Aspartic acid	3	10
Threonine	13	13
Serine	10	12
Glutamic acid	15	21
Proline	92	91
Glycine	287	269
Alanine	203	191
Valine	116	103
Methionine	—	2
Isoleucine	14	17
Leucine	40	47
Tyrosine	14	15
Phenylalanine	23	27
Desmosine (quarter)	—	10
Lysine	38	7
Histidine	—	1
Arginine	4	8

20 lysine residues that have disappeared during the formation of the mature elastin fibre. A partial answer to this is fairly easy to obtain, at any rate with small animals such as the chick. In some recent work carried out together with my colleague A. H. Whiting, pulse chase experiments with ^{14}C-lysine accounted for 36 moles of lysine per thousand residues as 18 different compounds. These lysine-derived compounds were not labile intermediates since they were still detectable nine weeks after the injection of the pulse label. Thus with chick aorta, nearly 90% of the total lysine in the chick tropoelastin can be detected as labelled lysine in apparently stable combination more than two months after the label is applied. There is thus no doubt that any lysine not used in the formation of intermolecular linkages can be accounted for as unchanged lysine or as an identifiable product of lysine.

In this connection the interactions between the ε-amino side chains of lysine and simple sugars or their aldehydic reaction products should not be forgotten. These reactions take place slowly, but under very mild conditions and, if steric conditions or availability are suitable, can give rise to complex condensation products showing yellow colour and fluorescence. With mixtures of lysine and free hexose, aldimine derivatives of hydroxymethylfurfural can be formed that can undergo further condensation to give rise to human-like coloured condensation products (Gottschalk and Partridge, 1950).

CONCLUSION

To conclude, it seemed possible to hope that the fairly large range of experiments I have described would now lead us to the possibility of visualising at least some features of the processes that take place during the biosynthesis of elastin. We know from the work of Anna Kadar and her colleagues (Kadar et al., 1971a, b) that with the concentric laminae of the large blood vessels the surface of the cell lies quite closely to the elastin surface that is gowing by accretion. We do not know in what condition the elastin precursor is transported in the cell. It may be protected by the addition of a 'Transport peptide' or it may rely on its own very high positive charge (due to the presence of some 40 or so lysine residues in an otherwise very non-polar molecule) to prevent close approach and precipitation while still within the cell.

Quite clearly however, at some point as the elastin precursor leaves the cell and enters the extracellular space, it must acquire a new property—the property of combining in a specific way with a growing surface of the elastin gel. This is to say that the soluble elastin precursor must not combine with itself during diffusion from the cell surface to a specific growing site on the new elastin structure: but here there must be a mechanism to bring about precipitation, correct apposition and final covalent link formation. Thus the precipitation might take place at a nucleation site perhaps on a pre-existing glycoprotein fibre as suggested by Ross and Bornstein (1970): or on a specific kind of elastin surface that by its conformational properties offers a suitable and proper site for further growth.

We do not know in detail how these processes take place, but I refer to Fig. 8 as an indication of one of the physical properties that are peculiar to elastin and a few other proteins and that may turn out to have some relevance in the matter. In the figure two of the U-shaped coacervation curves are shown. The curve on the right is for soluble α-elastin produced from fibrous bovine elastin by partial hydrolysis with oxalic acid. That on the left was produced from the same starting material by a route involving the action of an elastase isolated from a mould.

It will be seen that there is a considerable difference in the isoionic points of the two preparations (as indicated by the lowest point of the U-shaped curve). This difference in the isoionic point is clearly due to the difference in the way the protein has been prepared from fibrous elastin. An acidic isoelectric point represents the loss of amide ammonia and probably also the loss of some basic peptides due to the rather specific action of the particular microbiological enzyme used.

The point I really wish to make by showing these two curves is that it is now possible to visualize that by the action of a proteolytic enzyme with local activity at the site of fibre formation, it would be possible to arrange that coacervate formation occurs only at *that particular site*. Thus the placement of the three-dimensional isotropic elastin gel could be accurately confined to a specific growth site and the placement control would act before the new structure becomes permanent due to chemical crosslink formation.

If this kind of picture were true, it would lead to a rather similar mechanism to that now being envisaged for collagen particularly by Speakman (1971): however I must say that there seem to be so many quite fundamental differences between collagen and elastin that I am inclined to accept the view of the early histologists who insisted that the two fibrous proteins have a widely different evolutionary origin. Elastin seems to have a specific association with smooth muscle.

What I am now led to believe may be summarized as follows:

1. Although salt soluble elastin has an amino acid composition identical with fibrous elastin when allowance is made for its extra lysine content, it may exist as an extracellular protein in this form, only in copper deficiency and perhaps in BAPN lathyrism also.

2. In the normal physiological process the lysine residues available for oxidation are oxidized by copper-pyridoxal enzymes at the cell surfaces or in surface organelles of the cell.

3. The oxidized elastin precursor still contains some ten or so lysine residues: thus its isoelectric point is still too high to allow it to coacervate at physiological pH and ionic strength.

4. There may be some intervention of a regulatory proteolytic enzyme confined to the site of fibre formation, in order to adjust the isoelectric point of the precursor protein to an exact value. Accretion of elastin must take place exactly at predetermined growth points or points of repair. As we now know (Davis and Anwar, 1970) a coacervate of oxidized elastin precursor will need no catalysis by enzymes to bring about permanent polymerization: such a thing would surely be far too dangerous a material to allow loose in the extracellular space.

5. Finally it should be noted that the first process of crosslink formation—the formation of a reversible aldimine bond or Schiff base—is itself a mechanism for reducing the positive charge on a surface. The accretion of elastin would take place on the growing surface of the elastin fibre because, with each addition of a molecule, positive charges would be lost by the formation of aldimines from ε-amino residues. With a mechanism of this sort uncontrolled spacefilling with the protein elastomer would be avoided because the isoelectric point of the precursor protein is finely adjusted to ensure that, at the pH and ionic strength of the extracellular space, it lies just outside the coacervation range.

Finally I have to apologize if I seem to have laboured my point concerning the fine-trigger adjustment of the precipitation of the salt soluble elastin, but it seems that, although protein chemists have succeeded in making significant progress in the last decade with knowledge of the primary and secondary structures of some of the more important biopolymers, we still appear to be as far away as ever from the solution of the outstanding biological problems with which we are concerned: those concerning repair processes in vessel walls after damage; or the fine adjustment of growth of elastin formations in response to increasing load during early growth of the animal. These latter control mechanisms appear to be triggered by extracellular mechanisms and transmitted through fine changes in elastin tertiary structure as a result of stretching. As has been indicated throughout this chapter, the labile tertiary structure of elastin presents an ideal situation for transmitting a stretch response; either in the form of changed enzyme susceptibility or in the form of changes in the very exacting surface requirements for elastin growth points.

REFERENCES

Bailey, A. J., Perch, C. M. and Fowler, L. J. (1969). In "Chemistry and Molecular Biology of the Intracellular Matrix" (E. A. Balazs, ed.) Vol. 1, pp. 385–404. Academic Press, New York.
Balo, J. and Banga, I. (1949). *Nature* **164,** 491.
Barnes, M. J. and Partridge, S. M. (1968). *Biochem. J.* **109,** 883.
Cox, R. W. and O'Dell, B. L. (1966). *J. Roy. Microsc. Soc.* **85,** 401–409.
Davis, N. R. and Anwar, R. A. (1970). *J. Amer. Chem. Soc.* **92,** 3779–3782.
Davis, N. R. and Bailey, A. J. (1971). *Biochem. Biophys. Res. Commun.* **45,** 1416.
Franzblau, C., Sinex, M., Faris, B. and Lampidis, R. (1965). *Biochem. Biophys. Res. Commun.* **21,** 572–582.
Gelotte, B. J. (1960). *J. Chromatogr.* **3,** 330.
Gottschalk, A. and Partridge, S. M. (1950). *Nature* **165,** 684.
Greenlee, T. K. Jr., Ross, R. and Hartman, J. L. (1966). *J. Cell Biol.* **30,** 59.
Kadar, A., Gardner, D. L. and Bush, V. (1971a). *J. Path.* **104,** 253–260.

Kadar, A., Gardner, D. L. and Bush, V. (1971b). *J. Path.* **104,** 261–266
Laurent, T. C. (1964). *Biochem. J.* **93,** 106.
Lent, R. W., Smith, B., Salcado, L. L., Faris, B. and Franzblau, C. (1969). *Biochemistry* **8,** 2837.
Mammi, M., Gotte, L. and Pezzin, G. (1968). *Nature* **220,** 371.
Mechanic, G., Gallop, P. and Tanzer, M. (1971). *Biochem. Biophys. Res. Commun.* **45,** 644.
Mukherjee, D. P. (1969). Thesis. Massachusetts Institute of Technology, Boston, January, 1969.
Partridge, S. M. (1962). "Elastin" *Advances Protein Chem.* Vol. 17, pp. 227–297. Academic Press, New York.
Partridge, S. M. (1965). Eleventh Procter Memorial Lecture. *J. Soc. Leather Trades Chemists* **49,** 41.
Partridge, S. M. (1966). *Fed. Proc.* **25,** 1023–1029.
Partridge, S. M. (1967a). *Nature* **213,** 1123–1125.
Partridge, S. M. (1967b). *Biochim. Biophys. Acta* **140,** 132–141.
Partridge, S. M. (1970). *Advances Biol. Skin* **10,** 69–87.
Partridge, S. M., Davis, H. F. and Adair, G. S. (1955a). *Biochem. J.* **61,** 11–21.
Partridge, S. M. and Davis, H. F. (1955b). *Biochem. J.* **61,** 21–30.
Partridge, S. M., Elsden, D. F. and Thomas, J. (1963). *Nature* **197,** 1297.
Partridge, S. M., Elsden, D. F., Thomas, J., Dorfman, A., Telser, A. and Pei-Lee Ho (1964). *Biochem. J.* **93,** 30C.
Pikkarainen, J. and Kulonen, E. (1972). *Comp. Biochem. Physiol.* **41,** No. 4B, 705-713.
Richards, F. M. and Knowles, J. R. (1968). *J. Molec. Biol.* **37** 231.
Robert, L. and Poullain, N. (1963) *Bull. Soc. Chim. Biol.* **45,** 1317–1325.
Ross, R. and Bornstein, P. (1970). *In* "Chemistry and Molecular Biology of the Intercellular Matrix" (E. A. Balazs, ed.) Vol. 1, pp. 641–655. Academic Press, New York.
Sandberg, L. B., Weissman, N. and Smith, D. W. (1969). *Biochemistry* **8,** 2940.
Sandberg, L. B., Weissman, N. and Gray, W. R. (1971). *Biochemistry* **10,** 52–56.
Shotton, D. M. and Hartley, B. S. (1970). *Nature* **225,** 802.
Shotton, D. M. and Watson, H. C. (1970). *Nature* **225,** 811.
Siegel, L. M. and Monty, K. J. (1966). *Biochim. Biophys. Acta* **112,** 346.
Smith, D. W., Weissman, N. and Carnes, W. H. (1968). *Biochem. Biophys. Res. Commun.* **31,** 309–315.
Smith, D. W., Brown, D. M. and Carnes, W. H. (1972). *J. Biol. Chem.* **247,** 2427–2432.
Speakman, P. T. (1971). *Nature* **229,** 241–243.
Starcher, B. C., Partridge, S. M. and Elsden, D. F. (1967). *Biochemistry* **6,** 2425–2432.
Thomas, J., Elsden, D. F. and Partridge, S. M. (1963). *Nature* **200,** 651.
Urry, D. W., Starcher, B. C. and Partridge, S. M. (1969). *Nature* **222,** 795–796.
Weis-Fogh, T. (1961). *J. Molec. Biol.* **3,** 520.
Weis-Fogh, T. and Anderson, S. O. (1970a). *Nature* **227,** 718.
Weis-Fogh, T. and Anderson, S. O. (1970b). *In* "Chemistry and Molecular Biology of the Intercellular Matrix" (E. A. Balazs, ed.) Vol. 1, pp. 671–684. Academic Press, New York.

The Generation and Maintenance of Parallel Arrays in Cultures of Diploid Fibroblasts

TOM ELSDALE

Medical Research Council, Clinical and Population Cytogenetics Units
Western General Hospital, Crewe Road, Edinburgh 4, Scotland

INTRODUCTION

So useful is the concept of the cell in biology that it seems ungrateful to point out its limitations. Knowledge of nerve cells does not suffice for an understanding of the brain, although the cellular foundations of the brain are not in doubt. There are some intriguing situations at the borderland where the concept of the cells loses explanatory power. An example is provided by the simple morphogenesis that occurs in mass cultures of normal fibroblasts. The individual cells are distinguished their properties open to investigation. At the same time phenomena are encountered whose rationalization seems to leave the cell less sharply focussed, no longer quite in the centre of the stage.

METHODS AND BACKGROUND

The observations described were made on early subcultures of human, diploid fibroblasts obtained from therapeutic abortions. Lung fibroblasts were used routinely, occasional use was made of strains derived from skin, salivary gland, mammary gland, and kidney. See previous publications for culture methods (Elsdale, 1968; Elsdale and Foley, 1969).

Fibroblast form (Fig. 1)

Lone cells in sparse platings usually extend in the *ruffling membrane form*. The cell possesses one or more expanded, fan-like pseudopodia, the surfaces of which exhibit a characteristic 'ruffling' activity. A pseudopodium may apparently be produced anywhere on the cell; the succession of these transient extensions results in an interrupted and haphazard gait.

Even in sparsely initiated cultures, not all the cells extend in the *ruffling membrane form*, some will approximate to the *bipolar spindle form*. The cell exhibits a stable differentiation between pseudopodial and non-pseudopodial surface, the former confined to the two ends of the

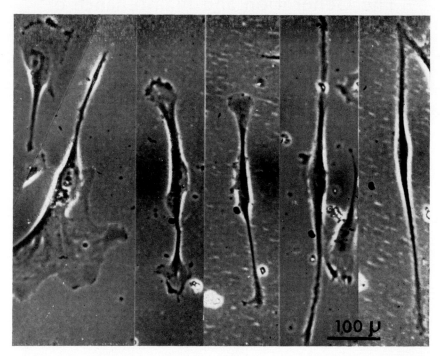

Fig. 1. A composite figure showing 6 cells observed in a single culture of fibroblasts a day after initiation. These illustrate the spectrum of forms from *ruffling membrane forms* on the left to *bipolar spindle forms* on the right. Phase contrast, living cells.

spindle. Ruffling membranes are not observed, the compact pseudopodia terminate in several fine prongs.

Intermediates are observed; the two forms as described can be regarded as the two extremes between which cell form modulates according to the prevailing conditions. *Ruffling membrane form* is accentuated by sparse culture (absence of crowding), pristine glass and plastic surfaces, and collagen deprivation. *Bipolar spindle form* is stabilised by the reverse of these conditions. As a culture approaches confluence there is a gradual assumption of the bipolar spindle form; virtually all the cells in a confluent culture are so stabilized.

The behaviour of fibroblasts is not the same in the two forms. Contact inhibition as described by Abercrombie (1970) and Trinkhaus et al. (1971) refers to cells in *ruffling membrane form*. This form is not present in the confluent cultures to be described.

Growth beyond confluence

Human, fetal, fibroblast strains normally grow beyond confluence to stationary densities characteristic for the tissue of origin, (Bard and Elsdale, 1972). Fetal lung strains grow to stationary densities around 10×10^6 cells/50 mm Petri plate. Taking confluence as $1.5 - 2.0 \times 10^6$ cells, those cultures achieve stationary populations around 5–6 monolayer equivalents.

DESCRIPTION OF THE DEVELOPMENT OF PARALLEL ARRAYS
Random initiation

Trypsinization is a scrambling procedure, the cells plated from a cell suspension extend themselves randomly arranged and distributed in the new Petri dish (Fig. 2).

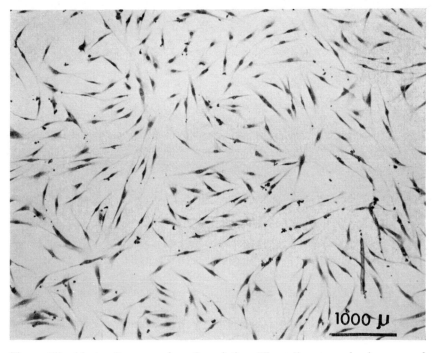

Fig. 2. Fibroblast culture one day after plating. The cells are randomly arranged and distributed. Fixed preparation.

The confluent patchwork

As a result of chance encounters between the cells in a sparsely initiated culture, side-by-side associations are formed. These associations grow by accretion (Fig. 3). Virtually all the cells in a confluent culture belong to one of these associations, which because they formed independently of one another give the confluent monolayer a patchwork appearance.

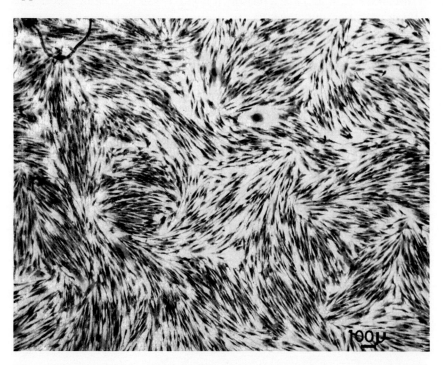

Fig. 3. Fibroblast culture approaching confluence. The monolayer is a patchwork of many, smaller, independently formed parallel associations. Fixed preparation.

These associations are arrays within which each cell lies parallel with its immediate neighbours. The angular difference in the orientation of the cells in adjacent arrays can be measured across the intervening discontinuity. The minimum angle so measured is in the order of 20°. Where the angular difference in the orientation of the cells in neighbouring arrays is less than this figure, the arrays merge at confluence and no discontinuity arises.

The confluent monolayer patchwork is but a transient stage in the development of pattern within a fibroblast culture. Post-confluent growth leads to the generation of multilayered stacks of cells built over and obscuring the underlying patchwork (Elsdale and Bard, 1972). It is possible to inhibit this development without inhibiting post-confluent growth. All the cells produced by subsequent proliferation then remain accommodated within the patchwork. The arrays become densely crowded and the discontinuities achieve an exceptional clarity (Fig. 4). This result is obtained by preventing the normal accumulation of collagen by adding low concentrations of bacterial collagenase to the medium, see Elsdale and Foley, 1969. This procedure provides a useful extension to observations on the evolution of the patchwork and the discontinuities therein.

Simplification

Provided the conditions for cellular motility are maintained by frequent changes of the medium, the patchwork gradually simplifies during prolonged cultivation. Superficially, the interaction of individual cells to generate arrays is repeated on a larger scale by the interaction of arrays moving so as to merge together according to the 20° rule previously mentioned. Cultures may eventually become dominated by a single large array, this being the end point onto which the morphogenetic process is homing. Furthermore the end point is the same whether multilayering is inhibited by collagenase or not. The generation and eventual dispersal of the multilayered stacks is but an interlude in the steady progression towards the global array.

The large array

The large array is perfectly stable; no multilayering takes place within it and no new discontinuities arise. Not only is the array self-maintaining, it is self-replacing too for cuts and blemishes inflicted upon it are effaced. Indeed if the whole pattern is scrambled by trypsinization and the cells set out again at random the whole process is repeated.

The large array therefore constitutes a simple situation exhibiting those properties of maintenance and repair which in the context of more complex systems are quite mysterious.

ANALYSIS

Status of the individual cell

The individual cell in a sparse culture is a unit of independent action entering into form generating interactions. This independence is lost

Fig. 4. Stationary culture maintained in the presence of collagenase. Multilayering is inhibited, the arrays are crowded. The discontinuities between arrays are especially clear. Fixed preparation.

when cells are incorporated into arrays within which their activities are concerted. It is the arrays that possess individuality as the units of interaction within the patchwork. The morphogenetic process is now to be understood in terms of the rules governing the formation, the stability and the eventual disappearance of the discontinuities within the patchwork. These rules are validated by observation and their internal consistency, they do not have to await a cellular or molecular explanation. As simplification proceeds the number of interacting units declines. Within the large array free of discontinuities no new generating interactions are possible, only those intimate interactions between the anonymous cellular constituents by which the array is dynamically maintained.

Cell motility

By motility is implied a scalar term providing a measure of displacement. This scalar term is to be distinguished from the corresponding vector term including a directional component for which the word movement will be used.

Time lapse films reveal high motility within arrays no matter how crowded. Movement is confined to motions up and down the array, movements, that is to say, that reaffirm and do not disrupt the array. Motility within the immediate vicinity of discontinuities is low; these are sites of semi-permanent contact inhibition. With the additional assumption that fibroblasts are inherently motile and move unless prevented, an overall view of the generation of large arrays can be constructed. We can envisage the cells in a fibroblast culture continuously exploring their immediate environment and taking advantage of the opportunities it offers for the exercise of motility. A cell halted at a discontinuity will eventually seek another avenue of movement. The summation of these movements will produce a resultant trend towards the configuration that allows the maximum exercise of motility. Motility is least inhibited in the absence of discontinuities. The resultant configuration will be therefore the large array free from discontinuities. It is no doubt true, but platitudinous to suggest that the large array is located in a trough or a well in the surface-free energy contours.

This overview, however, leaves an important consideration vague; how are the discontinuities eliminated during the simplification of the patchwork?

Discontinuities in the patchwork

It might be thought that the arrays in a confluent culture would come to interlock in an almost limitless number of ways, with the discontinuities taking a multiplicity of forms. Only two forms of discontinuity are observed, namely the meeting of two arrays and the meeting of three arrays. This restriction has important consequences.

Computing the index around the curve

This procedure provides the basis for classifying discontinuities.

Consider first the trivial case, where there is no discontinuity, provided by the parallel array. Sum in radians the angular difference between each element and the next as encountered in the course of making a clockwise circular tour starting and finishing at some arbitrary point. Divide this sum by 2π. The index is here 0.

Consider next a pattern of radially disposed elements around a central discontinuity.

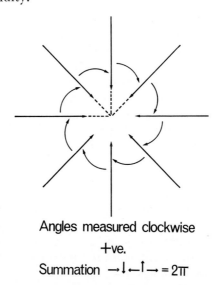

Angles measured clockwise

+ve.

Summation →↓←↑→ = 2π

By convention the tour is made in a clockwise direction. Note that the angle between one element and the next is in this case also measured in a clockwise direction. This defines the angles so measured as positive. The sum yields $+2\pi$. The index is $+1$. For brevity we will speak of a $+1$ discontinuity. Observe that if a half of this pattern were replaced by a parallel array the sum would be π and the index $+\frac{1}{2}$.

Consider next the following figure that represents the meeting of four arrays.

Notice that the angles between one element and the next are here measured in counterclockwise direction where the tour as before is

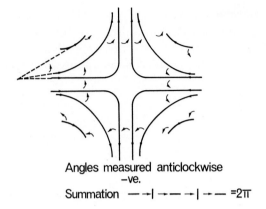

Angles measured anticlockwise −ve.
Summation — →| → — →| → — =2π

made in a clockwise direction. This defines the angles so measured as negative. The sum is −2π. The index is −1.

If half of this pattern is replaced by a parallel array the sum is −π, and the index is −½. This pattern constitutes the meeting of three arrays.

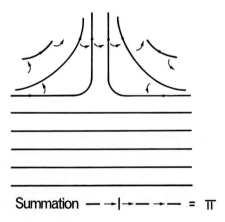

Summation — →|→—→— = π

There are two advantages of this treatment. First, it defines two sets of discontinuities, those with positive and those with negative indices. Forms of like sign are related, forms of unlike sign are unrelated and cannot be deformed one into another without making or eliminating discontinuities.

Second, this treatment is an aid to appreciating why only two, namely the $+\frac{1}{2}$ and $-\frac{1}{2}$ discontinuities are stable within the fibroblast patchwork.

Stability of discontinuities

Both the $+1$ and -1 discontinuities are inherently unstable and break up into two $+\frac{1}{2}$s and two $-\frac{1}{2}$s respectively.

(a) The $+1$ form. This is unstable along any diameter.

(b) The -1 form. This form is unstable along two axes.

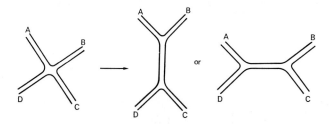

Higher members of the negative series, provided by the meeting of 5, 6 or more arrays are similarly unstable and break down into 3, 4 or more $-\frac{1}{2}$ discontinuities. The instability of those forms is also observed in the slippage between soap bubbles leading to hexagonal packing. $-\frac{1}{2}$ forms are observed in finger and palm prints (Fig. 5).

Discontinuities observed in the fibroblast patchwork

Regarding the $+1$ form it might be thought that this was an improbable formation to observe in fibroblast cultures grown up from a random initiation. Actually it results from fibroblasts doing their thing and forming side-by-side associations, it constitutes a single array with a central discontinuity. Quite good examples have occasionally been observed in cultures approaching confluence (Fig. 6a).

The -1 form is very rare, but has also been observed (Fig. 6b).

The stable discontinuities observed are the $+\frac{1}{2}$ and $-\frac{1}{2}$ forms only, Figs. 7 and 8.

The elimination of discontinuities during simplification of the patchwork

If the situation around discontinuities was entirely static simplification of the patchwork could not take place. The remarks above on

Fig. 5. Finger prints showing $-\frac{1}{2}$ discontinuities (triradii) and orthogonal forms of $+1$ and $+\frac{1}{2}$ discontinuities.

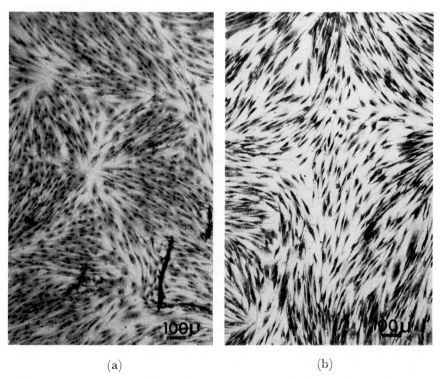

(a)   (b)

Fig. 6. Imperfect $+1$(a) and -1(b) forms appearing transiently in a culture approaching confluence. Fixed preparation.

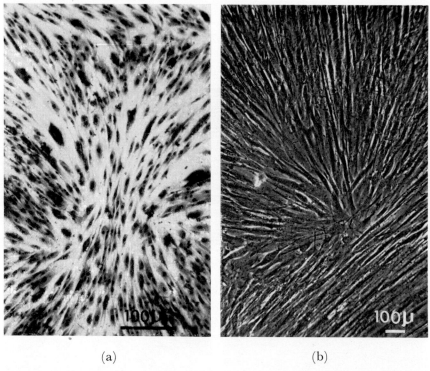

(a) (b)

Fig. 7. $+\frac{1}{2}$ discontinuities. (a) separation of two $+\frac{1}{2}$'s at an unstable $+1$, fixed preparation. (b) $+\frac{1}{2}$ discontinuity, phase contrast living cells.

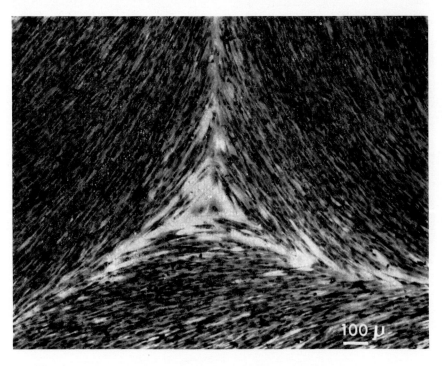

Fig. 8. $-\frac{1}{2}$ discontinuity at the meeting of three arrays. Fixed preparation.

motility imply no more than that a differential exists. The $+\frac{1}{2}$ discontinuities, in particular, are motile and migrate within the culture like a ripple through water. Fig. 9 illustrates how the collision between a $+\frac{1}{2}$ and a $-\frac{1}{2}$ discontinuity results in the disappearance of both. As the

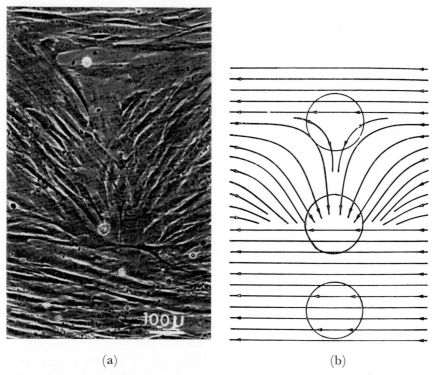

(a)   (b)

Fig. 9. Collision and mutual elimination of $+\frac{1}{2}$ and $-\frac{1}{2}$ discontinuities. (a) Collision, $-\frac{1}{2}$ above, $+\frac{1}{2}$ below, phase contrast living cells. (b) Diagrammatic representation of (a) The arrows define the $+\frac{1}{2}$ and $-\frac{1}{2}$ discontinuities. The cells in the centre more outside the figure, the horizontal array in the lower half of the figure grows by accretion and eventually merges with the similarly orientated array at the top of the figure.

cellular movements that engender this collision cannot themselves create new discontinuities, a mechanism is provided for the elimination of discontinuities within the patchwork.

Observation suggests that roughly equal numbers of $+\frac{1}{2}$ and $-\frac{1}{2}$ discontinuities arise in the initial patchwork, and equality is predicted as a consequence of Cauchy's theorem. Some $+\frac{1}{2}$ discontinuities are eliminated by knocking themselves out against the side of the dish, with the result that at the end of the day an excess of $-\frac{1}{2}$ discontinuities is present that cannot be got rid of.

SYNTHESIS

Generation and maintenance of arrays, an inherently precise process

A consideration of various ways of making and shaping things shows that certain methods possess the property of inherent precision. For example, using a lathe is not an inherently precise method because the precision of the work is limited by the tolerances to which the lathe was made. On the other hand, grinding two blanks together so that they bed even more snugly together is the basis of the inherently precise method for grinding spherical surfaces.

Given that the elements within a system move at random except in so far as they constrain one another according to some rule, the outcome of their activities, provided the elements remain articulated, will be that the system approaches a configuration that exemplifies displacement congruence. There is no limit in principle to the closeness of approach to this configuration—hence the designation is inherently precise. Notice that two spherical surfaces of the same radius articulated to form a joint are in contact at all points in the overlap—congruence, and remain so under displacement of the joint—displacement congruence. Similarly the movements of cells within an array, reaffirm the parallel arrangement; the array exemplifies displacement congruence. The principles of the inherently precise method, the application to fibroblast cultures, the implications for biological systems in general, and the manner in which an analog is provided for Waddington's concept of the chreod is discussed elsewhere (Elsdale, 1972). However, there is a consequence of the inherently precise hypothesis that is profoundly important in a discussion of the generation and maintenance of parallel arrays.

Consider, in particular, the maintenance of arrays. Arrays exhibit dynamic stability, the cells move to reaffirm the array. What controls the movement of the cells and prevents them from destroying the array? On the basis of the discussion so far, the idea can be developed that as motility is maximized the more precisely the cells adopt the parallel arrangement. Should a cell within an array find itself out of line, the constraints on motility will be increased, and it would find its way back to parallel where motility will be least constrained.

This type of control, no matter how precise the operation is taken to be, and one is free to conceive it as infinitely precise if you want, nevertheless allows the cells, in principle at least, a residual freedom to make small excursions from parallel. This follows because the control only becomes effective once a cell has so departed, although the necessary departures may be conceived as infinitesimally small. However, given this freedom the parallel array can no longer be considered as exemplifying displacement congruence, and the process of generation cannot therefore be taken as inherently precise. The inherently precise hypothesis predicts therefore that a quite different control will dominate movement within arrays.

The type of control considered above is an example of return control whose properties are diagrammed in Fig. 10. The inherently precise

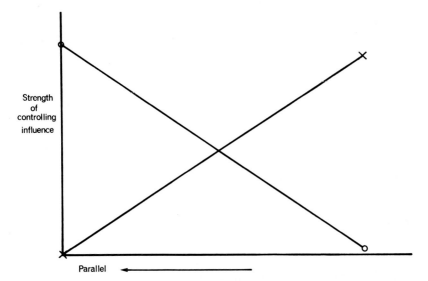

Fig. 10. Diagram contrasting return control x ──── x and initial departure control o ──── o. Return control does not become effective until a cell departs from parallel, the curve therefore rises from the origin. Initial departure control is maximally effective to restrain departures from parallel, the curve therefore declines from a high point above the origin.

hypothesis demands a control that is maximally effective to prevent initial departures from parallel, termed therefore initial departure control. The properties of this control are also diagrammed in Fig. 10 which demonstrates at a glance the contrasting feature of these two types of control.

Demonstration of initial departure control

It is necessary to contrive situations where the predictions based on the two types of control are different. The observations will be described in detail elsewhere, only a summary account is provided here. Consider the situation where a part of an array has been removed leaving a free edge parallel to the long axis of the cells, i.e. in analog with a ploughed field, a free edge along the furrows as opposed to across the furrows.

Return control implies that cells are free to make small departures from parallel. It predicts that the cells lining the edge will make gestures towards the free space provided. These gestures, however, encounter no restraint. A wholesale migration of cells from the free edge is predicted.

The existence of initial departure control carries a different prediction. The cells at the free edge are still incorporated in the array and are under a restraint preventing their departure. No wholesale migration from the free edge is predicted.

Observation reveals some loss of cells from a free edge in the course of a week, quite long stretches remain unchanged. There is no wholesale migration.

Further experiments demonstrate the growth of an edge by the trapping and incorporation of free cells seeded in the vicinity. This behaviour is positively predicted on the basis of initial departure control, whereas return control is neutral. These experiments are taken to confirm the existence of initial departure control.

The nature of the initial departure control

I suggest that the relevant mode of cellular interaction is lateral adhesion with slippage. Consider a rubber soap holder with suction feet of the type that might be used in a shower cabinet. If the walls of the cabinet are wet, the soap holder can be readily slid over the surface but it resists a straight pull off the wall. Applied to cells this means that the bipolar spindles are free to slip past one another but other movements involving a pulling apart of the cells will encounter resistance.

It is relevant that people interested in how cells move have explored the possibility that electrostatic forces are involved in locomotion. The difficulty, from their point of view, is that these forces provide for adhesion with slippage.

Return control and initial departure control fulfil complementary functions in accordance with the different ways they graph in the above diagram. Indeed the graph also represents diagrammatically the relative contributions of the two controls during stages in the progression from randomness to global array.

CONCLUSION

1. The generation of arrays is a stochastic process. The cells are individually and independently energized, a fibroblast normally moves

unless prevented from so doing. Movements are random except in so far as they are otherwise constrained by cellular interactions.

2. Two modes of cellular interaction are important.

(a) The maximizing of motility under mutual constraints, directs the movement of cells into parallel.

(b) Lateral adhesion with slippage results in the trapping of cells in side-by-side associations, and results in the dynamic stability of arrays.

These interactions by providing for both return control and initial departure control define the form generating method as an inherently precise process.

3. The two stable forms of discontinuities allow for mutual elimination.

4. Repair potential is a property of inherently precise systems. It results from the continued availability of the return control and initial departure control mechanisms, founded as they are, on basic cellular interactions.

5. The return control mechanism depends on specific cellular properties under close genetic control. The initial departure control mechanism depends on non-specific biophysical imperatives.

A gloss on contact inhibition

This term has been used by previous authors to refer to the cessation of movement observed following the pairwise collisions of fibroblasts in a sparse field. Contact inhibition so defined may be considered as the simplest special case within a more general scheme of fibroblast behaviour appropriately treated in terms of the discontinuities, their forms, stabilities, and interactions, observed in mass cultures. The simplest case because contact inhibition as previously described generalizes on pairwise interactions of cells, and thereby gives no inkling of the sophistications that arise within larger associations of cells. A special case because the cells are described in *ruffling membrane form*, a form that appears to be an artifact of an especially unnatural condition of culture.

ACKNOWLEDGEMENTS

I should like to thank Rae Fairholm and Maureen Atkinson for technical assistance, Norman Davidson, Sandy Bruce and Ian Wadell who did the drawings and photographic processing, Christopher Zeeman who saw topology in my photographs of the fibroblast patchwork, and showed me how to compute the indices around curves and Jonathan Bard for stimulating discussions throughout.

REFERENCES

Abercrombie, M. (1970). *In Vitro* **6**, 128–142.
Bard, J. and Elsdale, T. (1972). *In* "The Regulation of Growth in Cell Cultures" (J. Knight, ed.) pp. 187–206. CIBA Fnd. Symp. Churchill, London.
Elsdale, T. (1968). *Exp. Cell Res.* **51**, 439–450.
Elsdale, T. and Foley, R. (1969). *J. Cell Biol.* **41**, 298–311.
Elsdale, T. and Bard, J. (1972). *Nature* **236**, 152–155.
Elsdale, T. (1972). *In* "Towards a Theoretical Biology" Vol. 4. (C. H. Waddington, ed.) pp. 95–108. Edinburgh University Press.
Trinkhaus, J. P., Betchaku, T. and Krolikowski, L. S. (1971). *Exp. Cell Res.* **64**, 291–300.

II

Cytological Aspects and Differentiation

Some Overt and Covert Properties of Chondrogenic Cells

RICHARD MAYNE, JOHN R. SCHILTZ and HOWARD HOLTZER

Department of Anatomy, School of Medicine, University of Pennsylvania
Philadelphia, Pa. 19104, USA

Several criteria, both histological and biochemical, enable one to distinguish mature and functioning chondroblasts from functioning fibroblasts. Both *in vivo* and *in vitro* chondroblasts tend to be round or polygonal and relatively immotile in contrast to the amoeboid and more variably shaped fibroblasts (Holtzer and Abbott, 1968). Chondroblasts are embedded in a metachromatic matrix, generally a mixture of chondroitin 4- and 6-sulfates, and they synthesize little if any hyaluronic acid (Thorp and Dorfman, 1967; Shulman and Meyer, 1968). Also, chondroblasts synthesize a unique species of collagen designated α-1 (II) (Miller and Matukas, 1969; Trelstad *et al.*, 1970; Miller, 1970a, b), and which may be the only chain present in some types of cartilage (Strawich and Nimni, 1971). In contrast to chondroblasts, collagen extracted from skin (Miller *et al.*, 1971) or from bone (Toole *et al.*, 1972) or tendon (Bornstein, 1969) contains *at least* two different kinds of collagen chains, α-1 (I) and α-2, and it is likely that other tissues synthesize still other species of collagens (Kefalides, 1970; Morris and McClain, 1972). The predominant glycosaminoglycan synthesized by fibroblasts in culture is hyaluronic acid, though in addition they synthesize various amounts and kinds of chondroitin sulfate and other sulfated glycosaminoglycans. The quantities and species of sulfated glycosaminoglycans synthesized by fibroblasts either *in vivo* or *in vitro* varies greatly depending on their *in situ* location, their hormonal milieu and other environmental conditions (Schubert and Hamerman, 1968).

In addition to these criteria involving large populations of cells, there is also a powerful biological method to discriminate between the two phenotypes even at the level of a *single cell*, and that is the clonal test. Clones derived from a single chondroblast differ markedly both in their morphological and biochemical properties from a clone derived from a single fibroblast. Chondrogenic clones reared in the appropriate medium consist of immotile, epithelioid cells synthesizing and depositing chondroitin 4- and 6-sulfates, whereas fibroblasts in the same culture medium yield progeny of stellate or spindle-shaped, amoeboid cells synthesizing predominantly hyaluronic acid (Abbott and Holtzer, 1968; Abbott et al., 1972).

In this report we describe experiments relating to some covert genetic properties of terminally differentiated chondroblasts. Specifically, we will present data which suggests: (1) terminally differentiated chondroblasts can be induced to change into cells morphologically, and to a considerable degree biochemically, indistinguishable from some authentic fibroblasts; (2) movement of the induced, embryonic chondrogenic cell to the mature chondroblast involves DNA synthesis and cell division; and (3) presumptive chondrogenic cells from somites that incorporate 5-bromo-2-deoxyuridine (BrdU) into their DNA fail to mature into functional chondroblasts even though they exist as replicating cells for many generations.

THE TRANSFORMATION OF CHONDROBLASTS INTO FIBROBLASTS (OR FIBROBLAST-LIKE CELLS)

Glycosaminoglycan synthesis by non-replicating chondroblasts in intact trunks

A reasonably pure population of frank chondroblasts may be obtained by carefully stripping away the adhering connective tissue from the vertebral cartilages of 10-day chick embryos. Histological inspection of such cartilaginous trunks immediately after dissection revealed that approximately 2% of the cells were not surrounded by metachromatic matrix and of these many were probably perichondrial cells (Holtzer et al., 1960; Abbott and Holtzer, 1966). Equally convincing evidence that few fibroblasts were inadvertently included in such preparations is the biochemical data illustrated in Figs. 1 and 2. Ten-day cartilaginous trunks were grown for 2 days in F-10 supplemented with 10% fetal calf serum (henceforth called simply F-10 or control medium). ^{3}H-glucosamine was added for this period and the trunks removed and analyzed for total glycosaminoglycan by the methods described in Abbott et al. (1972). As shown in Fig. 1 virtually all the readily identifiable glycosaminoglycans synthesized by chondroblasts in the intact cartilaginous trunk is chondroitin sulfate. The relative absence of labeled hyaluronic acid demonstrates that few if any fibroblasts are present in the preparation. That the peak in Fig. 1 is authentic chondroitin sulfate is indicated by the following: (1) approximately

94% of the labeled hexosamine is galactosamine; (2) 92% of the labeled material is equally digested by chondroitinases ABC or AC (Saito et al., 1968; Yamagata et al., 1968).

It is known that embryo extract (EE) under certain conditions interferes with the protracted synthesis of chondroitin sulfate by replicating chondroblasts (Coon, 1966; Coon and Cahn, 1966; Nameroff and Holtzer, 1967). Accordingly, it was of interest to rear essentially non-

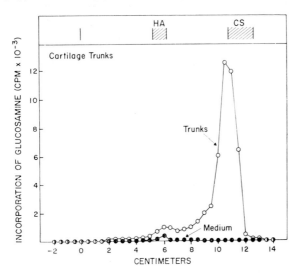

Fig. 1. Freshly isolated trunks of cartilage were incubated in F-10 for 48 h in the presence of ^{3}H-glucosamine. Glycosaminoglycans were isolated from both trunks and the medium, and then fractionated by high-voltage electrophoresis. HA, hyaluronic acid standard; CS, chondroitin sulfate standard.

replicating chondroblasts in intact 10-day trunks in F-10 plus 5% EE. As shown in Fig. 2 non-replicating trunk chondroblasts in medium with or without EE display the same synthetic profile of glycosaminoglycan synthesis; indeed in EE there is a net increase in ^{3}H-glucosamine incorporation into chondroitin sulfate (see also Chacko et al., 1969a). These results suggest there is no unique, large molecular weight, heat labile molecule in EE which selectively depresses chondroitin sulfate synthesis.

At appropriate concentrations BrdU suppresses the synthesis of chondroitin sulfate in liberated and replicating chondroblasts from 10-day trunks without grossly blocking rates of DNA synthesis, cell division, RNA synthesis or total protein synthesis (Abbott and Holtzer, 1968; Schulte-Holthausen et al., 1969; Holtzer et al., 1970; Holtzer et al.,

1973a). The effect of the analog can be prevented by adding excess TdR or by interfering with DNA synthesis with cytosine arabinoside, hydroxyurea or even cycloheximide (Mayne et al., 1973). The paradoxical effect of BrdU is intimately coupled to cell replication. Since only a modest number of chondroblasts replicate in 10-day trunks during these experiments, exposure to BrdU should not grossly impair the

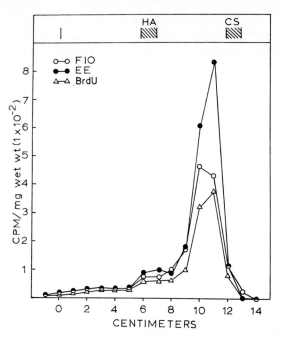

Fig. 2. Freshly isolated trunks of cartilage were incubated in F-10 with or without EE (5%) or BrdU (10 μg/ml). All cultures were exposed to ^{3}H-glucosamine for 48 h. The trunks were weighed, glycosaminoglycans isolated and fractionated by high voltage electrophoresis.

synthesis of glycosaminoglycans. Fig. 2 shows that, as anticipated, BrdU did not interfere with the synthesis of chondroitin sulfate by trunk chondroblasts. (For similar results where replication of chondroblasts was inhibited by high density see Mayne et al., 1973.)

In summary: essentially non-replicating chondroblasts in intact 10-day trunks grown for 2 days in F-10 with or without EE or BrdU continue to synthesize chondroitin sulfate. If these chondroblasts synthesize hyaluronic acid the quantities are below the resolution of the techniques used.

The glycosaminoglycan profiles of liberated, replicating chondroblasts

The great majority of chondroblasts liberated by trypsin from 10-day trunk cartilages are induced to synthesize DNA and subsequently divide (Holtzer et al., 1960; Abbott and Holtzer, 1966; Nameroff and Holtzer, 1967; Murison, 1972). Whether the progeny of such chondroblasts display the cartilage phenotype or transform into 'fibroblastic' or 'dedifferentiated' chondroblasts depends in part on the density at which they are grown, in part on the composition of the medium (Coon, 1966; Chacko et al., 1969a; Pawelek, 1969), and in part on the condition or kinds of cells with which they come into contact (Chacko et al., 1969b; Holtzer et al., 1970). The following experiments demonstrate that the addition of embryo extract to F-10, which had little effect on non-replicating trunk chondroblasts, dramatically alters the phenotypic behavior of *replicating* chondroblasts. Ten-day trunks were trypsinized and the liberated cells plated at a density of 1×10^5 cells/ml in 100 mm tissue culture dishes for 5 days. During this period many cells adhere to the dish; a few of these adherent cells are fibroblastic, but most are polygonal. The floating cells, however, are uniformly spherical and elsewhere have been shown by cloning experiments to constitute virtually a pure population of chondroblasts. Cultures of 'floaters' were grown for 5 days either in F-10, or F-10 plus 5% EE. During the fifth day of culture ^{3}H-glucosamine was added for 24 h and then the cells and medium were analyzed for glycosaminoglycans. It is worth stressing that the cell cycle is approximately 18–20 h in either medium and that replication occurs throughout the culture period although not all cells enter S with equal frequency (Murison, 1972).

As shown in Fig. 3a, the progeny of chondroblasts reared in F-10 synthesize chondroitin sulfate almost exclusively. This glycosaminoglycan is also released into the medium (Fig. 3b). In contrast, aliquots of the same floaters grown for the same period in F-10 plus EE yield a progeny which synthesizes hyaluronic acid in addition to chondroitin sulfate. That this peak is hyaluronic acid is based on the following: (*i*) it migrates on high voltage electrophoresis with the same mobility as a sample of authentic hyaluronic acid (*ii*) 93% of the labeled hexosamine is glucosamine (*iii*) 87% of the peak is sensitive to testicular hyaluronidase. The kinds and ratios of the glycosaminoglycans associated with the cells are also found in the medium; therefore, the changes of glycosaminoglycans with the cells are not due to failure to deposit or trap the extracellular material. If there is a differential degradation of the chondroitin sulfate by cells growing in EE then it must occur intracellularly. These results on the induction of hyaluronic acid synthesis in a population of replicating chondroblasts extend the earlier observations by Nameroff and Holtzer (1967). On the other hand these findings make untenable the claim of Bryan (1968) that the production of hyaluronic acid in a preparation of this kind is to be attributed to

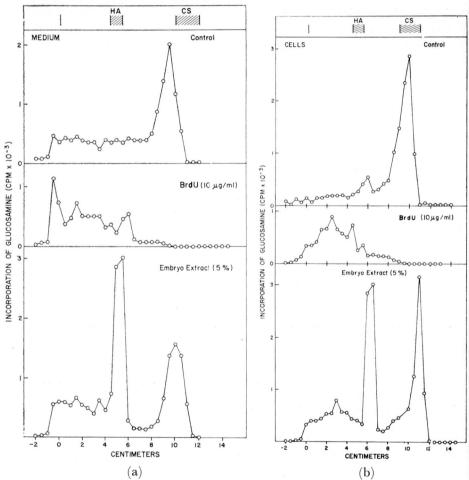

Fig. 3. Chondroblasts derived from 'floater' populations were plated at 1×10^5 cells/ml. After 24 h either embryo extract (5%) or BrdU (10 μg/ml.) was added to some dishes. After 5 days of growth in these conditions ^{3}H-glucosamine was added for 24 h, the glycosaminoglycans isolated and fractionated by high voltage electrophoresis. (a) cell layer; (b) medium. The results have been expressed per equal DNA content of replicate cultures grown in the same conditions.

small numbers of fibroblasts overgrowing the chondroblasts; virtually all cells replicate in these cultures and the cultures grow to comparable densities.

Experiments were carried out to determine if the hyaluronic acid synthesized by the progeny of chondroblasts grown in EE required protein synthesis. Table I shows that, contrary to the result obtained with other kinds of cells (Matalon and Dorfman, 1968; Smith and Hamerman, 1968) hyaluronic acid synthesis and chondroitin sulfate synthesis were equally depressed by cycloheximide (de la Haba and Holtzer, 1965; Telser et al., 1965).

Table 1
Inhibition by cycloheximide of hyaluronic acid and chondroitin sulfate synthesis in dedifferentiated chondroblasts

Chondroblasts from 'floater' populations were grown for 5 days in F-10 with 5% embryo extract, and then ^{3}H-glucosamine was added with and without cycloheximide (20 μg/ml) for 6 h. Glycosaminoglycans were isolated and separated by high voltage electrophoresis. Total counts in the hyaluronic acid and chondroitin sulfate peaks were determined.

Peak	−cycloheximide	+cycloheximide	% inhibition
Hyaluronic acid	2154	141	94
Chondroitin sulfate	2299	166	93

In another series of experiments floaters were grown in F-10 plus BrdU (10 μg/ml). As anticipated (Holtzer and Abbott, 1968; Holtzer et al., 1972) exposure of the replicating cells to BrdU led to a striking depression in the synthesis of chondroitin sulfate (Figs. 3a and b). In some experiments with BrdU a modest peak appeared in the hyaluronic acid region. Whether this peak is real or only more apparent because of the absence of the chondroitin sulfate peak, or whether the BrdU had a slight stimulatory effect on the production of hyaluronic acid is not clear.

The response to BrdU of cells synthesizing hyaluronic acid is not as simple as the response of cells synthesizing chondroitin sulfate. BrdU markedly depresses the syntheses of hyaluronic acid in amnion cells (Bischoff, 1971; Mayne et al., 1971) and in a hybrid line of mouse and hamster cells (Koyama and Ono, 1972). On the other hand, it has little apparent effect on the production of hyaluronic acid by somite cells (Abbott et al., 1972). Therefore, it was of interest to determine whether the hyaluronic acid synthesized by the descendants of chrondroblasts grown in EE was sensitive to BrdU. Table II shows there is a preferential inhibition of chondroitin sulfate synthesis with little inhibition of hyaluronic acid synthesis. The reason why the synthesis of hyaluronic acid is more or less resistance to the effects of BrdU is unknown but does suggest that; (1) there may be a 'constitutive' level of hyaluronic acid synthesis which is insensitive to BrdU (2) in different types of cells different mechanisms regulate the production of this glycosaminoglycan; or (3) there are subtle differences in the kinds of hyaluronic acid different kinds of cells synthesize and in different kinds of cells different structural genes may be involved.

Table II
Effect of BrdU on hyaluronic acid and chondroitin sulfate synthesis in chondroblasts grown in F-10 plus EE (5%)

Chondroblasts were grown for 5 days in F-10 plus EE (5%) with and without BrdU (10 μg/ml). ^{3}H-glucosamine was added for 6 h. The glycosaminoglycans were isolated and separated by high voltage electrophoresis. Total counts in the hyaluronic acid and chondroitin sulfate peaks were determined.

	F-10 plus EE	F-10 plus EE plus BrdU	% inhibition
Hyaluronic acid	1924	1667	13%
Chondroitin sulfate	2350	192	92%

In summary: the progeny of replicating chondroblasts grown in F-10 behave as non-replicating chondroblasts with respect to chondroitin sulfate synthesis. In contrast, equivalent cells in embryo extract display a glycosaminoglycan profile consisting of hyaluronic acid predominantly with variable amounts of chondroitin sulfate; this profile is indistinguishable from that of authentic fibroblasts obtained from the chick amnion (Mayne et al., 1971) or from muscle (Mayne et al., 1972), when grown to comparable densities. BrdU will suppress chondroitin sulfate synthesis in *replicating* chondroblasts irrespective of the presence of embryo extract.

Effect of control medium, embryo extract and BrdU on collagen synthesis

Since the kinds of glycosaminoglycans synthesized by chondroblasts can be readily influenced by the ambient medium, we have also asked whether such exogenous factors will influence either the amount or the kinds of collagen synthesized.

Ten-day intact trunks were pre-incubated for 2 h in control medium. After this time the following was added to each culture: ascorbic acid (50 μg/ml), β-amino-propionitrile fumarate (125 μg/ml) and ^{3}H-proline

(12·5 μC/dish). The trunks were incubated for 24 h and the soluble collagen extracted and characterized by fractionation on a carboxymethyl-cellulose (CMC) column by the procedure described in Schiltz *et al.* (1973). The results of these experiments are shown in Fig. 4. Virtually all the labeled collagen eluted with the α-1 subunit of the carrier lathyritic collagen. Essentially, no radioactivity was observed in the α-2 region. The lack of α-2 subunits

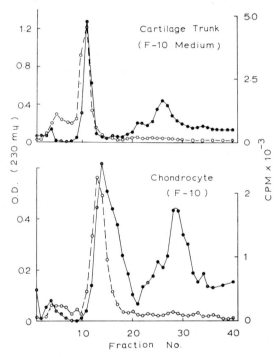

Fig. 4. Fractionation using carboxymethyl cellulose of NaCl-extracted, newly-synthesized collagen from (i) cartilage trunks incubated for 24 h in F-10 and ³H-proline (ii) chondroblast monolayer cultures grown for 5 days in F-10 and then for 24 h in ³H-proline. ●———● O.D. 230 mμ; o———o cpm/ml.

strongly suggests that the collagen synthesized *in vitro* by intact trunks is almost entirely of the cartilage type described by Miller and Matukas (1969). In additional experiments with either EE or BrdU present in the medium for the 24 h period only the synthesis of α-1 subunits was observed. Thus, the chondroblasts present in whole trunks do not change either the kind of glycosaminoglycans or the kind of collagen synthesized when exposed to medium containing either embryo extract or BrdU.

The next question was whether addition of either EE or BrdU to cultures of isolated and replicating chondroblasts would affect the rate of collagen synthesis. Floaters were plated in 100-mm Falcon plastic tissue culture dishes at an initial density of 5×10^5 cells/dish. Twenty four hours later, embryo extract (5%) or BrdU (10 μg/ml) was added to some of the cultures. All cultures were fed daily. After 5 days, the cultures were pulsed for 6 h with ^{3}H-proline. The cells and medium were separated and the content of ^{3}H-hydroxyproline and proline in TCA-precipitable material determined after acid hydrolysis. Values for hydroxyproline/(hydroxyproline + proline) × 100 are shown in Table III. It is concluded that EE or BrdU has no marked effect on total collagen synthesis.

Table III

Synthesis of collagen after growth of chondroblasts in embryo extract or BrdU

Chondroblasts were grown for 5 days in F-10 with EE (5%) or BrdU (10 μg/ml), and then pulsed for 6 h with ^{3}H-proline. Cells and medium were separated and incorporation of ^{3}H-proline as hydroxyproline and proline determined after acid hydrolysis of the TCA-precipitable material.

		$Hypro/(hypro + pro) \times 100$
Control	Cells	3·32
	Medium	3·88
	Culture	3·31
BrdU	Cells	2·69
	Medium	5·76
	Culture	3·08
EE	Cells	1·83
	Medium	8·17
	Culture	2·92

This analysis gives no information concerning the subunit composition of the collagen being synthesized. Accordingly, cultures were grown for 5 days as previously but were pulsed for 24 h with ^{3}H-proline, β-aminopropionitrile and ascorbic acid. Cells and medium were separated and analyzed for collagen subunits. The CMC elution profiles of the collagen from these cultures is shown in Fig. 5. In cultures grown in F-10 control medium, although the cells have replicated many times, they continue to synthesize only α-1 subunits. In contrast, the replicating chondroblasts in media with EE or BrdU have been induced to synthesize α-2 subunits (Fig. 5). α-1/α-2 ratios for BrdU-treated cultures are close to 2. On the other hand, the ratios for the embryo extract-treated cells were usually higher, in this case about 4·5. The reasons for this are not yet known, but the data suggest that from the standpoint of collagen synthesis not all cells have moved from the chondrogenic to the fibrogenic phenotype. Whether this will change with time and whether some individual cells have changed completely whereas others remain unchanged are questions for further experimentation. Similarly, it will be worth demonstrating that the α-1 collagen from the chondroblasts is in fact different in amino acid sequence from the α-1 from the altered cells.

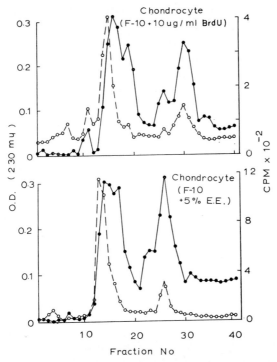

Fig. 5. Carboxymethyl cellulose column profile of NaCl-extracted, newly-synthesized collagen from chondroblasts grown for 5 days in either BrdU (10 μg/ml) or embryo extract (5%) and then pulsed for 24 h with ³H-proline. ●———●, O.D. 23 μ; o———o, cpm/ml.

In summary: non-replicating chondroblasts in trunks synthesize cartilage-type collagen in concert with chondroitin sulfate irrespective of EE or BrdU in the medium. However, EE which induces replicating chondroblasts to synthesize hyaluronic acid also induces the cells to synthesize fibroblast-type collagen. Nevertheless, that collagen and glycosaminoglycan synthesis in these cells can be uncoupled is demonstrated by the fact that BrdU does not block collagen synthesis in the very same cell in which chondroitin sulfate synthesis has been completely suppressed. Of even greater interest is the fact that BrdU induces a switch-over to the fibroblast-like collagen.

CHONDROGENESIS IN SOMITES

In the first section of this chapter we described some covert properties of definitive chondroblasts—mainly their capacity to be induced to yield progeny which operationally are remarkably similar to many kinds of fibroblasts. In this section we will describe experiments designed to learn more of the capacities of cells *ancestral* to functional chondroblasts—those precursor chondrogenic cells that arise in each somite of developing vertebrate embryos.

Vertebral chondrogenesis requires an obligatory inductive interaction between somite mesenchyme cells and adjacent spinal cord and notochord. As this literature has recently been reviewed (Holtzer and Matheson, 1970; Holtzer *et al.*, 1972; Holtzer and Mayne, 1973) only the unequivocal features of this system which are required for background will be described. The most reliable and comprehensive analysis of this inductive interaction at the biological level has been performed on salamander embryos (Holtzer and Detwiler, 1953; Holtzer, 1958). This work circumvents the trivial differences observed with chick material *in vitro* that stem from variations in culture medium. Total extirpation of the spinal cord and notochord early in development precludes the subsequent emergence of *all* vertebral chondroblasts. Grafting pieces of embryonic spinal cord into areas of the somite that normally yield only fibroblasts and myoblasts will induce the subsequent emergence of ectopic chondroblasts. The inductive activity of the spinal cord is confined to the ventral half of the spinal cord.

It is worth stressing that it is *not* the induced cell that transforms into a chondroblast; it is the *progeny* of the induced cell that displays the terminal characteristics of a chondroblast. It has been suggested (Holtzer, 1963; Holtzer *et al.*, 1972) that there is no mysterious 'cartilage inducing' molecule excreted by the inducing tissues. Rather the inductive interaction permits the requisite numbers of mitosis and movements in the proper microenvironment and the consequent playing out of the chondrogenic program initiated in the very early somite and pre-somite mesenchyme cells.

In the chick the inductive interaction begins in the late 2-day embryo and probably continues for the next 24–36 h. Accordingly, somites from Stages 14–20 contain varying numbers of induced cells and they may, in permissive medium, replicate, and after 4–6 days *in vitro* differentiate into chondroblasts without further interaction with the inducers (Holtzer, 1963, 1964, 1968; Strudel, 1967; Ellison and Lash, 1971). At earlier stages there are no cells in the somites that have been induced and consequently when they are cultured by themselves, vertebral chondroblasts will not 'spontaneously' appear. An intriguing question is: How many cells are induced in each somite? Elsewhere (Holtzer and Matheson, 1970; Holtzer and Mayne, 1973; Holtzer et al., 1973) we have suggested that normally this is probably of the order of 10^2 cells and that only after further interactions between the first or second generation progeny of the induced cells will chondroblasts emerge. There is an excellent correlation between histologically detectable chondroblasts and the synthesis and deposition of sizeable amounts of chondroitin sulfate. The modest amount of sulfated glycosaminoglycans synthesized by somite cells prior to overt chondrogenesis may be the product of authentic fibroblasts or myoblasts differentiating in this system. Thus, for example we have recently found both hyaluronic acid and a hydroxyproline-containing protein in 18-h chick embryos (Mayne, Abbott and Holtzer, unpublished observations). Until we determine which cells in the 18-h embryo synthesize these molecules their relationship to the vertebral chondrogenic lineage is unclear. Conceivably all or most embryonic cells, for purposes of mechanical cohesion, possess as a differentiated trait the capacity to synthesize hyaluronic acid-like or collagen-like molecules that may or may not require the same structural genes that later will be active in connective tissue cells (see for example Holtzer et al., 1973b, for similar arguments involving actin).

The following experiments demonstrate that subsequent to the inductive interaction between notochord and somites there is a further obligatory requirement for DNA synthesis and cell division if the cells are to differentiate into chondroblasts. Stage 17–18 somites alone, which contain a modest number of induced cells or Stage 17–18 plus notochord were grown in normal medium for 12 to 24 h and then exposed for 12 h to one of the following inhibitors: cytosine arabinoside, hydroxyurea, or colchimid. The cultures were washed and grown in normal medium for another 8 days. In approximately 80% of the cultures of somites alone nodules appeared containing between 10^2 and 10^3 chondroblasts. In 100% of the cultures of somites plus notochord nodules appeared with between 5×10^2 to 10^4 chondroblasts. In all the cultures where either DNA synthesis or cytokinesis was perturbed, chondrogenesis failed to occur. Failure of chondrogenesis cannot be attributed to generalized cytotoxic effects, for following the removal of the inhibitors replication resumed and the notochord, muscle and fibroblasts terminally differentiated.

A question of some theoretical interest is the time at which a chondroblast is clearly a chondroblast and not a precursor chondrogenic cell. In other words, when in each somite does a cell appear that has some or all of the properties of a definitive chondroblast? The claim that sulfated glycosaminoglycans are synthesized in early embryos (Franco-Browder et al., 1963), or by some somite cells (Marzullo and Lash, 1967) sheds no light on the kinds of glycosaminoglycans synthesized specifically by precursor chondrogenic cells, which constitute no more that 10% of the total somite population (Holtzer, 1968; Holtzer and Matheson, 1970). To learn more about the chondrogenic cells within the somite we have had recourse to cloning experiments. In addition to depositing chondroitin sulfates and α-1 (II) collagen, a single functional chondroblast will yield a characteristic chondrogenic clone when challenged in an appropriate environment (Coon, 1966; Abbott and Holtzer, 1968; Chacko et al., 1969a). Accordingly, we have determined the stage in development when a *single* cell that will give rise to a chondrogenic clone may be isolated from a somite; the number of cells within a somite varies from 5×10^3 to over 6×10^4 cells. Individual somite cells from as late as Stage 19 or 20 yield chondrogenic cells only very rarely. At this stage no cells within the somite have yet excreted and deposited significant amounts of chondroitin sulfate. Individual cells tightly adherent to the notochord, however, will give rise to chondrogenic cells when taken from Stage 21 or 22 somites. In another series of experiments Stage 17–18 somites were grown as organ cultures for 3 days or at a time when some sclerotome cells begin to excrete readily detectable quantities of chondroitin sulfate. These cultures were trypsinized and the resulting single cells challenged to clone. Large numbers of individual cells from these organ cultures gave rise to chondrogenic clones (Abbott et al., 1972). Apparently the capacity to form chondrogenic clones develops about the same time the cells commence excreting significant quantities of chondroitin sulfate.

To summarize the last two experiments: (1) there is not a *single* definitive chondroblast in most Stage 17–18 somites, and (2) after induction, the transition from the precursor chondrogenic cells to functional chondroblasts requires one or two more rounds of DNA synthesis and cell division.

In contrast to the more conventional view of developmental biologists (Zwilling, 1961; Searls, 1965; Medoff, 1967; Levitt and Dorfman, 1972) most of our experiments have stressed that the genetic programs of cells early in the chondrogenic lineage are unique and qualitatively different from those of the functional chondroblast. More recently we have added to this the notion that movement from one phenotypic compartment into the next within the chondrogenic lineage requires DNA synthesis and cell division. The response of somite cells incorporating BrdU provides yet another way of demonstrating that there are no cells in Stage 17–18 somites that possess all the properties of definitive chondroblasts.

Stage 17–18 somites with or without notochord were exposed to BrdU for days 0–3, 1–3, 2–5, 3–6. After any one of these 3-day exposures to the analog the cells were washed and grown in fresh medium until the cultures were 10 days old. They were then assessed for chondroblasts by staining with toluidine blue or trypsinized and challenged to yield chondrogenic clones. In these cultures it has been shown (Holtzer and Matheson, 1970) that each cell divides at least once during the first 24 h and that the great majority divide at least once more during the next 18 h. Cartilage nodules did not appear in any cultures exposed to BrdU during days 0–3 or 1–4. Modest sized nodules appeared in cultures exposed to BrdU during days 2–5 or 3–6. The presence of chondroblasts in cultures treated with BrdU after 48 h correlated with the marked decrease in numbers of somite cells synthesizing DNA after 48 h of culture. Of even greater interest, however, was the finding that the BrdU-suppressed somite cells when sub-cultured at high or low density consistently failed to yield frank chondroblasts.

BrdU-suppressed myogenic cells and amnion cells when allowed to replicate for several generations yield many normally terminally differentiated muscle and amnion cells respectively (Bischoff and Holtzer, 1970; Mayne et al., 1971). Although the data on 'reversibility' of BrdU-suppressed chondroblasts is not as hard as with other cell types, it is clear that many will yield frank chondroblasts. The apparent 'irreversible' effect of BrdU on very early cells in the chondrogenic lineage may well involve a genetic lesion that differs from that induced in frank chondroblasts. In this connection it is worth stressing that the incorporation of BrdU into early cells in the myogenic and erythrogenic lineages also precludes terminal differentiation (Holtzer et al., 1972; Holtzer et al., 1973a).

DISCUSSION

The recent recognition that the expression of the cartilage and fibroblast phenotypes involves the activities of different genes has made it possible to test directly for a close relationship between these two cell types; a relationship long suspected by the earlier histologists. The experiments reported in this paper reaffirm this notion, and hopefully will set the background for defining this relationship in terms of current concepts of cytoplasmic-nuclear interactions responsible for terminal differentiation.

Several types of cartilage are recognized—hyaline, fibro and elastic. Similarly, fibroblasts from different regions of the body often appear to be dissimilar; skin fibroblasts, for example, are able to synthesize dermatan sulfate (Matalon and Dorfman, 1966, 1968). What is not clear is which of these differences between kinds of cartilage cells and kinds of fibroblasts are relatively reversible modulations, and which are relatively irreversible genetic decisions. Some, and possibly all, of the differences between cartilage cells and fibroblasts may in fact be variable expressions of a 'basic connective tissue phenotype'; the particular form of expression varying with differences in the cell's microenvironment. In this chapter we have emphasized that in a particular *in vitro* microenvironment the progeny of vertebral chondroblasts may yield cells in many ways indistinguishable from some kinds of fibroblasts. The formation, following injury, of ectopic cartilage in such varied tissues as muscle, kidney or bladder (La Croix, 1950) demonstrates the converse, namely that at least one kind of widely distributed fibroblast may yield progeny that are indistinguishable from some kinds of chondroblasts.

The first detectable effect of the spinal cord-notochord interaction is a local burst of mitotic activity (Holtzer and Detwiler, 1953; Holtzer and Matheson, 1970). Further, it has been shown that perturbing the subsequent mitotic schedule of already induced cells will block terminal chondrogenesis. It will be of great interest to learn more of the obligatory role for DNA synthesis in moving cells through the successive compartments in the chondrogenic lineage during embryogenesis, and to compare this role of DNA synthesis with switching from the terminal chondroblast phenotype to the terminal fibroblast phenotype.

Original speculations concerning the effects of BrdU on differentiation have stressed that BrdU often has a greater inhibitory effect on the synthesis of 'terminal luxury' molecules than on the synthesis of essential molecules (Holtzer, 1970). More recently, it has been recognized that the presence of BrdU will often interfere with the final stages of terminal differentiation, but that this effect is reversible on removal of the BrdU (Holtzer *et al.*, 1973a, b). The introduction of BrdU in the early stages of a lineage, however, results in cells in which terminal differentiation is not readily realized (Abbott *et al.*, 1972; Holtzer *et al.*, 1973; Weintraub *et al.*, 1973). These observations may in part reflect the proposal of Lin and Riggs (1972) that BrdU interferes with the binding properties of regulatory proteins to DNA. The finding in this present work that BrdU stimulates the synthesis of α-2 collagen chains is a clear example of the way in which BrdU can interfere with *regulation*. It will be of considerable interest to learn more of the degree to which this regulation is reversible, and the ease with which the chondrogenic phenotype can be re-established on removal of BrdU.

ACKNOWLEDGEMENTS

Our researchers were supported by research grants from National Institutes of Health (5 ROI- HD 00189, 5 TOI- HD 00030), National Science Foundation (GB- 27933) and the American Cancer Society (VC- 45).

REFERENCES

Abbott, J. and Holtzer, H. (1966). *J. Cell. Biol.* **28**, 473–87.
Abbott, J. and Holtzer, H. (1968). *Proc. Natl. Acad. Sci. USA* **59**, 1144–1151.
Abbott, J., Mayne, R. and Holtzer, H. (1972). *Develop. Biol.* **28**, 430–442.
Bischoff, R. (1971). *Exp. Cell Res.* **66**, 224–236.
Bischoff, R. and Holtzer, H. (1970). *J. Cell Biol.* **44**, 134–150.
Bornstein, P. (1969). *Biochemistry* **8**, 63–71.
Bryan, J. (1968). *Exp. Cell Res.* **52**, 319–26.
Chacko, S., Abbott, J., Holtzer, S. and Holtzer, H. (1969a). *J. Exp. Med.* **130**, 417–442.
Chacko, S., Holtzer, S. and Holtzer, H. (1969b). *Biochem. Biophys. Res. Commun.* **34**, 183–189.
Coon, H. G. (1966). *Proc. Natl. Acad. Sci. USA* **55**, 66–73.
Coon, H. G. and Cahn, R. D. (1966). *Science* **153**, 1116–1119.
De la Haba, G. and Holtzer, H. (1965). *Science* **149**, 1263–1265.
Ellison, M. L. and Lash, J. W. (1971). *Develop. Biol.* **26**, 486–496.
Franco-Browder, S., De Rydt, J. and Dorfman, A. (1963). *Proc. Natl. Acad. Sci. USA* **49**, 643–647.
Holtzer, H. (1958). In "Regeneration in Vertebrates" (C. S. Thornton, ed.) pp. 15–33. Univ. of Chicago Press, Chicago, Illinois.
Holtzer, H. (1963). In "Induktion und Morphogenese". Springer Verlag, Berlin.
Holtzer, H. (1964). *Biophys. J.* **4**, 239–250.
Holtzer, H. (1968). In "Epithelial-Mesenchymal Interactions" (R. Fleischmajer and R. E. Billingham, eds.) pp. 152–164. Williams and Wilkins, Baltimore, Maryland.
Holtzer, H. (1970). In ISCH Symposium, "Control Mechanisms in Tissue Cells" (H. Padykula, ed.) pp. 69–88. Academic Press.
Holtzer, H. and Detwiler, S. R. (1953). *J. Exp. Zool.* **123**, 335–370.
Holtzer, H. and Abbott, J. (1968). In "The Stability of the Differentiated State" (H. Ursprung, ed.) pp. 1–16. Springer, Berlin.
Holtzer, H. and Matheson, D. W. (1970). In "Chemistry and Molecular Biology of the Intercellular Matrix" (E. A. Balazs, ed.) Vol. 3, pp. 1753–1769. Academic Press, New York.
Holtzer, H. and Mayne, R. (1973). In "AAPB Symposium on Developmental Pathobiology" (M. Finegold and E. Perrin, eds.) Williams and Wilkins, Baltimore, Maryland.
Holtzer, H., Abbott, J., Lash, J. and Holtzer, S. (1960). *Proc. Natl. Acad. Sci. USA* **46**, 1533–1542.
Holtzer, H. Chacko, S., Abbott, J., Holtzer, S. and Anderson, H. (1970). In "Chemistry and Molecular Biology of the Intercellular Matrix" (E. A. Balazs, ed.) Academic Press, New York.
Holtzer, H., Weintraub, H., Mayne, R. and Mochan, B. (1972). *Curr. Top. Develop.* (A. Moscona and A. Monroy, eds.), pp. 229–256. Academic Press, New York.
Holtzer, H., Mayne, R. and Weintraub, H. (1973a). (In preparation.)
Holtzer, H., Sanger, J. W., Ishikawa, H. and Strahs, K. (1973b). Cold Spring Harbor Symposium on Muscle, pp. 549–566.
Kefalides, N. A., In "Chemistry and Molecular Biology of the Intercellular Matrix" (E. A. Balazs, ed.) Vol. 1, pp. 535–573. Academic Press, New York.
Koyama, H. and Ono, T. (1971). *J. Cell. Physiol.* **78**, 265–272.
La Croix, P. (1950). "The Organization of Bones", Philadelphia, Blakiston Press.

Levitt, D. and Dorfman, A. (1972). *Proc. Natl. Acad. Sci. USA* **69**, 1253-1257.
Lin, S-Y and Riggs, A. D. (1972). *Proc. Natl. Acad. Sci. USA* **69**, 2574-2576.
Marzullo, G. and Lash, J. W. (1967). *In* "Experimental Biology and Medicine", Vol. 1, pp. 213-218. Karger, Basel.
Matalon, R. and Dorfman, A. (1966) *Proc. Natl. Acad. Sci. USA* **56**, 1310-1316.
Matalon, R. and Dorfman, A. (1968). *Proc. Natl. Acad. Sci USA* **60**, 179-185.
Mayne, R., Sanger, J. W. and Holtzer, H. (1971). *Develop. Biol.* **25**, 547-567.
Mayne, R., Abbott, J. and Schiltz, J. (1972). Abstract, *Cell Biology Meetings*, St. Louis.
Mayne, R., Abbott, J. and Holtzer, H. (1973). *Exp. Cell Res.* **77**, 255-263.
Medoff, J. (1967). *Develop. Biol.* **16**, 118-143.
Miller, E. J. (1971a). *Biochemistry* **10**, 1652-1659.
Miller, E. J. (1971b). *Biochemistry* **10**, 3030-3035.
Miller, E. J. and Matukas, V. J. (1969). *Proc. Natl. Acad. Sci. USA* **64**, 1264-1268.
Miller, E. J., Epstein, E. H. and Piez, K. A. (1971). *Biochem. Biophys. Res. Commun.* **42**, 1026-1029.
Morris, S. C. and McClain, P. E. *Biochem. Biophys. Res. Commun.* **47**, 27-34.
Murison, G. L. (1972). *Exp. Cell Res.* **72**, 595-600.
Nameroff, M. and Holtzer, H. (1967). *Develop. Biol.* **16**, 250-281.
Pawelek, J. M. (1969). *Develop. Biol.* **19**, 52-72.
Saito, H., Yamagata, T. and Suzuki, S. (1968). *J. Biol. Chem.* **243**, 1536-1542.
Schiltz, J., Mayne, R. and Holtzer. H. (1973). *Differentiation* Vol. 1. In Press.
Schubert, M. and Hamerman, D. A. (1968). "A Primer on Connective Tissue Biochemistry". Lea and Febiger, Philadelphia. Pa.
Schulte Holthausen, H., Chacko, S., Davidson, E. A. and Holtzer, H. (1969). *Proc. Natl. Acad. Sci. USA* **63**, 864-870.
Searls, R. L. (1965). *Develop. Biol.* **11**, 155-168.
Shulman, H. J. and Meyer, K. (1968). *J. Exp. Med.* **128**, 1353-1362.
Smith, C. and Hamerman, D. (1968). *Proc. Soc. Exp. Biol. Med.* **127**, 988-991.
Strawich, E. and Nimni, M. E. *Biochemistry* **10**, 3905-3911.
Strudel, G. (1967). *In* "Experimental Biology and Medicine" (E. Hagen, W. Wechsler and P. Zilliken, eds.) Vol. 1, 183-198.
Telser, A., Robinson, H. C. and Dorfman, A. (1965). *Proc. Natl. Acad. Sci. USA* **54**, 912-919.
Thorp, F. K. and Dorfman, A. (1967). *In* "Current Topics in Developmental Biology", (A. A. Moscona and A. Monroy, eds.) Vol. 2, pp. 151-190. Academic Press, New York.
Toole, B. P., Kang, A. H., Trelstad, R. L. and Gross, J. (1972). *Biochem. J.* **127**, 715-720.
Trelstad, R. L., Kang, A. H., Igarashi, S. and Gross, J. (1970). *Biochemistry* **9**, 4993-4998.
Weintraub, H., Campbell, G. L. and Holtzer, H. (1972). *J. Molec. Biol.* **70**, 337-350.
Yamagata, T., Saito, H., Habuchi, O. and Suzuki, S. (1968). *J. Biol. Chem.* **243**, 1523-1535.
Zwilling, E. (1961). *Advan. Morphogenesis* **1**, 301-330.

The Differentiation of Cartilage

DANIEL LEVITT and ALBERT DORFMAN
*Departments of Pediatrics, Biochemistry, and Biology
Joseph P. Kennedy, Jr., Mental Retardation Research Center
and the La Rabida-University of Chicago Institute
Pritzker School of Medicine
University of Chicago, Chicago, Illinois 60637, USA*

INTRODUCTION

Cartilage is distinguished from other connective tissues by the presence of relatively large amounts of matrix composed primarily of proteoglycans and collagen. The proteoglycans of various cartilages contain chondroitin, chondroitin 4-sulfate, chondroitin 6-sulfate and keratan sulfate II in various combinations and proportions depending upon species, tissue and age.

Chondroitin sulfates appear to be present in tissues other than cartilage, but there is as yet inadequate information to determine whether the protein core of the proteoglycan of cartilage is identical with that of other connective tissues. However, the quantity of chondroitin sulfate-containing proteoglycan in cartilage is markedly greater than that in other connective tissues. Miller and Matukas (1969) have shown that collagen of cartilage has a primary structure distinct from that of other connective tissues. This discovery has been confirmed by Trelstad et al. (1970). Since the pathway of biosynthesis of chondroitin sulfate chains is now reasonably well established and cartilage contains a unique collagen, the differentiation of cartilage can be chemically defined reasonably well.

The transformation of mesenchyme cells to cartilage is accompanied by development of a capacity for a high rate of synthesis of chondroitin sulfate proteoglycan and possibly by the development of a capacity to synthesize cartilage-specific collagen. Inadequate data are yet available to be certain that conversion of mesenchyme cells to chondrocytes is accompanied by a change in the nature of collagen synthesized.

Differentiation may be studied in various ways. The descriptive approach characterizes the distinctions between the particular differentiated cell type and precursor cells or other differentiated cells. Such studies may be carried out at levels of sophistication varying from morphological observations, which have been conducted for many years, to detailed description of biochemical parameters. The more profound approach is an attempt to understand the underlying mechanism of differentiation. Although this problem has been of interest to many investigators, no clearcut basis for the mechanism of differentiation has yet been established. The studies to be outlined in this chapter have been undertaken to establish a system for the study of cartilage differentiation which might be used for an investigation of mechanisms of differentiation.

The conversion of limb bud mesenchyme to cartilage was chosen because differentiation in this system is readily observed in cell culture and the problem is not dependent on 'inducers'. Previous studies by Zwilling (personal communication) demonstrated that Stage 24 chick limb bud mesoderm differentiates primarily into chondrocytes when grown in cell culture at high density (above confluency). Culture at lower density results in cells resembling fibroblasts. Searls and Janners (1969) have shown that up to Stage 25 chondrogenic properties of mesodermal chick limb bud cells are influenced by environmental factors, while after this stage commitment to chondrogenesis is stabilized. These findings suggested that study of Stage 24 limb bud mesodermal cells might offer a system for analysis of some of the events involved in the transformation of limb bud mesenchymal cells to chondrocytes. The methods and material used in these studies have been described in detail elsewhere (Levitt and Dorfman, 1972).

DIFFERENTIATION OF LIMB BUD MESENCHYME

The discovery by Zwilling that Stage 24 limb bud cells grown at densities above confluency readily differentiate to cartilage suggested that other environmental manipulations might promote or prevent differentiation. Previous studies in this laboratory have demonstrated that chondrocytes obtained from epiphyses of 13-day old chick embryos may be cultured in liquid suspension over agar (Horwitz and Dorfman, 1970; Nevo et al., 1972). Therefore, an attempt was made to determine whether culture on agar would promote differentiation of early embryonic limb bud mesenchyme when plated at densities below confluency. For this purpose the following experimental design, utilizing freshly liberated Stage 24 (Hamburger and Hamilton, 1951) limb bud cells was employed: (a) cells were cultured at a density of 1×10^7 cells per 100-mm dish over agar for 48 h and then plated at a density of 0·5

$\times 10^6$ cells per 60-mm dish, (b) cells were plated at 1×10^7 cells per 100-mm dish without agar and after 48 h subcultured at a density of 0.5×10^6 cells per 60-mm dish, and (c) cells were plated directly at 3.6×10^6 cells per 60-mm dish and were not subcultured. The initial plating density for each of the culture conditions was less than that required for a confluent monolayer. Sulfate uptake was measured on day 7 after subculture for (a) and (b) and on day 9 for (c).

The development of cartilage was monitored by phase microscopy and staining with 1% aqueous toluidine blue. Cultures first grown on agar (a) showed a large number of cells with typical epitheloid morphology and the presence of metachromatic matrix. The data presented in Table I demonstrate a striking increase in chondroitin sulfate synthesis by cultures derived from cells first grown over agar.

Table I

Effect of growth on agar on differentiation of limb bud mesenchyme

Experiment No.	(a) $cpm/10^6$ cells	(b) $cpm/10^6$ cells	(c) $cpm/10^6$ cells
1	16,500	348	—
2	6200	590	—
3	4630	595	428
4	53,750	5760	7050
5	245,000	4950	—
6	26,300	—	1610

$3.3~\mu$Ci $H_2{}^{35}SO_4$ per ml of medium was added to each culture 6 h before harvest. On the ninth day of culture, chondroitin sulfate was isolated.

The mechanism by which culture over agar promotes cartilage differentiation is not clear. Since it seemed possible that cell aggregation might be an important factor, experiments were performed utilizing the technique of Moscona (1961) for aggregating limb bud cells in a gyratory shaking flask. Three ml of a Stage 24 limb bud cell suspension (10^6 cells per ml) were incubated for 48 h in 25 ml Erlenmeyer flasks on a New Brunswick gyratory shaker rotating at 70 rpm at 38°C. The cells were then dissociated with 0.25% trypsin in EDTA and placed on plastic dishes at a density of 0.5×10^6 cells per 60-mm plate. After seven days of growth on the tissue culture plates, there was no recognizable cartilage formation based on morphology, metachromasia or increased chondroitin sulfate synthesis (as measured by increased $^{35}SO_4$ uptake). These results indicate that aggregation of limb bud cells is not in itself sufficient to promote chondrogenesis on subsequent monolayer culture.

PREVENTION OF DIFFERENTIATION OF CARTILAGE CELLS BY BrdU

Since previous studies (Abbott and Holtzer, 1968) have shown that 5-bromo-2-deoxyuridine prevents the expression of cartilage phenotype, experiments were performed to determine the effect of this analog on

Table II
Effect of BrdU on cartilage differentiation

Experiment No.	Control cpm/10^6 cells	BrdU cpm/10^6 cells
1	13,400	2380
2	53,750	8925
3	76,000	6250
4	245,000	65,200
5	26,300	3035
6	15,400	2540

Cells were grown over agar for 48 h and then plated at 0.5×10^6 cells per 60-mm dish. BrdU was added at a concentration of 10 μg ($3.2 \times 10^{-5} M$) per ml of medium during period of culture over agar. No BrdU was present on subculture. Cells were labeled with 3.3 μCi $H_2^{35}SO_4$ per ml medium 6 h before harvest on the seventh day of subculture.

the differentiation of dissociated limb bud cells. The data in Table II show the effect of BrdU added during the initial 48-h period when the mesenchymal cells were cultured over agar. In these experiments *no BrdU was present* during subsequent subculture on plastic dishes when the cartilage phenotype normally makes its appearance in untreated cells. These data indicate that the presence of BrdU in concentrations of $3.2 \times 10^{-5} M$ during the period of culture over agar inhibits the subsequent appearance of the cartilage phenotype. BrdU did not exert its effect simply as a result of toxicity since cells exposed to BrdU continued to exhibit normal rates of DNA, RNA, protein synthesis and cell division (Table III).

Table III
Effect of BrdU on cell multiplication and protein, DNA and RNA synthesis

	Control	BrdU
Cell number per dish	3.2×10^6	3.3×10^6
mg protein per dish	0.42	0.37
	cpm/10^6 cells	*cpm/10^6 cells*
^{3}H-thymidine incorporation	1420	990
^{14}C-uracil incorporation	139	126
^{14}C-leucine incorporation	13,600	14,500

All experiments were performed on cells exposed to 50 µg of BrdU per ml medium during growth over agar. Isotopes were added 6 h prior to harvest of cultures on the seventh day after subculture.

The effect of BrdU as measured by sulfate incorporation into chondroitin sulfate proteoglycan was dose dependent up to 10 µg/ml ($3.2 \times 10^{-5}M$). Doses larger than 10 µg/ml did not further depress radioactive sulfate incorporation. Inhibition was observed with as little as 1 µg of BrdU per ml.

In order to be certain that the effects observed were not simply on sulfation, another precursor of chondroitin sulfate, ^{3}H-acetate, was utilized. The results were identical with experiments utilizing $H_2^{35}SO_4$. In other experiments it was shown that exposure of cells over agar to BrdU for as little as 18 h was sufficient to prevent subsequent chondrogenesis.

Thymidine, at a concentration of $1.6 \times 10^{-4}M$, prevented the effect of $3.2 \times 10^{-5}M$ BrdU only if it was added simultaneously with the analogue. Subsequent addition of the same concentration of thymidine when cells previously exposed to BrdU were subcultured on plates, did not restore the cartilage phenotype. The inhibition of differentiation during this special period of determination is reversed neither by removal of BrdU nor by the subsequent addition of excess thymidine.

That the effect was irreversible was demonstrated by the decrease in the rate of chondroitin sulfate synthesis of BrdU-treated cells maintained through serial subculture in the absence of the drug. At higher doses of BrdU ($1.6 \times 10^{-4}M$ to $3.2 \times 10^{-5}M$), the diminution of the capacity to synthesize chondroitin sulfate was observed through 4 subcultures covering 31 days. Even at a dose of 1 µg/ml of BrdU ($3.2 \times 10^{-6}M$), the difference between BrdU and control cells was evident through 5 subcultures as illustrated by the data summarized in Table IV.

Table IV
Effect of BrdU after serial subculture

	Control cpm/10^6 cells	BrdU cpm/10^6 cells
First subculture (6 days)	15,400	8250
Second subculture (10 days)	9300	5260
Third subculture (21 days)	13,150	6975
Fourth subculture (27 days)	5025	2740
Fifth subculture (41 days)	20,900	11,400

$H_2{}^{35}SO_4$ was added 6 h before harvest of cells on day indicated. BrdU at a concentration of 3.2×10^{-6} M (1 µg/ml) was used only during initial growth over agar.

If Stage 24 limb bud cells which had undergone differentiation during the period of culture over agar were exposed to BrdU during subsequent subculture, a decrease in sulfate uptake occurred. However, further subculture of 6 days in the absence of BrdU resulted in restoration of high levels of sulfate incorporation. The inhibitory effect of BrdU on expression of cartilage phenotype in differentiated chondrocytes is reversible.

An additional confirmation of this conclusion was obtained when cartilage from Stage 32 hind limbs was first cultured in the presence of BrdU and then subcultured in its absence. No inhibition of sulfate incorporation was observed after subculture for 7 days in the absence of BrdU.

In view of the marked depression of $^{35}SO_4$ and ^{3}H-acetate incorporation, it was of interest to determine more precisely the exact spectra of glycosaminoglycans synthesized by control and BrdU-treated cells. The glycosaminoglycan composition of cells and media was determined separately. For this purpose cells were grown on 60-mm tissue culture dishes at high density (2×10^7 cells per culture) and labeled with both ^{3}H-acetate and $H_2{}^{35}SO_4$ 24 h prior to harvest. The cells were scraped from the plates, and cells and media were separated by centrifugation at 1000 rpm. Media was dialyzed against distilled water, then frozen, and cells were immediately frozen and stored at $-20°C$ until enough material had been collected for assay.

Cell-associated glycosaminoglycans were isolated after digestion with papain for 48 h. The non-trichloracetic acid-precipitable papain digest was brought to 0·03M salt concentration and was treated with cetyl pyridinium chloride (CPC) to precipitate acid polysaccharides. The CPC-precipitate was dissolved in 2·0M NaCl and the polysaccharide was precipitated with alcohol. After alcohol and ether washes, the precipitate was dissolved in 0·03M NaCl and reprecipitated with CPC, ethanol and ether. Total uronic acid, hexosamine and galactosamine/glucosamine ratios were measured on the final product.

Prior to papain digestion, the media was dialyzed against distilled water and 0·03M NaCl and was then treated with CPC. The precipitate which formed was papain digested and purified as was the cell-associated material.

By the use of streptococcal hyaluronidase, testicular hyaluronidase, chondroitinase ABC, and nitrous acid and Sephadex G-25 chromatography, it was possible to distinguish hyaluronic acid, chondroitin 4- or 6-sulfates, dermatan sulfate and heparan sulfate. Details of this method will be published elsewhere. A summary of the results of these experiments is presented in Table V.

Table V
Glycosaminoglycan synthesis by control and BrdU-treated cells

	Total GAG	Hyaluronic Acid	Chondroitin 4- and 6-SO_4	Dermatan Sulfate	Heparan Sulfate
	μM uronic acid/10^9 cells				
Control cells	4·12	0·40	3·73	0·16	0·05
BrdU cells	0·71	0·19	0·47	0·08	0·09
Control media	5·70	0·80	5·07	0	0·07
BrdU media	0·60	0·17	0·36	0	0·09
Control—total	9·82	1·20	8·80	0·16	0·12
BrdU—total	1·31	0·36	0·83	0·08	0·18

The data indicate that the chondroitin 4- and 6-SO_4 represent the major glycosaminoglycan synthesized by the control cells. It is suppression of synthesis of chondroitin sulfate which accounts for most of the effect of BrdU although some decrease of hyaluronic acid synthesis was also observed. The amounts of dermatan sulfate and heparan sulfate were so small that any difference indicated between control and BrdU-treated cells cannot be considered significant.

An attempt has been made to determine the mechanism by which BrdU interferes with formation of chondromucoprotein. Previous studies by Holthausen et al. (1969) have indicated that BrdU treatment of chondrocytes obtained from vertebral cartilage results in a marked decrease in the levels of UDP-glucose dehydrogenase, UDP-N-acetylhexosamine-4-epimerase and the enzyme system responsible for catalyzing 3'-phosphoadenosine-5'-phosphosulfate synthesis. We have initiated a series of experiments to examine the levels of enzymic activity of the specific transferases involved in formation of the carbohydrate chains of chondroitin sulfate-proteoglycan. Methods of enzyme assays have been previously described (Horwitz and Dorfman, 1968; Stoolmiller et al., 1972). For this purpose limb bud cells were cultured with or without BrdU for two days at a cell density which promotes chondrogenesis. The BrdU-containing media was removed and culture was continued for 7 additional days. Six hours before harvest for enzyme assays ^{14}C-acetate and $^{35}SO_4$ were added to replicate plates to determine the extent of glycosaminoglycan synthesis. The data in Table VI indicate

Table VI

Glycosaminoglycan synthesis and glycosyltransferase activities in control and BrdU-treated cells

		cpm/mg protein/h	cpm/10^6 cells
UDP-xylosyl transferase	Control	2060	—
	BrdU	1392	—
UDP-GalNAc transferase	Control	2420	—
	BrdU	1380	—
$^{35}SO_4$-incorporation	Control	—	13,600
	BrdU	—	1689
^{14}C-acetate incorporation	Control	—	7840
	BrdU	—	1155

a 7–8 fold inhibition of glycosaminoglycan synthesis but only a two fold inhibition of xylosyltransferase and N-acetylgalactosamine transferase activities.

Mechanism of action of BrdU

Although many studies have been concerned with the effect of BrdU on suppression of the differentiated function of cells, there is as yet little information concerning the mechanism of BrdU action. In view of the striking effects of the short exposure of cells to BrdU, it seemed appropriate to study some aspects of the metabolism of BrdU in these cells. Autoradiographs of limb bud cells cultured in ^3H-BrdU for

48 h over agar showed a dense accumulation of silver grains in 95% of the cells. However, after 7 days of subculture on plastic dishes, during which time approximately three cell divisions had occurred, silver grains were no longer detectable. An investigation of the kinetics of the loss of ³H-BrdU was accordingly undertaken. Similar kinetic studies utilizing ³H-guanosine and ³H-thymidine were also carried out. The

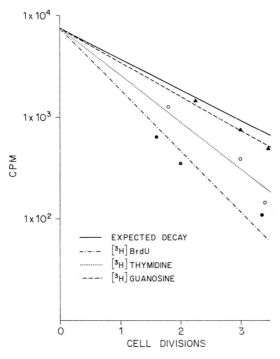

Fig. 1. Rate of disappearance of labeled bases as compared with expected rate.

results are indicated in Fig. 1. In the studies, preparations were treated with RNase to remove label incorporated into RNA. The rate of disappearance of the three labeled compounds is compared with the rate expected on the basis of cell division. The data show that ³H-guanosine disappears at the rate predicted on the basis of cell division while ³H-BrdU disappears at a more rapid rate and ³H-thymidine at an intermediate rate. These data indicate that both thymidine and BrdU turnover at a rate more rapid than is anticipated on the basis of DNA replication.

DISCUSSION

The results reported in this study indicate that early embryonic limb tissue is relatively unstable with respect to differentiation into mature cartilage. When freshly dissociated limb bud mesenchyme cells are grown at low density or are exposed to BrdU, there is permanent inhibition of chondrogenic expression. The effects of culturing the cells at low density can be alleviated by initially placing the cells in liquid suspension over agar, while the suppressive effects of early exposure to BrdU can only be overcome by simultaneous addition of high concentrations of thymidine.

Abbott *et al.* (1972) also have recently observed the irreversible inhibition of chondrogenesis in Stage 17–18 chick somites exposed to BrdU on days 0–3 of culture. When exposed to BrdU later in culture, chondrogenesis becomes increasingly evident. These investigators did not note suppression of hyaluronic acid synthesis. However, the methods used may not have been sufficiently sensitive to detect partial inhibition.

The studies reported in this communication indicate an approximately 90% inhibition of chondroitin sulfate synthesis. The site of defect of chondroitin sulfate synthesis is not clear. Biosynthesis of chondromucoprotein involves the synthesis of a core protein followed by the stepwise addition of monosaccharide units from uridine diphosphate sugars by six different glycosyltransferases and completion of the molecule by transfer of sulfate from 3′-phosphoadenosine-5′-phosphosulfate.

Decreased chondromucoprotein synthesis might then result from alterations of the synthetic machinery at various levels which may be enumerated as follows: (1) Failure to synthesize the core protein, (2) failure to synthesize the requisite sugar nucleotide precursors, (3) lack of requisite transferases, or (4) interference with as yet unknown control mechanisms. Complex glycoproteins are probably synthesized by multienzyme systems on cytoplasmic membranes. Failure of organization of such anatomical relationships might also interfere with orderly synthesis of such macromolecules. Anderson *et al.* (1970) have indicated that BrdU-treated chondrocytes have less prominent endoplasmic reticulum and Golgi complexes and more free ribosomes, microtubules and cytoplasmic filaments than do normal chick chondrocytes. Holthausen *et al.* (1969) have reported a decrease in UDP-GlcNAc epimerase, UDP-Glc dehydrogenase and the enzyme system responsible for synthesis of 3′-phosphoadenosine-5′-phosphosulfate in BrdU-treated cells. They, however, reported the presence of uridine diphospho-sugars in normal amounts. Data presented in this paper indicate a decrease in activity of two of the transferases involved in chondromucoprotein synthesis but the extent of decrease is not as striking as the decrease in chondroitin sulfate synthesis.

Chondroitin sulfate has been shown to be synthesized by a number of cell types other than chondrocytes. Whether the proteoglycan is identical with that of cartilage is unknown. It remains to be determined whether the chondroitin sulfate-proteoglycan produced by BrdU-treated cells is the same as that produced by chondrocytes. It is possible that as in the case of collagen, two types of molecules are produced only one of which is characteristic of cartilage, and the production of a cartilage-specific proteoglycan is suppressed by BrdU-treatment. Preliminary studies indicate that BrdU-treated chondrocytes synthesize collagen at a rate approximately equal to that of normal cells (Levitt and Dorfman, 1972). However, since it is now known that chondrocytes synthesize a characteristic type of collagen, it becomes of importance to determine the nature of collagen synthesized by BrdU-treated cells. Such studies are now under way.

The effect of BrdU on these 'protodifferentiated' cells (Rutter *et al.*, 1968) contrasts markedly with its influence on differentiated cells. The repression of differentiated expression by BrdU is both permanent and time-specific in the early embryonic cell type; however, it is reversible in mature cells. Miura and Wilt (1971) have recently demonstrated a suppressive effect of BrdU associated with a specific period of embryonic chick erythropoiesis. However, this inhibition could be reversed by subsequent addition of thymidine at twice the molarity of BrdU. Gontcharoff and Mazia (1967) have also shown a critical period during which BrdU perturbs sea urchin development, the effect being irreversible. The similarity between their finding with sea urchins and the one reported in this paper cannot be readily determined, although superficially the sequence of events appears comparable.

Recent studies with transplanted fragments of chick-limb mesenchyme indicate that a relatively fixed time period exists during which limb-bud cells are stabilized into a chondrogenic versus soft tissue phenotype (Searls and Janners, 1969). Before this crucial stage, cells in certain areas of the limb seem to be directed toward cartilage production but can become redirected by modification of their environment. Intrinsic cellular qualities appear to prevent environmental factors from determining cell expression after this important period. We have shown that the exhibition of the differentiated state by early limb mesenchyme cells (Stage 23–24) is exquisitely sensitive to their surroundings, while older embryonic cells (Stage 32) are only transiently affected by their milieu and, thus, possess the ability to revert to normal expressivity once their environment becomes favorable.

The mode of action of BrdU may allow the elucidation of mechanisms involved in cellular differentiation. On the basis of the specific activity and total incorporation of ^{3}H-BrdU, calculations indicate that BrdU is initially incorporated into DNA at a fairly low concentration (about 2% substitution for thymidine). It is then rapidly lost from the genome. Radioautographs revealed that 95% of the cells exposed to BrdU incorporate the analog into their DNA during the 48 h culture period over agar. It could not be determined whether inhibitors of DNA synthesis prevent BrdU suppression of differentiation, since previous studies with cytosine arabinoside, nitrogen mustard, and 5-fluorodeoxyuridine revealed that these inhibitors themselves disturbed the development of cartilage from limb-bud mesenchyme (Levitt and Zwilling, unpublished).

The possible importance of specific satellites of DNA or selective gene amplification in differentiation are only beginning to be appreciated. Walther et al. (1971) has recently shown BrdU to be selectively incorporated into a 'light' fraction of DNA. Others have postulated the involvement of cytoplasmic DNA molecules in influencing both tissue differentiation and gene expression (Bell, 1969). Since we have observed that even a relatively brief exposure to BrdU (18–24 h) during the initial period of cultivation of chick limb-bud cells will permanently repress chondrogenic differentiation, it appears plausible that some of the DNA synthesized by the limb-bud mesenchymal cell during this early period possesses characteristics not retained in mature cells. If this fraction of DNA is modified before differentiation is stabilized, the permanent loss of specialized function in affected cells might be expected. A possible explanation of these events is that BrdU is incorporated into a specific region of DNA, exerts its repressive action at a crucial time of differentiation, and then is rapidly lost from this now functionless region.

The unexpectedly rapid loss of BrdU requires further investigation. Since BrdU leads to strand breaks, it may be lost as a result of the excision and repair mechanism. This explanation would not, however, explain the difference in the rate of disappearance of thymidine and guanosine.

ACKNOWLEDGEMENTS

We are deeply indebted to the late Dr. Edgar Zwilling for several of the techniques and unpublished information in this report.

This study was supported by USPHS Grants AM-05996, HD-04583, and HD-00001, and a grant from the Illinois and Chicago Heart Association.

The assistance of Miss Pei-Lee Ho in these studies is deeply appreciated.

REFERENCES

Abbott, J. and Holtzer, H. (1968). *Proc. Natl. Acad. Sci. USA* **59**, 1144-1151.
Abbott, J., Mayne, R. and Holtzer, H. (1972). *Develop. Biol.* **28**, 430-442.
Anderson, H. C., Chacko, S., Abbott, J. and Holtzer, H. (1970). *Amer. J. Path.* **60**, 289-311.
Bell, E. (1969). *Nature* **224**, 326-328.
Gontcharoff, M. and Mazia, D. (1967). *Exp. Cell. Res.* **46**, 315-327.
Hamburger, V. and Hamilton, H. C. (1951). *J. Morph.* **88**, 49-92.
Holthausen, H. S., Chacko, S., Davidson, E. A. and Holtzer, H. (1969). *Proc. Natl. Acad. Sci. USA* **63**, 864-870.
Horwitz, A. L. and Dorfman, A. (1968). *J. Cell Biol.* **38**, 358-369.
Horwitz, A. L. and Dorfman, A. (1970). *J. Cell Biol.* **45**, 434-438.
Levitt, D. and Dorfman, A. (1972). *Proc. Natl. Acad. Sci. USA* **69**, 1253-1257.
Miller, E. J. and Matukas, V. J. (1969). *Proc. Natl. Acad. Sci. USA* **64**, 1264-1268.
Miura, Y. and Wilt, F. H. (1971). *J. Cell Biol.* **48**, 523-532.
Moscona, A. (1961). *Exp. Cell Res.* **22**, 455-475.
Nevo, Z., Horwitz, A. L. and Dorfman, A. (1972). *Develop. Biol.* **28**, 219-228.
Rutter, W. J., Kemp, J. D., Bradshaw, W. S., Clark, W. R., Ronzio, R. A. and Sanders, T. G. (1968). *J. Cell Physiol.* **72**, part 2, suppl. 1, 1-18.
Searls, R. L. and Janners, M. Y. (1969). *J. Exp. Zool.* **170**, 365-375.
Stoolmiller, A. C., Horwitz, A. L. and Dorfman, A. (1972). *J. Biol. Chem.* **247**, 3525-3532.
Trelstad, R. L., Kang, A. H., Igarashi, S. and Gross, J. (1970). *Biochemistry* **9**, 4993-4998.
Walther, B. T., David, J. D., Levine, S., Pictet, R. and Rutter, W. J. (1971). *Fed. Proc.* **30**, 1128 (Abst.).

Relationship between the Chick Periaxial Metachromatic Extracellular Material and Vertebral Chondrogenesis

GEORGES STRUDEL

Laboratoire d'Embryologie Expérimentale, Collège de France et Ecole Pratique des Hautes Etudes, 49bis, avenue de la Belle Gabrielle 94130 Nogent-sur-Marne, France

The role of extracellular material during developmental interactions has been studied for many years by numerous investigators. Since Konigsberg and Hauschka (cf. Konigsberg, 1970) obtained the differentiation of myoblasts into myotubes *in vitro*, the problem of developmental significance of intercellular material in cell or tissue interactions has gained a renewal of interest.

We have studied such a type of interaction for several years: namely the influence exercised by the neural tube and the notochord on the differentiation of chick vertebral cartilage. It is well known that the extirpation of the neural tube and the notochord in young embryos prevents the differentiation of the sclerotome cells into cartilage (Strudel, 1953; Watterson *et al.*, 1954; Holtzer, 1952; Grobstein and Parker, 1954).

The results of somite culture experiments *in vitro* (Strudel, 1963; Ellison *et al.*, 1969; Ellison and Lash, 1971; O'Hare, 1972) showed that chondrogenesis can occur without inducer or extract from the inducing organs, on condition that the *in vitro* environment favours chondrogenesis. For this reason the question arises as to whether the microenvironment provided by the extracellular metachromatic periaxial material, existing around the axial organs prior to vertebral cartilage differentiation, interferes with or provokes vertebral chondrogenesis or, at the very least, participates in the mechanisms of somite chondrogenesis *in vivo*.

The present paper deals with the experiments performed in order to study the relationships which may exist between the periaxial material and vertebral chondrogenesis. For a better understanding of the developmental significance of the role of the periaxial extracellular material in vertebral chondrogenesis we made a comparative study of limb and vertebral primordia chondrogenesis *in vivo* and in organ culture under normal and experimental conditions.

CHICK VERTEBRAL AND LIMB CHONDROGENESIS

Vertebral chondrogenesis *in vivo*

Vertebral cartilage cells originate from the sclerotome which is a part of the somitic mesoderm. In young chick embryos, prior to vertebral chondrogenesis, there exists an acellular area around the neural tube, the notochord and the somites containing extracellular material (Fig. 1). This extracellular material gives a true metachromasia with toluidine blue. Morphologically it is composed of flocculent masses, electron-dense granules, microfilaments and an electron-translucent ground substance. The chemical investigations of different workers (Thorp and Dorfman, 1967; Kvist and Finnegan, 1970; Frederickson and Low, 1971; Bazin and Strudel, 1972) demonstrated that this material contains glycosaminoglycans (hyaluronic acid, chondroitin 4- and 6-sulfate) collagen and sialoglycoproteins. The periaxial extracellular material can be completely hydrolysed by several unpurified enzymes. It is possible to hydrolyse selectively the different components by utilising purified protease-free collagenase, purified testicular or bacterial hyaluronidase and purified neuraminidase.

On the third day of incubation the periaxial material forms a dense three-dimensional network in the acellular area around the axial organs. This network can be considered as the stroma of the primary chick connective tissue. Between the third and the fourth day of incubation, the sclerotome cells emigrate from the sclerotome towards the axial organs. The meshes of the extracellular network seem to serve as a support for the emigrating sclerotome cells (Fig. 2). At the end of the fourth day, the emigrated cells have completely filled the periaxial area and are now embedded in the extracellular material. Between the fourth and the fifth day the cells located around the notochord differentiate into chondroblasts. Cartilage appears phenotypically first around the notochord (Fig. 3). Later chondrogenesis spreads out to the periphery. On the sixth day the vertebra is developed. While chondrogenesis occurs, the extracellular material is gradually transformed into cartilage matrix.

The extirpation in young embryos of the neural tube and the notochord not only results in the failure of vertebral cartilage differentiation but it provokes a total absence of any somitic mesenchyme differentiation in the operated area. Neither the dermatome and the myotome nor the sclerotome develops. The somitic mesenchyme cells remain phenotypically undifferentiated in spite of their genetic determination. A still more important fact has to be noticed: the extirpation of the axial organs, besides the consequences

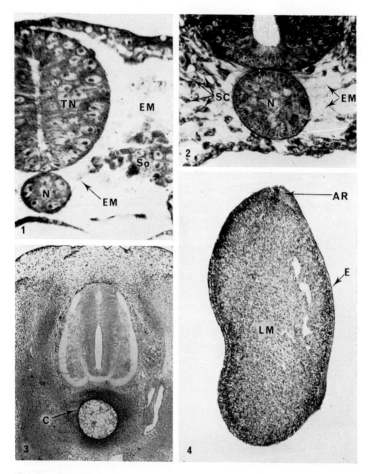

Fig. 1. Section through an intersomitic region of a 2-day old embryo. In the acellular area around the neural tube and the notochord exists metachromatic extracellular material. TN, neural tube; N, notochord; EM, extracellular material; So, somitic mesenchyme ($\times$ 640).

Fig. 2. Section through a 3-day old embryo. The meshes of the three-dimensional network of the extracellular material seem to support the emigrating sclerotome cells. SC, sclerotome cells; N, notochord; EM, extracellular material ($\times$ 360).

Fig. 3. Section through a 5-day old embryo. Cartilage has differentiated around the notochord. C, cartilage ($\times$ 56).

Fig. 4. Section through a limb primordium from a $4\frac{1}{2}$-day old embryo. The limb mesenchyme cells are identical throughout the mesenchyme. E, epithelium; AR, apical ridge; LM, limb mesenchyme ($\times$ 100).

already mentioned, causes the absence of the metachromatic extracellular material in the operated axial region. It is sufficient that either the neural tube or the notochord remain in place for the metachromatic material to appear and cartilage to differentiate. Furthermore a trypsinized neural tube/notochord system need only to be grafted into an operated embryo, for the metachromatic material to appear and for chondrogenesis to occur (Strudel, 1972a).

Vertebral chondrogenesis in organ culture

Different workers (Strudel, 1963; Ellison *et al.*, 1969; Ellison and Lash, 1971; O'Hare, 1972) showed that young somites cultured *in vitro*, without inducer can undergo chondrogenesis. This result led different authors to suppose that the determined sclerotome cells need only the contact with a microenvironment favouring chondrogenesis to differentiate into vertebral cartilage. However, this does not mean at all, that, *in vitro* and *in situ*, the action of the inducing organs is dispensable. At the present time, nobody has been able to provoke vertebral chondrogenesis *in vivo* and *in situ* in the absence of the inducing organs and nobody has been able to determine exactly the stage of the genetic determination of the sclerotome cells (see reviews: Thorp and Dorfman, 1967; Holtzer and Matheson, 1970; Strudel, 1972a).

Chondrogenesis of vertebral primordia taken from $2\frac{1}{2}$-4-day old embryos was studied in organ culture. A primordium is composed of a segment of the neural tube and of the notochord with the two adjoining rows of somites (Fig. 5). The primordia are cultured on an agar medium containing embryo extract, horse serum and Tyrode solution. Primordia cultured on such media undergo chondrogenesis. First, as *in vivo*, the metachromatic periaxial material appears and, after several days, cartilage differentiates especially around the notochord because the notochord develops better in organ culture than the neural tube.

Limb chondrogenesis *in vivo*

Limb chondrogenesis *in vivo* has been studied extensively by different authors. The results of our investigations are in accordance with the findings of Searls *et al.* (1972).

From the third to the fourth day of incubation limb mesenchyme cells appear to be uniform (Fig. 4). From the end of the 4th day onwards the cells of the limb cartilage-forming regions become distinguishable from the other mesenchyme cells. They form condensations or chondrogenic blastemas which do not stain metachromatically in spite of the existence of slight electron-dense extracellular material. Only at the beginning of the 5th day, do the chondrogenic regions acquire extracellular metachromatic materials. At the $5\frac{1}{2}$-day stage phenotypic cartilage is histologically recognizable; it possesses a typical cartilage matrix.

In conclusion, one can establish that limb chondrogenesis occurs *in vivo*, as does vertebral chondrogenesis, between the 4th and 5th day of incubation. However, there exists an important histological and biochemical difference

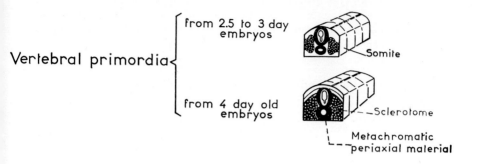

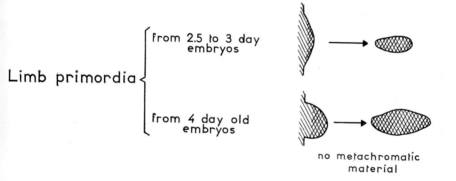

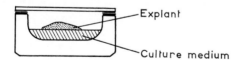

The primordia are cultured for 5 days on normal media or on media containing enzymes.

Fig. 5. Diagram illustrating the primordia organ culture technique.

between the two chondrogenic processes. In vertebral primordia we demonstrated the existence of periaxial extracellular metachromatic material before vertebral chondrogenesis is set in action. Nobody has been able to detect metachromatic extracellular material in limb chondrogenic mesenchyme before limb cartilage differentiation occurs.

Limb primordia chondrogenesis in organ culture

At the earliest stages at which limb primordia can be taken and transplanted on culture media, they develop and the chondrogenic mesenchyme undergoes chondrogenesis. The phenotypic events leading to limb chondrogenesis *in vitro* are similar to those occurring *in vivo*.

VERTEBRAL AND LIMB PRIMORDIA CHONDROGENESIS IN ORGAN CULTURE AFFECTED BY DIFFERENT FACTORS

Since no extracellular material exists prior to chondrogenesis in limb primordia we expect that the reaction of limb and vertebral primordia would be dissimilar if they are submitted to the same experimental conditions (Fig. 5). We cultured two types of each primordium: those taken between $2\frac{1}{2}$–3 days of incubation and primordia taken from 4-day old embryos. The primordia were cultured on control media and on media in which we incorporated enzymes (collagenase, testicular and bacterial hyaluronidase, neuraminidase). All the histological sections were stained by toluidine blue at pH 4. In the experimental media the concentration of the enzymes were chosen so that they neither kill the cells nor abolish their activity. The treated explants transplanted on control media recover and regenerate. At the present time, we do not know the mechanisms by which the enzymes act, probably by their primary action, but one cannot exclude secondary effects. The vitality of the treated explants was verified by culturing them on control media to which we added tritiated thymidine (20 μCi/dish). The uptake of the marker and the qualitative labeling were studied and compared with the incorporation by control explants for the same lapse of time (1 h and 20 h) by light microscopy examination of the radiographs. No noticeable difference could be detected between the labelling of the treated and the control explants (details are given elsewhere).

Results

Action of collagenase

The experiments were performed with purified protease-free collagenase purchased from Worthington Biomedical Corporation or generously provided by Dr. Kühn (Munich, Germany). The experimental primordia were cultured for 3–6 days in the presence of 10–20 μg of enzyme per sample.

The vertebral primordia taken from $2\frac{1}{2}$–3-day old embryos cultured on the experimental media contained neither metachromatic periaxial extracellular material nor cartilage. The mesenchyme remained undifferentiated (Fig. 6). Transplanted on control media the tissues of the treated explants recovered their normal histological aspect. First periaxial metachromatic extracellular material appeared and later cartilage differentiated, especially around the notochord.

If the vertebral primordia are taken from 4-day old embryos, cartilage differentiated around the notochord but during its differentiation, the cartilage is depleted by the enzyme. One may observe the holes within the cartilage matrix (Fig. 7).

If young limb primordia, either wing or leg primordia, taken from $2\frac{1}{2}$–3-day old embryos are cultured under the same conditions, the determined chondrogenic mesenchyme differentiated always into cartilage. However this cartilage, like that from 4-day old treated vertebral primordia, is depleted by the enzyme (Fig. 8). In 4-day old limb primordia, the depletion of differentiated cartilage is more intense (Fig. 9). When the 4-day treated vertebral primordia and all the treated limb primordia were transplanted on control media for regeneration they showed the disappearance of the holes between the chondrocytes and the cartilage acquired its normal structure.

The action of other enzymes

The results obtained by culturing the young vertebral primordia (from $2\frac{1}{2}$–3 days) in the presence of purified testicular hyaluronidase (Leo, Sweden), of bacterial hyaluronidase (Lederer, USA) or neuraminidase (Behring, Germany) were the same as those obtained with collagenase: neither metachromatic material nor cartilage appeared.

The action exercised by the above-mentioned enzymes on 4-day old vertebral and limb primordia from $2\frac{1}{2}$–4-day old embryos was also the same as that obtained with collagenase.

CONCLUSIONS

The obtained results can be summarized as follows: in young vertebral primordia treated with enzymes neither metachromatic material nor cartilage differentiated. In 4-day old vertebral primordia, as well as in $2\frac{1}{2}$–4-day-old limb primordia cartilage differentiated but it was depleted by the enzymes.

How may these dissimilar effects be explained? Our tentative explanation is the following. Since both types of primordia contain determined chondrogenic mesenchyme cells, and since one difference between them consists of the absence of metachromatic extracellular material in limb primordia, we believe that our results demonstrate that some relationship may exist between the extracellular periaxial material and vertebral

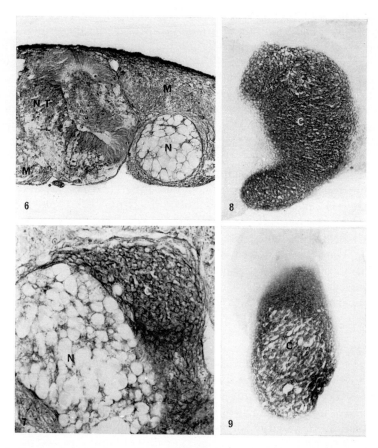

Fig. 6. Section through a vertebral primordium isolated from a 2½-day old embryo and cultured for 5 days on a medium containing protease-free collagenase. Neither cartilage nor metachromatic extracellular material differentiated. NT, neural tube; N, notochord; M, undifferentiated mesenchyme ($\times$ 138).

Fig. 7. Section through a vertebral primordium from a 4-day old embryo cultured with collagenase for 4 days. A cartilage nodule differentiated near the notochord but it is depleted by the enzyme. C, cartilage nodule; N, notochord ($\times$ 300).

Fig. 8. Section through a limb primordium from a 3-day old embryo cultured for 5 days with collagenase. A large nodule of cartilage differentiated and was depleted by the enzyme ($\times$ 115).

Fig. 9. Depleted cartilage nodule in a 4-day old limb primordium cultured for 4 days with collagenase. The depletion is more intensive than in the 3-day old primordium ($\times$ 115).

chondrogenesis. We suppose that in young vertebral primordia the enzymes hydrolyse either completely or partially the periaxial extracellular material. The determined sclerotome cells are then unable, in spite of the presence of the inducing organs, to differentiate according to their genetic determination. In 4-day old primordia the genetically coded mechanisms are already set in action. The enzymes are unable to act on the differentiating chondroblasts but they deplete the differentiated cartilage during its formation. In limb primordia, where no metachromatic material exists prior to cartilage differentiation, the enzymes act as they do in 4-day old vertebral primordia.

REFERENCES

Bazin, S. and Strudel, G. (1972). *C. R. Acad. Sci.* (Paris) **275**, 1167–1170.
Darzynkiewicz, Z. and Balazs, E. A. (1971). *Exp. Cell Res.* **66**, 113–123.
Ellison, M. L., Ambrose, E. J. and Easty, G. C. (1969). *J. Embryol. Exp. Morphol.* **21**, 331–340.
Ellison, M. L. and Lash, J. W. (1971). *Dev. Biol.* **26**, 486–496.
Frederickson, R. G. and Low, F. N. (1971). *Am. J. Anat.* **130**, 347–376.
Grobstein, Cl. and Parker, G. (1954). *Proc. Soc. Exp. Biol. Med.* **85**, 477–481.
Holtzer, H. (1952). *J. Exp. Zool.* **121**, 121–148.
Holtzer, H. and Matheson, D. W. (1970). *In* "Chemistry and Molecular Biology of the Intercellular Matrix" (E. Balazs, ed.) Vol. 3, pp. 1753–1769. Academic Press, New York.
Konigsberg, J. (1970). *In* "Chemistry and Molecular Biology of the Intercellular Matrix" (E. A. Balazs, ed.) Vol. 3, pp. 1779–1810. Academic Press, New York.
Kvist, T. N. and Finnegan, C. V. (1970). *J. Exp. Zool.* **175**, 221–240.
O'Hare, M. J. (1972). *J. Embryol. Exp. Morphol.* **27**, 215–228.
Searls, R. L., Hilfer, S. R. and Mirow, S. M. (1972). *Dev. Biol.* **28**, 123–137.
Slavkin, H. C. (1972). *In* "Developmental Aspects of Oral Biology" (H. C. Slavkin and L. A. Bavetta, eds.) pp. 165–201. Academic Press, New York.
Strudel, G. (1953). *Ann. sci. Nat. Zool.* **11**, 253–322.
Strudel, G. (1963). *J. Embryol. Exp. Morphol.* **11**, 399–412.
Strudel, G. (1967). *In* "Morphological and Biochemical Aspects of Cytodifferentiation" (E. Hagan, W. Wechsler and P. Zilliken, eds.) pp. 183–198. S. Karger, Basel.
Strudel, G. (1971). *C. R. Acad. Sci.* (Paris) **272**, 473–476.
Strudel, G. (1973). *Lyon Medical* **229**, 29–42.
Strudel, G. (1972a). *C. R. Acad. Sci.* (Paris). (In Press.)
Thorp, F. K. and Dorfman, A. (1967). *In* "Current Topics in Developmental Biology" (A. A. Moscona and A. Monroy, eds.) Vol. 2, pp. 151–190. Academic Press, New York.
Watterson, R. L., Fowler, I. and Fowler, B. J. (1954). *Am. J. Anat.* **95**, 337–400.

Studies, using Sponge Implants, on the Mechanism of Osteogenesis

GEORGE D. WINTER

Department of Biomedical Engineering
Institute of Orthopaedics (University of London)
Royal National Orthopaedic Hospital, Stanmore, Middlesex, England

The discovery that a synthetic sponge implanted into the skin of an animal can induce the formation of bone (Winter and Simpson, 1969) poses some interesting questions. Is this a unique foreign-body tissue reaction? Can the osteogenic potential of the sponge be put to practical use in surgery? Can the sponge be used to study the mechanism of osteogenesis itself? This paper describes the results of recent work providing partial answers to these questions.

MATERIALS AND METHODS

The sponge is made by solution polymerisation of glycol- monomethacrylate and 1% or less of glycoldimethacrylate in the presence of ammonium persulphate catalyst and an excess of water. The porous structure arises through the coalescence of the water phase droplets into interconnected channels (Wichterle and Lim, 1960). The polymer is not degraded by heating below 200°C and is resistant to hydrolysis and to attack by acids, alkalis and common organic solvents. It contains over 35% water, is hydrophilic and permeable to moderately large molecules so that when implanted, it may be expected to come into equilibrium with surrounding tissue fluids. It has poor tensile strength and low tear resistance. Before using the sponges for implant studies they are very thoroughly washed by repeated immersion in deionised water and centrifugation, then sterilised by autoclaving.

In the majority of experiments to be described, pieces of sponge in the shape of discs, rods or sheets, with volumes varying from about 3 cm³ to 12 cm³ were implanted into fat under the dermis in the back of 3–9 months old, pedigree Large White pigs under aseptic conditions. Variations in species, site of implantation and other experimental conditions are noted in the text where appropriate. Biopsy specimens have been obtained at frequent intervals up to 3 years and prepared for histological examination by conventional methods. Some specimens have been examined by electron probe analysis and by X-ray diffraction techniques.

RESULTS

Tissue reactions in the sponge leading to osteogenesis

Ingrowth of fibrous connective tissue

A typical polyhydroxyethylmethacrylate (polyHEMA, trade name: 'Hydron') sponge, crosslinked with 0·8% of the diester, has interconnected channels about 50 μm in diameter. Biopsy specimens taken within one hour of implantation show that the interstices rapidly fill with a protein rich exudate, replacing the water in the sponge. Blood vessels grow into this exudate (Fig. 1), accompanied by mononuclear cells and collagen fibres are formed. In four weeks the entire sponge is permeated by collagenous fibrous tissue and there are few inflammatory cells or giant cells and no indications of an adverse tissue reaction to the implant (Fig. 2).

Calcification of the polymer

The polymer brings about the deposition of calcium salt in some way not yet understood. Calcification can be detected histologically, usually within three weeks, by the characteristic basophilia with haematoxylin and by the von Kossa staining method. It starts at isolated small spherical foci near the perimeter of the implant (Fig. 3) but these centres of calcification grow progressively and many ultimately become confluent. In some implants it has been observed that the first calcium is deposited on the walls of small closed vesicles in the polymer, suggesting a membrane surface phenomenon (Fig. 4). Examination by glancing X-ray diffraction shows that particles in a sample of 'Hydron' implanted in the dermis of a pig for 22 days are poorly crystalline, but have the structure of hydroxyapatite (hexagonal $Ca_5(PO_4)_3OH$; $a_0 \simeq 9.42$ Å, $c_0 \simeq 6.88$ Å). A control sample of 'Hydron' not implanted, was essentially amorphous. Electron probe micro-analysis shows that the individual particles vary in size, the largest being about 50 μm across. The Ca:P ratio varies with the size of the particles. The average value for the atomic ratio was approximately 2·3 $\pm$ 0·5, based on calibrations with calcium phosphate powder dispersed in an electron microscope grid. Any errors due to particle size effects in the standard will tend to enhance the Ca:P ratio, so that the true value could well be slightly lower than 2·3:1. Calcium and phosphorous X-ray distribution images

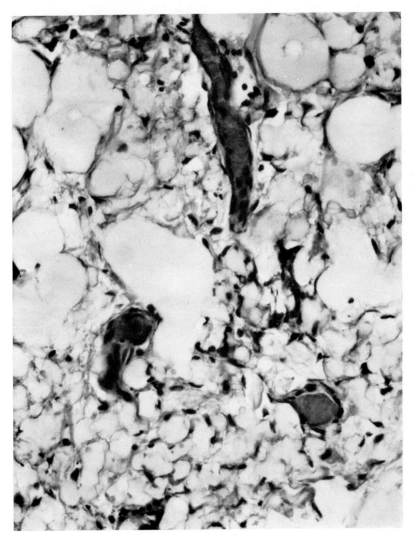

Fig. 1. 'Hydron' sponge penetrated by blood vessels (× 400).

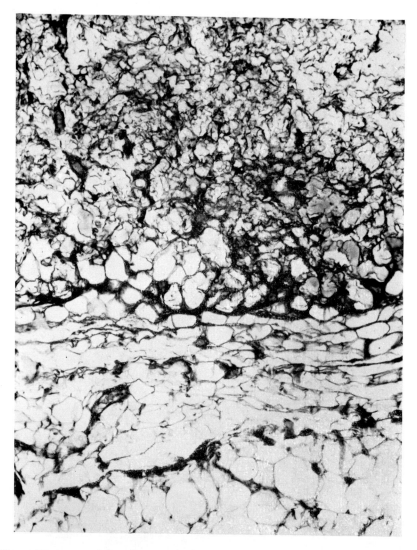

Fig. 2. 'Hydron' sponge (above) in fat under the dermis (below), interstices of sponge filled with fibrous tissue. Note absence of cellular foreign-body tissue reaction at 4 weeks ($\times$ 300).

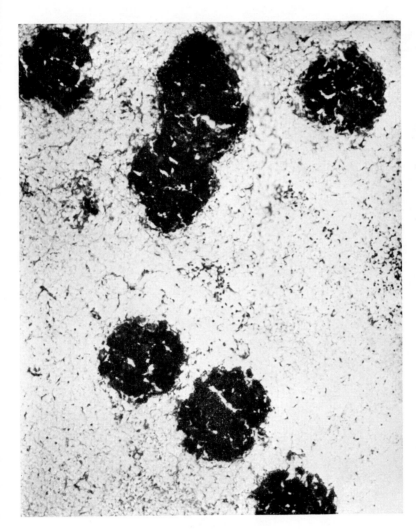

Fig. 3. Isolated calcified spherular patches at 3 weeks. Von Kossa stain ($\times$ 100).

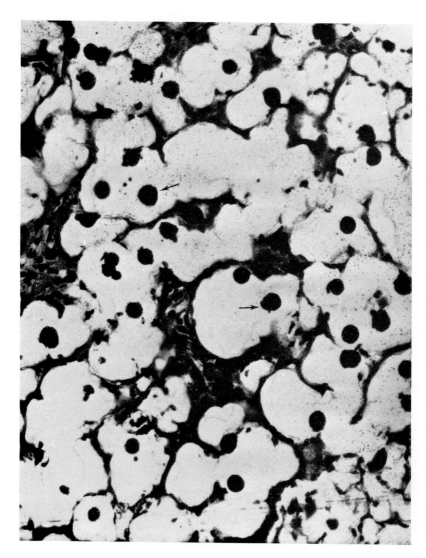

Fig. 4. Calcium deposits are first seen in closed vesicles in the polymer (arrows) at 7 days. Von Kossa stain ($\times$ 400).

have been obtained (Fig. 5) and demonstrate a good correlation in size, shape and distribution with similar areas tentatively identified as calcium deposits by the von Kossa method, which is not of itself a specific test for calcium. Such areas can be removed by soaking the sponge in ethylene diamine tetra-acetic acid (EDTA).

Calcification appears to induce some cellular reactions, notably an influx of mononuclear cells which penetrate the walls of blood vessels within the implant and are to be found interspersed with the connective tissue, filling the channels of the sponge and a moderate number of giant cells, possibly osteoclast type, associated with some of the calcified surfaces.

Bone formation

True bone has been seen, at the earliest, in 62 days and in all implants of 'Hydron', numbering over 100 specimens in at least 20 different pigs, after 100 days' implantation. Bone formation usually begins near the periphery of the implant adjacent to the initial areas of calcification (Fig. 6). The channels between calcified particles of the polymer fill with a homogeneous bone matrix enclosing cells in lacunae. The bone matrix closely invests the polymer, with no other tissue interposed between the polymer and the bone. Ossification rapidly leads to the formation of an extensive trabecular network of woven bone, but an approximately equal volume of the sponge occupies the spaces within the network and remains un-ossified (Fig. 7).

Low angle X-ray diffraction patterns have been obtained from woven bone in a 'Hydron' sponge implanted for ten weeks and are identical to the diffraction patterns obtained from bone in the animals' femoral cortex.

Following from its method of formation the bone is studded with embedded, interconnected particles of polyhydroxyethylmethacrylate (Fig. 8). Bone formation also extends into the fibrous capsule around the implant by a process of intramembranous ossification, the connective tissue cells being incorporated into lacunae to take on the function of osteocytes. Also it is observed that the sponge, being mechanically weak, is often split by the pressure of tissue growth within it. These splits become filled with bands of fibrous tissue which often ossify. Both these processes result in the formation of bony trabeculae which do not contain polymer particles (Fig. 9).

In one specimen only was a small nest of cartilage cells seen, in the centre of a mass of bone. Almost without exception heterotopic bone formation in these sponges takes place by direct ossification of fibrous connective tissue without intermediate cartilage formation. The spaces

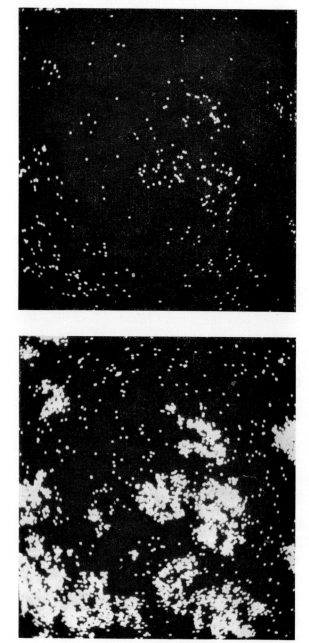

Fig. 5. X-ray distribution patterns in the electron probe microscope of calcium (left) and phosphorus (right) in calcified patches similar to those illustrated in Fig. 3.

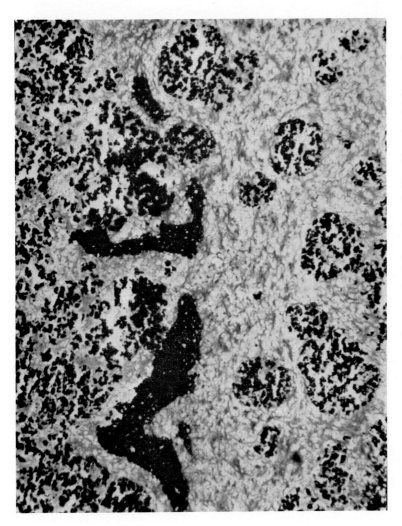

Fig. 6. Bone in relation to deposits of calcium mineral. Von Kossa stain ($\times$ 200).

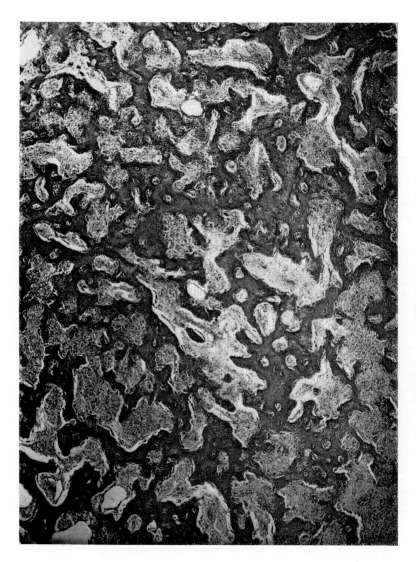

Fig. 7. Woven bone in 'Hydron' sponge at 77 days (× 25).

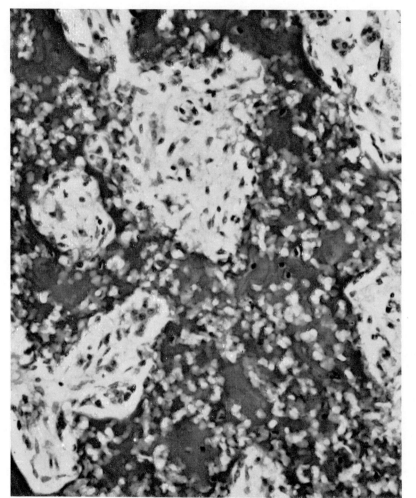

Fig. 8. Bone trabeculae with embedded particles of 'Hydron' (× 300)

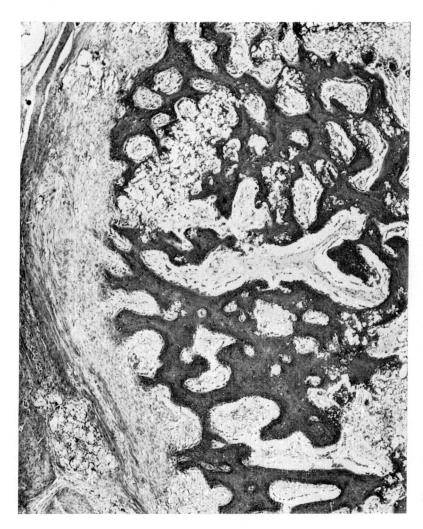

Fig. 9. Intramembranous ossification near periphery of 'Hydron' sponge implant showing trabeculae without polymer inclusions ($\times$ 40).

between the trabeculae become very well vascularised. Remodelling of the first formed bone begins immediately. The woven bone is lined with a typical palisade of osteoblasts and in some areas osteoclasts are active and characteristic resorption cavities are created. This remodelling activity leads to the creation of haversian systems. An ossicle is formed in the skin having a thin outer shell of bone, interrupted by fenestra through which pass bands of fibrous tissue and blood vessels and a trabeculated interior. The trabecular spaces are filled with a fatty marrow-like tissue but no certain identification of haematopoetic tissue has been made. Cutaneous implants have been studied so far for 953 days in the pig and although the trabeculae are thinner than at an earlier stage the ossicle still persists.

Experiments with other implants

Solid 'Hydron' sheet

PolyHEMA can also be cast in the form of a non-porous gel. When this is implanted as a solid sheet a fibrous capsule forms around the implant. Calcium mineral is deposited in the surface of the implant and the capsule responds by becoming more cellular. In studies extending over nine months in the skin of the pig no bone formation occurs in the capsule adjacent to the implant.

Solid 'Hydron' particles

Some solid polyHEMA was chopped into small particles 0·5 mm in width or less and several masses of these were implanted. In this way large pore sponges were simulated, having channels between the particles at least 0·5 mm in diameter. As expected the spaces between the particles became filled with fibrous tissue, the surfaces of the particles calcified and the tissue within the 'sponges' became more cellular, with some giant-cell reactions, but again no bone was formed within nine months.

PolyHEMA-Methacrylic acid copolymer sponge

Sponges made from a copolymer of polyHEMA and methacrylic acid have similar pore size and physical characteristics to the 'Hydron' sponges. Polymers containing 2%, 4%, 8% and 16% methacrylic acid were implanted as before. It was found that no calcification took place on this modified polymer and neither was there bone formation in implants studied for one year. In eighteen months there was scanty bone formation in one of the 2% polyHEMA—methacrylic acid copolymer sponges but the remainder contained only fibrous tissue.

PolyHEMA sponges employing different catalysts

The regular catalyst used in the manufacture of these sponges was ammonium persulphate and to investigate the possibility that a residue of sulphate ions, despite the rigorous washing procedures, was responsible for their calcification on implantation, a series of experiments was performed with sponges made from a reaction mixture in which an organic catalyst, di-isopropylperoxypercarbonate, was used in place of the persulphate. Bone occurred in these sponges as rapidly and as profusely as in the original 'Hydron' sponges.

Variation in cross-linking

Sponges made from polyHEMA crosslinked with 0·5%, 0·8% and 1% of the diester, all gave rise to bone with no detectable differences in the speed of its formation or the amount of bone formed.

Reactions in hydrophobic sponge

Several specimens of hydrophobic, polyether type polyurethane sponge ('Supersoft', Messrs. Harrison and Jones), having interconnected pores, were implanted under the dermis in pigs and studied histologically at 10, 35, 70 and 140 days. There was no calcification; the sponges were filled with collagenous fibrous tissue, there was a foreign-body giant-cell reaction and no bone formation.

Reactions in a sponge of polyvinyl alcohol

Formalised polyvinyl alcohol sponges ('Prosthex', Messrs. Ramer Chemical Company) were used as controls, for comparison with the 'Hydron' sponges in the majority of the experiments in the study. 'Prosthex' sponge, like 'Hydron' sponge is hydrophilic and it too becomes hardened by the deposition of calcium mineral on the polymer when implanted in the body. It differs in that the interconnected channels throughout the sponge are at least ten times larger in cross-sectional diameter than are the pores in 'Hydron' sponges. Implants of 'Prosthex' sponges have been studied at frequent intervals of time up to eighteen months and no bone formation has taken place. Giant cells line the calcified exposed faces of the polymer and the interstices of the sponges were filled with a well vascularised fibrous connective tissue (Fig. 10). There are signs of slow biodegradation of the polymer.

Bone formation in small-pore polyvinyl alcohol sponge

To investigate the possibility that pore size is a significant factor for the induction of heterotopic bone formation, it was decided to study the effects of implanting polyvinyl alcohol sponges with smaller pores, of the same order of size as those in the 'Hydron' sponges. None was available commercially and so attempts were made to modify the 'Prosthex' material by compressing it in water, reducing its volume to $\frac{1}{5}$th of the original and fixing it in this state by heating the water to boiling, then cooling to room temperature, whilst the sponge was kept compressed. But this collapsed the channels too much so that when the compressed sponges were implanted there was little tissue ingrowth and no bone formation. More success was obtained by chopping and rolling the dry 'Prosthex' sponge before implanting it. This broke open

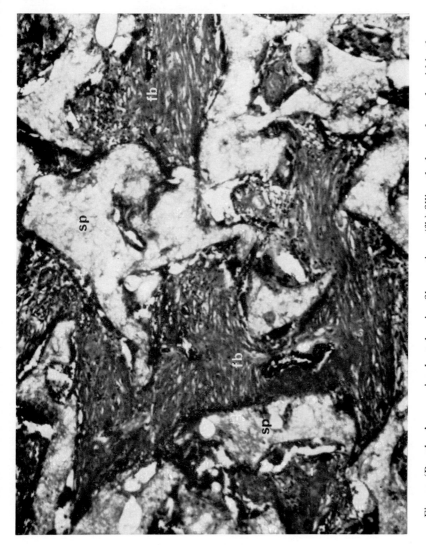

Fig. 10. 'Prosthex' sponge implant showing fibrous tissue (fb) filling the large channels within the sponge (sp) (× 70).

many of the small closed bubbles in the matrix of the sponge and it was in these cavities, measuring about 50 μm in diameter, that bone formation was initiated (Fig. 11). Ossification later spread to fill or partially occupy the much larger channels of the sponge. Bone formed in these modified 'Prosthex' sponges within 100 days (Fig. 12).

Implants in different tissues

Hydron sponges having 0·8% diester content have been implanted in the fat of the hypodermis on the back and subcutaneously on the abdomen in pigs. Other implants have been placed immediately beneath the epidermis, amongst the dense collagenous tissue of the reticular layer of the dermis and in skeletal muscle. Bone formation occurred in all these locations.

Implants in different species

Rats

Specimens of 0·5% and 1·0% diester 'Hydron' sponge were implanted into 12 adult male Wistar rats and biopsy specimens were obtained after 20 days, 30 days, 3 months and 6 months. Calcification occurred, but no bone formation (Simpson, personal communication). In a second experimental series sponges were implanted subcutaneously in 3- to 4-month old rats and biopsy specimens have so far been obtained at 5, 11, 20, 40 weeks, 1 year and 1·5 years. Again the sponge has calcified but no bone formation has taken place.

Mice

Similarly sponges implanted subcutaneously in the abdominal wall for up to 18 months in C_3H/Bi mice did not lead to bone formation.

Rabbits

'Hydron' and 'Prosthex' sponges have been implanted subcutaneously and intramuscularly in rabbits and studied for up to 13 months. After one year there is scanty bone formation in the 'Hydron' sponges.

Guinea pigs

Pieces of 0·8% diester 'Hydron' sponges were implanted subcutaneously and in the longer term biopsy specimens, at 9 and 12 months, there are changes indicative of bone matrix formation with cells incorporated in lacunae but by comparison with events in the pig the ossification is slight and no well marked trabeculae are formed.

Goat

'Hydron' and 'Prosthex' sponges were packed into holes drilled into the femur of a goat and other implants were inserted into skeletal muscle. Biopsy specimens at 120 days show that the 'Prosthex' sponges are filled with fibrous tissue. The 'Hydron' sponge implanted intramuscularly is also filled with

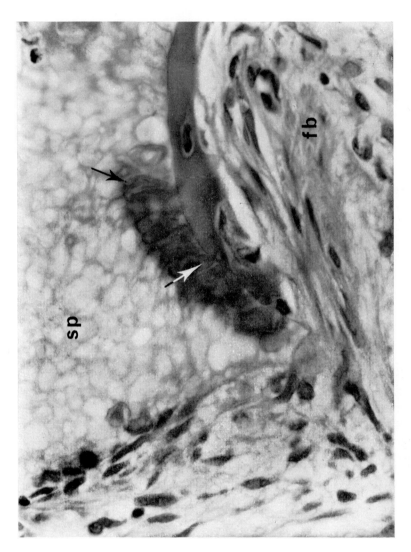

Fig. 11. Bone formation in 'Prosthex' sponge. Ossification occurred first in the smaller cavities (arrow) of the 'Prosthex' matrix (sp) and is spreading to replace the fibrous tissue (fb) in the channels of the sponge ($\times 400$).

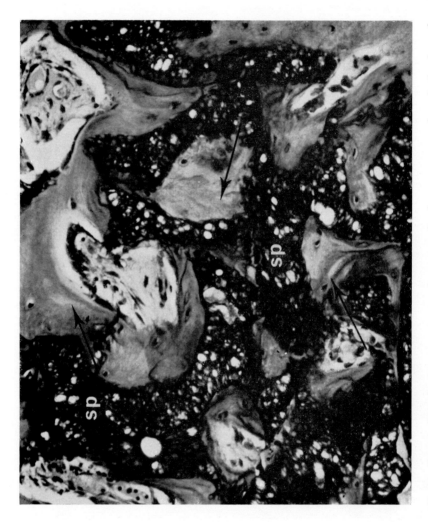

Fig. 12. Bone in 'Prosthex' sponge at 25 weeks. sp, matrix of sponge; arrows, bone in the channels of the sponge (× 120).

fibrous tissue but the specimen in the femur demonstrates heterotopic bone formation in the interior of the sponge.

DISCUSSION

It has been repeatedly demonstrated that implantation of 'Hydron' sponges into the tissues of a pig causes the formation of heterotopic bone within about ten weeks. Although heterotopic bone formation in humans (Keith, 1927) and other species occurs naturally in almost all tissues and organs of the body in certain morbid conditions, the only examples of heterotopic bone formation in response to foreign materials known to the writer are reported by Selye et al. (1960), who obtained bone in Pyrex glass cylinders inserted subcutaneously in rats and Blumberg et al. (1960) who mentioned that bone occurred in compressed 'Ivalon' sponges implanted subcutaneously in dogs for six months. Experimentally induced bone formation has intrigued many investigators (Bridges, 1959) and some have seen it as a model for the investigation of those factors which bring about ossification (Bassett, 1962). Modern work has focussed on the possibility that adult tissues may contain a specific osteogenic inductor and experiments with bone grafts have suggested that the alleged chemical messenger is a component of the organic matrix of bone (Burwell, 1966; Urist et al., 1967). Urist and Strates (1971) have postulated a 'bone morphogenic protein' and have tentatively identified it as part of the bone collagen fibril or the protein moiety of proteoglycans associated with the collagen. However, it may be seriously questioned whether these hypotheses are tenable in view of the evidence that osteogenesis in tissues remote from the skeleton can be provoked by such diverse procedures as injecting alcohol (Heinen et al., 1949), transplanting urinary bladder mucosa (Huggins, 1931), implanting certain foreign cells (Anderson and Coulter, 1967) and by implants of non-biological materials like glass, polyvinyl alcohol and polyhydroxyethylmethacrylate.

Uncertainty remains because the precise origin of the cells concerned with making heterotopic bone is in doubt and because the mechanism of calcification and of ossification has yet to be fully elucidated. In so far as the cells are concerned it is generally agreed that all the cells of the family of connective tissues, which includes bone, fibrous tissue and cartilage, belong to a common lineage and it is thought that there is a pool of relatively primitive connective tissue cells in the adult which are capable of becoming osteoblasts, fibroblasts or chondroblasts according to circumstances in their environment. Cells which have taken the decisive step and have begun to manufacture bone, fibrous tissue or cartilage and to look after the metabolism of these tissues are most probably end cells, never again to enjoy their former potential to make another kind of connective tissue or to reproduce themselves, but destined to die eventually. Connective tissue growth and its regeneration after injury relies upon the mobilisation and utilisation of the cells in the uncommitted pool. One example to illustrate this important conclusion comes from studies of wound healing in the skin where it is found that the resident fibrocytes play no part in the production of new connective tissue which originates exclusively from loose connective tissue around small blood vessels. It is believed that the perivascular loose connective tissue contains the primitive connective tissue cells which are activated by the injury, pass through several mitotic cycles, migrate to the wound and transform to fibroblasts. Evidently they become fibroblasts and not osteoblasts or chondroblasts because the dermis is a strongly 'fibrous-forming' environment and the challenge to research is to discover what are the signals from the environment that commit these cells to a specific behaviour. Some tentative conclusions about the nature of the signals can be deduced from the results of the experiments with sponges and other implants in the skin of the pig.

(*i*) In all implants fibrous tissue is formed first and it is some weeks before bone appears. Evidently the complete conditions for bone formation are not present initially but develop in time.

(*ii*) Bone formation is preceded by the deposition of calcium mineral which is, by 22 days, at least partially in the form of hydroxyapatite. In two otherwise similar sponges, namely 'Hydron' and 'Hydron'—methacrylic acid copolymer, bone formation occurs only in the calcifiable sponge—'Hydron'. But calcification alone is evidently not a sufficient signal for bone production, since solid 'Hydron' sheet, large-pore 'Prosthex' sponge and the simulated large-pore 'Hydron' sponge all calcify but do not ossify.

In this context it should be emphasised that calcification, involving as it does the precipitation of bone salt, is to be sharply distinguished from ossification, which requires the intervention of specific cells, the destruction of the existing fibrous tissue, the elaboration of a specific organic bone matrix or osseum and the calcification of this matrix.

Pathological heterotopic calcification is a familiar phenomenon and often mineral deposits in traumatised tissue consist of hydroxyapatite. Eisenstein et al. (1960) considering the pathogenesis of abnormal tissue calcifications point out that almost any matrix may calcify, even intracellular ones. The common denominator is injury to tissue but only those forms of tissue trauma which leave recognisable structure predispose to calcification. Pathological calcifications are frequently followed by bone formation. The hint that a structural or spatial factor may be significant is echoed in the usual explanation given for the osteogenic potential of bone grafts which are admitted not to supply a significant number of viable bone cells but rather their worth lies in providing a suitable superstructure for bone growth. Spatial factors seem to be important for the growth of bone in sponges.

(*iii*) No bone is formed adjacent to solid blocks of 'Hydron' gel nor to clumps of 'Hydron' particles having a large surface area which calcifies. Bone is formed in 'Hydron' sponges having channels of 50 μm diameter, but not in 'Prosthex' sponges with channels 0·5 mm in diameter. However, bone does occur in 'Prosthex' sponges that have channels 50 μm in diameter. The facts suggest that a spatial unit about 50 μm across is the critical dimension for bone formation. This is approximately the space occupied by one small blood vessel and satellite cells and intercellular substances. It is the optimum distance that can be efficiently serviced with oxygen and metabolites by diffusion and it can be looked upon as a histological unit of connective tissue.

Analogous factors can be recognised in the zone of bone growth at the ends of long bones. The cells of the cartilage in the epiphysis are aligned in rows and the longitudinal septa between them become calcified. It appears that the cartilage cells produce micro-vesicles, rich in enzymes that can increase the local concentration of orthophosphate (Ali *et al.*, 1970). The vesicles become separated from the cells and embedded in the matrix and in them the first crystals of hydroxyapatite are formed (Anderson, 1969). The cells die and the transverse septa are broken down and into the tunnels so formed grow blood vessels and cells from the medulla of the shaft of the bone. Ossification takes place on the surfaces of the calcified cartilage so that the initial trabeculae have calcified cartilage cores which are later resorbed when remodelling takes place. An almost exact parallel can be drawn between events leading to bone formation in 'Hydron' sponges in the skin and endochondral ossification in the epiphyses, where calcified 'Hydron' sponge can theoretically substitute for the sponge-like matrix of calcified cartilage. It may also be noted that subperiosteal and endosteal bone growth always takes place by ingrowth of connective tissue into small excavation cavities.

All these observations suggest that the critical factors in the environment that direct growing connective tissue to form bone are:

(a) *Spatial:* the connective tissue is divided into histological units about 50 μm in diameter by the porous structure of the matrix in which growth is taking place.

(b) *Chemical:* the contact of the growing units of connective tissue with deposits of calcium mineral.

It is possible that heterotopic bone formation in decalcified bone grafts can be explained in this way because the bone is certainly porous and has open channels of the critical size and it is undeniably a calcifiable matrix. Other factors which may be significant in determining the kind of connective tissue produced by activated primitive connective tissue cells are tension, predisposing to fibrous tissue formation, compression and high oxygen tension tending to produce bone and low oxygen tension which favours cartilage production (Bassett, 1962).

The suggestion that the presence of calcium mineral acts as a signal for bone growth is not new since Leriche and Policard in 1928 proposed that undifferentiated connective tissue cells elaborated bone matrix in response to a local 'calcific surcharge'. But the injection of calcium salts into muscle produces bone only occasionally (see Bridges, 1962) and modern theory has neglected the earlier observations.

It is also interesting to note that bone evolved as a dermal structure, probably originally as a means of storing phosphate and later as a dermal armour. The calcium phosphate will have been deposited around the walls of blood vessels, as discussed by Tarlo (1964) and hence the two conditions defined above—a microporous structure and deposits of calcium mineral were present in the skin of the earliest vertebrates known as fossils, the heterostracan ostracoderms and the evolution of this hard tissue can be traced in time towards true bone.

Having regard to potential clinical uses of 'Hydron' sponge it should be remarked that so far bone has been shown to occur, profusely and reproducibly, only in the pig, although there are indications that it happens in other animals. Possibly for each species there is a critical pore size. It is known that 'Hydron' calcifies in man and for this reason it is an unsuitable material from which to make prostheses for augmentation of the female breast because of the hardening and distortion that occurs with time. Sponge breast prostheses also have an unacceptably high incidence of infection.

No tumours have occurred in relation to 'Hydron' implanted for 18 months in rats and 3 years in pigs and there are no histopathological indications of other adverse reactions.

There is much interest at the present time in the possible uses of porous materials in orthopaedics and dentistry in an effort to obtain tissue ingrowth into prostheses with more thorough integration of implants with the tissues than has been achieved in the past. It is important to investigate the principles that govern the behaviour of tissues in porous materials and although the poor mechanical properties of 'Hydron' render it unsuitable for most orthopaedic applications, it is possible that as a filler in a macroporous prosthesis, made from a stronger material its bone-forming potential could be exploited.

ACKNOWLEDGEMENTS

I thank Dr. John T. Scales for his generous interest and support in this work. I am grateful to Messrs. Hydron Limited and Smith and Nephew Research Limited for supplying 'Hydron' for these studies, and to Dr. S. Chatterji, Crystallography Department, Birkbeck College, University of London, and B. P. Richards of the General Electric Company Limited, Hirst Research Centre, for the physico-chemical analysis. I am pleased to acknowledge skilful technical assistance by D. W. Clark and M. E. Wait.

I thank the European Research Office of the US Army for a grant, which enabled me to present the results of this work at the Sigrid Jusélius Foundation Symposium.

REFERENCES

Ali, S. Y., Sajdera, S. W. and Anderson, H. C. (1970). *Proc. Natl. Acad. Sci. USA* **67**, 1513–1520.
Anderson, H. C. and Coulter, P. R. (1967). *J. Cell Biol.* **33**, 165–177.
Anderson, H. C. (1969). *J. Cell Biol.* **41**, 59–72.
Basset, C. A. L. (1962). *J. Bone Joint Surg.* **44A**, 1217–1244.
Blumberg, J. B., Griffith, P. C. and Merendino, K. A. (1960). *Ann. Surg.* **151**, 409–418.
Bridges, J. B. (1959). *Int. Rev. Cytol.* **8**, 253–278.
Burwell, R. G. (1966). *J. Bone Joint Surg.* **48B**, 532–566.
Eisenstein, R., Trueheart, R. E. and Hass, G. M. (1960). *In* "Calcification in Biological Systems" (R. F. Sognnaes, ed.) Vol. 64, pp. 281–305. American Association for the Advancement of Science.
Heinen, J. H., Dabbs, G. H. and Masson, H. A. (1949). *J. Bone Joint Surg.* **31A**, 765–775.
Huggins, C. B. (1931). *A.M.A. Archs. Surg.* **22**, 377–408.
Keith, A. (1927). *Proc. Roy. Soc. Med.* **21**, 301–308.
Selye, H., Lemire, V. and Bajusz, E. (1960). *Wilhelm Roux' Arch. Entwickl.—Mech. Org.* **151**, 572–585.
Tarlo, L. B. (1964). *In* "Proc. of First European Bone and Tooth Symposium", pp. 3–15. Pergamon, Oxford.
Urist, M. R., Silverman, B. F., Büring, K., Dubuc, F. L. and Rosenberg, J. M. (1967). *Clin. Orthop.* **53**, 243–283.
Urist, M. R. and Strates, B. S. (1971). *J. Dent. Res.* **50**, 1392–1406.
Wichterle, O. and Lim, D. (1960). *Nature* **185**, 117–118.
Winter, G. D. and Simpson, B. J. (1969). *Nature* **223**, 88–90.

The Collagen Protein Synthesis in Long-term Cultures of Human Foetal Peripheral Blood

M. Macek,[*] J. Hurych[†] and K. Smetana[‡]

Hulliger (1956/1957) indicated the possibility of fibroblast growth in mammalian buffy coat cultures. Allgöwer and Hulliger (1960) revealed that fibroblast differentiation was accompanied by formation of collagen in rabbit buffy coat cultures. From their results rose the hypothesis that under special *in vitro* conditions either macrophage or fibroblast differentiation can occur. Paul (1958) was successful in initiating fibroblastoid cultures from human peripheral blood. Petrakis *et al.* (1960) found collagen during the *in vivo* transformation of human leucocytes in subcutaneous diffusion chambers. The question concerning the differentiation of hematogenous cells into fibroblasts is not yet definitively answered, especially because the sampling of the blood by cardiac puncture is connected with a high risk of contamination by tissue fibroblasts (Ross and Lillywhite, 1965).

The aim of our study was the investigation of the cell differentiation of the human foetal peripheral blood cells in order to elucidate their differentiation potencies in long term cultures.

MATERIAL AND METHODS

The peripheral blood cells were sampled from two male and five female foetuses, 16–25 weeks old, obtained by means of the sectio minor (Table I). The interruption of pregnancy was performed for various pathological affections of the mother. Immediately after the delivery, the umbilical

---

[*] Institute for the Child Development Research, Faculty of Pediatrics, Charles University, Prague, Czechoslovakia.
[†] Institute of Hygiene and Epidemiology, Centre of Industrial Hygiene and Occupational Diseases, Prague.
[‡] Laboratory for the Ultrastructure Research of Cells and Tissues, Czechoslovak Academy of Sciences, Prague.

Table I

Cultivation characteristics of long-term foetal peripheral blood cell cultures

Culture	Pregnancy (wks)	Epitheloid cells (days)	Fibroblastoid cells (days)	1st passage (days)	Length of cultivation (months)	Cultivation medium
E2$_F$	20	5–11	12	29–40	10	AS→BS→B
E3$_F$	22	9	22	—	4	BS
E4$_M$	20	10–29	24	—	2	BS
E6$_F$	17	—	—	—	—	BS
E8$_F$	16	18	35	70	9	BS→B
E9$_M$	19	27	27	49	18	BS→B
E10$_F$	25	13	26	35	12	BS→B

M, male; F, female; AS, Eagle's medium (MEM) with 10% of calf serum; BS, EPL medium with 10–20% of calf serum; B, EPL medium without calf serum.

cord was cut by scissors. The umbilical cord blood was sampled from the foetus and placenta. Peripheral blood (0·5–2·0 ml) was tapped by draining the umbilical cord vessels without their puncture or cannulation directly into the tissue culture medium. Any contact of the umbilical cord with the sampling bottle or medium was carefully avoided as well as the pressing of the placenta to accelerate the blood flow. Only in sample E2, an additional blood sample was taken by direct cardiac puncture. The medium for the blood sampling contained heparin (500 i.u./10 ml) in the sample E2, and 3·8% sodium citrate (1 ml/10 ml) in samples E3 and E4. In other samples, the anticoagulants were omitted. The whole blood was cultivated without any previous centrifugation and separation of the red and white blood cells. The cultures were started in Blake cultivation bottles with 30 ml of medium EPL (Michl, 1961) enriched by Eagle's MEM vitamins (Eagle, 1959) and 10–20% of calf serum. The medium was supplemented only at the beginning of cultivation by Phytohaemagglutinin M (PHA)(Difco)-0·3 ml/10 ml in all samples, except samples E6 and parallel culture of sample E2. The regular changes of the medium once or twice a week gradually purified the cell inoculum from disrupted erythrocytes and hemoglobin metabolites. Cell lines were subcultivated by 0·15% trypsin in PBS with a split ratio 1 : 2 when confluent monolayers were seen.

For electron microscopic examination, the cells were fixed in glutaraldehyde or osmium tetroxide, dehydrated in ethanol, containing uranyl acetate and embedded in Durcupan-epon mixture (Smetana, 1970).

The cultures for biochemical investigation were harvested in the stationary phase the 7th or 10th day after subcultivation. After discarding the medium, the cells were washed in the shortest possible time in 100 ml of PBS. After its immediate quantitative removal, 30 ml of 0·02% versene in PBS was applied for 30 min. Then the cells were shaken off the glass, resuspended in in the same solution and centrifuged (1000 rpm/10 min). Nitrogen (N) was determined after combustion by Nesslerization. Hydroxyproline (Hyp) was ascertained by the modified Prockop and Udenfriend's method (Kivirikko et al., 1967) after acid hydrolysis (6N HCl, 105°, 16 h). Hydroxyproline reflects the content of collagen in cultivated cells and in particular fractions.

The extraction of soluble collagen (4°C) was performed with 0·7 M NaCl, pH 7·2. Free and peptidyl hydroxyproline were removed by dialysis. The insoluble collagen was extracted from the sediment according to Fitch et al. (1955). Trichloroacetic acid was taken away by dialysis. A part of insoluble collagen ($\sim$15%) remained unextracted in the residue. Insoluble collagen refers to the trichloroacetic acid extracted and unextracted hydroxyproline. Total collagen is expressed as the ratio of hydroxyproline (μg) to cell nitrogen (mg). It corresponds to the sum of hydroxyproline in isolated collagen fractions and protein bound hydroxyproline in the supernate after centrifugation of cell suspension. Cell nitrogen refers similarily to the sum of nitrogen in the supernate after centrifugation of cell suspension, in isolated collagen fractions and in the residuum of cells after extraction with hot trichloroacetic acid. Nitrogen represents mainly cell proteins and nucleic acids because

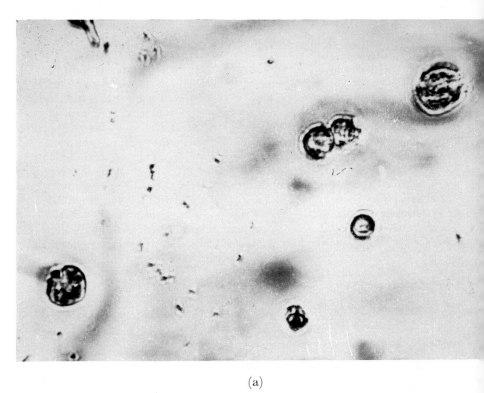

(a)

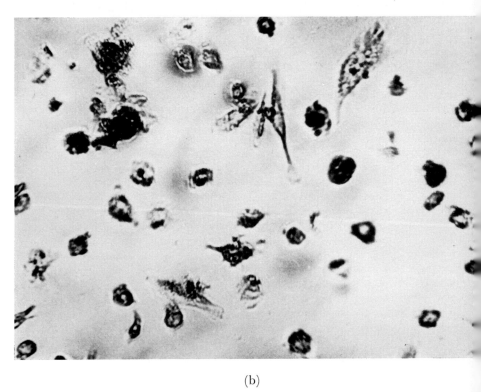

(b)

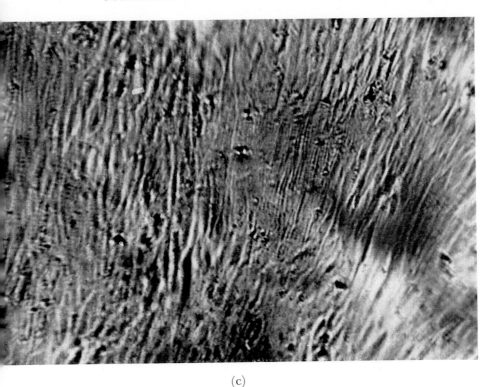

(c)

Fig. 1. Morphologic development of cultivated foetal peripheral blood cells growing on the glass in primoculture (objective 10×, ocular 4×). (a) Epitheloid cells; (b) appearance fibroepitheloid and fibroblastoid cells together with the epitheloid ones; (c) monolayer of fibroblasts.

low molecular nitrogen containing substances were removed by dialysis and by precipitation with trichloroacetic acid.

RESULTS

In the initial phase of cultivation (Table I, Fig. 1a) we observed epitheloid cells dispersed on the glass or growing in the small colonies. The fibroepitheloid cells seem to be formed by their elongation on one or both opposite ends. During further cultivation, the fibroepitheloid and fibroblastoid cells have appeared (Fig. 1b). The development of the cell population resulted in monolayers, formed by actively growing cells, characterized by fibroblast morphology (Fig. 1c).

The cultivation characteristic (Table I) demonstrates that the epitheloid cells dominated in individual samples at the beginning. The fibroblastoid cells have been observed in actively growing colonies between the 12th and 35th day of cultivation. The fibroblastoid differentiation was also revealed in cultures, derived by the experimental transfer of cultivated cell suspension not attached to the glass in the first two weeks of cultivation.

The fibroblastic differentiation was observed in 6/7 samples, in four samples there was possible to derive fibroblastoid lines, growing *in vitro* for 9–18 months after the first passage, performed between the 29th and 70th day of cultivation. The derived lines were able to grow actively in medium EPL without serum during the second phase of their *in vitro* proliferation (Table I).

The panoptic staining of the cells, cultivated on slides *in vitro*, documents their fibroblast morphology. In some cells the cytoplasma was filled by intensively stained granular material.

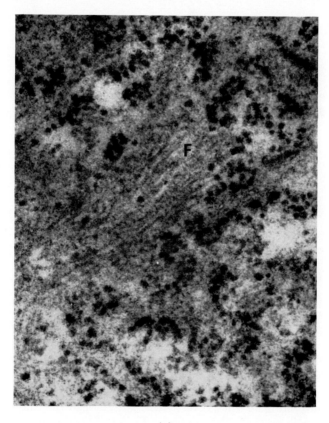

(a)

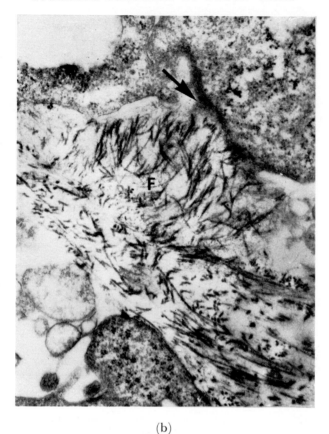

(b)

Fig. 2. Ultrastructure of cultivated foetal peripheral blood fibroblasts. (a) Fine cytoplasmic filaments (F) in a fibroblastoid cell. ($\times$ 88,000); (b) extracellular fibrils (F). Some of these fibrils seem to be in a close relationship with the surface of adjacent fibroblastoid cells (arrow). ($\times$ 23,000).

Ultrastructural investigation demonstrated developed rough endoplasmic reticulum. The fine intracytoplasmic (Fig. 2a) and extracellular (Fig. 2b) fibrils were clearly visible. Some of them were attached to the cell surface. Typical periodicity of mature collagen fibres was not detected.

The fibroblast differentiation was also proved by collagen synthesis in all fibroblast lines derived (Fig. 3). The ratio Hyp/N varied between 3·3–9·3. The highest value refers to the *in vitro* oldest E8 line.

The mutual proportion of soluble to insoluble collagen exhibits variability in particular lines (Fig. 4). In the E10 line, the ratio of both

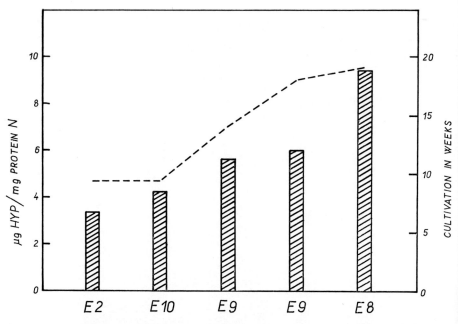

Fig. 3. Total collagen in cultivated foetal peripheral blood fibroblasts. Columns represent collagen (μg Hyp/mg protein N). Protein N corresponds to the sum of proteins and nucleic acids; low molecular substances containing N were removed by dialysis. Dotted line represents the length of *in vitro* cultivation.

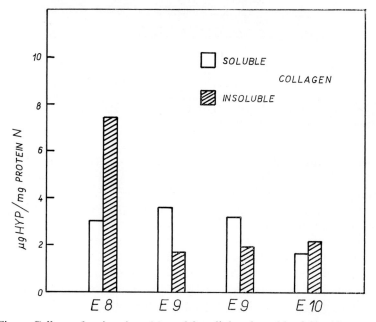

Fig. 4. Collagen fractions in cultivated foetalipheral per blood fibroblasts.

fractions was 0·75. In the E9 line, the average value from two examinations, differing in the length of *in vitro* cultivation, was 1·93. In the E8 line, the insoluble collagen forms the prevailing fraction (ratio 0·41).

DISCUSSION

The cultivation of the whole blood provided all blood cells with the same chance to be cultivated *in vitro* because it prevented any cell loss and their possible damage during the centrifugation before the start of cultivation.

The phase of primocultivation was longer than in other human diploid strains (Hayflick and Moorhead, 1961) due to the relatively slow differentiation of fibroblastoid cells. Its morphologic characteristic is similar to that in fibroblast cultures from human bone marrow (Berg and Rosenthal, 1961; Macek, 1967). Hentel and Hirschhorn (1971) brought evidence that some human bone marrow fibroblasts differentiate *in vitro* from hematopoetic cells as well. In the second phase of cultivation our fibroblast lines were able to grow in the EPL medium without serum as human diploids strains from other tissues (Macek and Michl, 1964).

Considering the origin of the cells responsible for the fibroblast differentiation, we have to take into account the fibroblast contamination from foetal tissues. Therefore we avoided puncture or cannulation and sampled the blood drops from the open umbilical vessels. The Wharton's jelly contains relatively small numbers of cells compared with the other tissues. We also found that it is very difficult to derive tissue cultures from the cells on the surface of the umbilical cord (Macek et al., 1972).

If we assume the fibroblast contamination, at least part of the fibroblasts would attach to the glass surface as do similarily the blood macrophages (Harris and Jahnz, 1958). They would form rapidly growing colonies with fibroblast morphology as is usual in cultivation of human embryonal and foetal fibroblasts. We did not see such colonies during the first weeks of cultivation. The relatively very low plating efficiency of human diploid cells (Pious *et al.*, 1964) and their inability to proliferate in suspension (Hayflick and Moorhead, 1961) decreases the probability of the proliferation of contaminating fibroblasts in our cultures. The gradual changes of the cell morphology with the length of *in vitro* cultivation do not indicate the growth of the tissue fibroblast either.

The late appearance of fibroblasts and their differentiation from cells remaining in suspension in the first two weeks of cultivation does not suggest the growth of tissue fibroblasts. The differentiation from the blood macrophages cannot be considered because the mammalian macrophages are not able to proliferate *in vitro* (Jacoby, 1965).

Despite the fact that it is impossible so far to eliminate the fibroblast contamination absolutely, there is the most plausible hypothesis that our fibroblast lines originated from the circulating pluripotential cell pool (Bond et al., 1958). The fibroblast differentiation of our cultures might be due to common mesenchymal origin of hematopoetic cells and is in agreement with the classical observation of Maximow (1926/1927).

It is impossible to compare our data with other experiments published so far, because they concern the cultivation of human peripheral blood cells from the postnatal period of life.

The negative evidence of fibroblast differentiation in human buffy coat cultures (Ross and Lillywhite, 1965; Kaluš et al., 1968) might be due among other things to a lower quantity of circulating undifferentiated cells in the postnatal period than during the foetal development. Recent observations indicate greater quantity of pluripotential cells in animal foetuses than in the postnatal period (Barnes and Loutit, 1967).

Fibroblasts derived from the foetal peripheral blood cells synthetize collagen in the same amount as the fibroblasts of human diploid cell strains from other tissues (Macek et al., 1967a; Macek et al., 1967b). The tendency to the slight increase of collagen with the length of cultivation (Fig. 3) was also seen in other fibroblast strains (Macek et al., 1967a). It can be probably related to the slight decrease of proliferation activity and thus better chance for collagen formation occurring mostly in the stationary phase (Green and Goldberg, 1964; Macek et al., 1967a).

The individual differences between the ratio of soluble to insoluble collagen in lines under investigation are not essentially different from variation range $x = 1\cdot 23 \pm 0\cdot 67$ found in human diploid fibroblast lines (Macek et al., 1966). The increase of insoluble collagen in the *in vitro* oldest E8 line was also observed in late passages of lines derived from patients with Marfan syndrome (Macek et al., 1966).

The tendency of the increase of total collagen and the decreased ratio of soluble to insoluble collagen with the length of cultivation suggests the possible relationship between ageing and collagen metabolism changes.

Therefore, the further study of human foetal peripheral blood cell differentiation *in vitro* may provide new information on the problem of the role of circulating cells in human tissue repair.

SUMMARY

The long-term cultivation of foetal peripheral blood cells of 7 foetuses 16–25 weeks old provides evidence of fibroblast differentiation in 6 of them. In four samples the fibroblast lines were derived. They were cultivated in medium EPL with 10–20% of calf serum. They have grown successfully in this medium without serum in the second phase of cultivation. Their *in vitro* development was similar as in human diploid cell strains, except the phase of primoculture. Epitheloid and fibroblastoid cells appeared and started to proliferate actively. The derived lines exhibit characteristic fibroblast morphology. The great majority of intra and extracellular fibrils did not exhibit regular periodicity. All lines investigated produced collagen. Its amount and ratio of soluble to insoluble collagen corresponded to the values in diploid fibroblast strains derived from different human tissues.

ACKNOWLEDGEMENTS

We should like to express great appreciation to Prof. Russell Ross of the University of Washington, Seattle, USA, for his helpful criticism. The sectio minors were kindly performed in the 3rd Department of Gynaecology and Obstetrics, Faculty of Pediatrics, Charles University, Prague.

REFERENCES

Allgöwer, M. and Hulliger, L. (1960). *Surgery* **47**, 603–610.
Barnes, D. W. H. and Loutit, J. T. (1967). *Lancet* **2**, 1138–1141.
Berg, R. B. and Rosenthal, M. S. (1961). *Proc. Soc. Exp. Biol.* **106**, 614–617.
Bond, V. P., Cronkite, E. P., Fliedner, T. M. and Schork, P. (1958). *Science* **128**, 202–203.
Eagle, H. (1959). *Science* **130**, 432–437.
Fitch, S. M., Harkness, M. L. R. and Harkness, R. D. (1955). *Nature* **176**, 136.
Green, H. and Goldberg, B. (1964). *Nature* **204**, 347–349.
Harris, H. and Jahnz, M. (1958). *Brit. J. Exp. Path.* **39**, 597–600.
Hayflick, L. and Moorhead, P. G. (1961). *Exptl. Cell Res.* **37**, 614–634.
Hentel, J. and Hirschhorn, K. (1971). *Blood* **38**, 81–86.
Hulliger, L. (1956/57). *Virchows Arch.* **329**, 289–318.
Hulliger, L. and Allgöwer, M. (1961). *Schweiz. med. Wschr.* **91**, 1201–1202.
Jacoby, F. (1965). *In* "Cells and Tissues in Culture, Methods, Biology and Physiology" (E. N. Wilmer, ed.) Vol. 2, pp. 1–93. Academic Press, London, New York.
Kalůs, M., Ghidoni, J. J. and O'Neal, R. M. (1968). *Pathologie et Microbiologie* **31**, 353–364.
Kivirikko, K. I., Laitinen, O. and Prockop, D. J. (1967). *Anal. Biochem.* **19**, 249–255.
Macek, M. (1967). Ph.D. Thesis, p. 22. Charles University, Prague.
Macek, M. and Michl, J. (1964). *Acta Univ. Carol. Med. (Prague)* **10**, 519–539.
Macek, M., Hurych, J., Chvapil, M. and Kadlecova, V. (1966). *Humangenetik* **3**, 87–97.
Macek, M., Hurych, J. and Chvapil, M. (1967a). *Cytologia (Tokyo)* **32** 426–443.

Macek, M., Hurych, J., Chvapil, M. and Dlouhá, M. (1967b). *Cytologia* (*Tokyo*) **32**, 308–316.
Macek, M., Řezáčová, D. and Kotásek, A. (1972). *Humangenetik*, **16**, 245–249.
Maximow, A. A. (1926/1927). *Proc. Soc. Exptl. Biol.* **24**, 570–572.
Michl, J. (1961). *Exptl. Cell Res.* **23**, 324–334.
Paul, J. (1958). *Nature* **182**, 808.
Petrakis, N. L., Davis, M. and Lucia, S. P. (1960). *Blood* **15**, 420.
Pious, D. A., Hamburger, R. N. and Mills, S. E. (1964). *Exptl. Cell Res.* **33**, 495–507.
Ross, R. and Lillywhite, J. W. (1965). *Lab. Invest.* **14**, 1568–1585.
Smetana, K. (1970). *In* "Methods in Cancer Research" (H. Busch, ed.) Vol. 5, p. 450. Academic Press, New York.

The Fibroblast as a Contractile Cell: The Myo-Fibroblast

GIULIO GABBIANI, GUIDO MAJNO and GRAEME B. RYAN
Department of Pathology, University of Geneva
Geneva, Switzerland

The various cellular elements composing connective tissue are classically divided into wandering cells and fixed cells. Typical examples of fixed cells are the fixed macrophages and the fibroblast. Upon different stimuli fixed macrophages can transform into actively moving cells; on the other hand, little attention has been given to the possibility that fibroblasts may become motile, their function is generally thought to be limited to the synthesis of fibers or ground substance. There are, however, some situations during which fibroblasts become not only motile (Ross, 1968a) but also contractile (Gabbiani *et al.*, 1971; Majno *et al.*, 1971; Hirschel *et al.*, 1971; Gabbiani *et al.*, 1972a) and in this way they influence the evolution of various pathologic processes (e.g. wound healing and granulation tissue contraction). Under such conditions, fibroblasts develop new morphologic, biochemical and pharmacological characteristics (Gabbiani *et al.*, 1971; Majno *et al.*, 1971; Hirschel *et al.*, 1971; Gabbiani *et al.*, 1972a) similar to smooth muscle—such cells are then able to fulfil new functional roles. The purpose of this chapter is to describe the features of such modified fibroblasts (which we have called 'myo-fibroblasts') and the dynamics of their formation.

The classic fibroblast

First, let us consider the characteristics of a typical 'normal' fibroblast. Since this cell has been the subject of several reviews (Ross, 1968a; Ross, 1968b), we will limit our description to the features most relevant to this discussion.

The fibroblast has long been identified and described histologically on the basis of its shape. The nucleus is generally oval and the cytoplasm is eosinophilic and possesses numerous long processes (Bloom and

Fawcett, 1968). The close association of these cells with bundles of collagen fibers was an early indication that fibroblasts are involved in fibrogenesis (Bloom and Fawcett, 1968). However only the use of modern techniques (such as autoradiography and electron microscopy) has clarified this problem. By means of electron microscopy it has been found that, during periods of increased collagen formation, fibroblasts assume the typical characteristics of actively synthesizing cells (i.e. large cisternae of rough endoplasmic reticulum and prominent Golgi apparatus) (Ross, 1968b). When incorporation of ^{14}C-labeled proline is followed autoradiographically in animals forming new collagen, the labeled amino acid is localized first in the endoplasmic reticulum of fibroblasts, then in the region of the Golgi apparatus and finally in the newly synthesized extracellular collagen fibers (Ross, 1968b; Ross and Benditt, 1965).

It appears, therefore, that the fibroblast is the connective tissue cell responsible for most collagen formation. However, there are other cellular elements capable of synthesizing collagen, such as smooth muscle cells, Schwann cells, chondrocytes and osteoblasts (Ross, 1968b). It should be noted that fibroblasts are probably also engaged in the formation of other extracellular components of connective tissue, such as elastin and reticulin (Ross, 1968b).

Other ultrastructural features of the fibroblast are: the relatively smooth contour of the nucleus which often contains a nucleolus, the occasional presence of small numbers of intracytoplasmic fibrils, peripheral vesicles and elongated cytoplasmic processes (Bloom and Fawcett, 1968). Sites of specialized cell contacts, such as desmosomes, have not been described between fibroblasts of adult animals. However, junctions have been described between fibroblasts of fetuses or newborn animals (Greenlee and Ross, 1967); it has been proposed that such junctions play a role in the maintenance of structural integrity in the embryologic development of tissue such as tendons (Greenlee and Ross, 1967).

The contractile fibroblast (myo-fibroblast)

Now let us look at the fibroblast of granulation tissue. Such cells show several remarkable features which clearly identify them as a distinct cellular type (Gabbiani *et al.*, 1971; Majno *et al.*, 1971; Hirschel *et al.*, 1971; Gabbiani *et al.*, 1972a). A meaningful differentiation from the normal fibroblasts has not been made on the basis of light microscopy; this probably explains why the spindle-shaped cells of granulation tissue have been generally considered as examples of classic fibroblasts (Ross 1968a). We first characterized myo-fibroblasts on ultrastructural grounds (Gabbiani et al., 1971); these studies were then complemented by chemical, pharmacological and immunological techniques (Majno et al., 1971; Hirschel et al., 1971; Gabbiani et al., 1972a).

Morphology

Electron microscopy shows that myo-fibroblasts possess several features which make them similar to smooth muscle:

(a) A *fibrillar system* within the cytoplasm (Gabbiani et al., 1971): not a few fibrils seen in normal fibroblasts, but bundles of parallel filaments measuring 40–80 Å in diameter, or sometimes 120–160 Å, which are usually arranged parallel to the long axis of the cell; many electron-dense areas are scattered among the bundles or located beneath the plasmalemma—these are similar to the 'dense bodies' or 'attachment sites' of smooth muscle (Figs. 1 and 2). Such fibrillar structures occupy a large part of the cytoplasm, but packed cisternae of rough endoplasmic reticulum (typical of normal fibroblasts) are still present (Fig. 1).

(b) *Nuclear deformations*. The nuclei consistently show multiple indentations or deep folds (Gabbiani et al., 1971), an appearance quite unlike that of normal fibroblasts, but very similar to that of smooth muscle and other cells undergoing contraction (Lane, 1965; Bloom and Cancilla, 1969; Majno et al., 1969) (Fig. 1).

(c) *Surface differentiations*. Numerous intercellular connections are present between myo-fibroblasts (Gabbiani et al., 1971); their structure identifies them as *maculae adhaerentes* or desmosomes, commonly with a clear-cut intermediate line (Fig. 2). In addition, part of the cell surface is often covered by a well defined layer of material having the structural features of basal lamina and separated from the cell membrane by a translucent layer (Fig. 2). Where it is covered by basal lamina, the cell usually shows dense zones in the fibrillar bundles immediately beneath the surface membrane: the resulting complex is reminiscent of the hemidesmosomes that bind smooth muscle cells, pericytes or endothelial cells to their basal lamina (Hogan and Feeney, 1963; Stehbens, 1966).

Having seen these ultrastructural features, we returned to our histologic sections and could then recognize differences between normal fibroblasts and myo-fibroblasts at the light microscopic level: in granulation tissues, and in the nodules of Dupuytren's disease (Gabbiani and Majno, 1972), the cells come to resemble smooth muscle in that they have bulky cytoplasm (containing red streaks with Masson's trichrome stain) and long nuclei that are often crossed transversely by a few basophilic lines, presumably corresponding to the infoldings seen electron microscopically.

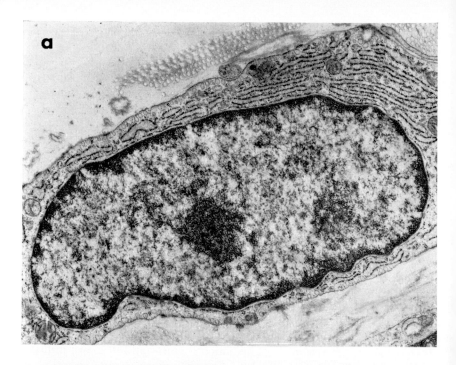

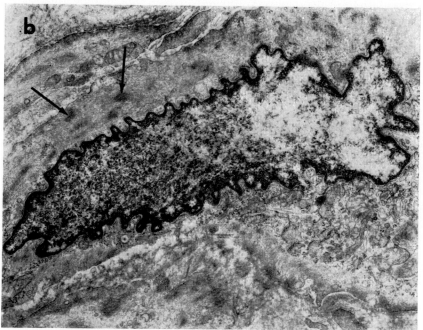

Fig. 1. Electron micrographs comparing the characteristics of fibroblasts and myo-fibroblasts. (a) Normal fibroblast from rat subcutaneous tissue. Note mitochondria, small peripheral vesicles and regular arrangement of endoplasmic reticulum ($\times$ 15,000). (b) Myo-fibroblast in a 21-day old granuloma pouch. A large part of cytoplasm contains bundles of densely packed fibrils, with attachment sites (arrows) typical of smooth muscle. Note the numerous nuclear indentations ($\times$ 15,400). (Reproduced from Gabbiani *et al.* (1972a) with the permission of the editors of the *J. Exp. Med.*)

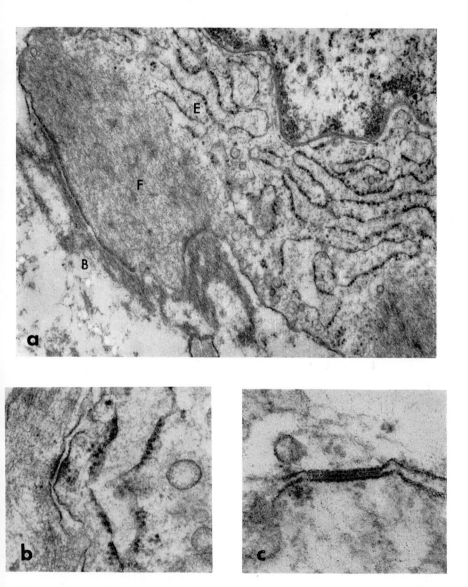

Fig. 2. Characteristics of human myo-fibroblasts. (a) A detail of the cytoplasm in a myo-fibroblast from human granulation tissue. Note a zone of basement lamina-like material (B) just outside the cell surface; beneath the cell membrane there is a thick layer of fibrils (F) with dense bodies; abundant rough endoplasmic reticulum (E) is situated more deeply ($\times 37{,}600$). (b) *Macula adhaerentes* between two myo-fibroblasts in a nodule of Dupuytren's disease. Note dense areas in adjacent cytoplasm ($\times 47{,}900$). (c) A well defined desmosome with a clear cut intermediate line between two myo-fibroblasts in a nodule of Dupuytren's disease ($\times 148{,}500$). ((b) and (c) are reproduced from Gabbiani and Majno (1972) with the permission of the editors of the *Amer. J. Path.*)

Chemistry

Our chemical estimations of actomyosin have been performed on the walls of Selye granuloma pouches (Selye, 1953). The yield of actomyosin obtained by extraction from such granulation tissue (4 mg of actomyosin/g wet weight of pouch wall) was comparable with that obtained in identically prepared extracts of pregnant rat uteri (3·5 mg/g wet weight) (Majno et al., 1971). The calcium-activated adenosinetriphosphatase activities of these extracts were similar, each splitting approximately 10 nmoles of adenosine triphosphate per mg of protein per min.

Pharmacology

We used granulation tissues from experimental animals in the form of strips (50–80 mm long) which were suspended from one end of a frontal lever and attached to the bottom of a bath containing 20 ml of Tyrode solution (Majno et al., 1971). The development of a micromethod made

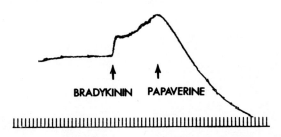

Fig. 3. Tissue from an 11-day old skin wound in the rat: contraction in response to bradykinin (1×10^{-5} g/ml) followed by relaxation due to papaverine (1×10^{-4} g/ml).

it possible to use shorter strips (about 10 mm) and to extend our investigations to specimens of human tissue (Cliff and Majno, in preparation).

Control strips of connective tissue (such as from normal skin or subcutaneous tissue) did not react to the various agents tested. On the other hand, tissues containing myo-fibroblasts (granuloma pouch, granulation tissue from a capsule (Ryan et al., 1971) that develops around intraperitoneally implanted blood clots, granulation tissue from wounds of rats and humans) did contract when exposed to substances that stimulate contraction of smooth muscle. Examples of active agents included serotonin (5-HT), angiotensin, vasopressin, bradykinin (Fig. 3), phenergan, nor-epinephrine and prostaglandin $F_1\alpha$. Papaverine caused relaxation (Fig. 3). The contraction induced by 5-HT was inhibited by anti-5-HT agents such as methysergide bimaleate or cyproheptadine. Cytochalasin B, which inhibits contraction due to microfilaments, caused a slight relaxation, and also inhibited the contraction induced by 5-HT; this inhibition could be partly reversed by washing. Finally, anoxia or potassium cyanide, a metabolic poison, relaxed strips of granulation tissue previously contracted by 5-HT and prevented contraction when applied prior to the stimulating agent.

Immunofluorescence

To investigate a possible relationship between the filaments of myo-fibroblasts and smooth muscle fibrils, we studied by means of immunofluorescence the labeling of such cells with human anti-smooth muscle sera (i.e. 'smooth muscle autoantibody' or 'SMA' sera) (Hirschel *et al.*, 1971) obtained from patients with chronic aggressive hepatitis (Johnson *et al.*, 1965). By means of the double layer technique, such sera stained smooth muscle cells from rat stomach, ureter, bladder and prostate; normal fibroblasts were never labeled. No labeling of smooth muscle or myo-fibroblasts was obtained with either normal serum or serum from a patient with myasthenia gravis. On the other hand myo-fibroblasts from different sources (e.g. rat or human skin wounds, Dupuytren's nodules or cultivated fibroblasts) always fixed SMA serum (Figs. 4 and 5).

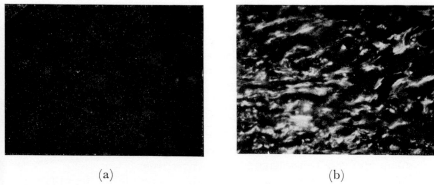

(a)　　　　　　　　　　(b)

Fig. 4. Immunofluorescent stain of myo-fibroblasts from human granulation tissue. (a) Control preparation treated with normal human serum, followed by fluorescent anti-human IgG: no labeling ($\times 400$). (b) Serial section of the same tissue treated with anti-smooth muscle serum followed by fluorescent anti-human IgG: intense labeling ($\times 400$).

In conclusion, these findings support the proposition that myo-fibroblasts have the features of contractile cells:

(a) *The cytoplasm contains a potentially contractile apparatus.* This consists of bundles of fibrils resembling those of smooth muscle. That this similarity is not simply morphologic is demonstrated by the high content of chemically extractable actomyosin in granulation tissue and by the selective localization of SMA serum to the cytoplasm of myo-fibroblasts. The nuclear shape of myo-fibroblasts, with indentations

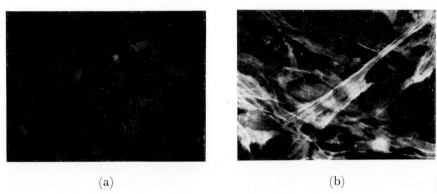

(a) (b)

Fig. 5. Immunofluorescent staining of cultured fibroblasts. (a) Control preparation treated with normal serum followed by fluorescent anti-human IgG: no labeling ($\times 400$). (b) Adjacent area on the same slide treated with anti-smooth muscle serum followed by fluorescent anti-human IgG: intense labeling. Note the specific staining in the form of long lines, probably corresponding to the distribution of microfilament bundles ($\times 400$).

and folds, provides indirect evidence of contraction. Moreover, the pharmacological tests indicate that strips of granulation tissue behave similarly to smooth muscle; the experiments with metabolic inhibitors or specific antagonists indicate that the contraction of the strips is a cell-mediated phenomenon.

(b) *The cell surfaces have devices for transmitting contraction to other cells or to the stroma.* These take the form of desmosomes and attachments to basal laminae. They are, of course, essential if cellular contraction is to explain an overall shrinkage of connective tissue.

We therefore suggest that, under certain conditions, a fibroblast can differentiate (as depicted in Fig. 6) into a cell type structurally and functionally similar to smooth muscle and that this cell, the myo-fibroblast, can play an important role in the contraction of connective tissue.

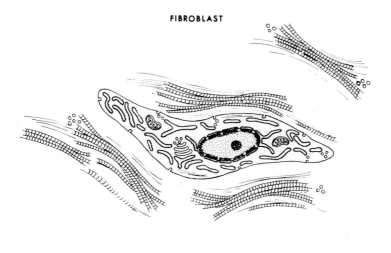

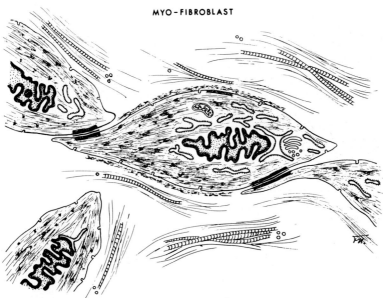

Fig. 6. Scheme comparing the characteristics of fibroblasts and myo-fibroblasts. The *upper part of the figure* shows a typical fibroblast with smooth contour of the nucleus which contains a nucleolus. The cytoplasm contains abundant cisternae of rough endoplasmic reticulum, mitochondria, a Golgi apparatus, peripheral vesicles, but only few intracytoplasmic fibrils. The extracellular tissue is mainly composed of collagen bundles. The *lower part of the figure* shows an area of granulation tissue: the cellular concentration is higher than in normal connective tissue. Myo-fibroblasts have a nucleus with numerous folds and indentations. The cytoplasm still has some cisternae of rough endoplasmic reticulum but its most striking feature is the presence of massive bundles of filaments, usually arranged parallel to the long axis of the cell; electron-dense areas are scattered among the bundles or located beneath the plasmalemma. Intercellular connections in the form of *maculae adhaerentes* or desmosomes are present between fibroblasts. In addition, part of the cell surface is often covered by a well defined layer of material similar to a basal lamina; in such regions, the cell commonly shows a dense zone (giving a hemidesmosome complex) in the fibrillar bundles immediately beneath the surface membrane. The extracellular tissue contains small unbanded fibrils as well as mature collagen fibers.

CONDITIONS LEADING TO MYO-FIBROBLAST DEVELOPMENT

We first observed myo-fibroblasts in experimental granulation tissues in rats (Gabbiani *et al.*, 1971). Later, these cells were seen in granulation tissues of other animals, including man, as well as in other pathologic conditions, which will be described here.

Granulation tissue

Experimental granulation tissue

We have studied, mostly in the rat, four experimental models:

(a) The retractive process after the removal from the chest, or the dorsum, of a square of skin (with the cutaneous muscle) measuring about 4 cm^2; such wounds begin to shrink about one week later and are fully closed at about the end of the second week. Wounds of the same size were also produced in rabbits, but leaving intact the cutaneous muscle; these take up to 30 days to heal.

(b) The shortening of a tail tendon (2 cm in length) homotransplanted into a pelvic fat body (Majno, 1958); this tendon becomes ensheathed by a layer of granulation tissue that progressively contracts, thus causing a decrease in the overall length of the tendon.

(c) The injury of the top surface of the liver (Ryan *at al.*, 1971) resulting from drying for 5 min with a gentle stream of air from a compressed air cylinder; this produces a thin layer of granulation tissue, the contraction of which causes a striking upward curling of the free edge of the liver, beginning at about 6 days.

(d) Selye's granuloma pouch (Selye, 1953), produced by the subcutaneous injection of 20 ml of air and 1 ml of croton oil in corn oil; the injected material becomes surrounded by a wall of granulation tissue; the resulting egg-shaped pouch slowly shrinks, starting at about 8 days, until it virtually disappears at about 3 months.

In all these experimental models, cells morphologically identifiable as myo-fibroblasts develop gradually (Fig. 1), being most prominent at the time of the maximal contractile activity of the granulation tissue.

Judged by immunofluorescence, the cytoplasmic labeling with SMA serum in the granuloma pouch appears first on the 7th day. It reaches a maximum between 20 and 30 days when fluorescent cells are widely distributed through the whole wall, with the exception of the innermost layer that contains mainly polymorphs and macrophages. Later, as the older granulation tissue is replaced by dense collagen, the outermost layer of the wall loses its fluorescence; by 50 days about half of the wall thickness is labeled (Hirschel *et al.*, 1971).

There are differences between the pharmacological reactivity of granulation tissues from different sources. Thus granuloma pouch strips are sensitive to 5-HT whereas wound strips do not respond to this agent under the same conditions of testing (Majno et al., 1971; Gabbiani et al., 1972a). This we could not explain—however, it is well known that smooth muscle cells from different organs show differing responses to certain pharmacological agents.

Our pharmacological experiments did not, by themselves, exclude the possibility that structures other than fibroblasts (e.g. blood vessels) are responsible for the contraction of the various granulation tissue strips. To clarify this point, we devised a new experimental model which consisted of an avascular 'granulation tissue' that develops as a capsule around intraperitoneally implanted blood clots (Ryan et al., 1971): about 8 ml of homologous blood are placed into the peritoneal cavity of 200–300-g rats. A proportion of these clots remain free in the peritoneal cavity particularly if the omentum and pelvic fat bodies are excised just before implantation—in such cases the coagulated blood becomes covered by a thin fibrous capsule which is completely avascular. The large majority of the cells composing this capsule have the morphologic and antigenic characteristics of myo-fibroblasts, and strips of the capsule are highly reactive to various agents (e.g. 5-HT, prostaglandin $F_1\alpha$ and papaverine) when tested pharmacologically. Thus, the contraction of such granulation tissue does not depend on the presence of vessels. Further evidence against the participation of vessels in granulation tissue contraction was obtained after producing granuloma pouches by injecting rats subcutaneously with 20 ml of air followed by 5 ml of a 3% solution of carrageenan in water (instead of the croton oil in corn oil mixture that is more usually used). After 2 weeks, these carrageenan pouches have a wall consisting of many macrophages and blood vessels, but with relatively few fibroblasts; such pouches remain soft, flabby and uncontracted *in vivo*, and do not contract *in vitro* in response to pharmacological agents that are active on the conventional croton oil pouch—this in spite of the fact that the number, distribution and age of the small vessels in the two types of pouches are virtually identical (Gabbiani et al., 1972a).

Human granulation tissue

We have studied specimens of granulation tissue obtained from several patients (Ryan et al., in press), e.g. during surgical correction of a pharyngostome; during adjustment of a displaced mammary prothesis; and during healing of an open abdominal skin wound. In all instances, we found myo-fibroblasts by means of electron microscopy (Fig. 2a) and immunofluorescence (Fig. 5). The electron microscope also showed cell-to-cell and cell-to-stroma connections. When provided with sufficient biopsy material, we were able to test these tissues pharmacologically: contractions were obtained with angiotensin, vasopressin and prostaglandin $F_1\alpha$, and relaxation occurred with papaverine and prostaglandin E_1. Thus, it appears that human granulation tissue during normal healing contains myo-fibroblasts analogous to those present in experimental animals.

Dupuytren's contracture and related conditions

Dupuytren's disease is one of the more spectacular examples of connective tissue shrinkage. Several authors have felt that the most significant changes take place in the nodules that appear in the palmar aponeurosis (Larsen 1966; Luck, 1959), but ultrastructural studies of Dupuytren's disease have usually focussed on the extracellular material, i.e. on the extra-nodular part of the aponeurosis (Jahnke, 1960; Patel, 1961; Dahmen, 1968). This was probably due to the long-established belief among pathologists that collagen can shorten *in vivo*. Since the nodules are essentially composed of masses of densely packed cells, they give a tumor-like appearance—indeed, this condition is now listed among the so-called fibromatoses, together with an analogous nodular lesion of the plantar aponeurosis (Ledderhose's disease), and the 'knuckle pads' that are sometimes seen over interphalangeal joints (Enzinger *et al.*, 1970). These last two conditions do not produce a contracture of the affected structures, probably because of the lack of appropriate connections with anatomic structures; on the other hand, in Dupuytren's disease, a shortening of the palmar aponeurosis, by its insertion into the first phalanges of the three last fingers, inevitably results in the typical traction deformity.

The studies performed on extracellular material, particularly collagen, have not shown definite abnormalities (Jahnke 1960; Patel 1961; Dahmen 1968). We have therefore examined the cells of Dupuytren's nodules in 6 patients and found in all cases that they have the morphological features of myo-fibroblasts, including cell-to-cell and cell-to-stroma connections (Gabbiani and Majno, 1972). Furthermore, such cells contain smooth muscle antigens, as shown by immunofluorescent labeling using SMA serum. It is worth noting that myo-fibroblasts have been found in some instances also in the dermis of the skin over the nodules, thus confirming the proposition of some surgeons that it is advisable to excise this overlying skin in order to reduce recurrence. Myo-fibroblasts were also found in the nodules of Ledderhose's disease, and in knuckle pads. We have tested, as described earlier, the pharmacological responses of small strips made from Dupuytren's nodules. To date, no contraction or relaxation has been observed with any agent. This is almost certainly due to a densely collagenous and inelastic stroma present in the nodules, a feature that would restrict detectable change in the overall length of a strip due to cellular contraction or relaxation.

These findings do not contribute any clue to the etiology of these fibromatoses. However, they indicate that myo-fibroblasts occur in certain non-inflammatory conditions where connective tissue contraction may be important.

Tissue culture

It is known that fibroblasts in culture can form collagen only when the culture consists of several layers of cells (Goldberg and Green, 1964). During the first outgrowth from explants, and during the phase of logarithmic growth, fibroblasts have been shown to possess intracellular filaments (40–80 Å in diameter) (Goldberg and Green, 1964; Abercrombie *et al.*, 1971; Goldman, 1971) which are believed to have no relationship to collagen (Goldberg and Green, 1964). Moreover it has been shown that such fibroblasts can pull and exert a tension quite similar to that observed in a contracting wound (James and Taylor, 1969). This is possible because the growing fibroblasts develop intercellular connections, mostly in the form of desmosomes (Devis and James, 1964). It appears likely that their intracellular filaments (which resemble those of smooth muscle) play a role in cellular movement or tension (Abercrombie *et al.*, 1971; Goldman, 1971; Wessels *et al.*, 1971).

We have tested the antigenic properties of cultivated fibroblasts during the first outgrowth and during the logarithmic phase of growth (these cells were derived from rat subcutaneous tissue or mouse dermis): bright fluorescence was obtained using SMA serum (Fig. 5)—this fluorescence was often in the form of multiple, narrow, parallel bundles (Farrow *et al.*, 1971). In preliminary studies of such fibroblasts obtained from normal rat dermis, we have observed that the addition of 5-HT (1×10^{-5} g/ml, final concentration in the bath) to the culture medium causes cellular contraction within 15 to 20 min, whereas tryptophan (1×10^{-5} g/ml) has no effect under the same conditions (Majno *et al.*, 1971). It appears therefore that, at least during some phases of *in vitro* growth, fibroblasts develop the typical features of myo-fibroblasts, i.e. they can no longer be considered 'normal' fibroblasts.

DISCUSSION AND CONCLUSIONS

All the data reported here show that fibroblasts, under certain conditions, can progressively assume ultrastructural, chemical, immunologic and functional characteristics similar to smooth muscle. Such an intermediate-type cell could well be responsible for the contraction of connective tissue *in vivo*—a process that is beneficial in wound closure but potentially harmful in other situations, e.g. wound contracture, Dupuytren's contracture, valvular deformity in chronic rheumatic heart disease, bowel obstruction following various inflammatory lesions. It should be stressed that the stimulus for the formation of myo-fibroblasts is not necessarily the same in every case—for example, it is generally believed that Dupuytren's disease is not the consequence of a local granulation tissue response (Enzinger *et al.*, 1970).

It remains to be seen whether the morphological and functional features of myo-fibroblasts are compatible with their proposed histogenetic origin from fibroblasts. At present most authors agree that fibroblasts of granulation tissues are formed locally from pre-existing cells of the same type, or possibly from more primitive mesenchymal cells (Ross, 1968a; Ross et al., 1970). Although the majority of the cells in our experimental models of granulation tissue are myo-fibroblasts, as judged by electron microscopy and immunofluorescence, it may be argued that these cells are derived from smooth muscle, e.g. from local blood vessels. We believe that this is unlikely because it would imply that the commonest connective tissue cell, the fibroblast, takes little part in the formation of granulation tissue. The relationship between fibroblasts and myo-fibroblasts receives further support from the fact that fibroblasts cultivated *in virto* normally develop extensive cytoplasmic fibrillar bundles (Goldberg and Green, 1964; Abercrombie et al., 1971; Goldman, 1971), intercellular connections (Devis and James, 1964), and can also fix anti-smooth antibodies. Finally, it appears that myo-fibroblasts can develop in areas with no blood vessels, such as in the capsule that surrounds an intraperitoneally implanted blood clot.

The pharmacological tests show some differences between the reactivity of granulation tissues strips and that of classical smooth muscle preparations. For example, granulation tissue fails to react to barium chloride or acetylcholine, agents which normally cause contraction of smooth muscle (Gabbiani et al., 1972a). Furthermore, the pattern of response is somewhat different, in that the peak of contraction is reached more slowly and maintained longer by the granulation tissue strips: in some instances (e.g. after stimulation by 5-HT), the contraction remained stable at the peak for more than 2 h. This 'spastic' behaviour suggests the presence of a contractile system similar to the 'catch-muscle' of invertebrates and falls well into place with the time-course of wound shrinkage or tissue contracture, which characteristically show slow but continuous progression.

It is of interest that cells ultrastructurally similar to myo-fibroblasts have been recently observed in certain normal tissues, such as chicken aorta (Moss and Benditt, 1970) and rat ovary (O'Shea, 1970) or pathologic tissues, such as ganglia of the wrist in man (Ghadially and Mehta, 1971), although their significance is not yet understood. Smooth muscle cells with fibroblastic features have been described in the uterus of rats treated with estrogens (Ross, 1968b) and in human or experimental arteriosclerotic lesions (Thomas et al., 1963; Parker and Odland, 1966). Furthermore, there is evidence that smooth muscle cells can produce collagen or elastin. Thus, it appears that smooth muscle can assume morphologic and functional characteristics of fibroblasts; on the other hand our studies indicate that the reverse is also true.

The relationship between intracytoplasmic microfilaments and cellular motion or development of tension has been proposed for a wide spectrum of cells, ranging from monocellular organisms to those of mammalian tissues (Wessels et al., 1971). These filaments consist of protein which probably represent variants of actomyosin. We have recently seen that several types of rat cells (different from smooth muscle) fix anti-smooth muscle serum under normal conditions (e.g. platelets, polymorphonuclear leucocytes, certain endothelia, the microvillous borders of intestinal epithelial cells, renal tubular brush borders) (Gabbiani et al., 1972b). In most of these cells, contractile properties and/or microfilaments have also been shown by various means (Gabbiani et al., 1972b).

The work reported here indicates that under emergency conditions fibroblasts can develop a system of contractile fibrils as well as junctions to transfer their shrinkage to the whole tissue. The question arises as to whether other mammalian cells can respond to appropriate stimulation by developing microfilaments. In preliminary experiments, we have seen that during healing of skin wounds and during liver regeneration the newly formed epithelial cells are rich in intracytoplasmic fibrils and show greatly increased specific binding of anti-smooth muscle antibodies. Thus, it would appear that the capacity for developing a prominent intracytoplasmic fibrillar system is not limited to fibroblasts, but represents a more general phenomenon of cellular adaptation. This probably takes place when cells of different embryologic origin face situations that require the enhancement of certain functional characteristics, such as the ability to move about or contract.

ACKNOWLEDGEMENTS

This work was supported in part by grants No. 3. 356. 70 and 3. 460. 70 from the Fonds National Suisse pour la Recherche Scientifique and a grant from Zyma S. A. Nyon, Switzerland.

We wish to thank Drs. A. Cruchaud and I. Nicod for kindly providing samples of antisera; Dr. J. E. Pike from the Upjohn Co., Kalamazoo, Michigan, USA for the samples of prostaglandins $F_1\alpha$ and E_1; Misses M. C. Clottu, M. C. Armand and Mrs. A. Fiaux for excellent technical help; and Messrs J. C. Rümbeli and E. Denkinger for photographic work. Mrs. Fiaux also drew Fig. 6.

REFERENCES

Abercrombie, M., Heaysman, J. E. M. and Pegrum, S. M. (1971). *Exp. Cell Res.* **67**, 359–367.
Bloom, S. and Cancilla, P. A. (1969). *Circ. Res.* **24**, 189–196.
Bloom, W. and Fawcett, D. W. (1968). "A Textbook of Histology". W. B. Saunders Co., Philadelphia, London.
Dahmen, G. (1968). *Z. Orthop.* **104**, 247–254.

Devis, R. and James, D. W. (1964). *J. Anat. (London)* **98**, 63–68.
Enzinger, F. M., Lattes, R. and Torloni, H. (1970). "Types histologiques des tumeurs des tissus mous". World Health Organisation, Geneva.
Farrow, L. J., Holborrow, E. J. and Brighton, W. D. (1971). *Nature* **232**, 186–187.
Gabbiani, G. and Majno, G. (1972). *Amer. J. Path.* **66**, 131–146.
Gabbiani, G., Ryan, G. B. and Majno, G. (1971). *Experientia* **27**, 549–550.
Gabbiani, G., Hirschel, B. J., Ryan, G. B., Statkov, P. R. and Majno, G. (1972a). *J. Exp. Med.* **135**, 719–734.
Gabbiani, G., Ryan, G. B., Badonnel, M-C. and Majno, G. (1972b). *Path. Biol.* Suppl. **20**, 6–8.
Ghadially, F. N. and Mehta, P. N. (1971). *Ann. Rheum. Dis.* **30**, 31–42.
Goldberg, B. and Green, H. (1964). *J. Cell Biol.* **22**, 227–258.
Goldman, D. (1971). *J. Cell Biol.* **51**, 763–771.
Greenlee, T. K. Jr. and Ross, R. (1967). *J. Ultrastruct. Res.* **18**, 354–376.
Hirschel, B. J., Gabbiani, G., Ryan, G. B. and Majno, G. (1971). *Proc. Soc. Exp. Biol. Med.* **138**, 466–469.
Hogan, M. J. and Feeney, L. (1963). *J. Ultrastruct. Res.* **9**, 47–64.
Jahnke, A. (1960). *Zbl. Chir.* **85**, 2295–2303.
James, D. W. and Taylor, J. F. (1969). *Exp. Cell Res.* **54**, 107–110.
Johnson, G. D., Holborrow, E. J. and Glynn, L. E. (1965). *Lancet* **2**, 878–879.
Lane, B. P. (1965), *J. Cell Biol.* **27**, 199–213.
Larsen, R. D. (1966). "Hand Surgery" (J. E. Flyn, ed.) pp. 922–952. Williams & Wilkins Co., Baltimore.
Luck, J. V. (1959). *J. Bone Joint Surg. (Am.)* **41**, 635–664.
Majno, G. (1958). *Lancet* **2**, 994.
Majno, G., Shea, S. M. and Leventhal, M. (1969). *J. Cell Biol.* **42**, 647–672.
Majno, G., Gabbiani, G., Hirschel, B. J., Ryan, G. B. and Statkov, P. R. (1971). *Science (Washington)* **173**, 548–549.
Moss, N. S. and Benditt, E. P. (1970). *Lab. Invest.* **22**, 166–183.
O'Shea, J. D. (1970). *Anat. Rec.* **167**, 127–131.
Parker, F. and Odland, G. F. (1966). *Amer. J. Path.* **48**, 451–481.
Patel, J. C. (1961). *Presse Med.* **69**, 793.
Ross, R. (1968a). *Biol. Rev.* **43**, 51–96.
Ross, R. (1968b). *In* "Treatise on Collagen" (G. N. Ramachandran, ed.) Vol. 2, part A, pp. 1–82. Academic Press, New York.
Ross, R. and Benditt, E. P. (1965). *J. Cell Biol.* **27**, 83–106.
Ross, R., Everett, N. B. and Tyler, R. (1970). *J. Cell Biol.* **44**, 645–654.
Ryan, G. B., Grobéty, J. and Majno, G. (1971). *Amer. J. Path.* **65**, 117–148.
Ryan, G. B., Cliff, W. J., Gabbiani, G., Irlé, C., Montandon, D. and Majno, G. (1973). *Human Path.* (In Press.)
Selye, H. (1953). *JAMA* **152**, 1207–1213.
Stehbens, W. E. (1966). *J. Ultrastruct. Res.* **15**, 389–399.
Thomas, W. A., Jones, R., Scott, R. F., Morrison, E., Godale, F. and Imai, H. (1963). *Exp. Molec. Path.* suppl. **1**, 40–61.
Wessels, N. K., Spooner, B. S., Ash, J. F., Bradley, M. O., Luduena, M. A., Taylor, E. L. A., Wrenn, J. T. and Yamada, K. M. (1971). *Science (Washington)* **171**, 135–143.

Density Dependent Regulation of Growth and Differentiated Function in Suspension Cultures of Mouse Fibroblasts

ANDRÉ D. GLINOS

Department of Cellular Physiology, Walter Reed Army Institute of Research
Walter Reed Army Medical Center, Washington, D.C.
and Department of Zoology, University of Maryland
College Park, Maryland, USA

Although density dependent regulation of growth occurs in attached cultures of all normal euploid mammalian cells and of some established cell lines, it is of particular significance for connective tissue cells because of its resemblance to the behavior of fibroblasts in the healing wound. Based on this resemblance, the suggestion was made that the factors controlling fibroblast proliferation in these two situations are the same, and that, therefore, one of the most effective ways to study wound healing is through the application of cell and tissue culture methods (Ross, 1968).

Compared to conventional wound healing studies, the much simpler cell and tissue culture models would indeed appear to be the most promising, once present difficulties concerning divergent views on how increased cell density inhibits growth are resolved. One of these views assumes the transmission of inhibitory signals necessitating physical cell-to-cell contact (Dulbecco, 1971), while a second considers decreased availability of media components essential for growth. The latter could be due to bulk media depletion of serum factors (Holley and Kiernan, 1971), while for other media constituents it could involve the cellular micro-environment (Rubin and Rein, 1967) and/or reduced transport through the surface membrane (Castor, 1970; Cunningham and Pardee, 1971; Griffiths, 1972).

For obvious reasons, experimental dissociation of the variables involved in these interpretations is extremely difficult in crowded cell populations attached on solid surfaces. The question therefore arises whether this task could be facilitated through the use of suspension cultures. This would in turn depend on whether suspension cultures of cells of fibroblastic origin, grown to high cell densities through regular medium renewal, exhibit the characteristics typical of attached dense cultures. These are: maintenance of intact viability for prolonged periods of time; marked inhibition of DNA synthesis and mitosis, which however is reversed when population density is decreased; preferential synthesis of collagen; and finally, regulation of the activity of specific enzymes. In the following, experiments bearing on this question are briefly described and correlations between suspension and attached cell cultures discussed in terms of the regulatory features exhibited by dense populations in both types of cultures.

GENERATION OF HIGH-DENSITY STABLE SUSPENSION CULTURES

The origin and quantitative karyotype of the L-929, subline WRL-10A cells and details of the culture methods used have been described elsewhere (Glinos, 1967; Glinos and Hargrove, 1965; Glinos and Werrlein, 1972). To obtain high-density populations the culture medium was renewed daily by centrifuging the cells at 225 g for 15 min and resuspending them in fresh medium. During these manipulations, proper care was taken to keep cell loss to a minimum. The range of the daily cell counts obtained in 17 suspension cultures grown in this fashion, in a series of experiments extending over a period of four years, is shown in Fig. 1. The range of the initial and final cell densities in these cultures was $3 \cdot 5 – 5 \times 10^5$ cells/ml and $6 – 11 \times 10^6$ cells/ml, respectively. Doubling time during the exponential-growth phase decreased from 28 h in the earlier experiments to 18 h in the more recent ones; there was no correlation among initial cell density, doubling time during the exponential phase, and final cell density. Duration of the exponential and near exponential growth phase varied from three to four days and of the ensuing retardation phase from four to five days. Stabilization of the cell population occurred at about nine to ten days after setting up of the cultures. Such dense stable cell populations have been studied in our laboratory for periods varying from 16 to 168 days, their discontinuation being due either to utilization of the cells for experimental purposes or to accidental causes.

During the progression of the cultures from the exponential-growth phase to the retardation phase and throughout the duration of the dense stable populations, the fraction of cells excluding nigrosin remained within a range of $0 \cdot 94 – 1 \cdot 00$, with mean values varying from $0 \cdot 97$ to

1·00, indicating that there was no decrease of the viability of the cells in any of the populations investigated.

These results are entirely comparable to findings with attached cultures and contrast sharply with all previous studies on suspension cultures. This is so because in the latter, limitation of growth was obtained by omission of medium renewal (Eidam and Merchant, 1965; Littlefield, 1962; Tobey and Ley, 1970; Ward and Plagemann,

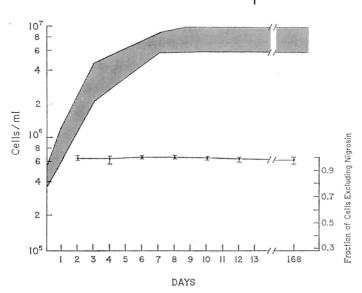

Fig. 1. Cell population kinetics and viability. Shaded area represents the range of daily cell counts (left ordinate) obtained in 17 suspension cultures grown without dilution for periods varying from 16 to 168 days. Curve represents cellular viability in terms of the fraction of cells excluding nigrosin (right ordinate). Each point represents the mean of at least four determinations and the vertical lines indicate the range around each mean.

1969). The population densities achieved with that procedure were two to three times lower than the ones reported here and in no case was viability retained beyond three days.

REGULATION OF DNA SYNTHESIS AND MITOSIS

To determine the fraction of cells synthesizing DNA, 9-ml samples were removed from the experimental cultures at desired time intervals and pulse labeled for ten minutes with ^{3}H-thymidine added to the samples at a final concentration of 0·25 μCi/ml. The cells were then centrifuged, washed with cold Earle's balanced saline, fixed and used in the preparation of conventional autoradiographs.

To determine the fraction of the cell population in mitosis at desired time intervals following medium renewal, appropriate samples of the cell suspension were fixed and stained with crystal violet; the mitotic cells were then counted on wet mounts prepared from these samples. To determine the fraction of the cells entering mitosis over an extended period of time, colchicine at a final concentration of 6 μg/ml was added to cultures two hours after renewal of the medium, when dispersal of the cells was almost complete. Eighteen hours later, i.e. 20 h following medium renewal, wet mounts were prepared and mitotic cells counted as just described.

The daily kinetics of the cells synthesizing DNA and undergoing mitosis in cultures of varying densities is shown in Figs. 2 and 3.

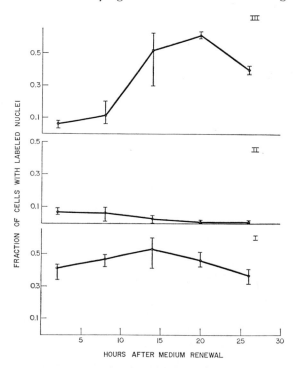

Fig. 2. Effects of culture density on DNA synthesis. At the indicated times, cell samples from (I) low-density (4–8 × 10⁵ cells/ml), (II) high-density (6–10 × 10⁶ cells/ml), and (III) high-density cultures following dilution to the density level of (I) at 0 time, were pulse labeled with tritiated thymidine as described. Nuclei with more than five grains were considered labeled. Each point represents the mean of at least three cultures, and vertical lines indicate the range around each mean.

Diurnal synchronization effects due to medium renewal and differences in the range of values around the mean, between the curves and the bar graphs of Fig. 3 have been discussed elsewhere (Glinos and Werrlein, 1972). Comparison of curves I and II of Fig. 2 shows that the fraction of cells synthesizing DNA in the high-density cultures exhibits a reduction ranging from 87% for the 6-h period immediately following medium renewal, to 97% during the last 6-h period between 20 and 26 h. The corresponding reduction of the mitotic index obtained by

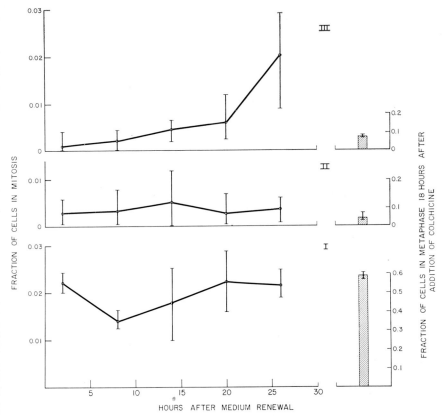

Fig. 3. Effects of culture density on mitosis. At the indicated times, cell samples from (I) low-density ($4-8 \times 10^5$ cells/ml), (II) high-density ($6-10 \times 10^6$ cells/ml), and (III) high-density cultures following dilution to the density level of (I) at 0 time, were taken and the fraction of cells in mitosis determined as described. Each point represents the mean of at least four cultures and vertical lines indicate the range around each mean. The bar graphs on the right indicate the fraction of cells in metaphase, in samples obtained at 20 h from a group of cultures which had received 6 μg/ml of colchicine 2 h after medium renewal. Each bar graph represents the mean of at least three cultures and vertical lines indicate the range around each mean.

comparing curves I and II in Fig. 3 varied between 75% at 14 h and 85% at 26 h. Calculation of the mitotic flow, based on the fraction of cells arrested in metaphase by colchicine during a period of 18 h (Fig. 3, bar graphs I and II), yields a value of approximately 0·003 cells/h for high-density cultures as compared to 0·033 for low-density cultures, or a reduction of over 90%.

The slight discrepancy in the degree of reduction when calculated on the basis of (a) the fraction of DNA synthesizing cells, (b) the mitotic index, and (c) the mitotic flow is an indication that the duration of mitosis is prolonged in the high-density cultures. Taking this into consideration it may be calculated that the maximum fraction of cells expected to complete mitosis per day in the high-density cultures is approximately 0·05 (Glinos and Werrlein, 1972). This is nearly identical with the maximum value of 0·06 for the fraction of nonviable cells (Fig. 1). Growth in these cultures appears thus to be limited to the compensation of the small degree of cell death, encountered also in low-density exponentially growing cultures, most probably because of the manipulations associated with medium renewal.

In terms of growth inhibition, therefore, these high-density suspension cultures are entirely comparable to dense attached cultures where, also, DNA synthesis and mitosis do not cease completely (Kruse et al., 1970) and respond to medium renewal with a sudden small rise (Todaro et al., 1965; cf. Fig. 2–II). As in these attached cultures, the inhibition of DNA synthesis and mitosis is reversible and growth resumes readily following dilution of the high-density suspension cultures to the density level of exponentially growing populations (Figs. 2-III, 3-III). It can be seen that the fraction of cells synthesizing DNA under these conditions reaches a maximum at 20 h. On the basis of the L cell life cycle, a partially synchronized population such as the one shown in Fig. 3-III would be expected to exhibit maximum mitotic activity at approximately 12 h after the peak of DNA synthesis, i.e. at 32 h after dilution. This has been confirmed (Glinos and Werrlein, 1972) and explains the absence of a maximum in the mitotic activation illustrated in Fig. 3-III where observations did not extend beyond 26 h. Detailed analysis of the kinetics of post-dilution activation of DNA synthesis and mitosis shows that a large fraction of the cells is in a G zero or early G one state (Glinos and Werrlein, 1972). In this respect, therefore, high-density suspension cultures are similar to growth-inhibited, dense attached cultures.

REGULATION OF COLLAGEN SYNTHESIS AND RELEASE

To determine general protein and collagen synthesis and release, 3,4-^{3}H-L-proline was added to paired low-density, exponentially growing and high-density, stable cultures, one hour after medium renewal to a final concentration of 2 μCi/ml. At desired time intervals, 5-ml samples were withdrawn and the cells separated from the medium by centrifugation. The cell pellets were washed twice with 5 ml of Eagle's medium and the washings added to the original supernatant. The combined culture medium and washings were then incubated for 15 min at 35°C in the presence of added 8×10^{-2} M cold proline to minimize artefacts resulting from the exchange of unbound labeled proline with serum proteins contained in the medium. Total radioactivity in the acid-soluble and acid-precipitable fractions of the medium and the distribution of the latter among its proline and hydroxyproline residues were determined as described elsewhere (Glinos et al., 1973).

The uptake of ^{3}H-proline and the appearance of labeled extracellular, high molecular weight compounds in exponentially growing and high-density, stable suspension cultures are shown in Table I. The decline of the level of radioactivity of the acid-soluble portion of the medium during 4 h of incubation with labeled proline, from the 1st to the 5th h after medium renewal, showed that the cells in both the growing and the stable cultures readily utilized the administered isotope. In the case of the low-density cultures, the counts decreased from 4.7×10^6 dpm/ml at 30 sec after addition of the isotope to 3.5×10^6 dpm/ml 4 h later, which is equivalent to a 25% reduction. During the same interval, half of the radioactivity initially present in the acid-soluble fraction of the medium was removed by the cells in the high-density culture, the counts decreasing from 4.6 to 2.3×10^6 dpm/ml. As the determinations conducted 5 h following medium renewal did not take into account the possible presence of labeled, low molecular weight compounds metabolically derived from tritiated proline, the figures representing radioactive uptake (Table I, Column C) must be considered as minimal estimates of actual proline utilization.

Incorporation of ^{3}H-proline into high molecular weight compounds released from the cells into the medium, as measured by the radioactivity present in the PCA-insoluble portion of the medium, is shown in Table I, Column D. The fraction of precipitate radioactivity resulting from nonspecific exchange of tritiated, free proline with the serum proteins of the medium was determined by incubating serum containing medium with ^{3}H-proline but without cells for a 24 h period. Prior to PCA fractionation, 8×10^{-2} M cold proline was added to the medium and allowed to incubate for 15 min at 35°C. The magnitude of the exchange artefact in the acid-insoluble precipitate treated in this fashion was found to be 6644 dpm/ml, or, less than 5% of the value obtained in the high molecular weight fraction of the medium from an exponentially growing culture exposed to the isotope for 4 h and treated in a similar fashion (Table I, Column D). This error was considered to be inconsequential and was therefore ignored in calculating

Table I

Uptake of tritiated proline and appearance of labeled extracellular high molecular weight compounds in low- and high-density L cell suspension cultures

Type of culture	Cell concentration (10^5 cells/ml)	Radioactivity in medium (dpm/ml)			Acid insoluble	
		Acid soluble				
		A $T = 1$ h	B $T = 5$ h	C Difference $(A - B)$	D $T = 5$ h	$D/C \times 100$ %
Low-density exponentially growing	6·55	4,679,495	3,517,017	1,162,478	133,413	11·5
High-density stable	77·60	4,630,320	2,313,171	2,317,149	233,730	10·1

Radioisotope was added to the cultures one hour after medium renewal (T = 1) and aliquots of the cultures withdrawn 30 sec and 4 h later (T = 1 and 5 h, respectively) and processed as described. Cell population densities were determined one hour after medium renewal, immediately before isotope addition.

incorporation of label into PCA-precipitable material of the medium and relative rates of collagen and protein synthesis.

After a 4 h incubation in the presence of labeled proline, the amounts of radioactivity incorporated into high molecular weight compounds released into the medium were approximately 233 and 133 $\times$ 10^3 dpm/ml for the high- and low-density cultures, respectively. Comparison of these values with the quantity of radioactivity removed from the medium (Table I, Column C) reveals that in both phases of growth about 10–12% of the label utilized reappeared in the high molecular weight fraction of the culture medium. This is in good quantitative agreement with the findings of Halpern and Rubin (1970) on the release of newly synthesized proteins by attached diploid chick fibroblasts.

Table I also shows that approximately twice as much radioactivity was incorporated into the high molecular weight compounds released into the medium by stable rather than by growing cultures, although, cell density in the former was approximately ten times higher than in the latter. This indicates that the rate of release of newly synthesized proteins by high-density cultures was inhibited by approximately 80%. Determination of the distribution of radioactivity between the proline and the hydroxyproline residues of the acid-precipitable material of the medium revealed that this inhibition affected general protein synthesis to a greater extent than collagen synthesis (Glinos et al., 1973).

As demonstrated by Manner (1971) and Priest and Davies (1969), this is precisely the type of regulation which operates also in dense attached cultures of fibroblasts and accounts for the preferential synthesis of collagen in such cultures, first described by Green and Goldberg (1964). The expression $\Delta C/\Delta P$, first introduced by these authors, may be conveniently used to indicate the fraction of labeled collagen relative to total labeled protein released into the media of low- and high-density L cell suspension cultures following addition of radioactive proline. The results are shown in Fig. 4 where it can be seen that the $\Delta C/\Delta P$ ratio obtained from actively growing cultures 3 h after medium renewal was 1·2%, increasing subsequently to 1·5% at 5 h and 1·9% at 7 h. The values from the high-density, stable cultures were 1·6% at 3 h following medium renewal, sharply increasing to 4·3% at 7 h, an almost 3-fold increase. This progressive increase of the ratio as a function of labeling time, in both low- and high-density cultures, is due to the time needed for maximum labeling of the newly synthesized collagen prior to its release from the cells. This increase was significantly greater in the high-density stable cultures, thus indicating preferential synthesis and release of collagen, comparable to the one described previously in attached dense fibroblastic populations (Green and Goldberg, 1964; Manner, 1971; Priest and Davies, 1969).

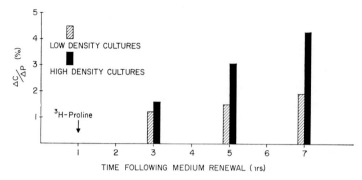

Fig. 4. Fraction of labeled collagen relative to total labeled protein appearing in the media of low- and high-density L cell suspension cultures following addition of radioactive proline. As indicated by the arrow, the isotope was added to paired low- and high-density cultures one hour after medium renewal and samples withdrawn for analysis at 3, 5, and 7 h. The bar graph shows the mean $\Delta C/\Delta P$ ratios (cf. Green and Goldberg, 1964 and text) obtained in two experiments from two different pairs of high- and low-density cultures.

REGULATION OF ENZYME ACTIVITY

Acetylcholinesterase represents one of the best documented examples of enzyme activity regulation in growth inhibited dense attached cultures. In addition to the recent detailed description of such regulation in neuroblastoma cells by Blume et al. (1970), there is evidence that the activity of this enzyme is inversely related to the rate of cell division in attached cultures of cells of mesenchymal origin as well (Goodwin and Sizer, 1965). On this basis, it would appear that the question as to whether growth inhibited dense suspension cultures are capable of enzyme activity regulation comparable to dense attached cultures could be approached experimentally using acetylcholinesterase as an indicator.

Accordingly, acetylcholinesterase activity was determined at various time intervals during the progression of L cell suspension cultures from exponential-growth to high-density stable populations (cf. Fig. 1). All samples were obtained at 5 h after medium renewal. The cells were separated from the medium by centrifugation at 5°C, washed three times with cold Earle's balanced salt solution and resuspended in a 50 mM potassium phosphate buffer, pH 6·8, containing 1 mM EDTA, potassium salt. The samples were then assayed for acetylcholinesterase activity using the method of Wilson et al. (1972).

By comparing the results shown in Table II with the cell population kinetics curve in Fig. 1, it can be seen that the specific activity of acetylcholinesterase in L cell suspension cultures is relatively stable during exponential growth up to the 6th day. There is a moderate increase

Table II
Specific activity of acetylcholinesterase (AChE) in L cell suspension cultures

Age of culture (Days)	Density of culture (10^5 cells/ml)	AChE activity (pmoles ^{14}C-acetate/mg/min)
1	5.9	20
4	23.5	21
5	38.2	36
6	51.2	64
7	63.7	123
12	91.5	1140
16	76.0	1610
23	73.2	3650

Figures in last column express activity of 'true' acetycholinesterase inhibited by BW 284C51; residual, non-specific, esterase activity for the 1-7 days, low-density cultures, was 30-70 pmoles, and for the 12-33 days, high-density cultures, 50-140 pmoles ^{14}C-acetate/mg/min.

during the period of growth retardation between 6 and 10 days which culminates in a dramatic 10- to 100-fold rise upon establishment of the high-density stable populations around the 12th day. The order of magnitude of this increase is the same as that reported for dense attached cultures of neuroblastoma, although basal acetylcholinesterase activity levels in those cells are approximately 100-fold higher than in L-929 cells (Blume et al., 1970). Also, like in neuroblastoma, the regulatory mechanism operating in L cells is specific for acetylcholinesterase, the level of nonspecific esterases increasing only by a factor of 2 in the high-density cultures (Table II). These data strongly suggest that regardless of cell type and initial levels the activity of acetylcholinesterase in growth inhibited dense suspension cultures is regulated in essentially the same fashion as in similarly inhibited attached cultures.

DISCUSSION AND CONCLUSIONS

In the preceding sections evidence was presented that in suspension cultures of L-929 cells increased cell density leads to inhibition of growth with regulation of DNA synthesis, mitosis, collagen synthesis and enzyme activity. This is in sharp contrast to the behavior of L-929 cells and other virally or chemically transformed fibroblasts in attached cultures. In the latter, the absence of growth inhibition leads to the formation of very crowded, multilayered populations, exhibiting extensive cell death and detachment in certain areas while active mitosis continues in others (Castor, 1970; Green and Goldberg, 1965).

Transformed cell lines appear thus to respond to high density in opposite ways depending on whether their cells are attached or suspended. The most likely explanation for this is that as the density of suspension cultures increases, availability of essential media constituents decreases uniformly for all cells in the population eliciting the changes described in this paper in regard to DNA synthesis, mitosis, collagen synthesis and enzyme activity. The evidence presented indicates that the end result is the establishment of a viable, stable, functionally active cell population in equilibrium with the available essential constituents of the media. In attached cultures, on the other hand as the cell density increases, crowding and multilayering would cause wide variations with regard to the availability of essential media components. Depending on local topography, excess or deprivation of the latter would result in maintaining parts of the cell population in a state of active growth while simultaneously the viability of other parts would decrease. Regulation of growth would thus be impossible in attached cultures of transformed cell lines unless, as Burger and Noonan (1970) have shown, specific cell surface binding agents are used.

Differences in availability of essential media constituents, therefore, appear to account for the presence of growth regulation in suspension and its absence in attached cultures of transformed cells. Since cell surface treatments are capable of restoring regulation in such cultures, the function of the normal cell membrane in growth control may be conceived as modulating the availability of essential media constituents in relation to population size. The surface membrane could be instrumental in (a) restricting the transport of substances essential for growth from the media into the cells (Castor, 1970; Cunningham and Pardee, 1971; Griffiths, 1972), (b) transmitting to the nucleus signals inhibiting DNA synthesis and mitosis (Dulbecco, 1971), and (c) transmitting similar inhibitory signals from cell to cell (Bard and Elsdale, 1971). These responses of the surface membrane could be triggered by: (a) cell-to-cell contact (Castor, 1970; Dulbecco, 1971), (b) diffusible inhibitors (Burk, 1967; Froese, 1971; Pariser and Cunningham, 1971; Rubin and Rein, 1967; Yeh and Fisher, 1969), and (c) local microenvironmental depletion of essential media constituents (Bard and Elsdale, 1971; Rubin and Rein, 1967).

In all cases the end result would be prevention of overgrowth and of the ensuing extreme topographical variation of the availability of essential media constituents, thus rendering possible the establishment of viable, stable, functioning fibroblastic populations.

These hypotheses are not mutually exclusive and each—with the notable exception of local microenvironmental depletion—is supported in part by experimental findings. Conclusive evidence, however, for any of these processes is lacking.

As it was shown in this chapter, even without cell surface treatment, dense suspension cultures of transformed cells show regulation of DNA synthesis, mitosis, collagen synthesis and enzyme activity, similar in all aspects so far examined to growth inhibited dense attached cultures. The question therefore arises whether the same media constituents are responsible for growth regulation in both cases, their decrease involving the bulk of the media in the case of suspensions while in attached cultures it would involve the micro-environment and/or cell surface transport. This possibility is currently under investigation in our laboratory, with a comparison of the kinetics of oxygen availability in the bulk media of suspension cultures and in the micro-environment of attached cultures as the first step in this direction.

ACKNOWLEDGEMENTS

The work discussed represents the collective effort of a team involving Drs. Bartos and Vail and Messrs. Werrlein and Taylor, my close associates and co-workers.

REFERENCES

Bard, J. and Elsdale, T. (1971). *In* "Growth Control in Cell Cultures" (G. E. W. Wolstenholme and J. Knight, eds.) pp. 187–206. Churchill Livingstone, London.
Blume, A., Gilbert, F., Wilson, S., Farber, J., Rosenberg, R. and Nirenberg, M. (1970). *Proc. Natl. Acad. Sci. USA* **67**, 786–792.
Burger, M. M. and Noonan, K. D. (1970). *Nature* **228**, 512–515.
Burk, R. R. (1967). *In* "Growth Regulating Substances for Animal Cells in Culture" (V. Defendi and M. Stoker, eds.) pp. 39–50. The Wistar Institute Press, Philadelphia.
Castor, L. N. (1970). *J. Cell Physiol.* **75**, 57–64.
Cunningham, D. D. and Pardee, A. B. (1971). *In* "Growth Control in Cell Cultures" (G. E. W. Wolstenholme and J. Knight, eds.) pp. 207–220. Churchill Livingstone, London.
Dulbecco, R. (1971). *In* "Growth Control in Cell Cultures" (G. E. W. Wolstenholme and J. Knight, eds.) pp. 71–87. Churchill Livingstone, London.
Eidam, C. R. and Merchant, D. J. (1965). *Exp. Cell Res.* **37**, 132–139.
Froese, G. (1971). *Exp. Cell Res.* **65**, 297–306.
Glinos, A. D. (1967). *In* "Control of Cellular Growth in Adult Organisms" (H. Teir and T. Rytömaa, eds.) pp. 41–53. Academic Press, New York.
Glinos, A. D. and Hargrove, D. D. (1965). *Exp. Cell Res.* **39**, 249–258.
Glinos, A. D., Vail, T. M. and Taylor, B. (1973). *Exp. Cell Res.* **78**, 319–328.
Glinos, A. D. and Werrlein, R. J. (1972). *J. Cell Physiol.* **79**, 79–90.
Goodwin, B. C. and Sizer, I. W. (1965). *Develop. Biol.* **11**, 136–153.
Green, H. and Goldberg, B. (1964). *Nature* **204**, 347–349.

Green, H. and Goldberg, B. (1965). *Proc. Natl. Acad. Sci. USA* **53,** 1360–1365.
Griffiths, J. B. (1972). *J. Cell Sci.* **10,** 515–524.
Halpern, M. and Rubin, H. (1970). *Exp. Cell Res.* **60,** 86–95.
Holley, R. W. and Kiernan, J. A. (1971). *In* "Growth Control in Cell Cultures" (G. E. W. Wolstenholme and J. Knight, eds.) pp. 3–15. Churchill Livingstone, London.
Kruse, P. F., Jr., Keen, L. N. and Whittle, W. L. (1970). *In Vitro* **6,** 75–88.
Littlefield, J. W. (1962). *Exp. Cell Res.* **26,** 318–326.
Manner, G. (1971). *Exp. Cell Res.* **65,** 49–60.
Pariser, R. J. and Cunningham, D. D. (1971). *J. Cell Biol.* **49,** 525–529.
Priest, R. E. and Davies, L. M. (1969). *Lab. Invest.* **21,** 138–142.
Ross, R. (1968). *Biol. Rev.* **43,** 51–96.
Rubin, H. and Rein, A. (1967). *In* "Growth Regulating Substances for Animal Cells in Culture" (V. Defendi and M. Stoker, eds.) pp. 51–66. The Wistar Institute Press, Philadelphia.
Tobey, R. A. and Ley, K. D. (1970). *J. Cell Biol.* **46,** 151–157.
Todaro, G. J., Lazar, G. K. and Green, H. (1965). *J. Cell. Comp. Physiol.* **66,** 325–333.
Ward, G. A. and Plagemann, P. G. W. (1969). *J. Cell Physiol.* **73,** 213–231.
Wilson, S. H., Schrier, B. K., Farber, J. L., Thompson, E. J., Rosenberg, R. N., Blume, A. J. and Nirenberg, M. W. (1972). *J. Biol. Chem.* **247,** 3159–3169.
Yeh, J. and Fisher, H. W. (1969). *J. Cell Biol.* **40,** 382–388.

DISCUSSION

Dr. Bard:

Dr. Glinos, why do you refer to your cultures as 'high density'? If, as I assume, your cells are rounded up with a diameter of about 10 microns, then a density of 10^7 per ml means that the cells occupy only about one per cent of the total volume. This does not seem to me to be a very high density, particularly when compared to the gross crowding of normal confluent cultures.

Dr. Glinos:

Since we are working with suspension cultures, 10^7 cells/ml is high density relative to the order of 10^5 cells/ml usual for this type of culture. Of course, cell crowding comparable to attached confluent cultures cannot possibly occur in suspended cell populations. This is precisely the reason we used such populations as we wanted to see whether density dependent inhibition of growth could occur without cell-to-cell contact and thus uncover the role of essential media constituents, as I have just been discussing.

Phases of Experimental Granulation Tissue Formation and Nucleic Acid Metabolism

J. AHONEN, M. VASTAMÄKI and PEKKA H. MÄENPÄÄ
Department of Medical Chemistry, University of Turku, Finland
Department of Medical Chemistry, University of Helsinki, Finland

The formation of granulation tissue is a basic phenomenon in mammalian repair processes. Thus the understanding of its development is medically important. There are several methods to produce granulation tissue experimentally for detailed studies; a commonly used method is implantation of viscose cellulose sponges subcutaneously (Viljanto and Kulonen, 1962). Three phases can be distinguished in the development of granulation tissue so induced: during the first week an intensive cellular proliferation, rapid synthesis of collagen during the second and third weeks, and finally a phase of involution to a scar (Ahonen, 1968).

Our investigations on the developing granulation tissue have been directed to reveal the changes in nucleic acid metabolism that accompany the developmental phases. These studies have also indicated that during its development the growing granulation tissue affects liver function.

RIBONUCLEIC ACIDS OF GRANULATION TISSUE AT DIFFERENT PHASES OF DEVELOPMENT

Methods of extraction and fractionation

The problem with the extraction of nucleic acids from tissues is to free them from associated proteins and lipoprotein membranes and simultaneously protect them against nucleases. Since the binding of nucleic acids and the contents of various nucleases appears to vary from tissue to tissue, one must find in each case optimal conditions of extraction. We have found the phenol method, introduced by Kirby (1956), supplemented with detergents and lipophilic salts, suitable for granulation tissue. Fractionating was performed using salt-precipitation, gel-filtration and variation of the extraction temperature.

Granulation tissue formation was induced by implanting viscous cellulose sponges subcutaneously in rats. The tissue was harvested at given intervals and homogenized in a mixture containing a phenol phase (freshly distilled, water-saturated) and an aqueous phase containing 0·5% sodium deoxycholate and 0·5% sodium naphthalene-1·5-disulphonate in distilled water. An equally good preparation was obtained when 1% tri-isopropylnaphthalene disulphonate-6% *para*-aminosalicylate-6% 2-butanol was used as an aqueous phase. When sodium dodecylsulphate was used the preparation invariably contained RNAase.

After phenol-extraction nucleic acids were precipitated with ethanol, dissolved again and a high molecular weight RNA fraction (HMW-RNA 20°) was precipitated with 3–4 M sodium acetate. DNA that remained in solution together with RNA of low molecular weight (LMW-RNA) was separated from it by gel-filtration in Sephadex G-200 (Ahonen and Kulonen, 1966).

When the phenol-extraction residue was re-extracted at 65 °C a further RNA-fraction of high molecular weight was obtained (HMW-RNA 65°).

Table I

Some properties of RNA-fractions obtained from 14-day granulation tissue

Fraction	$S_{20°, w}$	5-Ribosyl-UMP content % of alkali-liberated nucleotides	Amino acid-accepting capacity
HMW-RNA 20°	16·1	0	—
	27·9		
HMW-RNA 65°	12	0	—
	16		
	28		
LMW-RNA	4·6	1·8	+

Properties of the ribonucleic acid fractions

The various RNA-fractions were obtained from 14-day granulation tissue. Their properties are summarized in Table I and II.

HMW-RNA 20° consists for the most part of ribosomal RNA. HMW-RNA 65° contains most rapidly labelled and DNA-like RNA as shown in Table II. According to Georgiev *et al.* (1963) the RNA extracted at an elevated temperature consists of nucleolar and chromosomal RNA, that is both ribosomal precursor and messenger-RNA. To illustrate further the origin of this RNA-fraction subcellular fractions were prepared from granulation tissue by differential and zonal

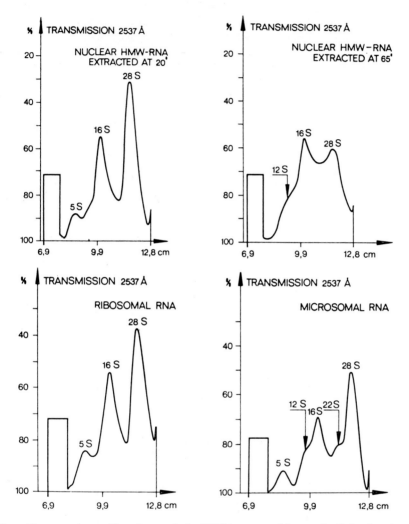

Fig. 1. Rate zonal centrifugation analysis of RNA extracted from subcellular fractions. Sucrose density gradients (5–20%) in 0·1 M sodium acetate, 30,000 rpm at 4°C, MSE 50 SS-centrifuge. Sedimentation from left to right, the distance from the center of rotation is indicated. Transmission was continuously monitored using a gradient evaluator described by Ahonen (1968).

Table II
Specific activities of RNA-fractions and hybridization of radioactive RNA with homologous DNA

Fraction	Specific activity cpm/mg	% hybridized* with DNA
HMW-RNA 20°	155,800	5.2
HMW-RNA 65°	1,401,200	18.6
LMW-RNA	363,600	10.4

The labelling of the various RNA-fractions was studied by injecting the rats with 10 mCi ^{32}P-phosphate intraperitoneally, RNA was isolated 2 h after the injection.
* The experiment was performed as described by Armstrong and Boezi (1965).

centrifugation (Ahonen 1968). Nuclei were first extracted at 20°C and then at 65°C, and the HMW-RNA-fractions thus obtained were compared with those obtained from unfractionated tissue. The rate zonal centrifugation patterns are shown in Figs. 1 and 2. It can be

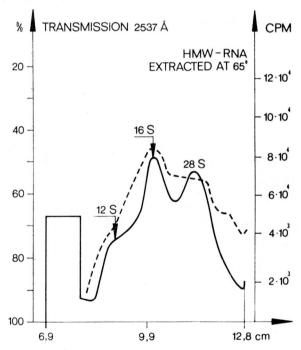

Fig. 2. Rate zonal centrifugation analysis of HMW-RNA 65° extracted from unfractionated granulation tissue. Details as in Fig. 1.

seen that HMW-RNA 65° is similar to that obtained from nuclei at 65°C, whereas that obtained from nuclei at 20°C is similar to the ribosomal RNA.

High molecular weight ribonucleic acids at different phases of development

The two HMW-RNA-fractions were then isolated from 6-day (proliferation phase) and 18-day (collagen synthesizing) granulation tissue, and their nucleotide compositions and rate zonal centrifugation patterns were compared.

There appears to be no differences in the HMW-RNA 20°, which indicates that the nucleotide composition of ribosomal RNA does not reflect the pattern of proteins synthesized. This is an agreement with

Table III
Nucleotide compositions of HMW-RNAs from 6- and 18-day granulation tissue

	CMP	AMP	UMP	GMP	$\dfrac{GMP + CMP}{AMP + UMP}$
HMW-RNA 20°					
6-day	32.0	18.0	17.5	32.5	1.8
18-day	31.9	18.2	17.4	32.5	1.8
HMW-RNA 65°					
6-day	30.0	19.7	20.1	30.2	1.5
18-day	30.7	18.2	17.5	33.8	1.8

Nucleotide compositions were determined as described by Ahonen (1968).

the data on the nucleotide composition of total RNA from granulation tissue (Ahonen, 1968). In the HMW-RNA 65° there appears to be present more CMP and GMP during the period collagen synthesis than before it. At the same time centrifugation analysis indicates only one peak sedimenting at 22S that was uniformly labelled (Fig. 3). Since collagen contains much glycine and proline (+hydroxyproline), together about 55%, one would expect that the corresponding mRNA contains much GMP and CMP, since the codes for these amino acids are GGX and CCX. The size of a collagen mRNA would be such that sedimentation coefficient would be about 22S. Our studies thus suggest that HMW-RNA 65° from 18-day granulation tissue would contain RNA that have some characteristics of collagen mRNA. To show this we should indicate that it can direct cell-free synthesis of collagen and it would also contain nucleotide sequences that correspond to the amino acid sequences in collagen. Despite much effort our trials on cell-free synthesis of collagen have not yet succeeded. Work to reveal the

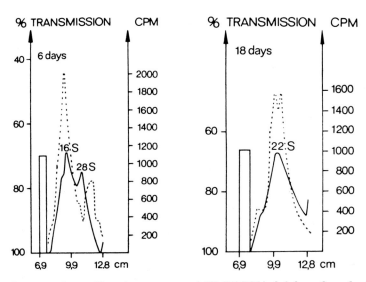

Fig. 3. Rate zonal centrifugation patterns of HMW-RNA 65° from 6- and 18-day granulation tissue. Details as in Fig. 1. The broken line indicates radioactivity, ^{32}P-labelling for 2 h *in vivo*.

nucleotide sequences of further fractionated HMW-RNA 65° is in progress.

Transfer-ribonucleic acids at different phases of development

The amount and nature of specific transfer-RNAs has been shown to vary among specialized tissues and cells. In some cases the amounts of specific tRNAs reflect the amino acid pattern in the major protein produced by the cells. (For a review, see Sueoka and Kano-Sueoka, 1970). This should also be valid for granulation tissue: during the phase of collagen synthesis one would expect increased contents of tRNAs specific for glycine and proline, especially, as we know that the amino acid composition of noncollagenous proteins does not change during granulation tissue development (Ahonen, unpublished).

As shown earlier the LMW-RNA-fraction contains tRNA. The nucleotide composition of this fraction at various phases of granulation tissue development is shown in Table IV.

Table IV
Nucleotide composition of LMW-RNA during granulation tissue development

	\multicolumn{5}{c}{Age of granulation tissue (days)}				
	7	14	21	27	60
CMP	29·6	31·7	34·1	34·1	29·9
AMP	17·1	14·7	14·0	15·4	20·6
UMP	16·8	16·7	14·0	15·4	20·6
GMP	33·7	35·0	35·4	32·7	27·7
5-ribosyl-UMP	2·4	1·8	2·5	2·2	1·0
$\frac{\text{GMP} + \text{CMP}}{\text{AMP} + \text{UMP}}$	1·7	2·0	2·3	2·0	1·4

The nucleotide compositions show a clear increase in the ratio (G + C)/(A + U) during the period of maximal collagen synthesis. Whether this is due to increased amounts of glycine and proline specific tRNAs required further studies since the RNA-fraction may contain other components than tRNAs, for instance broken pieces of collagen mRNA. Thus, the amino acid acceptances of tRNA-preparations were also studied.

Amino acid acceptances of tRNA isolated from granulation tissue at different phases of development

(a) *Methods.* Transfer-RNA was isolated from 6- and 15-day granulation tissue with a slightly modified method as previously described. The major difference was that DNA was separated from tRNA by DNAase digestion (Mäenpää and Bernfield, 1969). The contents of tRNAs accepting glycine, proline, lysine, serine and leucine were determined by measuring the limiting amounts tRNA in the reaction mixture containing among others Mg^{++}, ATP, 19 non-radioactive and one radioactive amino-acid, tRNA and amino acyl-tRNA synthetase obtained from 15-day granulation tissue (for details see Mäenpää and Ahonen, 1972). Isoaccepting species of tRNAs were studied using chromatography on benzoylated DEAE-cellulose as described by Mäenpää and Bernfield (1969).

(b) *Amino acid acceptances.* The acceptances measured are shown in Table V. Interestingly enough the results show about 30% higher content of tRNAs accepting 'collagen amino acids' during the period

Table V

Amino acid acceptances of unfractioned tRNAs obtained from 6- and 15-day granulation tissue

Amino acid	6-day preparation	15-day preparation	Ratio 15-day/6-day
		$nmoles/A_{260}$	
Glycine	10·7	14·3	1·34
Proline	5·9	7·6	1·29
Lysine	16·2	21·1	1·30
Serine	48·4	48·6	1·00
Leucine	28·6	29·5	1·03

of collagen synthesis in granulation tissue, than during the preceding period. They also indicate that the variation in the nucleotide composition shown in Table IV is at least partially due to changes in tRNA accompanying collagen synthesis. Our results confirm those reported by Lanks and Weinstein (1970) regarding proline, but also show an increased acceptance of glycine, which they did not detect. This may be due to the fact that we used homologous synthetase enzyme, whereas they used a liver preparation.

(c) *Chromatographic comparisons of specific transfer-RNAs obtained from 6- and 15-day granulation tissue.* To see whether the observed changes in acceptances in relation collagen synthesis were reflected in patterns of isoaccepting species of respective tRNAs, the tRNAs were chromatographed on benzoylated DEAE-cellulose. The 6-day preparation of tRNA was aminoacylated with ^{3}H-amino acids and the 15-day with ^{14}C-amino acids. Proline and lysine tRNAs were resolved into two fractions the patterns being similar in both preparations. Glycine-tRNAs were resolved into three fractions of which the firstly eluted was significantly increased in 15-day preparation as compared with the 6-day preparation (Fig. 4). This difference was independent of the isotope used. Thus there appears to be a specific glycine-tRNA that is used in the synthesis of collagen. It would be most interesting to see which codon it recognizes.

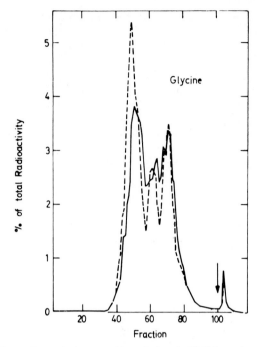

Fig. 4. Comparison of the elution profiles of glycyl-tRNA's on benzoylated DEAE-cellulose. ^{3}H-glycyl-tRNA (solid line) from 6-day granulation tissue was cochromatographed with ^{14}C-glycyl-tRNA (interrupted line) from 15-day granulation tissue.

NUCLEIC ACID METHYLATION DURING GRANULATION TISSUE DEVELOPMENT

A small proportion of purine and pyrimidine bases of both RNA and DNA are methylated at the polynucleotide stage of the biosynthesis (Borek and Srinivasan, 1966). The methylated purines are excreted in mammalian urine, presumably because there exist no enzymes for their degradation (Mandel *et al.*, 1966). The methylation is related to growth and differentiation (Craddock, 1970), for instance an increased urinary excretion of methylated purines has been observed in animals and man carrying malignant tumours. Our studies indicate changes in the methylation of nucleic acids during the development of granulation tissue.

Methods

Granulation tissue formation was induced by implanting viscous cellulose sponges in male rats. Urinary excretion of methylated purines was studied after an intraperitoneal injection of (Me-^{3}H)-methionine. The first injection was performed at least 3 days before the implantation of the sponges. The injections were repeated several times at 4–7-day intervals during the granulation tissue development. Urine was collected for three days after the injection and analyzed for methylated purines as described by Ahonen et al. (1972).

To study the incorporation of (Me-^{3}H)-methionine into nucleic acids of granulation tissue and liver the tracer was injected 18–40 h before killing the animals. The tissues were removed and homogenized, and the homogenate fractionated by differential centrifugation into subcellular fractions. RNA was then isolated from the fractions and its specific activity determined.

Urinary excretion of radioactive methylated purines during granulation tissue development

The following radioactive methylated purines were detected: 1-methylhypoxanthine, 8-hydroxy-7-methylguanine, 1-methylguanine and 7-methylguanine. 99% of the radioactivity was excreted during the first 48 h following the injection of the tracer. When the excretion of

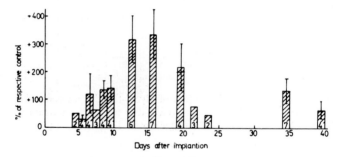

Fig. 5. The effect of granulation tissue formation on the urinary excretion of radioactive methylated purines during 24–48 h period after injection of (Me-^{3}H)-methionine. Values are as average % over preimplantation level. Means, SEM and number of experiments are indicated.

radioactive purines were measured for 48 h after the injection of the tracer an increase of about 100% was observed during the 10–20th days of granulation tissue development. A more pronounced effect was, however, noticed when the excretion during 24–48 h after the tracer injection was studied. The increase at day 15 was over 300%, as shown in Fig. 5.

The time of maximal excretion corresponds to the time in granulation tissue development when the most active collagen synthesis starts and DNA-synthesis ceases. Most of the label was incorporated into RNA. Apparently there exists a RNA-fraction related to collagen synthesis, whose metabolism causes this increased excretion of the radioactive methylated purines.

To clarify the origin of increased excretion of radioactive methylated purines we isolated RNA from liver and granulation tissue 30 h after the injection of the tracer. The specific activities of ribosomal and soluble RNAs are shown in Fig. 6.

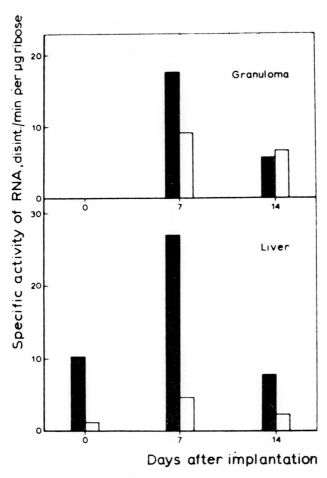

Fig. 6. Specific activities of liver and granulation tissue RNA's 30 h after injection of (Me-^{3}H)-methionine. Shaded columns: sRNA, open columns: rRNA.

These results show that the changed pattern of the excretion of the radioactive methylated purines is partly due to a change in liver RNA-methylation, especially during the first week of granulation tissue development. But the increase observed later was derived largely from the metabolism of the nucleic acids in the granulation tissue itself.

Table VI

Effect of granulation tissue formation on some liver microsomal enzyme activities

	Hexobarbital metabolizing capacity μ moles hexobarbital metabolized in 30 min/g liver	Glucose-6-phosphatase μ moles P_i liberated/ 20 min/g liver	ATPase n moles liberated min/mg protein	Cytochrome P-450 $E_{450-500}$/mg protein
Sham-operated controls	1·36 ± 0·21	71·6 ± 6·1	73·1 ± 4·6	0·037 ± 0·003
Rats bearing 6 18-day granulomas each	0·63 ± 0·15	71·4 ± 11·8	54·0 ± 5·7	0·019 ± 0·006
Rats bearing 12 18-day granulomas each	0·40 ± 0·06	62·1 ± 9·6	34·2 ± 4·1	0·013 ± 0·003

Values are means ± SEM n = 8–10. Determinations were performed as described by Ahonen and Auranen (1972).

LIVER MICROSOMAL CHANGES DURING GRANULATION TISSUE DEVELOPMENT

The experiments on the methylation of nucleic acids during granulation tissue development showed that liver nucleic acid metabolism was affected by the growth of the granulation tissue. We thought that it would be most interesting to see whether this is a general phenomenon, since the practical consequences are most obvious. First we directed our studies on liver microsomes, some of the results obtained are reported in this connection.

Granulation tissue formation was again induced by implanting viscous cellulose sponges, sham-operated animals served as controls. After a given time of granulation tissue development animals were bled, livers removed, microsomes isolated and studied as described by Ahonen and Auranen (1972).

The effect of granulation tissue formation on some enzymes is presented in Table VI. The results show that the effect on enzyme activities is selective and also 'dose-dependent', the more granulation tissue is growing in the animal, the more are enzyme activities affected. Since glucose-6-phosphatase has been reported to be very sensitive to increased levels of adrenal corticosteroids (Cohn et al., 1969), we can exclude the possibility that the effect of granulation tissue formation would be caused by unspecific stress.

Table VII

Effect of granulation tissue formation on some liver microsomal constituents

	Protein mg/g liver	RNA mg/g liver	Phosholpipids mg/g liver
Sham-operated controls	19·5	3·5	7·6
Rats bearing 6 granulomas each	21·7	5·1	6·4

Granuloma formation time was 18 days.

Table VII shows that protein and RNA contents of liver microsomes were increased during granulation tissue formation. Preliminary data also indicates that incorporation of radioactive glycine into microsomal proteins was increased. There was a slight reduction in the amounts of microsomal phospholipids, isotope studies indicate an increased rate of turnover of microsomal phospholipids.

The data show that during the formation of granulation tissue, especially when collagen synthesis is most active in the repair tissue, liver functions are specifically altered, probably by some humoral mechanisms.

ACKNOWLEDGEMENTS

The studies reported in this chapter have been supported by the Sigrid Jusélius Foundation, Emil Aaltonen Foundation and the National Research Council for Medical Sciences, Finland.

REFERENCES

Ahonen, J. (1968). *Acta Physiol. Scand.* Suppl. **315.** (A Thesis.)
Ahonen, J. and Auranen, A. (1972). *Eur. Surg. Res.* (In Press.)
Ahonen, J., Kivisaari, J., Vastamäki, M., Välimäki, M. and Kulonen, E. (1972). *Biochim. Biophys. Acta* **262,** 233–238.
Ahonen, J. and Kulonen, E. (1966). *J. Chromatog.* **24,** 197–198.
Armstrong, R. L. and Boezi, J. A. (1965). *Biochim. Biophys. Acta* **103,** 60–69.
Borek, E. and Srinivasan, P. R. (1966). *Ann. Rev. Biochem.* **35,** 275–296.
Craddock, V. M. (1970). *Nature* **228,** 1264–1268.
Cohn, R. M., Herman, R. H. and Zahim, D. (1969). *Am. J. Clin. Nutr.* **22,** 1204–1210.
Georgiev, G. P., Samarina, O. P., Lerman, M. I., Smirnov, M. N. and Severtzov, A. N. (1963). *Nature* **200,** 1291–1294.
Kirby, K. S. (1956). *Biochem. J.* **64,** 405–408.
Lanks, K. W. and Weinstein, I. B. (1970). *Biochem. Biophys. Res. Commun.* **40,** 708–710.
Mäenpää, P. H. and Ahonen, J. (1972). *Biochem. Biophys. Res. Commun.* **49,** 179–184.
Mäenpää, P. H. and Bernfield, M. R. (1969). *Biochemistry* **8,** 4926–4934.
Maudel, L. R., Srinivasan, P. R. and Borek, E. (1966). *Nature* **209,** 586–588.
Sueoka, N. and Kano-Sueoka, T. (1970). *Progr. Nucl. Acid. Res. and Mol. Biol.* **10,** 23–48.
Viljanto, J. and Kulonen, E. (1962). *Acta Path. Microbiol. Scand.* **56,** 120–126.

Mechanism for Cellular Ageing in Long-term Culture

JAN PONTÉN, BENGT WESTERMARK and ULF BRUNK
Department of Pathology, University of Uppsala
751 22 Uppsala, Sweden

Fibroblasts from all higher organisms grow luxuriantly in tissue culture with preserved normal morphology and proliferation control. However, after a certain time, they either die out or transform into sarcoma cells. Hayflick (1966) first proposed that the former event may be an *in vitro* expression of ageing. Absence of transformation and subsequent ageing is typical of so called stable species (e.g. human, bovine, avian) (Pontén and Lithner, 1966; Pontén, 1970).

We have extended these observations to human glia cells (Pontén and Macintyre, 1968). Altogether 85 lines from adult non-neoplastic human brain biopsies have been cultivated in Eagle's MEM with 10% calf serum. The cultures are composed of astrocyte-like, diploid and well monolayered cells with virtually perfect contact inhibition of movement and division at a critical density of about 80,000 cells per cm^2 solid support (Westermark, 1971). Because of their stability, absence of spontaneous transformation, well regulated growth behaviour and suitability for ultrastructural studies (Brunk *et al.*, 1971) we have explored their use for ageing studies.

The glia cells reach the end of their finite life span *in vitro* after an estimated 15–20 mitotic divisions. In their senescent stage they display numerous residual bodies principally similar to the alterations seen in old nerve and liver cells *in vivo* (Brunk and Ericsson, 1972; Essner and Novikoff, 1960). It is not known if the accumulations of residual bodies with their content of hydrolytic enzymes, potentially capable of destroying the cell from the inside, is a cause or an effect of ageing although the former alternative has been the most favoured one (Toth, 1968).

Another unsolved question is whether senescence is caused by the passage of time *per se* or by continuous cell division. Several recent reports (Hay *et al.*, 1968; McHale *et al.*, 1971) have favoured the former view.

We have devised a system which should differentiate between various alternative mechanisms for senescence *in vitro*. Monolayer cultures were started from single cells. Due to the differential effect of contact inhibition of growth a gradient is created with respect to number of cell cycles. Cells in the periphery are constantly stimulated to replicate in contrast to centrally placed cells which only undergo a few divisions before they become contact inhibited. The former therefore have a longer *cell cycle age* than the latter under conditions of identical *chronological age*. After an incubation time of about 1 to 2 months a *cell cycle age gradient* has been created with young cells at the centre and senescent glia in the periphery.

A cell-free strip ('wound') was created along a diameter of colonies with well developed cell cycle age gradients. The sudden abolishment of contact inhibition induced cells along the edge of the strip to multiply. The proliferative potential along the defect was followed autoradiographically (^{3}H-TdR) and compared with that of undisturbed parts. A direct correlation between distance from the centre and diminished regenerative power was found (Fig. 1).

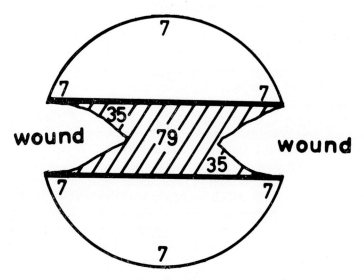

Fig. 1. Schematic representation of a wounding experiment with a circular cell clone having a cell cycle age gradient along its radii (see text). % cells incorporating thymidine (^{3}H-TdR: 48 h) indicated by numbers at the respective spots where cells were counted. The striped area corresponds to the zone into which cells migrated during the first two days after wounding.

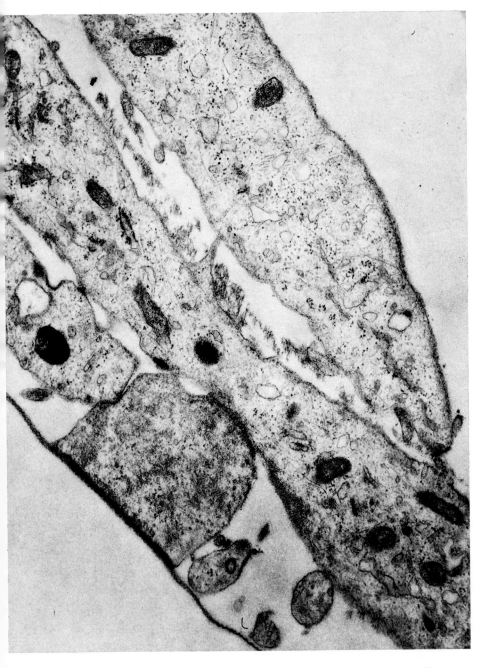

Fig. 2. Normal 'young' human glia cells after 2 weeks of confluent culture. Note scanty dense bodies (secondary lysosomes) ($\times 22{,}000$).

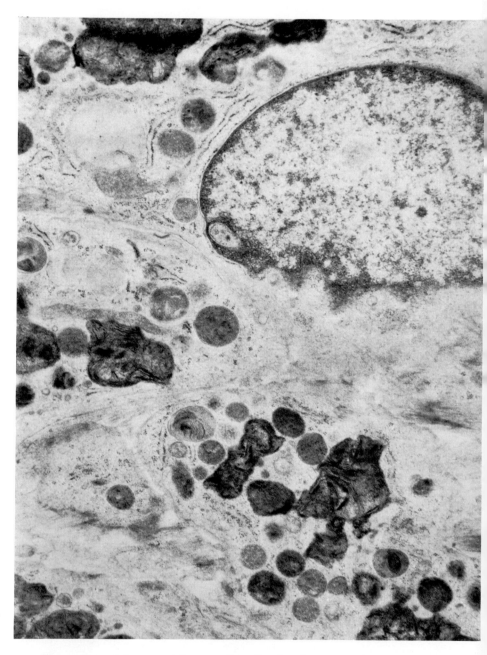

Fig. 3. Normal 'senescent' human glia cells after 6 months of confluent culture (same experiment as in Fig. 2). Note great amount of highly irregular dense bodies, many of which are 'lipofuscine-like' ($\times 22{,}000$).

In the same experiment the capacity of the cells to migrate into the wound during the labelling period of 48 h was also measured. It is seen from Fig. 1 that migration was directly correlated to capacity to undergo DNA-synthesis.

As a mirror image of the above experiment *chronologic age* was varied under conditions of a *constant cell cycle age*. Contact inhibited cells were kept stationary for many months and investigated by electron microscopy after various periods of time. A gradual accumulation of secondary lysosomes was seen (Figs. 2 and 3). Apparently accumulation of lysosomes and residual bodies was primarily influenced by chronologic age. Cell viability was largely unimpaired indicating that the lysosomes did not play an essential role for experimental ageing *in vitro*. Stabilization of lysosomes by cortisone only gave a marginal positive effect on the cellular life span.

Our model of a cell cycle age gradient is based on a cell clone. The effects can therefore not be explained by the preexistence of genetically unequal cells.

Since lysosomes and residual bodies would accumulate as a function of chronologic time and nevertheless such cells were capable of proliferation it seems unlikely that accumulation of these bodies in the cytoplasm is the cause of senescence.

The actual cause of senescence has been debated recently. Holliday and Tarrant (1972) have tried to verify Orgel's hypothesis (1963) of ageing based on a cytoplasmic protein synthesis error catastrophe. Essentially the theory says that errors in protein synthesis will—when they have reached a critical threshold—rapidly escalate to a situation where cell viability is lost. Defect proteins such as nucleic acid polymerases and ribosomal proteins will by their very presence promote further errors in transcription and translation and thereby create a vicious circle with senescence as its phenotypic expression.

It is difficult to see how a protein error catastrophe can be caused by repeated cell divisions. It is still possible that the finding of a large percentage of 'malformed' proteins in the cytoplasms of senescent cells (Molliday and Tarrant, 1972) is a secondary phenomenon. We hope to be able to elucidate whether the protein error catastrophe is cause or effect of ageing by the present experimental model which seemed to exclude lysosomal accumulation as a cause of senescence *in vitro*.

CONCLUSIONS

1. Accumulation of lysosomes and/or residual bodies takes place as a result of increasing chronologic rather than cell cycle age.
2. The capacity to respond by proliferation after severing of inter-

cellular contacts is reduced in proportion to the length of the cell cycle age. Chronologic age seems to be relatively unimportant in this respect.

3. The capacity to migrate into a wound is affected by cell cycle age similarly to capacity to proliferate. It does not seem to be reduced significantly by chronologic age *per se*.

REFERENCES

Brunk, U. and Ericsson, J. L. E. (1972). *J. Ultrastruct. Res.* **38**, 1–15.
Brunk, U., Ericsson, J. L. E., Pontén, J. and Westermark, B. (1971). *Exp. Cell Res.* **67**, 407–415.
Essner, E. and Novikoff, A. B. (1960). *J. Ultrastruct. Res.* **3**, 379–391.
Hay, R. J., Menzies, R. A., Morgan, H. P. and Strehler, B. L. (1968). *Exp. Gerontol.* **3**, 35–44.
Hayflick, L. (1966). *Perspect. Exp. Gerontol.* **14**, 195–211.
Holliday, R. and Tarrant, G. M. (1972). *Nature* **238**, 26–30.
McHale, J. S., Mouton, M. L. and McHale, J. T. (1971). *Exp. Gerontol.* **6**, 89–93.
Orgel, L. E. (1963). *Proc. Natl. Acad. Sci. USA* **49**, 517–521.
Pontén, J. (1970). *Int. J. Cancer* **6**, 323–332.
Pontén, J. and Lithner, F. (1966). *Int. J. Cancer* **1**, 589–598.
Pontén, J. and Macintyre, E. (1968). *Acta Path. Microbiol. Scand.* **74**, 465–486.
Toth, S. E. (1968). *Exp. Gerontol.* **3**, 19–30.
Westermark, B. (1971). *Exp. Cell Res.* **69**, 259–264.

Nucleolus Specific Antigens in Human Fibroblasts and Hybrid Cells Studied with Patient Autoantibodies

NILS R. RINGERTZ, THORFINN EGE and STEN-ANDERS CARLSSON
*Institute for Medical Cell Research and Genetics, Medical Nobel Institute
Karolinska Institutet, 104 01 Stockholm 60, Sweden*

Patients suffering from lupus erythematosus develop autoantibodies against a variety of nuclear components. The nuclear autoantibodies react with a variety of components, e.g. single- and double-stranded DNA, single- and double-stranded RNA, nucleoproteins and also some other components of unknown composition. We have found such autoantibodies to be of use in the analysis of the postmitotic reassembly of nucleolar components and in studies of intracellular protein migration in interspecific (man-chick) cell hybrids. The human antinucleolar antibodies used in our work were obtained from a 52-year old female patient suffering from polymorphic manifestations of lupus erythematosus. The binding of the antibodies was studied on fixed cells and cell hydrids using the indirect immune fluorescence reaction. Quantitation of the immune fluorescence was carried out by microfluorimetry.

THE CELL CYCLE

HeLa cells and human fibroblasts showed a clear nucleolar localization of the antigens (Figs. 1a and 2a) throughout interphase. The nucleolar localization was changed with the onset of mitosis. During prophase the nucleolar staining pattern broke up together with the disaggregation of the nucleolus and at the end of prophase the antigens were found throughout the cytoplasm, but appeared to be absent from the metaphase plate (Fig. 1b).

In anaphase. the antigens concentrated along the chromosomes, first diffusely, but later as small grains along the chromosomes (Fig. 1c). These structures seemed to merge and grow during telophase (Fig. 1d), and in early interphase they had fused into nucleoli.

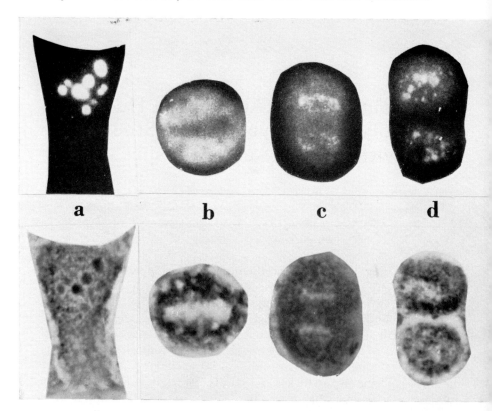

Fig. 1. Phase contrast and fluorescent microphotographs, showing the changes in the localization of the nucleolus specific antigen during mitosis in HeLa cells: a, interphase; b, metaphase; c, anaphase; d, telophase. During metaphase the nucleolar antigens are scattered throughout the cytoplasm. The antigens begin to bind to the chromosomes already in anaphase and in telophase nucleolus-like bodies have reassembled.

ANTIGENICITY OF ACTINOMYCIN SEGREGATED NUCLEOLI

Actinomycin D (AMD), a drug which binds specifically to double-stranded DNA and inhibits RNA synthesis, is known to induce a phenomenon known as *nucleolar segregation*. This phenomenon consists of both condensation of the chromatin fibrils and segregation of the fibrillar and granular ribonucleoprotein components into different regions of the nucleolus. Since normal nucleoli when observed in the light or fluorescence microscope do not show any clearly definable substructures, it appeared that immune staining of AMD-segregated nucleoli might provide information about the antigenicity of different nucleolar subcomponents. When AMD was added to the growth medium of HeLa cells in a concentration of 5 µg/ml, structural changes were observed in the nucleolus both by phase contrast microscopy and by fluorescence microscopy after immune staining. About 2 h after the addition of AMD the first changes were observed. In the phase contrast microscope small dark nucleolar caps were found to appear at the periphery of the nucleolus. These grew and after 6–8 h they reached maximal size. At that time the caps tended to be greater than the central mass. At all stages examined the central mass was strongly fluorescent while the nucleolar caps only gave a very weak reaction (Fig. 2b). More extensive treatment with AMD led to disaggregation of the

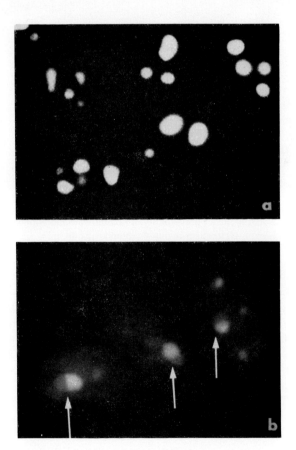

Fig. 2. Nucleolar fluorescence of 5 normal HeLa cell nuclei (a). (b) Shows the antibody binding to cells treated with actinomycin D. The nucleoli (arrows) have segregated into two parts, one of which binds the antibodies.

nucleolus and death of the cell. After short exposure to moderate doses of actinomycin (0·04 μg/ml) the nuclei of some cells contained numerous small bodies (Fig. 3). These cells occurred in pairs and therefore probably represented postmitotic cells where the reorganization of nucleoli had been impaired by the actinomycin treatment.

As the nucleoli in the majority of the cells underwent segregation, and starting at about the same time as the nucleolar caps appeared,

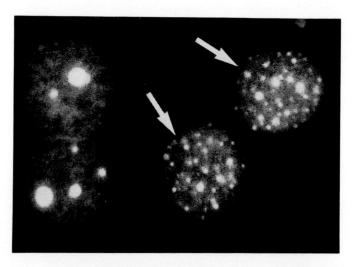

Fig. 3. Increased nucleoplasmic fluorescence in actinomycin treated HeLa cells. Arrows indicate a pair of nuclei presumably of postmitotic cells which show numerous small nucleolar bodies. Cytoplasm is non-fluorescent in all cells.

antigenic material also accumulated in the extra-nucleolar chromatin (Fig. 3). The intensity of the immune fluorescence reaction in extra-nucleolar regions of the nucleus and in the cytoplasm was measured by microfluorimetry as shown in Fig. 4. The fluorescence of extra-nucleolar regions in the nucleus began to increase within a few hours after administration of 5 μg/ml of actinomycin and reached a maximum after about 5 h. Throughout the observation period, the cytoplasm showed a negligible fluorescence.

First the antigen was evenly distributed but later the fluorescence structure became more granular parallelling the gradual increase in chromatin condensation. When the nucleolus was completely disintegrated, the nuclear chromatin was condensed into numerous small grains which gave a positive fluorescence reaction. Throughout the observation period cytoplasms appeared negative.

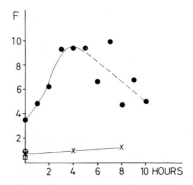

Fig. 4. Microfluorimetric measurements of the increase in nucleoplasmic fluorescence (●) after actinomycin treatment. The cytoplasm of both actinomycin treated (x) and normal cells (o) is nonfluorescent. Cytoplasm of untreated control stained with buffer instead of serum (□).

MIGRATION OF HUMAN NUCLEOLAR ANTIGENS IN HUMAN-CHICK HETEROKARYONS

Human autoantibodies to nucleolar antigens have been used as an instrument to study the formation of a nucleolus in chick erythrocyte nuclei undergoing reactivation in heterokaryons. This application is based on the observation that the human autoantibodies react only with nucleolar antigens in human, mouse, rat and calf cells but *not* with nucleoli in chick cells. The chick erythrocyte nuclei are inactive with respect to RNA, DNA and protein synthesis but can be reactivated by fusion of chick erythrocytes with human fibroblasts or HeLa cells (Bolund *et al.*, 1969; Ringertz *et al.*, 1971; Ege *et al.*, 1971). This offers unique possibilities to analyse gene regulatory mechanisms when a large number of genes are suddenly switched on in a diploid eukaryotic cell. During reactivation in heterokaryons the erythrocyte nucleus enlarges and synthesizes RNA. New chick specific proteins are, however, not synthesized until rather late in the reactivation process. Since proteins make up most of the dry weight of the nucleolus, this raised questions as to how the nucleolus in the chick erythrocyte nucleus was formed. The nucleolus could either be made up of chick proteins specified by the RNA produced by the chick erythrocyte nucleus or the nucleolus could be formed from material provided by the human cell, i.e. from human proteins. In order to examine these problems, heterokaryons formed by Sendai virus induced fusion of human cells with chick erythrocytes were examined at various time-points after fusion. As illustrated in Fig. 5, the chick erythrocyte nucleus was at first found to be negative with respect to human nucleolar antigens. As the nucleus enlarged and nucleoli formed, these were clearly positive with respect to human nucleolar antigens. These observations show that human macromolecules migrate into the chick erythrocyte nucleus during its

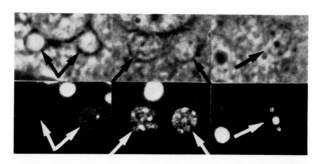

Fig. 5. Migration of human nucleolar antigens into chick erythrocyte nuclei undergoing reactivation in the cytoplasm of a HeLa cell. Running left-right the arrows indicate the appearance of chick erythrocyte nuclei representing different stages of reactivation. The fully activated chick erythrocyte nuclei have formed nucleoli which show human nucleolar antigens (extreme right) while chick erthrocyte nuclei which have not begun to enlarge are negative with respect to human nucleolar antigens (extreme left).

reactivation and play a role in the formation of the nucleolus (Ringertz et al., 1971).

CONCLUSIONS

Our observations suggest that human autoantibodies to nucleolar components can be used as a powerful tool in studying the formation of nucleoli. The behaviour of nucleolar antigens studied by us suggest that during mitosis nucleolar components are dispersed throughout the cytoplasm during metaphase and then reaggregate to form a new nucleolus in telophase. The behaviour of the human nucleolar antigen in human-chick heterokaryons suggests that the antigen can migrate relatively rapidly between cytoplasm and nucleus and that within the nucleus the antigen becomes selectively concentrated in the nucleolus organizing chromatin. The nature of the antigen is not yet known but it is interesting to note that the antibodies only react with certain regions in actinomycin segregated nucleoli.

REFERENCES

Bolund, L., Ringertz, N. R. and Harris, H. (1969). *J. Cell Sci.* **4**, 71–87.
Ege, T., Carlsson, S-A. and Ringertz, N. R. (1971). *Exp. Cell Res.* **69**, 472–477.
Ringertz, N. R., Carlsson, S-A., Ege, T. and Bolund, L. (1971). *Proc. Natl. Acad. Sci. USA*, **68**, 3228–3232.

Putrescine as a Growth Factor

PIRKKO POHJANPELTO
Department of Virology, University of Helsinki, Helsinki, Finland

A marked increase in the synthesis of putrescine in connexion with stimulation of cell proliferation has been demonstrated *in vivo* by several workers (Snyder and Russell, 1970; Raina and Jänne, 1970). Together with A. Raina I have been able to show that growth factor activity (GF) found in the conditioned medium from cultures of human fibroblasts is mainly due to putrescine (diaminobutane) (Pohjanpelto and Raina, 1972). In this chapter studies on the growth stimulating effect of putrescine are enlarged upon.

PRODUCTION OF GROWTH FACTOR

Maximal production of GF parallels maximal cell proliferation. When the rate of cell proliferation was modified by using different temperatures, serum concentrations and cell densities, most GF was found in the medium from cultures with the largest absolute increase in the cell number, the total number of cells being less important (Pohjanpelto, 1973a). The results suggest that proliferating cells are mainly responsible for the production of GF. If the growth factor activity found in the medium from cultures of human fibroblasts reflects the overall synthesis of putrescine by the cells, this is consistent with the findings *in vivo*, which suggest that synthesis of putrescine is connected with cell proliferation.

REQUIREMENT FOR EXTRA PUTRESCINE BY HUMAN FIBROBLASTS

It seems that cells *in vitro* are unable to synthesize enough putrescine to ensure maximal cell proliferation probably because of the partial loss of putrescine in the medium. Hence extra putrescine is required. The concentrations of putrescine that have a stimulating effect on cell proliferation fall within a narrow range. The optimal concentration of putrescine varies in different cultures, and it seems to depend on the absolute increase in the cell number during incubation with putrescine (Pohjanpelto, 1973a). It may be, that with proliferating cells the requirement for putrescine rises more than production of GF, and therefore a larger quantity of extra putrescine is needed to compensate.

UPTAKE OF PUTRESCINE BY HUMAN FIBROBLASTS

Attachment of putrescine to human fibroblasts is not simply due to electrostatic binding, because labelled putrescine cannot be removed

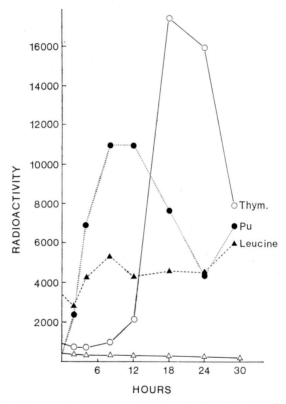

Fig. 1. Uptake of putrescine by human embryonic fibroblasts after stimulation of cell proliferation by serum. The cells were starved of serum for 3 days. At time 0 the medium was replaced with fresh medium containing either 0% or 10% serum. 30-min pulses of ^{14}C-putrescine and ^{3}H-thymidine, and 10-min pulses of ^{3}H-leucine were given at different time intervals, and radioactivity in the cells was determined. △, 0% serum, ^{14}C-putrescine cpm/10^6 cells; ●, 10% serum, ^{14}C-putrescine cpm/10^6 cells; ○, 10% serum, ^{3}H-thymidine cpm/6×10^5 cells; ▲, 10% serum, ^{3}H-leucine cpm/2×10^5 cells.

by washing with excess of unlabelled putrescine. If cells are killed by heating or incubating in water, no attachment of putrescine to the cell remnants occurs. The process is dependent on temperature: at 37°C about 30 times more putrescine is taken up by the cells than at 4°C (Pohjanpelto, 1973b).

When cells are stimulated to proliferate by adding serum to cultures starved of serum, there is a striking increase in the uptake of putrescine (Fig. 1) (Pohjanpelto, 1973b). Two hours after the addition of serum, uptake of putrescine had increased sixfold, and after 8 h 25-fold. Incorporation of ^{3}H-thymidine into DNA did not start to increase until at 12 h, and it reached maximum at about 18 h. Thus increase in the uptake of putrescine preceded DNA synthesis by 8–12 h.

To find out whether RNA and protein synthesis were necessary for the increase in putrescine uptake, the effect of actinomycin D and cycloheximide was tested. In the presence of actinomycin D, which blocks RNA synthesis, there was almost complete inhibition of increase in the uptake of putrescine after addition of serum. Cycloheximide in a concentration preventing 90% of the incorporation of labelled aminoacids into acid insoluble fraction, unexpectedly, caused only partial inhibition in the increase of the uptake of putrescine. This gives an impression that protein synthesis would be less important for the increase in the uptake of putrescine than RNA synthesis, and might be required only indirectly, in as far as it is necessary for RNA synthesis. However, unspecific effects of actinomycin D were not excluded, and therefore no definite conclusions can be drawn (Pohjanpelto, 1973b).

EFFECT OF PUTRESCINE ON DNA SYNTHESIS

When cell proliferation is determined by counting the cell number, the growth stimulating effect of putrescine is clearly visible only after 3–4 days incubation. This gave rise to an assumption that the growth stimulating effect of putrescine was based on the shortening of the cell cycle rather than on the triggering of cell division in quiescent cells. To find out whether this was true, the length of the cell cycle in the presence and absence of putrescine was determined by counting labelled mitoses at different time intervals after ^{3}H-thymidine pulse (Baserga and Malamud, 1969). The results are presented in Fig. 2. It appears, that putrescine shortens the cell cycle from 20 h to 16 h, and that this is almost entirely due to reduction in the length of the S-period. Without putrescine synthesis of DNA takes 13·4 h and in the presence of putrescine 9·4 h. Thus putrescine shortens the S-period by 30% (Pohjanpelto, 1973b). This result is consistent with the findings of Dion and Cohen (1972), who showed that DNA synthesis in an *E. coli* mutant, unable to accumulate putrescine in the presence of arginine, was heavily retarded when the cells were grown in an arginine containing medium. It seems that availability of putrescine is a prerequisite for the normal synthesis of DNA in organisms as far apart as *E. coli* and man.

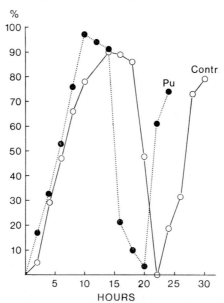

Fig. 2. Effect of putrescine on the cell cycle. Putrescine ($3 \times 10^{-8}M$) was added to sparsely populated cultures of human embryonic fibroblasts. Two days later the cell cycle in cultures incubated with and without putrescine, was determined by counting the labelled mitoses after ^{3}H-thymidine pulse (Baserga and Malamud, 1969).

REFERENCES

Baserga, R. and Malamud, D. (1969). "Autoradiography". Harper and Row, New York, Evanston and London.
Dion, A. S. and Cohen, S. S. (1972). *Proc. Natl. Acad. Sci. USA* **69**, 213–217.
Pohjanpelto, P. and Raina, A. (1972). *Nature* **235**, 247–249.
Pohjanpelto, P. (1973a). *Exp. Cell Res.* (In Press.)
Pohjanpelto, P. (1973b). In preparation.
Raina, A. and Jänne, J. (1970). *Fed. Proc.* **29**, 1568–1571.
Snyder, S. H. and Russell, D. H. (1970). *Fed. Proc.* **29**, 1575–1582.

Determination of Proteolytically Most Active Cell Population in Experimental Granuloma and Purification of its Cathepsins

Š. Stražiščar and V. Turk

Department of Biochemistry, J. Stefan Institute
University of Ljubljana, Ljubljana, Yugoslavia

Several physiological and pathological processes characterized by a rapid breakdown of connective tissue components are known. High levels of tissue hydrolases can be detected in the course of such processes (Dingle, 1971; Woessner, 1968). Evolution of granuloma is accompanied by continuous remodelling of tissue components and in the phase of involution their degradation occurs.

Our work is oriented to the study of proteolytic enzymes in experimental granuloma. The evolution of granulomas was induced by implantation of sterile plastic sponges into the subcutis of rats according to the method of Boucek and Noble (1955). After 1, 3, 8, 16, 30 and 60 days the animals were killed, sponges withdrawn and cut into small pieces. Tissue was homogenized and centrifuged. DNA by the Dishe's method, nitrogen according to the modified Kjeldahl's method and proteolytic activity by the modified Anson's method were measured. Proteolytic activities were increasing through all the phases of granuloma evolution. This increase was parallel to that of activity to nitrogen ratios, as soluble nitrogen did not vary very much in the course of evolution. In initial phases of granuloma evolution the nitrogen belongs not only to tissue components but also to those of inflammatory exudate. Therefore the ratios of activity to DNA were taken as an additional criterion to check whether the concentrations of proteolytic enzymes increase in chronic granulomas. The results obtained are shown in Table I.

For the purification of proteolytic enzymes the 60-day old granulomas, being the richest, were taken as a starting material. A greater amount of

Table I
Some biochemical parameters indicating the levels of proteolytic enzymes in granulation tissue

	Age of granuloma (days)					
	1	3	8	16	30	60
Proteolytic activity*	0·105	0·173	0·378	0·770	1·047	1·126
Soluble N mg/ml of supernatant	1·415	1·414	1·537	1·748	1·701	1·790
DNA mg/g of tissue	0·950	0·791	0·980	1·190	1·610	1·580
Activity/N	0·074	0·122	0·246	0·440	0·615	0·629
Activity/DNA	0·110	0·219	0·386	0·647	0·650	0·713

* Hydrolysis of haemoglobin at pH 3·5, incubation time 30 min. Activities are given in E_{750}.

granulation tissue was obtained by the implantation of eight sponges through two incisions on the back of each animal. Twenty animals were used in each experiment. All purification procedures were done in the cold room at $+3°C$. Preparations of supernatants were the same as already described. The pH of supernatant was adjusted to 3·5 with 3N HCl, centrifuged and the resulting precipitate was discharged. The obtained supernatant was fractionated with solid ammonium sulphate to 30% and 60% of saturation. The fraction 30–60% contained the bulk of proteolytic activity and was used for further purification. The precipitate was dissolved in 0·05 M phosphate buffer pH 8·0 and chromatographed through DEAE-cellulose under the same conditions. A proteolytically active first peak was collected and concentrated by ultrafiltration through UM-10 membranes, Amicon. (The second peak probably corresponds to cathepsin E and was eluted with the addition of 0·5 M NaCl in the same buffer.) Concentrate was applied to the Sephadex G-100 column in 0·1 M NaCl. It is possible to remove a great deal of protein impurities in this way (Fig. 1). Fractions with high enzyme activity were collected and concentrated by ultrafiltration.

The last purification step was acrylamide gel electrophoresis using 7% acrylamide gel in tris-glycine buffer, pH 8·4. Some gels were stained with amido black in 7% acetic acid and some remained unstained. Unstained gels were sliced into sections using stained gels as markers and proteins were extracted with water. Proteolytic activities were measured in the extracted fractions.

Fig. 2 represents disc electrophoresis patterns through different purification steps. Disc electrophoresis of Sephadex G-100 fraction gave three main bands: the second band had low activity on haemoglobin as substrate but the third band showed a high activity on haemoglobin as substrate. From the wideness of the active band it can be seen that it is

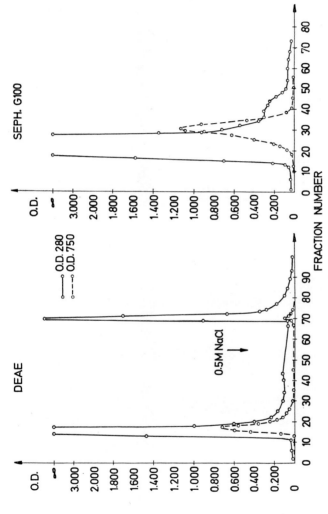

Fig. 1. Column chromatography of granuloma extract on DEAE cellulose and gel filtration on Sephadex G-100.

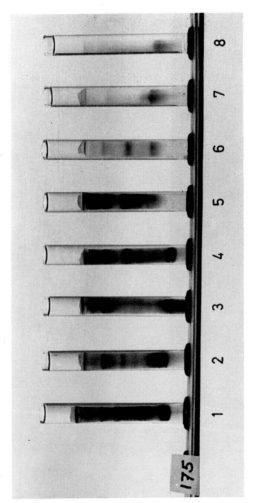

Fig. 2. Disc electrophoresis patterns through different purification steps: 1, neutral supernatant; 2, acid supernatant; 3, 0–30% ammonium sulphate precipitation; 4, 30–60% ammonium sulphate precipitation; 5, chromatography on DEAE cellulose; 6, gel filtration on Sephadex G-100; 7 and 8, extracted active fractions after disc electrophoresis as last purification step.

still inhomogeneous and contains at least two components (7th and 8th gels on Fig. 2). The extracted enzymes from these two bands were used for further investigation. Specific activity was 110 Anson's units/mg of protein, which corresponds to about 150-fold purification of the crude extract.

The study of the pH optimum was made with haemoglobin and bovine serum albumin as substrates. pH optimum is at pH 3·2 for both substrates (acetate buffer). The hydrolysis of bovine serum albumin is only about 1/20th of that of haemoglobin. Results are characteristic for cathepsin D (Press et al., 1960; Lebez et al., 1968).

The molecular weight was determined by gel filtration method on Sephadex G-100 in 0·1 M NaCl. Blue dextran marked the void volume and cytochrome c, chymotrypsinogen and bovine serum albumin served as weight markers. Elution volume of cathepsin D corresponds to 38,000, which is very similar or the same with those of in bovine uterus (37,000–42,000) as reported by Sapolsky and Woessner (1972), in human and chicken liver (45,000) as reported by Barrett (1971), and in chicken muscle (36,000) as reported by Fukushima (1971) and others.

Bazin and Delaunay (1971) found cathepsin D and collagenolytic cathepsin in experimentally induced subcutaneous granulation tissue in rats. Our experimental results support and extend the findings of French authors regarding cathepsin D. We can conclude that cathepsin D exists in multiple forms (at least two) in granulation tissue. In further experiments we will study some other biochemical characteristics of cathepsin D and possible collagenolytic cathepsin, which will lead to better knowledge of the mechanism of action of these enzymes in the catabolic processes.

ACKNOWLEDGEMENTS

Our research was supported by Boris Kidrič Foundation and Grant GF-31389 from the US National Science Foundation.

REFERENCES

Barrett, A. J. (1971). *In* "Tissue Proteinases" (A. J. Barrett and J. T. Dingle, eds.) pp. 109–128. North-Holland Publishing Co., Amsterdam.
Bazin, S. and Delaunay, A. (1971). *Ann. Inst. Pasteur* **120**, 50–61.
Boucek, R. J. and Noble, N. L. (1955). *A.M.A. Arch. Pathol.* **59**, 553–558.
Dingle, J. T. (1971). *In* "Tissue Proteinases" (A. J. Barrett and J. T. Dingle, eds.) pp. 313–324. North-Holland Publishing Co., Amsterdam.
Fukushima, K., Gnoh, G. H. and Shinano, S. (1971). *Agr. Biol. Chem.* **35**, 1495–1502.
Lebez, D., Turk, V. and Kregar, I. (1968). *Enzymologia* **34**, 344–348.
Press, E. M., Porter, R. R. and Cebra, J. (1960). *Biochem. J.* **74**, 501–514.
Sapolsky, A. J. and Woessner, Jr., J. F. (1972). *J. Biol. Chem.* **247**, 2069–2076.
Woessner, Jr., J. F. (1968). *In* "Treatise on Collagen" (G. N. Ramachandran and B. S. Gould, eds.) Vol. 2, pp. 253–330. Academic Press, New York.

Isolation of Cells from Experimental Granulation Tissue

K. Ivaska* and H. Pertoft

*Department of Medical Chemistry, University of Uppsala
Uppsala, Sweden*

The aim of the present work was to develop a method, which would yield connective tissue cells with preserved functions in amounts sufficient for biochemical studies. Granulation tissue was tried as the source of cell suspensions because its soft and cellular character seems to promise a good and easy yield of cells and because the tissue can be produced experimentally in large quantities.

The principles of the cell isolation procedure are presented in the flow sheet of Fig. 1. The tissue, which was produced by the cellulose sponge implantation technique (Viljanto and Kulonen, 1962), was sliced and immersed in the medium. The tissue was recovered by filtration, washed with the medium and agitated for one hour in the medium containing collagenase and hyaluronidase (Yamada and Ambrose, 1966). The material brought into the medium (Fraction B in the flow sheet) was separated from the residual tissue and cellulose sponge by filtration and put on the top of a density cushion prepared from two solutions of colloidal silica having densities of 1·050 g/ml and 1·090 g/ml (Pertoft et al., 1968). The choice of densities was based on the preliminary observation, that the viable cells banded at densities higher than 1·050 g/ml and the non-viable cells and cell debris at lower densities. After centrifugation the material at the interface was removed, incubated for half an hour in the presence of trypsin and deoxyribonuclease, and rebanded in the density cushion to obtain the cells for the final cell suspension (Fraction BB in the flow sheet).

The final cell suspension from granulomas at different ages consisted of rounded single cells. Over 90% of the cells in the preparations were viable, when tested with the trypan blue exclusion test. Suspensions containing up to 530 million cells were obtained from twelve granulomas, and the preparations accounted for up to 20% of DNA present in the granulomas. The

* Present address: Department of Medical Chemistry, University of Turku, Turku, Finland.

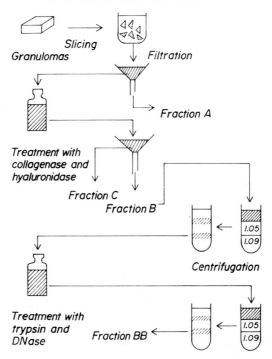

Fig. 1. Flow sheet of the preparation procedure. All operations were performed at room temperature, except the enzyme treatments, which were carried out at 37°C. Eagle's minimum essential medium was used throughout the procedure. Fraction A contained damaged cells and cell debris. Fraction C contained the material not brought into the medium by the enzyme treatment. These fractions were not examined further.

buoyant density (Fig. 2) and the cell size (Fig. 3) distributions of the preparations from 19- and 26-day old granulomas were identical and suggest the presence of one main population of cells. The preparations from younger granulomas were more heterogeneous.

When the cells were plated in plastic culture dishes in Eagle's minimum essential medium supplemented with 10% of calf serum, primary cultures of fibroblast-like cells were derived. The results of the incorporation experiments (Fig. 4) indicate the preparations to have active synthesis of proteins, including collagen, and glycosaminoglycans. The preparations from young granulomas seem to be biosynthetically more active than the preparations from older granulomas.

Large amounts of cells of connective tissue origin can be prepared by the present method. The cells are viable and capable of synthesizing

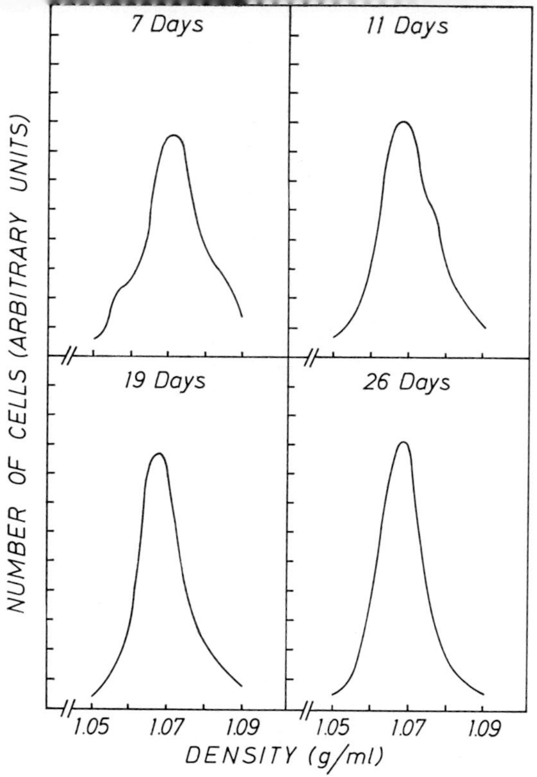

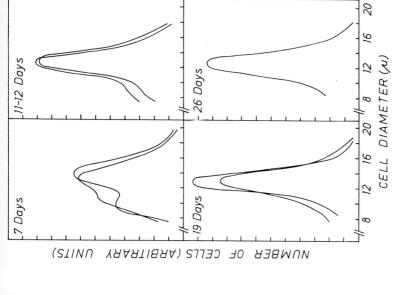

Fig. 3. The diameter of cells isolated from granulomas at indicated ages. The measurements were made with an electronic cell counter.

Fig. 2. The buoyant density of the cells isolated from granulomas at indicated ages. The results were obtained from banding the cells in a linear density gradient from 1·050 g/ml to 1·090 g/ml.

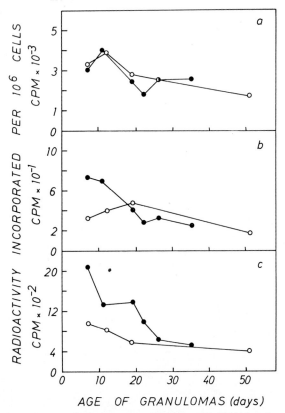

Fig. 4. a, incorporation of ^{14}C-proline into TCA-precipitable proteins; b, conversion of ^{14}C-proline into ^{14}C-hydroxyproline in TCA-precipitable proteins; c, incorporation of ^{35}S-sulfate into CPC-precipitable glycosaminoglycans. The two series of experiments were performed with two different concentrations of radioisotopes.

matrix components, e.g. collagen and glycosaminoglycans. Preparations with different properties can be obtained from granulomas at different ages, and probably also from granulomas produced under various experimental conditions and treatments.

ACKNOWLEDGEMENTS

Our research was supported by grants from the Swedish Medical Research Council (99F-3274; 13X-4) and the Swedish Cancer Society (53).

REFERENCES

Pertoft, H., Bäck, O. and Lindahl-Kiessling, K. (1968). *Exp. Cell Res.* **50**, 355–368.
Viljanto, J. and Kulonen, E. (1962). *Acta Path. Microbiol. Scand.* **56**, 120–126.
Yamada, T. and Ambrose, E. J. (1966). *Exp. Cell Res.* **44**, 634–636.

The Three-dimensional Culturing of Fibroblasts in Collagen

JONATHAN BARD and TOM ELSDALE

Medical Research Council, Clinical and Population Cytogenetics Unit, Western General Hospital, Crewe Road, Edinburgh EH4 2XU, Scotland

It is almost a truism to say that the object of studying cells *in vivo* is to obtain perceptions valid *in vivo*. Culturing fibroblasts on plastic or glass substrata involves removing cells from their natural three-dimensional environment of matrix and other cells and putting them onto an unyielding artificial surface for which they are not prepared and to which they adapt as best they can. To improve on this, we have tried to produce a more normal environment in which to study cells; one that is simple and cheap to use and involves no special skills. The essence of the technique is to prepare routine medium containing collagen in the form of native fibrils so that the mix has the appearance and feel of a gel, although it is in fact an hydrated collagen lattice (HCL) rather than a true gel. The cells that we have used are early subcultures of human diploid fibroblasts grown from lung fragments of 12–14 week foetuses.

This chapter gives the essential details for making HCL's and presents some observations on the behaviour of fibroblasts within them (see also Elsdale and Bard, 1972). It complements earlier work where gels have been used as a substratum on which to grow cells (e.g. Ehrmann and Gey, 1956).

PREPARATION AND PROPERTIES OF HYDRATED COLLAGEN LATTICES

A solution of collagen from rat-tail tendons is prepared essentially by the technique of Wood and Keech (1960) but dialysed twice against one-tenth strength MEM Eagle's medium unit and then centrifuged at 20,000 rpm overnight in a Spinco ultracentrifuge (which both removes debris and sterilises the solution). Precipitation of the collagen is caused by simultaneously raising the ionic strength and the pH of the ice-cold collagen solution to physiological values. This is done by preparing syringes of appropriate quantities of serum, 10 × Eagle's medium and enough $M/7$ NaOH to raise the final pH to about 7·6. (A cell concentrate can be added at this stage if required.) The ingredients are added and the mix rapidly transferred to Petri dishes (3 ml/5-cm dish) where it sets in a few minutes.

These HCL's contain only about 0·1% collagen in the form of 640 Å-periodic native fibrils and, when sectioned and examined in the electron microscope, show little solid material in spite of their superficial firmness. For photographic purposes, the HCL's are transparent.

CELL BEHAVIOUR IN HYDRATED COLLAGENS LATTICES

Fibroblasts treat the HCL's as solid substrata in the sense that the cells within the mix spread and extend rather than stay rounded up (Fig. 1). The fibroblasts tend to show extended bipolar rather than

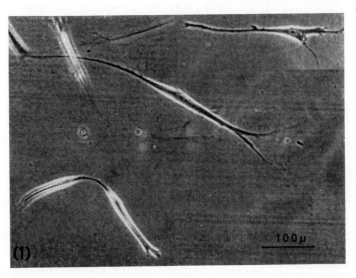

Fig. 1. Fibroblasts in an HCL. Composite of three cells photographed under phase contrast (Elsdale and Bard, 1972).

ruffling-membrane morphology. Time-lapse films show rapid cellular motility in the gels and occasional mitoses (Fig. 2). Thymidine incorporation rates are less for the cells in gels than on plastic. The adhesions between cells and HCL are far stronger than between cells and plastic: cells which have spread on top of an HCL cannot be trypsinised off and cells on plastic move into an HCL overlay (even if this contains as little as 0·005% collagen—a most insubstantial gel) within about 12 hours.

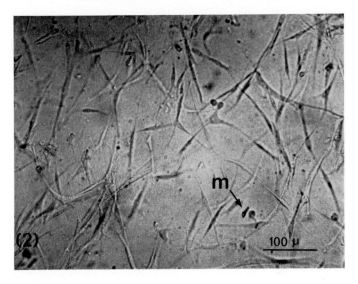

Fig. 2. Fibroblasts in an HCL. From 16-mm time-lapse film taken with a stopped-down 5 × objective to improve depth of focus. Cell density: 2×10^5/ml (m, mitosis).

USES OF THE TECHNIQUE

HCL's have proven helpful in studying other phenomena:

1. They provide a useful substrata for organ cultures: fragments are placed on the surface with an additional few drops of medium to maintain moistness.

2. In measuring fibroblast membrane potentials, one wastes fewer electrodes if the cells are on a yielding substratum.

3. Adult human skin fragments do not adhere well to plastic substrata; when placed within an HCL, the cells will migrate from the fragments and can readily be harvested by breaking up the HCL with bacterial collagenase.

4. Abnormal cells show properties on HCL's which differ from those of more normal cells: for example, SV40-transformed rat fibroblasts cluster on the HCL surface rather than spreading evenly and make unexpectedly weak adhesions to the collagen.

REFERENCES

Ehrmann, R. L. and Gey, G. O., (1956). *J. Natl. Cancer Inst.* **16**, 1375–1403.
Elsdale, T. E. and Bard, J. B. L., (1972). *J. Cell Biol.* (In Press.)
Wood, G. C. and Keech, M. K., (1960). *Biochem. J.* **75**, 588–597.

III

Extracellular Space and Cell Surface

Diffusion Properties of Model Extracellular Systems

B. N. Preston and J. McK. Snowden

Department of Biochemistry, Monash University
Clayton, Victoria, Australia

It is generally agreed that the immediate environment of the cells in connective tissues is the 'ground substance'—a histological term referring to the non-fibrillar extracellular region. It is this region through which nutrients, metabolites, ions and water must pass between the cells and the circulation. The functions of this medium are thought to include a mechanism for homeostasis of the cellular environment, a defence mechanism against the intrusion of micro-organisms and macromolecules, the control of the degree of hydration and the control of the transport and exchange processes of various molecules.

One of the major components of the ground substance is the high molecular weight glycosaminoglycans and it is these materials which are considered to play an important role in the determination of the mechanical and biological properties of the extracellular space. Although there are considerable variations in the amount and type of glycosaminoglycans found in the various connective tissues (Schubert and Hamerman, 1968), the glycosaminoglycans themselves have a number of common features. They all occur as negatively charged essentially linear macromolecules which are largely composed of regularly repeating disaccharide units linked glycosidically to form the polymer. The polyanionic nature of these materials arises from the regularly distributed carboxyl and O-sulphate ester groups which are fully ionized, with Na^+ as the main counterion, under physiological conditions. Of particular interest in present studies are hyaluronic acid—a non-sulphated extremely high molecular weight linear chain molecule and chondroitin sulphate, normally associated with a protein core to form a complex 'bottle brush' type structure (proteoglycan) having molecular weights of the order of millions (Mathews and Lozaityte, 1958). As pointed out by Ogston (1970) both these macromolecules can be regarded as flexible chain-like structures of rather indefinite configuration occupying in solution very large domains. Arising from their diffuse nature, the macromolecules probably exist at physiological concentrations as an interpenetrating mesh of entangled chains (for review see Laurent, 1970) having some of the properties of a microreticular ion-exchange network. In the tissues the polysaccharide molecules are further enmeshed within a collagenous framework and are thus immobilized (Fessler, 1960). It is the complex inter- and intra-molecular interactions that occur within such a network system that may determine the physiological properties of the extracellular space.

Our present interest in the behaviour of the proteoglycan solutions is as a medium for the transport of various types of solutes. In considering the interactions that may occur between the polyanionic chains and diffusing solutes, it is evident that two or more distinct effects may occur. These effects may depend on the one hand on the specific properties of the proteoglycan solution—namely (1) its ability to form a polymeric network and (2) its charged nature. Similarly, the observed effects will also be dependent upon the molecular size and charged nature of the diffusant molecules.

Let us first comment upon the nature of the diffusional process within polymeric networks. In general the polymer matrix has been regarded as giving rise to an 'obstruction' effect. A number of mechanistic models have been proposed by various workers (Pappenheimer et al., 1951; Renkin, 1954; Wang, 1954; Mackie and Meares, 1955; Lauffer, 1961; Newsom and Gilbert, 1964) which attempt to relate the measured diffusion coefficient within a matrix $\bar{D}$, to the diffusion of that solute in the absence of the meshwork D_0.

In the absence of specific binding of the solute to the matrix material, the factors considered can be summarized by the general relationship,

$$\frac{\bar{D}}{D_0} = \frac{(A/A_0) \ (f/f_0)}{h} \qquad (1)$$

where (A/A_0) represents the limitation of the total area available for diffusion; (f/f_0) is a frictional drag factor given by the ratio of the frictional interactions experienced by the solute within the matrix, f, to that experienced in the absence of the matrix f_0; h is a tortuosity factor which allows for the increase in the effective diffusional path length in the presence of the meshwork. The various models differ in respect to which factors are considered to be operative depending upon the size and nature of the diffusing species and upon the model adopted for the diffusion process.

An alternative approach to the problem of diffusion through polysaccharide solutions has been suggested by the recent review articles of Lieb and Stein (1969, 1971). These authors have viewed the permeability properties of biological membrane in the context of diffusion across polymeric networks. They have pointed out the well established, but not widely appreciated fact that diffusion within a complex media such as a polymeric network can be markedly and qualitatively different from that in free solution. They recall the application of Eyring's 'hole' theory of diffusion (1941) to explain the diffusion of hydrocarbons through polyisobutylene membranes (Prager and Long, 1951). This theory considers the probability that a diffusional jump will take place is proportional to the probability that a 'hole' of sufficient size is adjacent to the diffusing species. Furthermore it is possible to consider that arising from the local fluctuations in density in liquids, that even if an individual 'hole' is not large enough to accommodate the diffusant, a co-operative motion may allow two or more 'holes' to merge into a single 'hole' large enough for a diffusional jump to occur. The work of Prager and Long (1951) also indicated that the shape of the diffusing molecule was extremely important to its transport in a network. The studies suggested that an elongated molecule will diffuse almost exclusively along the direction of its major axis since the 'hole' required for movement in this direction is so much smaller than that for diffusion normal to its long axis. Since the concentration of the polymeric networks in connective tissue is dilute compared to those found in biological and synthetic membranes, the usefulness of the 'hole' concept may be greatest in the discussion of the diffusion of macromolecular solutes.

The changes in diffusion brought about by polymer networks are obviously dependent upon the concentration of the meshwork and it is evident that the observed rate of transport of solute *across* a membrane does not represent a simple diffusion process. The overall transport process across a membrane, a permeation process, is usually expressed in terms of the operational quantity, the permeability coefficient P, which is related to the actual diffusional process by

$$P = \lambda \overline{D} \tag{2}$$

where λ is the molar distribution coefficient of the solute between the membrane phase and the external solution.

Let us now consider some specific examples of the diffusional behaviour of solutes through polysaccharide media.

DIFFUSION OF LOW MOLECULAR WEIGHT SOLUTES ACROSS AND WITHIN PROTEOGLYCAN NETWORKS

The diffusion of a number of low molecular weight solutes *across* a gelatin-agarose membrane containing proteoglycan networks was measured using a steady state technique (Preston and Snowden, 1972). It has earlier been shown that the gelatin gel system seems to behave as an inert constraint merely serving to immobilize the proteoglycan network (Preston and Meyer, 1971; Meyer *et al.*, 1971). Preliminary studies indicated that the 'obstruction' effect due to the gelatin-agarose matrix alone caused a slight reduction ($\sim 1.5\%$) in the diffusion of glucose and glycine. However, to eliminate this effect from consideration in

Table I
Diffusion of low molecular weight solutes through gel membranes

Measurements were carried out in $0.1\,M$ NaCl $-$ $1.2 \times 10^{-4}\,M$ NaHCO$_3$, pH 7.0. Results have been corrected to 25°C.

Solute	Conc. in membrane		$P_{25°C} \times 10^7$ $(cm^2 sec^{-1})$	Published $D_{25°C} \times 10^0$ $(cm^2 sec^{-1})$
	Hyaluronate proteoglycan (mg/ml)	Sulphated proteoglycan (mg/ml)		
Glucose	—	—	67.5	67.5*
	1.0	—	69.0	
	2.0	—	64.8	
	—	5.7	66.8	
	—	8.5	65.5	
Sucrose	—	—	53.0	52.0*
	2.0	—	51.8	
	—	5.7	50.0	
	—	8.5	51.6	
Ribose	—	—	85.9	—
	2.0	—	87.7	
	—	8.5	82.4	
Glucosamine hydrochloride	—	—	74.0	76.4**
	—	5.7	75.2	
	—	8.5	80.0	
Glycine	—	—	101.0	106.0*
	2.0	—	98.3	
	—	5.7	101.5	
	—	8.5	96.3	
Caffeine	—	—	80.0	—
	2.0	—	77.4	
	—	5.7	77.4	
	—	8.5	82.2	
Water	—	—	229.0	244.0†
	2.0	—	228.0	
	—	5.7	235.0	
	—	8.5	233.2	

* From Hand book of Chemistry and Physics, 47th Edition. The Chemical Rubber Co., Cleveland, Ohio (1966).
** Corrected to 25°C for the value given at 20°C by Paulson et al. (1951).
† From Wang et al. (1953).

subsequent studies, the membranes were calibrated by measuring the diffusion of glucose or tritiated water assuming a value of $\lambda = 1$ and $\bar{D}$ given by the published diffusional values in water. The results obtained with a number of solutes are given in Table I. Further characterization of the system was achieved by measuring the distribution coefficients of several of the solutes between solutions of proteoglycan (10 mg/ml) and solvent. These are given in Table II. For a limited number of solutes their diffusion *within* the proteoglycan solution was measured in the ultracentrifuge using standard techniques. These results are shown in Table III.

Table II

Distribution of low molecular weight solutes between solutions of sulphated proteoglycan and 0·1 M NaCl

All experiments were performed with an original solute concentration of $2 \cdot 0 \times 10^{-4}$ g/ml and a concentration of the sulphated proteoglycan of 10 mg/ml.

Solute	Distribution coefficient
Glucose	0·98
Sucrose	0·97
Glycine	0·97
Ribose	0·98
Water (^{3}HHO)	0·99
Glucosamine hydrochloride	1·11

Table III

Diffusion coefficients of low molecular weight solutes in the presence of sulphated proteoglycan

All measurements carried out in 0·1 M NaCl at a solute concentration of 5 mg/ml using the synthetic boundary cell in the ultracentrifuge. Results have been corrected to 25°C.

Solute	Proteoglycan conc. (mg/ml)	Diffusion coeff. $\times\ 10^7$ ($cm^2 sec^{-1}$)	D/D_0
Glucose	—	67·4	
	10·0	65·4	0·97
Sucrose	—	51·8	
	10·0	50·8	0·98
Glycine	—	103·0	
	10·0	97·9	0·95

It is evident that in most cases the presence of the polysaccharide network had a negligible effect on the distribution and diffusive behaviour of the solutes. Some of the results require specific comment.

Glucose. The results are in disagreement with earlier reports (Ogston and Sherman, 1961; Buddecke *et al.*, 1967). The explanation of the discrepancy may be associated with possible interactions between the proteoglycans and the millipore membranes used by these workers to contain the polysaccharide solution.

Caffeine. A more puzzling discrepancy is that between the present observations with caffeine and that recently reported by McCabe (1972). Using a system very similar to the present one, he observed a marked reduction in the permeability of this solute from and into agar gels containing hyaluronate.

Glucosamine hydrochloride. The slight increase in the diffusion rate of this solute with increase in concentration of proteoglycan and its distribution in favour of the polysaccharide phase is real and can be understood in terms of a Donnan type distribution (see below) with the solute behaving as a uni-univalent electrolyte.

Water. It is interesting to note that whereas the bulk flow of water is markedly resisted by a polysaccharide network (Preston *et al.*, 1965) the self-diffusional flow appears to be unaffected.

DIFFUSION OF MICRO-IONS ACROSS AND WITHIN PROTEOGLYCANS

Using a tracer steady state technique, the permeability of the gelatin-agarose-proteoglycan membrane to a number of ions have been investigated (Preston and Snowden, 1972). A typical set of results for the diffusion of Na^+ and Cl^- through a sulphated proteoglycan containing membrane is given in Table IV. It is apparent that there are marked effects on the behaviour of the counter- and of the co-ion. There is a systematic increase in the permeability coefficients for Na^+ and a corresponding decrease for Cl^- as the concentration of the external salt is lowered. This behaviour can be understood in terms of a Donnan equilibrium existing with regards to the distribution of *mobile* ions between the membrane and external solution. In this case the value of λ for Na^+ is greater than and that for Cl^- less than unity. If the Donnan equilibrium condition exists then

$$\left[\frac{P}{D}\right]^{1/z} = \text{constant} \qquad (3)$$

where z is the algebraic charge on the diffusible ion. It is evident from Table IV, cols. 4 and 8 that eqn. 3 is seen to hold. However, since the

Table IV

Self-diffusion of Na^+ and Cl^- through gel membranes

D is value obtained in absence of added proteoglycan; P is value obtained by incorporation of sulphated proteoglycan at a concentration of 8.5 mg/ml (charge density of 22.3 m equiv./l.). Solutions were unbuffered at pH 5–7 at 20°C. Results have been corrected to 25°C.

Membrane equilibrated in NaCl (mM)	Na^+			Cl^-				Ideal value $(P_i/D_i)_{1/z}$
	$P \times 10^5$ $cm^2 sec^{-1}$	$D \times 10^5$ $cm^2 sec^{-1}$	P/D	$P \times 10^5$ $cm^2 sec^{-1}$	$D \times 10^5$ $cm^2 sec^{-1}$	P/D	D/P	
100	1.46	1.34	1.09	1.93	2.08	0.93	1.08	1.12
50	1.57	1.34	1.17	1.71	2.04	0.84	1.19	1.25
25	1.82	1.34	1.36	1.57	2.08	0.75	1.32	1.54
10	2.46	1.36	1.81	1.08	2.04	0.53	1.89	2.63
5	3.00	1.38	2.17	0.91	2.04	0.45	2.24	4.67
2.5	3.06	1.34	2.28	0.85	2.05	0.41	2.41	9.03

concentration in the membrane and the chemical composition of the proteoglycan is known, we can predict the changes in the diffusional behaviour of the ions (Table IV, col. 9). It appears that not all the counter-ions of the polyanionic salt can be regarded as 'mobile'. Indeed the behaviour of the Na^+ and Cl^- ions can be explained quantitatively in terms of a certain fraction (about 40%) of the counter-ions being 'bound' and not able to take part in the diffusional process. Subsequent distribution studies have confirmed this counter-ion fixation (Preston et al., 1972). It is interesting to note that the ion fixation acts so as to reduce the effect of the polyanionic network on the diffusional behaviour of the ions. Although the results in Table IV would indicate that the changes in the diffusion rates are small at physiological ionic strength, it should be realised that their magnitude is dependent upon the concentration and charged nature of the polysaccharide material within the meshwork. Calculations suggest that in tissues having a charge density of 0·2 equiv./l and under physiological conditions, the diffusion of Na^+ ion would be approximately doubled. Changes of this order have been observed by Maroudas (1968) on the diffusion of Na^+ through slices of articular cartilage.

The effect of counter-ion fixation on the diffusion process was further investigated by measurements on the self-diffusion of Na^+ ions *within* a polyanionic network. (Preston and Kitchen, unpublished observations), using a modification of the capillary technique of Anderson and Saddington (1949). Since with self diffusion, there is neither net back flow of solvent nor variation in the activity coefficients of the components along the diffusional path, the situation becomes theoretically much simpler. The recent polyelectrolyte theory of Manning (1969a, b) predicts a reduction in the self diffusion of both counter and co-ions in the presence of the charged polymer. This can be understood in that a fraction of the ions are 'partially immobilized' to the polymeric network and do not contribute to the overall rate of self diffusion of the ions. The results of such a study for Na^+ within a dextran sulphate solution is seen in Fig. 1. There is excellent agreement between the observed reduction in the self diffusion of the ions and the theoretical values based upon a counter-ion fixation of between 70–80%, a value suggested from earlier studies (Preston and Snowden, 1972; Preston et al., 1972).

It is evident that the presence of a charged polysaccharide matrix within the interstitial space could have considerable effects on the transport of charged species through it. In the extreme case, a densely charged polysaccharide medium could act as an impermeable barrier to an anionic solute.

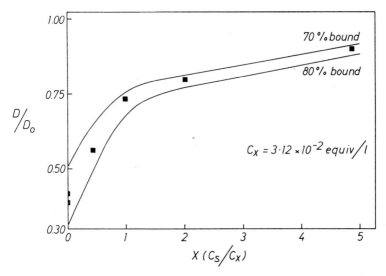

Fig. 1. The relative decrease in the self diffusion of Na$^+$ (■) in the presence of dextran sulphate at varying values of X. X is given by the ratio of concentration of polyion, C_x, expressed in equiv./l to the concentration of added uni-univalent electrolyte (NaCl) C_s. Solid lines calculated according to eqn. 39 of Manning (1969b) with a counter-ion fixation corresponding to 70% and 80%.

DIFFUSION OF MACROMOLECULAR SOLUTES WITHIN PROTEOGLYCAN NETWORKS

Some early observations (Johnston, 1955; Platt et al., 1959; Fessler, 1960) indicated that hyaluronate could decrease the sedimentation rate of some proteins. Laurent and his co-workers (Laurent and Pietruszkiewicz, 1961; Laurent et al., 1963; Laurent and Persson, 1964) carried out a detailed investigation of these effects on the sedimentation of a number of proteins and colloidal particles and concluded that the hyaluronate was acting as a molecular sieve. Only a few attempts have been made (Ogston and Sherman, 1961; Laurent et al., 1963) to study the diffusion of macromolecular solutes in the presence of connective tissue polysaccharides and these investigations again suggested the material was acting as a molecular filter.

From the extensive sedimentation studies Laurent and co-workers demonstrated that the sedimentation of the various solutes in the presence of hyaluronate could be described by the empirical relationship

$$\frac{S}{S_0} = A\, e^{-B\sqrt{C_{HA}}} \qquad (4)$$

where S and S_0 are the sedimentation coefficients measured in the presence and absence of the hyaluronate respectively; C_{HA} is the concentration of the hyaluronate (gm/ml) and A and B are constants for a given solute. The limited diffusional measurements appeared to obey the same empirical relationship. A particular feature of the results was the dependence of the reduction in sedimentation and diffusion upon the square root of the polysaccharide concentration. Furthermore, it was shown that the constant B was directly proportional to the molecular diameter of the diffusing solute.

Our present investigation is an extension of this work by Laurent. We have studied the effect of two different connective tissue polysaccharides—a bovine synovial hyaluronate (molecular weight $\sim 10^7$) and a sulphated proteoglycan from bovine nucleus pulposus (molecular weight $\sim 5 \times 10^5$)—on the diffusional behaviour of a number of globular proteins. In addition, limited studies have been undertaken to observe the effect of the overall shape of the diffusing solute, on its diffusivity in the polymeric network.

The main technique employed in these studies was that used by Laurent et al. (1963) of measurement in the ultracentrifuge with schlieren optics, of the spreading of a boundary formed between a solution containing the polysaccharide and diffusing solute and a solution of equal polysaccharide concentration. It should be noted that in these experimental measurements, the diffusion coefficient of the macromolecular solute is computed by expressing the concentration of the solute in terms of the total volume and by taking the macroscopic dimensions of the diffusion apparatus irrespective of whether or not the diffusion path is blocked microscopically. Thus the steric exclusion effects of polysaccharide networks (for reviews see Laurent, 1970; Ogston, 1970) in limiting the volume within which diffusion can take place, will not explain any changes in the diffusion rates that may be observed since this limited volume factor would apply both to estimates of the flux and of the concentration gradient. We shall discuss the results in terms of the molecular shape and configuration of the diffusing macromolecule.

Globular proteins. The results for the three proteins ovalbumin, serum albumin and γ-globulin are seen in Fig. 2 where a plot of D/D_0 against polysaccharide concentration is made. It is evident that a marked reduction in the diffusion of the proteins occurred in the presence of the network. In agreement with the work of Laurent and his co-workers, the diffusional behaviour of the proteins was compatible with the equivalent empirical relationship,

$$D/D_0 = A^* e^{-B^* \sqrt{C_p}} \qquad (5)$$

where C_p is the concentration of the polysaccharide and A^* and B^* are constants. This can be seen from Fig. 3. The numerical values obtained

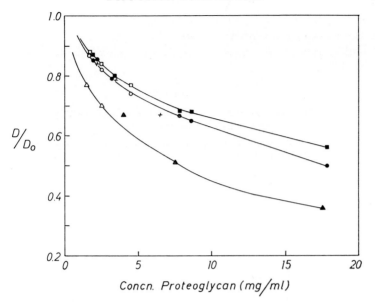

Fig. 2. The relative decrease in diffusion rate of ovalbumin (□, ■), albumin (○, ●) and γ-globulin (△, ▲) in the presence of the hyaluronate (open symbols) and the sulphated proteoglycan (closed symbols). The lines shown were drawn according to eqn. 5 using the values of A* and B* given in Table V. Also included in this figure are values for albumin (+) in hyaluronate (molecular weight $1 \cdot 7 \times 10^6$) from the data of Laurent et al. (1963).

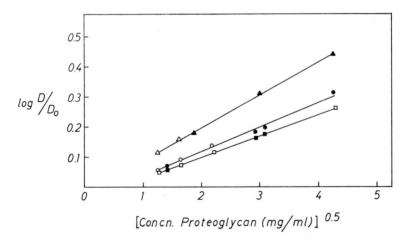

Fig. 3. Plot of log (D/D_0) versus square root of the proteoglycan concentration for ovalbumin (□, ■), albumin (○, ●) and γ-globulin (△, ▲) in the presence of hyaluronate (open symbols) and of the sulphated proteoglycan (closed symbols). The lines were fitted by least means square assuming a linear regression.

for A* and B* are given in Table V together with comparable data from the work of Laurent et al. (1963). From Figs. 2 and 3 it is evident that both polysaccharide preparations were equally effective in restricting the diffusion of these solutes.

Table V
The constants A* and B* for the solutes in the empirical relationship (eqn. 5)

The values given correspond to the proteoglycan concentration being expressed in terms of g/ml to conform with Laurent et al. (1963).

Solute	A*	B*
Ovalbumin	1·08	4·96
Bovine serum albumin	1·13	5·98
	1·03†	6·05†
Human γ-globulin	1·01	8·70
	1·20†	9·94†

† Data of Laurent et al. (1963) from use of sedimentation data and eqn. 4.

Laurent et al. (1963) had discussed these effects in terms of an increased viscous frictional resistance in the polysaccharide media to the translational movement of the macromolecules. However, recent work (Laurent and Öbrink, 1972; Preston et al., 1973) has shown that the rotational movement of albumin in polysaccharide networks is only slightly suppressed, thereby indicating that the frictional drag factor is not an important one in determining the motional behaviour of the protein within dilute network structures.

The particular form of the hindered transport equations (eqns. 4 and 5) has remained unexplained since their formulation some ten years ago. However, Ogston (personal communication), using a random walk model for the diffusional process in which particles move by random steps taken at a certain mean frequency, has shown that the dependence of the restricted diffusion upon the square root of the concentration of the polymeric matrix is consistent with a 'hole' concept. In this treatment the restriction can be regarded as a reduction in the stepping frequency experienced by the solute molecule dependent upon the availability of a 'hole of suitable size' adjacent to it. This model suggests that the hindered diffusion relationship takes for form

$$D/D_0 = e^{-kR\sqrt{C_P}}$$

where R is the radius of the diffusing solute; a relationship which is in good agreement with the above experimental data and that of Laurent et al. (1963).

Random coil molecules. The diffusional behaviour of two neutral random coil macromolecules, a polyethylene glycol (molecular weight 20,000) and a polyvinylalcohol (molecular weight 46,000) as measured in the ultracentrifuge in a range of polysaccharide media was investigated and the results are given in Table VI. An obvious and striking feature

Table VI
Diffusion coefficients of random chain polymers within solutions of polysaccharide material

Measurements were carried out in $0 \cdot 1 M$ NaCl $- 1 \cdot 2 \times 10^{-4} M$ NaHCO$_3$, pH 7·0.

Polysaccharide	Polysacc. conc. (mg/ml)	$D \times 10^7$ $(cm^2 sec^{-1})$	D/D_0
Polyethyleneglycol (Mol. wt. 20,000)			
—	—	4·20	1·0
Sulphated proteoglycan	8·15	10·7	2·5
	16·3	17·0	4·1
	22·0	14·8	3·5
Hyaluronate proteoglycan	3·43	5·76	1·4
Dextran sulphate	5·0	8·80	2·1
	10·0	14·7	3·5
Dextran	5·0	5·88	1·4
	10·0	9·28	2·2
Polyvinylalcohol (Mol. wt. 46,000)			
—	—	2·24	1·0
Sulphated proteoglycan	5·5	4·62	2·1
	11·0	7·67	3·4
Dextran sulphate	10·0	9·34	4·2

of these results is that the polysaccharide matrix appears to cause a very marked increase in the diffusion of these two solutes. Although the data are as yet limited it seems that the increase in diffusion rate is a function of network concentration although the results of the polyethylene glycol-sulphated proteoglycan system may suggest a reversal of the effect has occurred at high proteoglycan concentrations.

In view of the surprising nature of these results and the non-specificity of the measurement of the concentration gradient in the schlieren optical system, a reinvestigation of the diffusional processes using a capillary technique was undertaken. According to this method uniform capillaries, with lower ends sealed, are filled with polysaccharide solution containing radioactively labelled solute and held vertically in a large circulating bath of polysaccharide solution of the same concentration. The labelled solute was allowed to diffuse upward and the concentration of this solute at the upper end of each capillary was kept at zero by stirring the contents of the bath. The results of these studies (Preston and Kitchen, unpublished observations) have confirmed the previous observations.

By analogy with the behaviour of asymmetric molecules in synthetic membranes (Lieb and Stein, 1971), it is tempting to suggest that these random coil structures are moving along their long axis within the polysaccharide network. This may in part explain their diffusional behaviour, but perhaps a point of equal importance is that the molecules would be oriented by the polymeric matrix. The observed increase in diffusion rate appears to be difficult to explain, but if the absence of marked frictional interactions between the diffusants and the polysaccharide matrix (Laurent and Öbrink, 1972) is a general phenomena, the undoubted loss of entropy by the flexible molecules by confining them within a matrix may be converted to an increase in kinetic energy (an increase in the stepping frequency in the random walk model)—the macromolecule tending to 'worm' its way from 'hole' to 'hole' through the network.

It must be emphasized that we are discussing the increased rates of diffusion of solute within a polymeric network. This should not be confused with transport from a solution free of obstacles into and across a layer filled with obstacles such as a membrane. In this latter case the diffusion rates of all types of solutes will be reduced by a reduced area or steric exclusion factor (see eqn. 1).

If these findings are substantiated as being generally applicable to the transport of asymmetric molecules within polymeric networks then such an effect could have considerable biological implications. It would add a new role for the gel-like structures in being able to control the transport rate and degree of orientation of asymmetric molecules. This may be important in explaining such phenomena as the vectorial synthesis of collagen fibres.

From this brief review of the ground substance, it is evident that it will have a considerable effect on the general problem of the transport of materials to and from the connective tissue cells. The behaviour of the low molecular weight solutes is fully understood in view of the known properties of the glycosaminoglycan structures. However, the apparent effect of the polysaccharide matrix on the transport of macromolecules allows for a number of interesting biological mechanisms to be considered.

ACKNOWLEDGEMENTS

The authors are greatly indebted to Professor A. G. Ogston and Professor T. C. Laurent for the many lengthy discussions on this subject; to Professor T. C. Laurent and to the University of Uppsala for facilities in the Institute of Medical Chemistry; to the Swedish Medical Research Council and the Swedish Cancer Society for support; to the Australian Research Grants Committee under whose support much of the experimental studies were carried out.

REFERENCES

Anderson, J. S. and Saddington, K. (1949). *J. Chem. Soc.* 5381.
Buddecke, E., Kroz, W. and Tittor, W. (1967). *Hoppe-Seyler's Z. Physiol. Chem.* **348**, 651.
Eyring, H. (1941). "The Theory of Rate Processes" Ch. 9, S. Glasstone, K. J. Laidler and H. Eyring. McGraw-Hill Book Co., Inc. New York. N.Y.
Fessler, J. H. (1960). *Biochem. J.* **76**, 124.
Johnston, J. P. (1955). *Biochem. J.* **59**, 620.
Lauffer, M. A. (1961). *Biophys. J.* **1**, 205.
Laurent, T. C. (1970). In "Chemistry and Molecular Biology of the Intercellular Matrix". (E. A. Balazs, ed.) Vol. 2, p. 703. Academic Press, London.
Laurent, T. C. and Öbrink, B. (1972). *Eur. J. Biochem.* **28**, 94.
Laurent, T. C. and Persson, H. (1964). *Biochim. Biophys. Acta* **78**, 360.
Laurent, T. C. and Pietruszkiewicz, A. (1961). *Biochim. Biophys. Acta* **49**, 258.
Laurent, T. C., Björk, I., Pietruszkiewicz, A. and Persson, H. (1963). *Biochim. Biophys. Acta* **78**, 351.
Lieb, W. R. and Stein, W. D. (1969). *Nature* **224**, 240.
Lieb, W. R. and Stein, W. D. (1971). *Nature* [*New Biol.*] **234**, 220.
McCabe, M. (1972). *Biochem. J.* **127**, 249.
Mackie, J. S. and Meares, P. (1955). *Proc. Roy. Soc. Lond.* **A232**, 498.
Manning, G. S. (1969a). *J. Chem. Phys.* **51**, 924.
Manning, G. S. (1969b). *J. Chem. Phys.* **51**, 934.
Maroudas, A. (1968). *Biophys. J.* **8**, 575.
Mathews, M. B. and Lozaityte, I. (1958). *Arch. Biochem. Biophys.* **74**, 158.
Meyer, F. A., Comper, W. D. and Preston, B. N. (1971). *Biopolymers* **10**, 1351.
Newsom, B. G. and Gilbert, I. G. F. (1964). *Biochem. J.* **93**, 136.
Ogston, A. G. (1970) In "Chemistry and Molecular Biology of the Intercellular Matrix" (E. A. Balazs, ed.) Vol. 3, p. 1231. Academic Press, London.
Ogston, A. G. and Sherman, T. F. (1961). *J. Physiol.* **156**, 67.
Pappenheimer, J. R., Renkin, E. M. and Borrero, L. M. (1951). *Am. J. Physiol.* **167**, 13.
Paulson, S., Sylven, B., Hirsch, C. and Snellman, B. (1951). *Biochim. Biophys. Acta* **7**, 207.

Platt, D., Pigman, W. and Holley, H. L. (1959). *Arch. Biochem. Biophys.* **79,** 224.
Prager, S. and Long, F. A. (1951). *J. Am. Chem. Soc.* **73,** 4072.
Preston, B. N. and Meyer, F. A. (1971). *Biopolymers* **10,** 35.
Preston, B. N. and Snowden, J. McK. (1972). *Biopolymers* **11,** 1627.
Preston, B. N., Davies, M. and Ogston, A. G. (1965). *Biochem. J.* **96,** 449.
Preston, B. N., Öbrink, B. and Laurent, T. C. (1973). *Eur. J. Biochem.* **33,** 401.
Preston, B. N., Snowden, J. McK. and Houghton, K. T. (1972). *Biopolymers* **11,** 1645.
Renkin, E. M. (1954). *J. Gen. Physiol.* **38,** 225.
Schubert, M. and Hamerman, D. A. (1968). "A Primer on Connective Tissue Biochemistry". Lea and Febiger, Philadelphia.
Wang, J. H. (1954). *J. Am. Chem. Soc.* **76,** 4755.
Wang, J. H., Robinson, C. V. and Edelman, I. S. (1953). *J. Am. Chem. Soc.* **75,** 7666.

Factors Affecting the Biosynthesis of Sulphated Glycosaminoglycans by Chondrocytes in Short-Term Maintenance Culture Isolated from Adult Tissue

O. W. WIEBKIN and HELEN MUIR

*Kennedy Institute of Rheumatology, Bute Gardens
Hammersmith, W6 7DW, England*

Adult cartilage has a limited capacity for repair even though there is a continuous turnover of proteoglycans. Bossman (1968) and Fitton-Jackson (1970) have shown that the chondrocytes of embryonic chicken cartilage in organ culture respond to depletion of the matrix by a rapid resynthesis, suggesting that the local environment of the cells affected their synthetic processes. Depletion of 80% of the chondroitin sulphate of the matrix by hyaluronidase, induced five times the normal rate of synthesis of proteoglycans by such cultures (Hardingham et al., 1972). Since adult cartilage continues to synthesise proteoglycans, the possibility that the synthetic processes of adult chondrocytes may be influenced by macromolecules in their vicinity was studied. Initial results are presented here.

MATERIALS AND METHODS

Serum and all media except Tyrode's solution were obtained from Flow Laboratories, Ayrshire, Scotland. Collagenase, protamine and lysozyme were obtained from Sigma Chemical Co., St. Louis, USA. Hyaluronic acid and Dextran sulphate (M.W. 500×10^3) were obtained from B.D.H. Biochemicals, Poole, Dorset, England. Dextrans of molecular weight 40×10^3, 70×10^3 and 250×10^3 were obtained from Pharmacia Ltd., Uppsala, Sweden. Phytohaemagglutinin was obtained from Miles-Yeda Ltd., Rehovat, Israel, and purified concanavalin A was a gift from Dr. B. Mackler. Chondroitin sulphate was prepared from proteoglycan (extracted from pig laryngeal cartilage) which was digested exhaustively with activated papain, then isolated and purified (Muir, 1958).

Cartilage from pigs about 25 weeks old was used, the tissue from the thyroid plate of one larynx providing sufficient cartilage for several control and experimental cultures. Thus pairs of controls and test cultures could be compared using cells from the same animal to account for variation between animals.

Using sterile precautions, the cartilage was dissected and minced and then digested at 37°C for 12–18 h with crude bacterial collagenase at pH 6·8. The enzyme (2·5 mg/ml) was dissolved in Hank's balanced salt solution containing 8% foetal calf serum, streptomycin and penicillin. About 4–5 ml was used for each 10 g of cartilage. The cells were separated at room temperature from debris and most of the enzyme, by using a Ficoll gradient (Zembala and Asherson, 1970), made up in Hank's solution. The cells were exposed to Ficoll for no more than 1.5 h. They were then washed free of Ficoll with Tyrode's solution by centrifuging at low speed. The cells were incubated for 3–8 days in Leibovitz L-15 medium containing 8% foetal calf serum and streptomycin and penicillin, the medium being replaced once after 24 h. The cells were then transferred to Tyrode's solution, in which $MgCl_2$ replaces $MgSO_4$, but which contained no additives other than 0·25–0·5 mc of carrier free $^{35}SO_4^=$. The polymers to be tested were dissolved in the Tyrode's solution and added in varying amounts to the experimental tubes, while equivalent volumes of Tyrode's solution were added to the controls. The cell suspensions containing 10^5 cells/ml were incubated at 37°C for 2 h while being gently agitated. The cells were then centrifuged and washed twice by resuspending in 1 ml of cold Tyrode's solution containing no $^{35}SO_4^=$. The washings were added to the supernatants. The cells were resuspended in Tyrode's solution, and using an aliquot of the suspension, cell viability was tested by dye exclusion using a haemocytometer. Cells showing less than 90% viability were rejected.

Carrier chondroitin sulphate and an equal volume of 2% cetyl pyridinium chloride (CPC) were added separately to the supernatant and to the cells, together with some Kieselguhr and each stirred vigorously at intervals during 2 h at 25–30°C. Complete cell disruption was confirmed by microscopic examination of an aliquot of the suspension. The CPC precipitates and Kieselguhr were centrifuged and washed five times by resuspending in 5 ml of 0·5% CPC. The precipitates were then redissolved in either $1·25 M$ $MgCl_2$ or propanol and the radio-activity determined using a scintillation counter. The radioactivity of the solution was not reduced by dialysis, and there was negligible radioactivity when the cells had been killed by freezing and thawing before incubating with $^{35}SO_4^=$.

Previous experiments had shown that when the radioactive material after treatment with $0·33 N$ NaOH was examined by micro-column fractionation as CPC complexes (Antonopoulos et al., 1964), the radioactivity was mainly eluted by $1·25 M$ $MgCl_2$ and a little by 1% CPC, suggesting that most of the radioactivity was incorporated into chondroitin sulphate and a little into keratan sulphate.

RESULTS

The results show (Table I) that after 2 h about one third of radioactivity incorporated by the cells was still present within the cells. This

Table I

The effect of hyaluronic acid and chondroitin sulphate in the medium on the incorporation of [^{35}S]-sulphate into glycosaminoglycans by adult chondrocytes *in vitro*

Glycosaminoglycan in medium	Conc. µg/ml	Total radioactivity incorporated % of control	% of total radioactivity cell bound	No. of experiments
None (control)	0	100	32·8 ± 3·5	61
Hyaluronic acid	8·0	33·8 ± 15·6	50·5 ± 7·2	8
	0·8	44·8 ± 7·4	48·5 ± 5·3	4
	0·08	62·9 ± 3·4	39·6 ± 4·7	4
Chondroitin 4-sulphate	6000	135·8 ± 19·0	7·0 ± 1·2	7
	600	102·5 ± 7·7	6·6 ± 2·3	7
	2–300	137·2 ± 20·1	7·6 ± 0·7	15
	1–200	103·2 ± 6·8	19·7 ± 6·5	12
	80	115·8 ± 5·8	79·0 ± 10·0	9
	40	161·5 ± 9·1	86·0 ± 14·0	6
	20	100·0 ± 10·0	39·0 ± 5·0	4
	10	110·0 ± 1·2	37·7 ± 9·3	6

agrees with results of pulse/chase experiments *in vitro* using cartilage slices from a similar source (Hardingham and Muir, 1972a), when radioautography showed that after a 3 h chase with unlabelled sulphate, some of the radioactivity was still within the cells. The rate of export of labelled material into the medium from isolated cells cultured for several days, was therefore comparable with that observed after incubation of fresh cartilage slices. Previous experiments had shown that cells in contact as a pellet, or in suspension, did not affect the proportion of radioactivity retained by isolated cells during 2 h incubation period.

The presence of small amounts of various polysaccharides including dextrans of different molecular weight increased somewhat the proportion of material retained by the cell without significantly affecting the total incorporation. Dextran sulphate had a similar effect as the uncharged dextrans.

Free chondroitin sulphate chains had a noticeable effect between 40–80 μg/ml by increasing the amount of labelled material retained by the cell to over 80%, without affecting significantly the total incorporation (Table I). At higher levels from 200 μg/ml upwards however, the amount retained by the cell was greatly reduced, while total incorporation was unaffected.

In contrast with all other material tested, hyaluronic acid markedly inhibited the total incorporation of radioactive sulphate, even at very low concentrations (Table I), and it also decreased the export of labelled material from the cell. This effect may be connected with the specific interaction of hyaluronic acid with chondroitin sulphate proteoglycans reported recently (Hardingham and Muir, 1972b), and with the association of hyaluronic acid with cell surfaces (Kraemer, 1971).

However, purified concanavalin A and phytohaemagglutinin, although known to interact with cell surfaces, did not inhibit total incorporation, indeed concanavalin A had the opposite effect and markedly increased total incorporation at high concentration, and reduced the amount of cell bound radioactivity (Table II). Phytohaemagglutinin did not have the same effect, however (Table II).

There is a considerable amount of lysozyme in cartilage (Kuettner *et al.*, 1968), but its function is unknown. The addition of egg white lysozyme to the cultures had little effect except at high concentrations, when there was some increase in total incorporation (Table II). As this effect was not shown by similar concentrations of protamine (Table II), which is also a basic protein, it may be a specific effect of lysozyme which might be more evident with the lysozyme of cartilage itself.

It would appear from these results that adult chondrocytes in culture

Table II

The effect of mitogens and basic proteins in the medium on the incorporation of [^{35}S]-sulphate into glycosaminoglycans by adult chondrocytes *in vitro*

Protein or mitogen in medium	Conc. µg/ml	Total radioactivity incorporated % of control	% of total radioactivity activity cell bound	No. of experiments
None (control)	0	100	32·8 ± 3·5	61
Lysozyme	17,000	215 ± 22·1	20·3 ± 4·3	4
	2–6000	166 ± 12·5	28·8 ± 6·7	6
	1–200	115 ± 14·3	24·5 ± 2·3	6
	10–20	164 ± 9·1	42·8 ± 2·9	6
	1	112 ± 7·2	39·0 ± 7·0	2
Protamine	14,000	83 ± 13·0	42·6 ± 3·4	4
	8–9000	72 ± 18·2	46·7 ± 6·3	4
	130	106 ± 7·3	35·6 ± 1·2	4
	1·3	117 ± 3·9	45·0 ± 2·0	4
Concanavalin A	12,500	275 ± 15·2	17·2 ± 2·0	4
	6200	225 ± 17·1	21·4 ± 7·0	4
	25	170 ± 19·1	26·8 ± 5·0	4
	2·5	98 ± 8·5	30·1 ± 5·2	4
Phytohaemagglutinin	200	120 ± 7·5	14·8 ± 1·4	4
	20	110 ± 6·8	17·7 ± 3·7	4

respond in a specific way to various macromolecules in the culture medium, and that the inhibitory effect of hyaluronic acid may have some function in controlling synthesis of proteoglycans, since the cartilage used in these experiments contains about 0·05% wet weight of hyaluronic acid (Hardingham and Muir, 1973).

REFERENCES

Antonopoulos, C. A., Gardell, S., Szirmai, J. A. and van Bouen-De-Tyssonsk, E. R. (1964). *Biochim. Biophys. Acta* **83,** 1–19.
Bosmann, H. B. (1968). *Proc. Roy. Soc. Lond. Biol.* **169,** 399–425.
Fitton Jackson, S. (1970). *Proc. Roy. Soc. Lond. Biol.* **175,** 405–453.
Hardingham, T. E. and Muir, H. (1972a). *Biochem. J.* **126,** 791–803.
Hardingham, T. E. and Muir, H. (1972b). *Biochim. Biophys. Acta* **279,** 401–405.
Hardingham, T. E. and Muir, H. (1973). *Biochem. Soc. Trans.* **1,** 282.
Hardingham, T. E., Fitton Jackson, S. and Muir, H. (1972). *Biochem. J.* **129,** 101–112.
Kraemer, P. (1971). *Biochemistry* **10,** 1437–1451.
Kuettner, K. E., Guenther, H. L., Ray, R. D. and Schumacher, G. F. B. (1968). *Calcif. Tissue Res.* **1,** 298–305.
Muir, H. (1958). *Biochem. J.* **69,** 195–204.
Zembala, M. and Asherson, G. L. (1970). *Immunology* **19,** 677–681.

The Effect of Hyaluronic Acid on Fibroblasts, Mononuclear Phagocytes and Lymphocytes

E. A. BALAZS and Z. DARZYNKIEWICZ

Department of Connective Tissue Research
Boston Biomedical Research Institute
20 Staniford Street, Boston, Mass. 02114, USA

In the multicellular organism cells move and multiply on the solid surfaces of highly hydrated macromolecular structures such as: (a) the cell coatings of other cells, (b) basal laminae, (c) network or bundles of collagen fibrils and (d) fibrin coagulums in inflammatory and repair processes.

All the surfaces on which cells move in nature are highly hydrated unlike the plastic, glass or denatured protein membranes used as solid support in tissue culture methods. Most of these surfaces are engulfed by a viscoelastic liquid which contains proteoglycans and hyaluronic acid. Therefore, the question may be asked whether or not the polyanionic macromolecules of the intercellular matrix exert any influence on the cells moving and multiplying on these solid surfaces. The effect of heparin and other polyanions of high negative surface charge on cell activities is well known (Balazs and Jacobson, 1966; Regelson, 1968).

The cells of the lymphomyeloid system: lymphocytes, mononuclear phagocytes and granulocytes pass through several compartments of the intercellular matrix of the connective tissue during their life cycle. It is this matrix which surrounds these cells when they are stored, when they undergo mitosis and when they carry out their biological function. In inflammation, these cells pass through the perivascular intercellular matrix to reach the target area and, in most cases, the granulation process is located in the matrix. Thus, the question arises whether or not some of the macromolecular components of the matrix exercise any control over the activity of these cells.

Since the structural specificity of the intercellular matrix of the connective tissue originates primarily from the qualitative and quantitative variations of its proteoglycan and glycosaminoglycan content, one may suspect that these polyanionic macromolecules are instrumental in some control processes.

This paper presents a brief summary of the experimental work carried out in our laboratory in the past few years to test the effect of hyaluronic acid and some other extracellular glycosaminoglycans on some of the biological activities of the cells of the lymphomyeloid system.

HYALURONIC ACID AS A CELL IMMOBILIZING AGENT OF THE LYMPHOMYELOID SYSTEM*

Cells of the lymphomyeloid system and lymphoid tumor cells move very fast *in vitro* on solid plastic or glass surfaces or on top of each other. When these cells are suspended in tissue culture media and packed by centrifugation into a hematocrit tube (non-heparinized) and this tube is placed into a small plastic chamber which contains tissue culture maintenance media with or without calf serum, the cells migrate rapidly from the tube into the chamber covering a semicircular migration area. This migration is the result of a cellular motility and does not occur at low ($<8°C$) temperatures, or when a large percentage of the packed cells are dead. Within 10 to 16 h an equilibration is reached and a quantitative measure of the cell migration can be given by counting the number of cells in the chamber. The effect of macromolecules present in the medium on cell migration can be quantitatively measured and expressed as cell motility index which is the ratio of the number of cells migrated into the chamber in the presence and in the absence of the macromolecule tested.

The motility index of various lymphocytes and lymphosarcoma cells (YAC) as a function of the concentration of hyaluronic acid in the medium is shown in Fig. 1. Among these cells the most sensitive are the lymphocytes from thymus and spleen and the least sensitive, the blood lymphocytes.

The *in vitro* movement of mononuclear phagocytes (macrophages) collected from the surface of lung alveoli and from the peritoneal cavity, and blood granulocytes are also inhibited by hyaluronic acid, but in order to obtain the inhibition, as with lymphocytes, a higher hyaluronic acid concentration must be used.

* These data are taken from the unpublished work of E. A. Balazs, S. Friberg and M. Freeman.

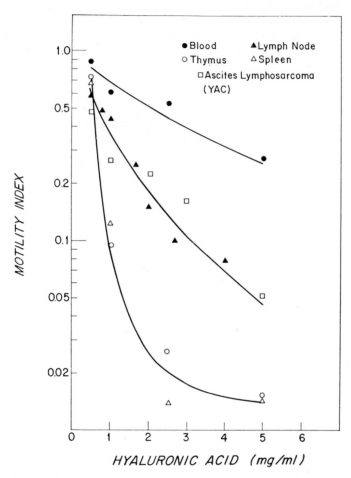

Fig. 1. The motility index of various lymphoid cells as the function of the concentration of hyaluronic acid in the medium. The medium used is Eagle's Basal Medium in Earle's Balanced Salt Solution supplemented with 10% calf serum and containing penicillin and streptomycin.

Only high polymeric hyaluronic acid acts as a cell immobilizing agent. Table I shows that the oligosaccharides of hyaluronic acid and the low molecular weight ($<50,000$) polymer are not effective. Similarly, other low molecular weight ($2 - 4 \times 10^4$) glycosaminoglycans such as chondroitin-4 and 6-sulfates, keratan sulfate, dermatan sulfate, heparan sulfate, and heparin do not influence the motility of the lymphomyeloid cells *in vitro*.

Table I

The motility index of mononuclear phagocytes and lymphocytes in the presence of N-acetylglucosamine and hyaluronic acid of various molecular size

Addition (4 mg/1ml media)	Mononuclear phagocytes	Lymphocytes (rat lymph node)
N-acetyl glucosamine	1·10	0·90
Hyaluronic acid:		
Tetrasaccharide	0·95	0·98
Hexasaccharide	1·00	0·99
Polysaccharide (MW 1×10^4)	0·98	0·95
Polysaccharide (MW 2×10^6)	0·01	0·02

Motility index is the ratio of the number of cells migrated from the capillary into the chamber in the presence of the test substance to those migrated in the absence of the test substance. Cell migration experiments were carried out at 37°C for 16–24 h.

Since these glycosaminoglycans have a higher charge density than hyaluronic acid one can rule out the possibility that the mechanism of the effect of hyaluronic acid is based on its polyanionic character.

The viscosity of the medium in which these cells move is an important factor. Viscous solutions of high polymeric DNA also inhibit the movement of the cells of the lymphomyeloid system. This suggests that the movement of these cells can be regulated by the viscosity of the medium. Indeed, the movement of these cells is completely inhibited when they are embedded in a fibrin or collagen gel.

Not only pure hyaluronic acid solutions, but natural fluids containing hyaluronic acid inhibit the movement of the lymphomyeloid cells *in vitro*. We found that synovial fluid and liquid vitreous (free of collagen) both containing high polymeric hyaluronic acid inhibit the movement of the cells of the lymphomyeloid system. When the hyaluronic acid is degraded by hyaluronidase or by oxidation-reduction systems (ascorbic acid + Cu^{++} + O_2) the inhibition effect is abolished (Fig. 2).

Testing the cell inhibition effect of synovial fluids and liquid vitreous we could demonstrate that the inhibition effect of these fluids is much greater than one would expect from the concentration of hyaluronic

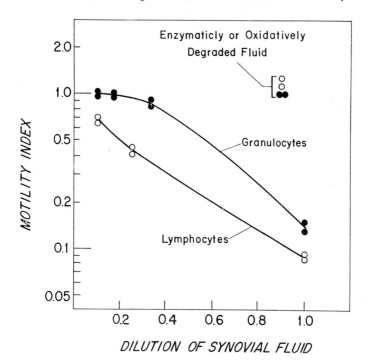

Fig. 2. The effect of horse synovial fluid on the motility index on blood lymphocytes and granulocytes obtained from the same animal. Enzymatic degradation of the hyaluronic acid in the synovial fluid was carried out by $\beta 1 \rightarrow 3$ hyaluronic acid hydrolase (prepared from leeches). Oxidative degradation of hyaluronic acid was carried out by incubation with copper sulfate and ascorbic acid in the presence of atmospheric oxygen. After degradation, the fluid was dialyzed against the medium.

acid in them (Table II). Since this effect is completely abolished when the hyaluronic acid is degraded with a specific hydrolytic enzyme ($\beta 1 \rightarrow 3$ hyaluronic acid hydrolase, prepared from leeches) one has to assume that the hyaluronic acid in the natural state has a specific quality which is partially altered during purification (conformational change) or a specific cofactor is present in these fluids which enhances the effect of hyaluronic acid.

In conclusion, hyaluronic acid is a cell immobilizing agent of lymphomyeloid cells and this effect is specific in the sense that only these

Table II
Inhibition of the *in vitro* motility of lymphocytes and mononuclear phagocytes by the liquid vitreous of the Owl Monkey

Addition	Hyaluronic acid $[\eta]/cm^3 g$	mg/ml	Mononuclear phagocytes (rat peritoneal)	Lymphocytes (rat lymph node)
Vitreous*	4500	0·46	0·06	0·09
Hyaluronic acid	3400	0·50	0·70	0·60

* The liquid vitreous of the owl monkey contains 0·06–0·09 mg/ml proteins also.

types of cells are affected and various cell types of this system show different sensitivity. The effect is specific also from the point of view of hyaluronic acid because it depends on the molecular size and probably the conformation of the large polysaccharide molecule. Proteoglycans in this system have not been tested systematically but preliminary experiments indicate that some proteoglycans may have similar effects to those of hyaluronic acid. Since hyaluronic acid-containing fluids such as the vitreous and synovial fluid under normal conditions contain very few cells and are extremely potent inhibitors of *in vitro* cell motility, one is tempted to speculate that hyaluronic acid is a natural regulator of the migratory movement of lymphomyeloid cells in the intercellular matrix.

EFFECT OF HYALURONIC ACID ON FIBROBLASTS

When fibroblasts from embryonic tissue as first explants or established cell lines are seeded over a very viscous tissue culture medium, the viscosity of which is from its hyaluronic acid content (0·5–5 mg/ml) the cells slowly settle on the bottom of the plastic or glass dish. Once the cells reach the solid surface and adhere to it they migrate and multiply and the number of clones formed (plating efficiency) is the same in the presence and absence of hyaluronic acid. The plating efficiency of fibroblasts seeded on the top of collagen gels is also unaffected by viscous hyaluronic acid solutions present in the culture medium. Very viscous ($MW > 1 \times 10^6$, concentration > 5 mg/ml) hyaluronic acid solutions, however, inhibit the movement and multiplication of fibroblasts by preventing them from reaching the solid surface. Unadhered, floating, these cells slowly die.

Thus, hyaluronic acid can assert an important control over fibroblast activity provided it can prevent the cells from adhering to a solid tissue surface; but once a fibroblast is attached to a solid surface, hyaluronic acid in these model systems does not seem to affect their motility and multiplication.

INFLUENCE OF HYALURONIC ACID ON LYMPHOCYTE STIMULATION (Darzynkiewicz and Balazs, 1971)

Human lymphocytes obtained from peripheral blood were separated from erythrocytes and polymorphonuclear leukocytes, and were suspended in tissue culture media and stimulated with phytohemagglutinin (PHA), pokeweed mitogen (PWM), streptolysin O (SLO) or purified protein derivative of tuberculin (PPD). The modulation of the lymphocytes to lymphoblasts was assayed by morphological criteria (percent of lymphoblasts) and by autoradiographic detection of ^{3}H-thymidine incorporation into the nucleus of the divided cells (percent of DNA-synthesizing cells). When hyaluronic acid is added to the incubation media of the stimulated cells, the blastoid modulation of the cells can be prevented completely. This inhibition of the blastoid modulation depends on the concentration and on the size of the hyaluronic acid molecule.

Hyaluronic acid of high molecular weight shows much greater effect than the lower molecular weight polymers. Hyaluronic acid with a molecular weight of 2.0×10^6 can suppress blastoid modulation at concentrations as low as 0·25 mg/ml (Fig. 3).

Other factors also play a role in the hyaluronic acid caused inhibition of blastoid modulation. Namely, the inhibitory effect of hyaluronic acid depends on the distance between the cells adhered to the glass and the type of mitogen used. In dense cultures the inhibitory effect of hyaluronic acid is less than in cultures of low cell densities. It appears that crowding of lymphocytes counteracts the suppressive effect of hyaluronic acid.

At the same cell densities the suppressive effect of hyaluronic acid is more pronounced in cultures stimulated by PWM, SLO or PPD than by PHA. This probably can be explained by the high leukoagglutinating properties of PHA. In the presence of PHA, aggregates of lymphocytes (local cell crowdings) occur and it is known that the frequency of blastoid modulation in these agglutinates is higher than among unagglutinated cells.

The suppression of lymphocyte stimulation induced by many factors (i.e. corticosteroids) is related to their toxic effect on lymphocytes. This, however, is not the case for hyaluronic acid, since the cells' viability remains unchanged in cultures containing high concentrations of hyaluronic acid. Also, lymphocytes preincubated in hyaluronic acid and then washed and transferred into media which do not contain hyaluronic acid respond to mitogen stimuli as well as lymphocytes that never have been in contact with hyaluronic acid.

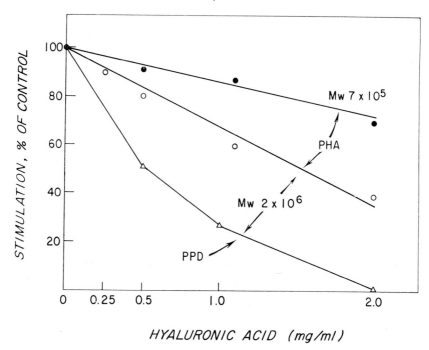

Fig. 3. Effect of hyaluronic acid of different molecular weight on blastoid modulation of lymphocytes. ●, lymphocytes stimulated by PHA, cultured in the presence of hyaluronic acid of low molecular weight; ○, PHA-stimulated lymphocytes, cultured in the presence of hyaluronic acid of high molecular weight; △, stimulation in cultures treated with PPD, in the presence of hyaluronic acid of high molecular weight. The stimulation was assayed autoradiographically by counting the cells which incorporated ^{3}H-thymidine.

The influence of hyaluronic acid on lymphocyte stimulation is not related to the possible protection of cells against binding mitogens. The suppression of lymphocyte stimulation may be induced by hyaluronic acid after the initial events of stimulation (binding of mitogen to cell membrane, 'genome activation') have occurred.

Once the stimulated lymphocytes advanced to the stage of DNA synthesis, hyaluronic acid is not effective. Hyaluronic acid was found to be totally ineffective on G_2, S and M phases of stimulated lymphocytes.

The effect of hyaluronic acid seems to be limited only to the 'inductive phase' of lymphocyte stimulation ($G_0 - G_1$ phase of the first generation cycle); the duration of this phase is markedly extended in the presence of hyaluronic acid.

The mechanism of interaction between hyaluronic acid and stimulated lymphocytes by which the stimulation is inhibited remains unknown. We presume that hyaluronic acid of high molecular weight interferes with cell-cell contacts since it impedes lymphocyte movement as it was shown in experiments on lymphocyte migration. The active movement of lymphocytes in cultures is probably a necessary factor for establishing cell-cell contacts in the absence of cell agglutination. Abundant experimental evidence indicates that cell-cell contacts (interaction) are essential in the stimulation of lymphocytes (Peters, 1972).

EFFECT OF HYALURONIC ACID ON GRAFT VERSUS HOST REACTION*

Since lymphocyte migration and proliferation are essential in graft versus host reaction, and since hyaluronic acid appears to suppress both types of lymphocyte activities *in vitro* we tested the effect of hyaluronic acid on the development of this reaction in mice. Graft versus host reaction was produced by intraperitoneal injection of spleen cells from A/Jax adult mice into $(C_{57}G_1/6J \times A/Jax)F_1$ male hybrids.

Before injection the donor's cells were suspended in tissue culture medium (Minimum Essential Medium) or in the same medium containing 7 mg hyaluronic acid per 1 ml medium. $2 \cdot 5 \times 10^7$ living cells suspended in 0·7 ml medium was injected intraperitoneally. Four different hyaluronic acid preparations were used with limiting viscosity numbers of 193, 1100, 2300 and 3400 cm^3/g, which correspond to molecular weights of 6×10^4, 5×10^5, $1 \cdot 4 \times 10^6$ and $2 \cdot 2 \times 10^6$, respectively. Ten days after injection of the cells, the mice were killed, the body weight and spleen weight were measured and the relative spleen weight (ratio of spleen weight to body weight) was calculated. The spleenic index was calculated for each experimental animal by dividing its relative spleen weight by the mean relative spleen weight of 6 non-injected mice of the same strain, age and sex. The results presented in Fig. 4 are expressed as a ratio of spleenic index of animals which received cells suspended in hyaluronic acid to the spleenic index of mice injected with cells without hyaluronic acid. Hyaluronic acid of low molecular weight seems to enhance graft versus host reaction, whereas

---

* These data are taken from the unpublished work of E. A. Balazs, E. Skopinska and Z. Darzynkiewicz.

hyaluronic acid of high molecular weight ($> 1.4 \times 10^6$) has a very distinctive suppressive effect. The increase of spleen weight was completely prevented with high molecular weight hyaluronic acid

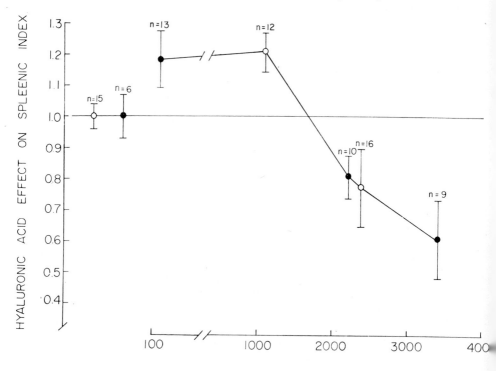

Fig. 4. The effect of hyaluronic acid preparations of various limiting viscosity numbers on the graft versus host reaction. The ordinate represents the ratio of the spleenic index of mice which received parental cells with hyaluronic acid to the spleenic index of mice which received cells without hyaluronic acid. The abscissa is the limiting viscosity number of the various hyaluronic acid preparations used. The vertical bars represent the standard error of the mean and n the number of animals used in each group. The filled and open circles indicate two different experimental series with two different control groups (the two points on the left).

(2.2×10^6). These results were verified by histological examination of the spleens and livers of the host animal. All animals which were inoculated with cells suspended in this hyaluronic acid preparation had perfectly normal spleens and livers. The control mice, receiving the same number of cells in the absence of hyaluronic acid or in the presence of hyaluronic acid of low molecular weight (6×10^4; 5×10^5) had extensive perivascular infiltration of mononuclear and pyroninophilic cells in the spleen and in the liver.

EFFECT OF HYALURONIC ACID ON LYMPHOCYTE CYTOTOXICITY*

Lymphocytes from individuals immunized with allogeneic cells destroy such (target) cells *in vitro*. Also, lymphocytes stimulated by non-specific mitogens are cytotoxic to allogeneic cells provided that contact between lymphocytes and target cells is established. The active movement and the modulation of lymphocytes are believed to be essential in this phenomenon. Since both of these lymphocyte activities are suppressed by hyaluronic acid, we tested the cytotoxic effect of lymphocytes in the presence of hyaluronic acid. Two types of experiments were performed. In the first type, the monolayer of allogeneic fibroblasts were used as target cells and PHA-stimulated lymphocytes were the effector cells. In the absence of hyaluronic acid, PHA-stimulated lymphocytes destroyed cell monolayers during 2 days of culturing. The destruction of monolayers was completely inhibited when PHA-stimulated lymphocytes were applied on monolayers in the presence of high molecular weight hyaluronic acid at concentrations of 2 mg/ml and higher.

In the other type of experiments, the effect of haluronic acid on cytotoxicity of sensitized mouse spleen cells against tumor cells was tested quantitatively by using ^{51}Cr-labeled target cells and measuring the release of ^{51}Cr. In the test tube, the effector cells or the target cells were placed in suspension. Over this cell suspension hyaluronic acid dissolved in tissue culture medium was layered. On the top, suspension of target, or effector cells were placed and the test tube was incubated without mixing or shaking. Thus, the effector and target cells were separated by the layer of hyaluronic acid. The destruction of the target cells was completely prevented when high molecular weight hyaluronic acid in 1·3 mg/ml or higher concentration was present.

EFFECT OF HYALURONIC ACID ON SURVIVAL OF SKIN ALLOGRAFTS†

In light of the observation that hyaluronic acid interferes with lymphocyte migration, proliferation and cytotoxicity we studied the effect

* These data are taken from the unpublished work of E. A. Balazs, S. Friberg and Z. Darzynkiewicz.

† These data are taken from the unpublished work of E. A. Balazs, E. Skopinska and Z. Darzynkiewicz.

of hyaluronic acid on survival of skin allografts. Balb/c male, 2–3-month old mice were used as skin graft recipients and $C_{57}BL/6J$ adult male mice were skin donors. Hyaluronic acid was administered either topically: 1–2 mg hyaluronic acid on the graft bed at the time of transplant and then 1–3 mg of hyaluronic acid injected subcutaneously under and around the graft area; or intraperitoneally: 5 mg of hyaluronic acid every second day after transplantation. Hyaluronic acid of high molecular weight (2.2×10^6) was used. Table III shows that the

Table III
The effect of hyaluronic acid treatment on
the survival time of allogeneic skin grafts in mice

Hyaluronic acid treatment	No. of mice used	Mean survival time (days ± SEM)	Significance (t-test)
None	11	10.8 ± 0.4	
Topical	14	12.9 ± 0.7	<0.05
Topical	13	13.0 ± 0.4	<0.001
Intraperitoneal	10	11.1 ± 0.5	n.s.

topically administered hyaluronic acid significantly extended the time of survival of skin allografts. Intraperitoneally administered hyaluronic acid was ineffective indicating that the effect is local and probably based on the same mechanism as *in vitro* systems, viz. on the lymphomyeloid cell immobilizing effect.

THE EFFECT OF HYALURONIC ACID ON SOME METABOLIC ACTIVITIES OF FIBROBLASTS AND MONONUCLEAR PHAGOCYTES *in vitro**

The effect of hyaluronic acid on the metabolism of cell surface glycoproteins (trypsin removable and trichloroacetic acid-insoluble), extracellular glycoproteins and hyaluronic acid was tested in fibroblast cultures (from monkey subcutaneous tissue). The metabolism of these macromolecular fractions was studied by incubating the cells adhered to plastic surfaces for 23 h with 6-^{3}H-glucosamine in the culture media, and separating these fractions according to standard methods. When the culture media contained relatively low concentration (0.1 mg/ml) of hyaluronic acid the specific activity of ^{3}H in all these three fractions

* These data are taken from the unpublished work of B. Jacobson and E. A. Balazs.

was several-fold higher than in cell cultures incubated under the same conditions without hyaluronic acid. When the same hyaluronic acid preparation was added to the fibroblast cultures in a concentration of 2 mg/ml all three fractions showed a considerably lower specific activity than the controls.

When the same experiments were carried out with mono-nuclear phagocytes obtained from the rat peritoneum or rabbit alveolar space the results obtained were different in some respects. In the presence of 2 mg/ml hyaluronic acid in the media the incorporation of the ^{3}H into the cell surface glycoproteins was suppressed. At the same time the specific activity of the hyaluronic acid in the medium was significantly higher. The presence of hyaluronic acid did not cause an increase in the number of cells dead during incubation and control experiments also showed that the 6-^{3}H-glucosamine uptake by the cells was not impaired by the presence of hyaluronic acid.

The presence of radioactively labeled hyaluronic acid in the culture media of mononuclear phagocytes was confirmed by cetylpyridinium chloride fractionation, hyaluronidase treatment and identification of ^{3}H-glucosamine. Furthermore, using UDP-^{14}C-glucuronic acid, the homogenate of mononuclear phagocytes contained glucuronosyl transferase activity and the undialyzable labeled product was susceptible to degradation with $\beta 1 \rightarrow 3$ hyaluronic acid hydrolase (from leeches).

These experiments show that hyaluronic acid present in the environment of cells has a specific effect on the metabolism of cell surface glycoproteins and extracellular glycoproteins and on hyaluronic acid synthesis itself. Furthermore, these findings provide evidence that mononuclear phagocytes can synthesize hyaluronic acid *in vitro*. It was reported previously that the mononuclear phagocytes of the vitreous (hyalocytes) can also synthesize hyaluronic acid (Österlin and Jacobson, 1968).

It is possible that the extracellular hyaluronic acid exerts a regulatory function on the regeneration of cell coating glycoproteins and by this action influences the movement and activity of the cells of the lymphomyeloid system. These experiments also suggest the possibility of a feedback mechanism in the regulation of the formation of extracellular hyaluronic acid by fibroblasts and macrophages.

EFFECT OF HYALURONIC ACID ON CARTILAGE, TENDON AND FASCIA WOUNDS (Rydell and Balazs, 1971)

Connective tissue surfaces which are not covered by epithelium and are subject to displacement during the movements of the musculoskeletal systems are covered with hyaluronic acid molecules and separated by an elastoviscous fluid that contains hyaluronic acid. Such surfaces (synovial membranes, articular cartilage, fasciae, tendon sheaths and tendons) when wounded have a tendency to support heavy granulation tissue formation (pannus) which can develop to fibrous tissue coupling of two adjacent tissue surfaces (adhesions) and thereby interfere with free movement and proper function. In view of the *in vitro* and *in vivo* biological activities of hyaluronic acid described above, we proposed the hypothesis that the presence of hyaluronic acid on the connective tissue surfaces is not to ensure less friction, but to prevent the growth together of these surfaces and, after injury, to control cell invasion and prevent the formation of excess fibrous tissue. To test this hypothesis the dorsal fascia of rabbits, guinea pigs and owl monkeys, the articular cartilage of the knee joints of dogs and owl monkeys and the long extensor tendon of the leg of rabbits were traumatized by light mechanical damage and the effect of high molecular weight sterile, pyrogen-free hyaluronic acid applied directly on the wounded surfaces was studied. In all of these experimental model systems the granulation tissue and subsequent fibrous tissue formation was considerably less in the hyaluronic acid-treated wounds than in the controls. A development of adhesions between the wounded tendon and tendon sheath was significantly less when hyaluronic acid was applied to the wounded surface in the form of viscous paste (10 mg/ml) or dried membranes.

To test the effect of hyaluronic acid on the development of granulation tissue around foreign body implants, polyethylene tubing filled with hyaluronic acid paste was placed subcutaneously in rats and guinea pigs and the same paste was placed around the tubing. The presence of hyaluronic acid delayed and decreased the cellular invasion inside the tubes as well as the development of the granulation tissue and newly formed capillaries around it.

These experiments suggest that by increasing the hyaluronic acid concentration in a traumatized connective tissue compartment by injecting pure high molecular weight sterile, pyrogen-free hyaluronic acid one can regulate the regeneration process. Thus hyaluronic acid applied locally to traumatized connective tissue surfaces can decrease the formation of scar tissue and adhesions.

Hyaluronic acid preparations

In all the experiments described the sodium salt of hyaluronic acid prepared from human umbilical cord or rooster comb was used. The sterile, pyrogen-free solution containing 10 mg hyaluronic acid of $1-2 \times 10^6$ molecular weight in one ml physiological buffer solution (Healon) was obtained from Biotrics,

Inc. (24 Beck Road, Arlington, Mass. 02174 USA). The lower molecular weight hyaluronic acid was obtained by treating these samples with hyaluronidase, oxidation-reduction systems or heat.

CONCLUSION

In *in vitro* model systems hyaluronic acid inhibits the migration and proliferation of lymphocytes. This inhibition is dependent on the molecular size and probably conformation of the hyaluronic acid molecules, but not on charge density since other glycosaminoglycans with higher surface charge density do not show this effect. The various cells of the lymphomyeloid system exhibit different sensitivity to hyaluronic acid; therefore, we conclude that this effect is highly specific. This specificity probably relates to both the receptor of the cell surface and to the hyaluronic acid molecule.

This *in vitro* effect of hyaluronic acid is concentration dependent and the concentration range of the effect is that of the physiologically occurring concentration of hyaluronic acid in various connective tissue matrices. Furthermore, two hyaluronic acid containing fluid matrices, the synovial fluid and the liquid vitreous exhibit a very strong biological activity in terms of cell movement control.

The suppressive effect of hyaluronic acid on the graft versus host reaction and its effect on the skin allograft rejection and on the adhesion formation between tendon and tendon sheaths suggests that this cell regulatory effect of hyaluronic acid is also operative in *in vivo* conditions.

The same sterile, pyrogen-free high molecular weight hyaluronic acid preparations used in the experiments reported in this paper were recently used in various pathological conditions as therapeutic agents. It was reported that this hyaluronic acid, when injected into the vitreous space to replace the fluid vitreous during retinal detachment surgery, influenced favourably the healing of vitreo-retinal wounds (Algvere, 1971; Klöti, 1971; Regnault, 1971; Balazs *et al.*, 1972). Hyaluronic acid promoted the healing of chronic joint inflammation (osteoarthritis) developed after traumatic arthritis in race horses (Rydell *et al.*, 1970).

Intraarticular injection of this hyaluronic acid preparation into knee and hip joints of patients with chronic arthritis caused considerable improvement of the clinical symptoms (personal communication from Rydell, Helfet and Peyron).

It is possible that the injection of hyaluronic acid into the vitreous or the joint space, when these connective tissue compartments have less than normal concentrations and molecular size hyaluronic acid, changes the course of the chronic inflammation by acting as a cell immobilizing agent of the lymphomyeloid system.

With these experiments we attempted to draw attention to the possible cell regulatory role of some components of the intercellular matrix, and to explore new avenues which may provide an explanation of the biological role, and suggestions for the possible therapeutic application of hyaluronic acid.

REFERENCES

Algvere, P. (1971). *Acta Ophthalmol.* **49,** 975–976.
Balazs, E. A. and Jacobson, B. (1966). In "The Amino Sugars" (E. A. Balazs and R. W. Jeanloz, eds.) Vol. 11B, p. 386. Academic Press, New York and London.
Balazs, E. A., Freeman, M. I., Klöti, R., Meyer-Schwickerath, G., Regnault, F. and Sweeney, D. B. (1972). In "Modern Problems in Ophthalmology" Vol. 10, pp. 3–21. S. Karger, Basel.
Darzynkiewicz, Z. and Balazs, E. A. (1971). *Exp. Cell Res.* **66,** 113–123.
Klöti, R. (1971). *Ophthalmologica* **165,** 351–359.
Österlin, S. and Jacobson, B. (1968). *Expt. Eye Res.* **7,** 497–510.
Peters, J. H. (1972). *Exp. Cell Res.* **74,** 179–186.
Regelson, W. (1968). In "Advances in Chemotherapy" (A. Goldin and F. Hawking, eds.) Vol. III, p. 303. Academic Press, London and New York.
Regnault, F. (1971). *Bull. mem. Ste Fse. Ophtalmo.* **84,** 106–112.
Rydell, N. W., Butler, J. and Balazs, E. A. (1970). *Acta Vet. Scand.* **11,** 139–155.
Rydell, N. and Balazs, E. A. (1971). *Cl. Orthop.* **80,** 25–32.

Aggregation of Feline Lymphoma Cells by Hyaluronic Acid

BENGT WESTERMARK and ÅKE WASTESON
*The Wallenberg Laboratory and The Institute of Medical Chemistry
University of Uppsala, Uppsala, Sweden*

Cells *in vivo* show complex and specific interactions which *inter alia* keep them arranged in organs and tissues of fixed composition and topography. Such interactions may be reproduced *in vitro*, where cells will often organize non-randomly to form aggregates of similar cells with the exclusion of unlike elements. The nature of the aggregating factors is largely unknown.

The present report is concerned with a different kind of aggregation phenomenon, expressed by a single cell type (cat lymphoma) in response to a factor produced by a number of non-related cells. The factor has been identified as hyaluronic acid, suggesting a hitherto unrecognized biological function for this polysaccharide. These results confirm and extend those of Pessac and Defendi (1972), who observed a similar phenomenon in a murine system.

RESULTS AND DISCUSSION

Aggregation of feline lymphoma cells

When dispersed feline lymphoma cells, FL 74 (Theilen *et al.*, 1969) were mixed with used medium from cultures of a variety of human cells, aggregation of the lymphoma cells was induced. Apparently some component released to the medium from the former cells was responsible for this phenomenon. An assay for cell aggregating factor was developed with FL 74 cells as target cells. The cells were brought into single cell suspension in their growth medium by gentle pipetting. The cell density was adjusted to about 400,000 cells/ml and 0·5 ml portions were transferred to the wells of a microtiter plate. The test samples were added in portions of 10 μl to 500 μl; the final volume was adjusted to 1 ml by the addition of phosphate-saline buffer. After 20 min at room temperature the aggregation had reached its maximal extent. The degree of aggregation was then estimated semiquantitatively in an inverted microscope on a 0–4+ scale.

FL 74 cells clumped equally well at 4°C, 20°C and 37°C. Incubation in Ca^{2+}-free medium, obtained by dialysis against 0.5 mM EDTA in phosphate-saline buffer did not impair the formation of aggregates. The target cells retained their clumping ability after treatment with 1% glutaraldehyde for 1 h.

After incubation at room temperature in ordinary growth medium with no cell aggregating factor added, spontaneous aggregation appeared after 45 min. This clumping was not sensitive to hyaluronidase treatment but was almost completely abolished when assessed in Ca^{2+}- and Mg^{2+}-free medium. An enhanced spontaneous clumping was evident after glutaraldehyde fixation. This type of aggregation was not influenced by Ca^{2+} or Mg^{2+}.

Target cells for cell aggregating factor

The cell species tested as targets for the aggregating factor are seen in Table I. Apart from the feline lymphoma cell line, they include

Table I

Ability of various cell species to serve as target cells for cell aggregating factor

Designation of cell species	Derivation of cell species	Aggregation
FL 74	Feline lymphoma	+
255 Bm	Human lymphoblastoid	–
517 Mj	Human lymphoblastoid	–
P3	Burkitt lymphoma	–
Penina	Burkitt lymphoma	–
Namalwa	Burkitt lymphoma	–
266 Bl	Human myeloma	–
Human lymphosarcoma		–
138 MG	Human malignant glioma	–
587 CG	Human normal glia	–
Human peripheral lymphocytes		–
Human erythrocytes		–

various human lymphoid cell species, such as lymphoblastoid cells, Burkitt lymphoma, lymphosarcoma and peripheral blood lymphocytes. So far, the feline lymphoma is the only agglutinable species which we have found. The reason for this specificity is not known. However,

Table II
Cell aggregating activity of media sampled from cultures of various human cell species

Designation of cell line	Derivation of cell line	Cell aggregating activity*
87 MG	malignant glioma	0
118 MG	malignant glioma	4+
105 MG	malignant glioma	3+
138 MG	malignant glioma	3+
178 MG	malignant glioma	3+
178 FSV	malignant glioma	1+
251 MG	malignant glioma	3+
251-0	malignant glioma	2+
AG Cl-1	malignant glioma	2+
343 MG	malignant glioma	3+
333/343	malignant glioma	0
373 MG	malignant glioma	2+
489 MG	malignant glioma	2+
495 MG	malignant glioma	4+
495 MG G	malignant glioma	0
587 CG	normal glia	0
618 CG	normal glia	0
551 CG	normal glia	0
587 FSV	virus transformed glia	3+
309 SV40	virus transformed glia	3+
2T	sarcoma	2+
4T	sarcoma	4+
393T	sarcoma	3+
2S	skin fibroblast	1+
559S	skin fibroblast	2+
Hel 16	lung fibroblast	3+
377 Hel	lung fibroblast	2+
P3	Burkitt lymphoma	0
Namalwa	Burkitt lymphoma	0
Penina	Burkitt lymphoma	0
517 Mj	lymphoblastoid	2+
255 Bm	lymphoblastoid	3+
605 T	lymphoblastoid	3+

* According to the aggregation assay, described in the text.

according to a recent report by Pessac and Defendi (1972), murine lymphoma cells can also be aggregated, and possibly there are still other species with the same property. The feline and murine cell lines were both transformed by oncorna virus, and it is possible that the virus-carrier state *per se* is responsible for the affinity to the aggregating factor. Alternatively, the aggregation property may represent a stage of lymphoid differentiation, the expression of which may be detectable in a variety of perhaps even normal cells. Attempts to 'unmask' such a property in various lymphoid cells e.g. by enzymatic treatment are in progress.

Production of cell aggregating factor

Some of the cell species tested for their ability to produce the cell aggregating factor are seen in Table II. It is evident from this table that most of the malignant gliomas were effective as producers of the cell aggregating factor. Whereas several normal glia lines lacked the ability to secrete the factor, this property was revealed in a number of virus-transformed glia cultures. Although more cell lines have to be tested the present data suggest a coupling of the production of the cell aggregating factor to the malignant state of the cells. It is worthy of note that one of the normal glia cell lines acquired the ability to produce aggregating factor on transformation with tumour virus. Several sarcoma cells and fibroblasts not unexpectedly proved to secrete cell aggregating factor. Also, a number of other mesenchymal cells were producers of the aggregating factor.

A very marked aggregation of the FL 74 cells was induced by media from certain glioma and sarcoma lines, particularly those designated 4 T and 178 MG. Since they produced the factor equally well in serum-free medium, serum could be omitted in the production procedure to simplify further handling.

Properties of cell aggregating factor

The characterization of the properties of the cell aggregating factor was performed with 4 T cell supernatants. Since the cell aggregating factor was non-dialyzable, 4 T medium was readily concentrated by ultrafiltration through dialysis tubing. Chromatography of such media on Sepharose 2 B (Fig. 1) revealed a high molecular weight of the factor, since the cell aggregating activity was eluted with the void volume of the column. A standard preparation of hyaluronic acid was eluted at a significantly later position on the same column. Furthermore, chromatography on DEAE-cellulose indicated an anionic character of the aggregating factor. The anionic forces were sufficiently strong to retain the factor on the column at physiological pH and ionic strength, but were abolished when the pH of the eluant was lowered to 1·5. It should be noted that carboxyl groups become non-dissociated by this pH shift.

Further, the susceptibility of the aggregating factor to various treatments was investigated. The factor was heat-stable and resistant towards digestion with trypsin. It was therefore reasonable to assume that protein was no integral part of the aggregating factor. On the other hand

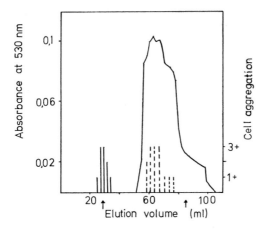

Fig. 1. Chromatography of 4T sarcoma cell culture medium and of hyaluronic acid on Sepharose 2 B: ——, absorbance of hyaluronic acid in the carbazole reaction; ┃┃┃, cell aggregating activity of 4T medium; ┊┊┊, cell aggregating activity of hyaluronic acid. The arrows indicate V_0 and V_t, respectively, of the column. Carbazole analysis of 4T medium showed too little absorbance to permit quantitation, therefore not included.

it was completely inactivated by treatment with polysaccharide-splitting enzymes. Whereas testicular hyaluronidase, chondroitinase ABC and AC and streptococcal hyaluronidase, respectively, are capable of degrading more than one species of polysaccharide. including hyaluronic acid, leech hyaluronidase is known to be specific for hyaluronic acid. The finding that the factor was inactivated by treatment with this enzyme strongly suggests that hyaluronic acid was essential for the aggregating activity.

Cell aggregating activity of standard polysaccharides

Several polysaccharide species were tested for their ability to induce aggregation of FL 74 cells. The polysaccharide preparations included hyaluronic acid, chondroitin sulphate, chondroitin sulphate proteoglycan dermatan sulphate, heparan sulphate and heparin (Table III).

Table III
Cell aggregating activity of standard polysaccharides

Polysaccharide	Final conc. of polysaccharide	Aggregation activity
Hyaluronic acid	>30 ng/ml	+
Hyaluronic acid	<30 ng/ml	−
Chondroitin sulphate	30 ng/ml–5 mg/ml	−
Chondroitin sulphate proteoglycan	30 ng/ml–5 mg/ml	−
Dermatan sulphate	30 ng/ml–5 mg/ml	−
Heparan sulphate	30 ng/ml–5 mg/ml	−
Heparin	30 ng/ml–5 mg/ml	−

As can be seen from the table, only hyaluronic acid aggregated the cells. Hyaluronic acid was efficient in concentrations as low as 30 ng/ml. Below this point the sensitivity of the aggregation assay was too low to permit reproducible measurements. The structural basis for the observed specificity of hyaluronic acid is unknown. However, it is interesting to note that neither the highly anionic character of heparin nor the high molecular weight of the chondroitin sulphate proteoglycan conferred any cell aggregating properties to these species.

As expected, hyaluronic acid was not effective after it had been degraded with testicular hyaluronidase or other polysaccharide-splitting enzymes. Since the effect of treatment of hyaluronic acid with testicular hyaluronidase is to reduce the number of repeating disaccharide units per polysaccharide fragment, it was reasonable to assume that the enzyme had split the polysaccharide beyond the minimal chain length, necessary for cell aggregation. Therefore, a preparation of partially hyaluronidase-degraded hyaluronic acid was tested for cell aggregating activity before and after separation on a column of Sepharose 4 B (Fig. 2). All the activity was found in the earlier part of the chromatogram. On the other hand, the carbazole analysis revealed that most of the uronic acid-containing material was eluted at a significantly later position. This discrepancy indicates that the cell aggregation phenomenon is dependent on the molecular weight of the hyaluronic acid.

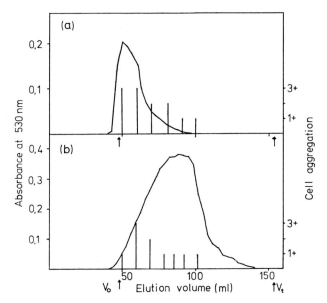

Fig. 2. Chromatography on non-digested (a) and partially degraded (b) hyaluronic acid on Sepharose 4 B. The column effluents were analysed for uronic acid (———) and cell aggregating activity (| | |). The arrows indicate V_0 and V_t, respectively, of the column.

Inhibition of cell aggregation

After separation of partially degraded hyaluronic acid on Sepharose 4 B, it was found to contain cell-aggregating components, emerging in the earlier part of the chromatogram. In contrast, before separation, the same preparation of hyaluronic acid did not induce cell aggregation. Apparently, the high molecular weight chains of hyaluronic acid, remaining after the partial degradation, were not effective in aggregating the cells in the presence of those of low molecular weight. A further test of this assumption was performed by preincubating the FL 74 cells with known amounts of hyaluronic acid oligosaccharides, prior to assay for aggregation with hyaluronic acid. Preexposure of the target cells to a concentration of 1 mg/ml of hyaluronic acid hexa- or tetrasaccharides was sufficient to abolish the aggregation induced by 'polymeric' hyaluronic acid. The inhibition was reversible, since additional amounts of hyaluronic acid reestablished the aggregation. For comparison, similar experiments were carried out with the corresponding chondroitin sulphate oligosaccharides. They proved to have approximately the same inhibitory effect on the aggregation as did the hyaluronic acid oligosaccharides.

That only a small part of the hyaluronic acid molecule binds to the receptor is suggested by the reversible abolishment of aggregation by oligosaccharides from digests of hyaluronic acid. However, since oligosaccharides from chondroitin sulphate (which does not aggregate FL 74 cells as intact molecules) were equally effective the interpretation of the inhibitory effects is uncertain. In view of the observation that high molecular weight hyaluronic acid was the only polysaccharide with aggregating activity it is reasonable to assume that the latter should be ascribed to the unique chain length of hyaluronic acid.

Receptor studies

Digestion of FL 74 cells with 0·2% crystalline trypsin in phosphate-saline buffer for 30 min at 37°C totally abolished the clumping ability. The reaction between hyaluronic acid and the target cat lymphoma cell therefore seems to involve a trypsin-sensitive receptor-like structure on the cell membrane. The reappearance of aggregation in trypsinized FL 74 cells was studied at intervals after the enzymatic treatment. The resynthesis of the receptor required approximately 24 h as judged from the reestablishment of full aggregation ability.

The receptor-hyaluronic acid interaction was peculiarly resistant to changes in temperature and even unaltered after treatment of the target cells with glutaraldehyde. This finding would seem to suggest that the interaction is not dependent on an intact cell metabolism.

ACKNOWLEDGEMENTS

This work was supported by grants from the Swedish Medical Research Council (13X-2309), the Swedish Cancer Society (55 and 53), Konung Gustaf V:s 80-årsfond and the Faculty of Medicine, University of Uppsala. The skilful technical assistance of Miss A. Bäfwe and Mrs. F. Karlsson is gratefully acknowledged.

REFERENCES

Pessac, B. and Defendi, V. (1972). *Science* **175**, 898–900.
Theilen, G. H., Kawakami, T. G., Rush, J. D. and Munn. R. J. (1969). *Nature* **222**, 589–590.

The Assembly of Fibroblast Plasma Membranes

C. A. PASTERNAK
Department of Biochemistry, University of Oxford
and
J. M. GRAHAM
Imperial Cancer Research Fund, Lincoln's Inn Fields, London W.C.2, England

Knowledge of the biosynthesis and assembly of the components of the plasma membrane is crucial to an understanding of the role of the cell surface in biological phenomena. A useful approach is to analyse the plasma membrane at different stages of the life cycle of cells. We have developed a selection technique, based on separation of cells according to size (Bergeron *et al.*, 1969; Warmsley and Pasternak, 1970; Pasternak, 1973; Pasternak and Warmsley, 1973) to achieve this and have used it to measure the appearance of several membrane associated enzymes in P815Y mastocytoma cells (Warmsley *et al.*, 1970). These cells contain particularly low levels of 5'-nucleotidase (an enzyme characteristic of the plasma membrane) and our initial studies were restricted to membrane markers of mitochondria and the endoplasmic reticulum. Our results indicated that such enzymes are synthesized gradually throughout interphase, in concert with general cell protein and phospholipid (Warmsley *et al.*, 1970).

In order to test the generality of this result, we have now extended these studies to cultured fibroblasts. Since such cells contain adequate amounts of 5'-nucleotidase and Na^+/K^+-stimulated Mg^{++} ATPase we have been able to assess the appearance of specific plasma membrane markers. Moreover the use of an efficient technique for isolating plasma membranes (Graham, 1972, 1973) has enabled us to analyse non-enzymic components throughout the cell cycle.

It appears that plasma membrane protein, phospholipid and 5'-nucleotidase are synthesized in concert with total cell protein but that Na^+/K^+-stimulated Mg^{++} ATPase rises dramatically in late interphase.

METHODOLOGY

Transformed mouse fibroblasts (NIL 2 HV obtained from Dr. I. A. MacPherson) were cultured as previously described (Graham, 1972). Exponentially growing cells (total 10^9) were exposed to a pulse of ^{3}H-thymidine (2mCi) for 1 h at 37°, harvested, washed and centrifuged through 2–10% Ficoll in an MSE zonal rotor A (Warmsley and Pasternak, 1970; Pasternak, 1973; Pasternak and Warmsley, 1973). The separated fractions (25 ml) were analysed for incorporated ^{3}H-thymidine by filtration after the addition of 5% trichloroacetic acid to delineate the S period, and for cell number and volume with a Coulter counter. Fractions were pooled to give sufficient material for isolation of plasma membranes.

Cells (approx 2×10^8 in 30 ml of $0.25M$ sucrose/$0.2mM$ $MgCl_2$/$5mM$ tris-HCl pH7.4) were disintegrated by nitrogen cavitation (Graham, 1973). A sample of the homogenate was kept for analysis and the remainder spun at $1000 \times g$ for 10 min to remove debris, unbroken cells and nuclei. The supernatant fraction was centrifuged through a discontinuous gradient of sucrose and dextran as described by Graham (1973). Fractions corresponding to plasma membrane, endoplasmic reticulum and mitochondria were pooled for enzymic and chemical analyses.

Enzyme activities were analysed as follows: 5'-nucleotidase (E C 3.1.3.5), Na^+/K^+-stimulated Mg^{++} ATPase (E C 3.6.1.3. and NADH oxidase (E C 1.6.99.3) as described by Avruch and Wallach (1971); succinate—cytochrome C reductase (E C 1.3.99.1) as described by Mackler et al. (1962).

Protein was estimated by the method of Lowry et al. (1951) and phospholipid phosphorus, after extraction with chloroform: methanol (Folch et al., 1957) and digestion at 300°C with 72% perchloric acid, by the method of Bartlett (1959).

RESULTS AND DISCUSSION

Fig. 1a shows that zonal centrifugation of NIL cells effectively separates G_1 from S cells, with less clear-cut separation at the S-G_2 boundary. This may be due partly to rather few cells actually in G_2 (i.e. a relatively short G_2 period) and partly to a smaller size difference between S and G_2 cells, than between G_1 and S cells. Measurement of cell volume confirms that the major increase in size is between G_1 and S.

Estimation of succinate—cytochrome c reductase (mitochondria) and NADH oxidase (endoplasmic reticulum) shows that specific activity remains fairly constant throughout the cell cycle, with possibly a small rise during S in the case of succinate—cytochrome c reductase (Fig. 1b). Since the specific activity of these enzymes in the isolated membrane fractions, relative to that in the total homogenate ('enrichment factor'), does not vary significantly during the cell cycle, it follows that the synthesis of the enzymes parallels that of total protein.

In the case of plasma membrane markers, the specific activity of 5′-nucleotidase also remains fairly constant throughout the cell cycle; Na^+/K^+-stimulated Mg^{++} ATPase, on the other hand, increases markedly towards the end of interphase (Fig. 1c). Interestingly enough, the pattern for Mg^{++} ATPase in the plasma membrane resembles that for 5′-nucleotidase rather than that for Na^+/K^+-stimulated Mg^{++}

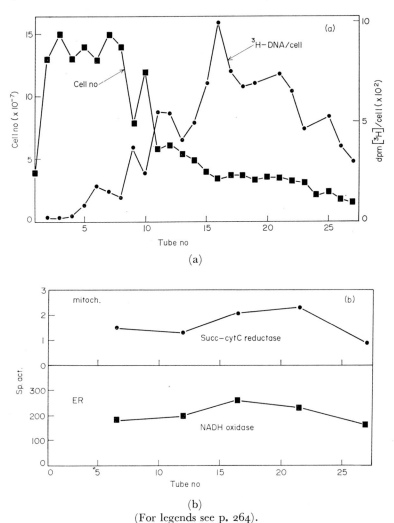

(For legends see p. 264).

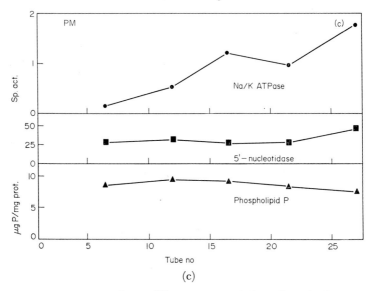

Fig. 1. Analysis of NIL cells at different stages of the cell cycle. ^{3}H-thymidine-labelled cells were fractionated through a Ficoll gradient and samples analysed as described in "Methodology". Pooled fractions (3–10, 11–14, 15–19, 20–24 and 25–30) were homogenized, separated into plasma membrane (PM), endoplasmic reticulum (ER) and mitochondria (mitoch.) and analysed as described in "Methodology".
(a) Total cell number, ■——■; ^{3}H-DNA/cell, ●——●, (b) Succinate-cytochrome c reductase, ●——●; NADH oxidase, ■——■; (c) Na$^+$/K$^+$—stimulated Mg^{++} ATPase, ●——●; 5′-nucleotidase, ■——■; phospholipid phosphorus, ▲——▲. All enzyme activities are expressed in μmoles substrate metabolized/h/mg protein.

ATPase. The 'enrichment factor' for 5′-nucleotidase and Na$^+$/K$^+$-stimulated Mg^{++} ATPase does not alter significantly during the cell cycle. Thus 5′-nucleotidase appears to be synthesized concomitantly with total cell protein and this is true of bulk plasma membrane protein and phospholipid also (Fig. 1c). Na$^+$/K$^+$-stimulated Mg^{++} ATPase, on the other hand, displays a 'peak' pattern (Mitchison, 1971); this is particularly significant since the other enzymes investigated in this study are all 'continuous' (Mitchison, 1971). As yet we have not distinguished between changes in *activity* and *amount* of Na$^+$/K$^+$-stimulated Mg^{++} ATPase. The physiological significance of the later increase in this enzyme may be related to the fluctuations in intracellular K$^+$ and Na$^+$ observed during the life cycle of L51787 cells (Jung and Rothstein, 1967).

The characteristic distribution between plasma membrane, endoplasmic reticulum and mitochondria of each of the enzymes studied in this chapter remains essentially the same throughout the cell cycle. This indicates that insertion of the enzyme into membranes probably follows fairly closely on its biosynthesis and that large precursor pools, e.g. in the endoplasmic reticulum, are not involved.

ACKNOWLEDGEMENTS

We are grateful to Dr. I. A. MacPherson, Mrs. P. Davies, Mrs. S. J. Picking, Mr. M. C. B. Sumner, Mr. K. J. Micklem and Mr. D. Harvey for help at various stages and to the Medical Research Council for financial support.

REFERENCES

Avruch, J. and Wallach, D. F. H. (1971). *Biochim. Biophys. Acta* **233**, 334–347.
Bartlett, G. R. (1959). *J. Biol. Chem.* **234**, 466–468.
Bergeron, J. J. M., Warmsley, A. M. H. and Pasternak, C. A. (1969). *FEBS Letters* **4**, 161–163.
Folch, J., Lees, M. and Sloane–Stanley, G. H. (1957). *J. Biol. Chem.* **226**, 497–509.
Graham, J. M. (1972). *Biochem. J.* **130**, 1113–1124.
Graham, J. M. (1973). *In* "Methodological Developments in Biochemistry" (E. Reid, ed.), Vol. 3. Longmans, London. (In Press.)
Jung, C. and Rothstein, A. (1967). *J. Gen. Physiol.* **50**, 917–932.
Lowry, O. H., Rosebrough, N. J., Farr, A. L. and Randall, R. J. (1951). *J. Biol. Chem.* **193**, 265–275.
Mackler, B., Collipp, P. J., Duncan, H. M., Rao, N. A. and Huennekens, F. M. (1962). *J. Biol. Chem.* **237**, 2968–2974.
Mitchison, J. M. (1971). "The Biology of the Cell Cycle", pp. 159–180. Cambridge University Press, Cambridge.
Pasternak, C. A. (1973). *In* "Methods in Molecular Biology" (A. I. Laskin and J. A. Last, eds.) Vol. 3. Marcel Dekker, New York. (In Press.)
Pasternak, C. A. and Warmsley, A. M. H. (1973). *In* "Methodological Developments in Biochemistry" (E. Reid, ed.), Vol. 3. Longmans, London. (In Press.)
Warmsley, A. M. H. and Pasternak, C. A. (1970). *Biochem. J.* **119**, 493–499.
Warmsley, A. M. H., Phillips, B. and Pasternak, C. A. (1970). *Biochem. J.* **120**, 683–688.

Cell Surface and Initiation of Proliferation

ANTTI VAHERI*, ERKKI RUOSLAHTI†, TAPANI HOVI*
and STIG NORDLING‡
University of Helsinki, Haartmaninkatu 3, SF-00290 Helsinki 29, Finland

For some time it has been recognized that the state of the cell surface plays a key role in the regulation of cell proliferation and the expression of malignant transformation. Some of the earliest changes detected after re-initiation of cell growth occur in the membrane. Normal and transformed cells differ in several surface properties in physical, chemical, immunological or physiological character (Burger, 1971a; Pardee, 1971).

Normal cells in culture stop dividing after growing to confluent monolayers. Transformation by oncogenic viruses removes this *in vitro* blocking of cell division by density dependent inhibition (DDI). Normal fibroblasts can be released from DDI by mild proteolytic treatment (Burger, 1970; Sefton and Rubin, 1970). We have extended these findings and have shown that modulation of the cell surface by certain treatments stimulates growth in density-inhibited chick fibroblasts. It is especially noteworthy that normal cells when released from DDI, assume some of the surface properties characteristic of transformed cells.

We prepared density-inhibited chick embryo fibroblasts according to Sefton and Rubin (1970) from primary cultures in medium 199, containing 2% tryptose phosphate broth and 2% chicken serum. After 48 h at 38°C the cultures were confluent, had a low rate of DNA synthesis and a density of about 1.5×10^6 cells per 20 cm^2 (Vaheri *et al.*, 1972).

SURFACE MEDIATED STIMULATION BY INSULIN, NEURAMINIDASE AND PROTEOLYTIC ENZYMES IN DENSITY-INHIBITED CULTURES

Microgram quantities of insulin, proteolytic enzymes or neuraminidase will initiate proliferation in density-inhibited chick embryo cell

---
* Antti Vaheri and Tapani Hovi: Department of Virology.
† Erkki Ruoslahti: Department of Serology and Bacteriology.
‡ Stig Nordling: Department of Pathology III.

cultures (Table I). Under these conditions no insulin was taken up by the cells (Vaheri et al., 1973), indicating that an action on the cell surface is sufficient to trigger the sequence of events leading to cell

Table I

Stimulation of DNA synthesis in density-inhibited fibroblast cultures by certain substances acting on the cell surface

Substance	Conc. ($\mu g/ml$)	Rate of DNA synthesis at 16 h: ratio to control*
Insulin	1	20
Trypsin	3	16
Papain	3	12
Neuraminidase	1	8

* Assayed by measuring incorporation of ^{3}H-thymidine during 60-min pulse (cpm/10^6 cells).

proliferation. This is consistent with the observation that interaction of insulin with receptors at the plasma membrane surface, may suffice to induce an increase in glucose transport as well as other metabolic changes in isolated fat cells (Cuatrecasas, 1969).

STIMULATION OF EMBRYONIC CELLS BY A CARCINO-EMBRYONIC SERUM PROTEIN

It is conceivable that any protein present in large quantities during embryogenesis may be involved in regulation of cell proliferation or differentiation. α-Fetoprotein (AFP) is present in concentrations up to several mg/ml in the fetal serum of a wide range of vertebrates ranging from man to chicken (Abelev, 1971). During adult life the serum concentration is only a few ng/ml (Ruoslahti and Seppälä, 1972) but it may increase to fetal levels in the presence of certain malignancies (Abelev, 1971). It was of special interest to find that purified AFP, in microgram quantities, stimulates density-inhibited embryonic fibroblasts (Tables II and III). Further study is needed, however, to establish association of this activity with the unmodified AFP molecule.

In very large doses other serum proteins, possibly any protein, (Table II, Vaheri et al., 1973) will also stimulate incorporation of ^{3}H-thymidine, but the doses needed are about a hundred times higher than that of AFP. In fact the stimulatory effect of chicken and bovine serum appeared to correlate with the serum protein content (Table II).

Table II
Stimulation of density inhibited chick embryo fibroblasts by serum protein

Substance	Protein conc. ($\mu g/ml$)	Rate of DNA synthesis at 16 h: ratio to control*
Calf serum 10%	6000	17
1%	600	4
0·1%	60	1
Bovine serum albumin	500	3
	100	1
Human serum albumin	500	5
	100	1
Bovine fetuin	500	4
	100	1
Human α-Fetoprotein	5	8

* Assayed by measuring incorporation of ^{3}H-thymidine during 60-min pulse (cpm/10^6 cells).

Table III
Sequence of events after stimulation of density-inhibited chick embryo fibroblasts by insulin, trypsin, neuraminidase, α-Fetoprotein or serum

Change	Time (h)
Increase in sugar uptake	1–3
Increase in cell volume	1–3
Increase in ^{3}H-thymidine incorporation Start	5
Maximum	8–9
Peak of mitosis	12

INCREASE OF SUGAR UPTAKE—AN EARLY CHANGE IN PLASMA MEMBRANE

The uptake of sugars, but not of amino acids or nucleosides, is increased in virus transformed cells, (Hatanaka et al., 1970; Isselbacher, 1972). Normal cells, stimulated to proliferate, share this property (Sefton and Rubin, 1971; Vaheri et al., 1972, 1973, Table III). Within 1–3 h after addition of the test substances the uptake of, for example, 2-deoxy-^{3}H-glucose or ^{3}H-glucosamine increased 2–3 fold, but that of ^{3}H-leucine or ^{3}H-thymidine did not. Another early change in stimulated cells is the rapid increase in cell volume after re-initiation of growth. The increase in ^{3}H-thymidine incorporation starts after about 5 h, reaches a peak at 8–9 h and is followed by increase in the mitotic rate and cell division.

STIMULATED NORMAL CELLS AND TRANSFORMED CELLS SHARE SEVERAL SURFACE PROPERTIES

It appears significant that rapidly growing cells, or cells released from DDI, transiently share properties that are characteristic of transformed cells (Table IV). The increase in sugar uptake indicates that early changes in membrane function precede proliferation. Protein synthesis is needed for this membrane modulation in both stimulated normal cells (Sefton and Rubin, 1971) and virus-transformed cells (Kawai and Hanafusa, 1972). Cinematographic analyses of chick fibroblasts show that reduction in cell contacts and increased cell movement occur very early after the addition of serum (Baker and Humphreys, 1971). The change in lectin agglutinability is a later event. This surface alteration requires DNA synthesis, and is characteristic of normal mitotic and transformed cells (Burger, 1971b). Certain sialosylglycolipids are synthesized as cells grow to high density. After transformation their concentration decreases sharply (Hakomori et al., 1971). It is unknown as yet, whether the changes in glycolipids or in the enzyme activities involved are the result of loss of contact inhibition in transformed cells, or whether the changes cause loss of contact inhibition. Kinetic analysis of normal cells during the re-initiation of proliferation may prove helpful.

The level of cyclic AMP is transiently reduced when density-inhibited mouse embryo fibroblasts (3T3) have been exposed to serum, trypsin or insulin (Sheppard, 1972). Insulin inhibits adenyl cyclase activity and decreases the level of cyclic AMP in adipose cells (Illiano and Cuatrecasas, 1972). Addition of cyclic AMP or its derivatives, and phosphodiesterase inhibitors such as theophylline appears to alter the morphology and growth properties of some, but not all, transformed cells towards those characteristic of normal cells (Otten et al., 1972). In chick fibroblasts, transformed by a temperature sensitive mutant of RSV, the levels of cyclic AMP are elevated at the higher, nonpermissive temperature, at which the cells assume normal morphology and growth characteristics. These data suggest that in transformation a viral function is involved in the controls of the level of cyclic AMP (Otten et al., 1972) and that cyclic AMP may be a mediator between the triggering modulation at the cell surface and the events leading to cell proliferation.

Table IV

Certain properties of normal resting, normal growing and transformed cells

Property	Cell type*	Normal resting	Normal growing, or normal released from DDI	Transformed	References
Enhanced sugar uptake	CEF	−	+	+	Sefton and Rubin, 1971; Vaheri et al., 1972, 1973
Reduction of cell contacts and increase in cell movement	CEF	−	+	+	Baker and Humphreys, 1971
Increased lectin-effected agglutinability	3T3	−	+†	+	Burger, 1971; Fox et al., 1971
	BHK21	−	+	+	Hakomori, 1970
Decreased amount of certain glycolipids	CEF				Hakomori et al., 1971
Decreased cAMP level	3T3	−	+	+	Sheppard, 1972
	CEF				Otten et al., 1972

* 3T3: a line of mouse embryo fibroblasts, BHK21: a line of baby hamster kidney fibroblasts, CEF: chick embryo fibroblasts.
† During mitosis only.

ACKNOWLEDGEMENTS

This research was supported by grants from the Sigrid Jusélius Foundation and the National Research Council for Medical Sciences.

REFERENCES

Abelev, G. I. (1971). *Adv. Cancer Res.* **14,** 295–354.
Baker, J. B. and Humphreys, T. (1971). *Proc. Natl. Acad. Sci. USA* **68,** 2161–2164.
Burger, M. M. (1970). *Nature* **227,** 170–171.
Burger, M. M. (1971a). *Curr. Top. Cell. Regulation* **3,** 135–193.
Burger, M. M. (1971b). *In* "Growth Control in Cell Cultures" (G. E. W. Wolstenholme and J. Knight, eds.), Ciba Fdn. Symp., pp. 45–63. Churchill Livingstone, London.
Cuatrecasas, P. (1969). *Proc. Natl. Acad. Sci. USA* **63,** 450–457.
Fox, T. O., Sheppard, J. R. and Burger, M. M. (1971). *Proc. Natl. Acad. Sci. USA* **68,** 244–247.
Hakomori, S. (1970). *Proc. Natl. Acad. Sci. USA* **67,** 1741–1747.
Hakomori, S., Saito, T. and Vogt, P. K. (1971). *Virology* **44,** 609–621.
Hatanaka, M., Augl, C. and Gilden, R. V. (1970). *J. Biol. Chem.* **245,** 714–717.
Illiano, G. and Cuatrecasas, P. (1972). *Science* **175,** 906–908.
Isselbacher, K. J. (1972). *Proc. Natl. Acad. Sci. USA* **69,** 585–589.
Kawai, S. and Hanafusa, H. (1971). *Virology* **46,** 470–479.
Otten, J., Bader, J., Johnson, G. S. and Pastan, I. (1972). *J. Biol. Chem.* **247,** 1632–1633.
Pardee, A. B. (1971). *In Vitro* **7,** 95–104.
Ruoslahti, E. and Seppälä, M. (1972). *Nature* **235,** 161–162.
Sefton, B. and Rubin, H. (1971). *Proc. Natl. Acad. Sci. USA* **68,** 3154–3157.
Sheppard, J. R. (1972). *Nature (New Biol.)* **236,** 14–16.
Vaheri, A., Ruoslahti, E. and Nordling, S. (1972). *Nature (New Biol.)* **238,** 211–212.
Vaheri, A., Ruoslahti, E., Hovi, T. and Nordling, S. (1973). *J. Cell Physiol.* (In Press.)

The Nature of the Surface of Normal and Virus-Transformed Tissue Culture Cells

L. WARREN, J. P. FUHRER and C. A. BUCK

*Department of Therapeutic Research, School of Medicine
University of Pennsylvania, Pa 19104
and Division of Biology, Kansas State University
Manhattan, Kansas 66502, USA*

The surface of the cell is now a major concern in our thinking about cancer and the peculiar behavior of the cancer cell. For our thinking to become more meaningful, however, we must answer two questions: does the surface of a malignant cell differ chemically from that of its normal counterpart? And, if surface differences are, in fact, found are these responsible for the malignant behavior of the cell or are they secondary effects to be added to a long list of almost random changes manifested by malignant cells?

From a functional point of view there is good reason to believe that the surface of the malignant cell differs from that of the normal. Malignant cells adhere to one another less tightly (Coman, 1944, 1961) and they appear to have lost contact inhibition of motion (Abercrombie and Heaysman, 1954). Both of these processes undoubtedly involve the surface of the cell. Furthermore, there are clear differences in the agglutinability of normal and malignant cells by various plant lectins (Burger, 1968; Inbar and Sachs, 1969).

Our laboratory is, at the present time, concerned with establishing chemical and structural differences between cell surfaces of normal and virus-transformed (malignant) cells in tissue culture. We have, in fact, found differences which will be described below but whether they are of prime importance in malignancy remains to be seen. Interesting comparative biochemical studies have also been carried out by Meezan et al. (1969), Wu et al. (1969), Sheinin and Onodera (1970) and Onodera and Sheinin (1970).

In our laboratory we have been comparing the carbohydrate components of glycoproteins residing on the surfaces of control and virus-transformed cells (Buck et al., 1970, 1971a, b; Warren et al., 1972a, b, 1973).

We have used a double-label technique in which control cells are grown in the presence of ^{14}C-L-fucose while the corresponding transformed cell is cultured in the presence of ^{3}H-L-fucose. The radioactivity of L-fucose is incorporated into macromolecules and essentially all bound radioactivity can be recovered as L-fucose. After 3 days during which the cells have been in log phase of growth they are treated with trypsin. This treatment detaches the cells from the glass surface on which they had grown and releases glycopeptides which contain approximately 20% of the bound fucose (Buck et al., 1970). This fraction is called the trypsinate. Surface membranes of the cells are also isolated by the zinc ion procedure (Warren and Glick, 1969). Trypsinates from both normal and transformed cells are mixed and are treated with pronase to digest virtually all of the polypeptide component leaving intact polysaccharides bearing the amino acid which formerly anchored the carbohydrate to the polypeptide. Similarly the two sets of surface membranes are mixed, dissolved in sodium dodecyl sulfate solution and are exhaustively digested with pronase. The digests, containing chains of carbohydrates, ^{14}C-labeled from control and ^{3}H-labeled from transformed cells are then co-chromatographed on a column of Sephadex G-50. Each tube is counted for ^{14}C and ^{3}H, the data is processed and plotted by computer (Figs. 1a–f). It can be seen (Fig. 1a, c, e) that there is a large, relatively early-eluting material which comes from the surface of the virus-transformed cell. A shoulder or very small peak is usually observed in this region in plots of radioactivity from control cells (Buck et al., 1970). Essentially the same pattern is seen using trypsinates (probably from the more superficial layers of the cell surface) or using isolated surface membranes—the deeper elements of the cell surface. The results are the same if the labels are reversed, i.e. control cells are grown in the presence of ^{3}H-L-fucose and transformed cells with ^{14}C-L-fucose.

This early-eluting material has been found in mouse, chick and hamster cells transformed by both DNA and RNA-containing oncogenic viruses (Buck et al., 1971b). It is growth-dependent for it disappears in both control and transformed cells in plateau phase of growth (Buck et al., 1971a). Early-eluting material is found on the surface of chick embryo fibroblasts transformed by T5 virus and grown at 35°C (permissive temperature) but not at 40°C (Warren et al., 1972a). T5 is a temperature-sensitive mutant of Rous sarcoma virus (Martin, 1970). Cells transformed by this mutant and grown at 35°C appear to behave as malignant cells, while those grown at 40°C are indistinguishable from controls. It is of interest that cells transformed by T5 show no significant differences from the control in their glycolipid content nor are there changes in shifts between permissive and non-permissive temperatures (Warren et al., 1972a).

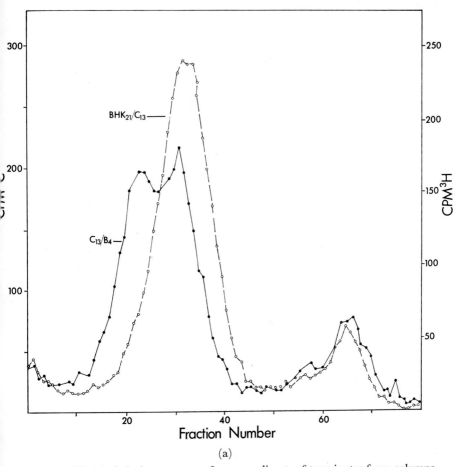

(a)

Fig. 1. Double-label elution patterns of pronase digests of trypsinates from columns of Sephadex G-50 (fine). The procedure has been described in detail previously and in the text. (a and b) Material derived from BHK_{21}/C_{13} labeled in log phase of growth for 3 days with ^{14}C-L-fucose and from C_{13}/B_4 labeled with ^3H-L-fucose. In (b) the pronase digests had been treated with 5 units of neuraminidase for 1·5 h at pH 5·2 in 3 mM $CaCl_2$ prior to application to the column. Controls were also incubated but without enzyme. (c and d) Material derived from BHK_{21}/C_{13} and its polyoma virus transformed counterpart, PyY. Note that the pattern, though slightly different from that of C_{13}/B_4 is almost identical to that previously described (Buck et al., 1971b). (d) Material treated with neuraminidase before application to the column. (e and f) Material derived from chick embryo fibroblasts before and after transformation with the Schmidt-Ruppin (SR) strain of Rous Sarcoma virus. (f) Material treated with neuraminidase.

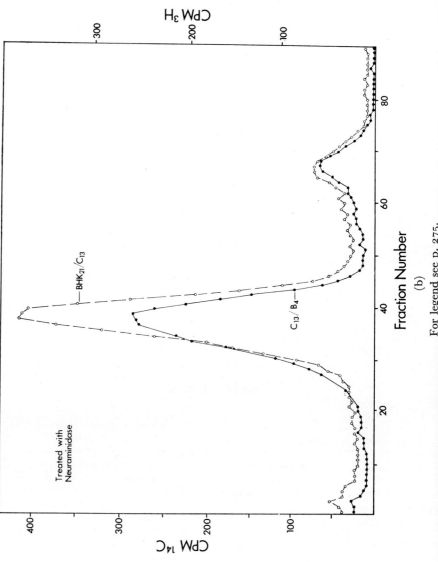

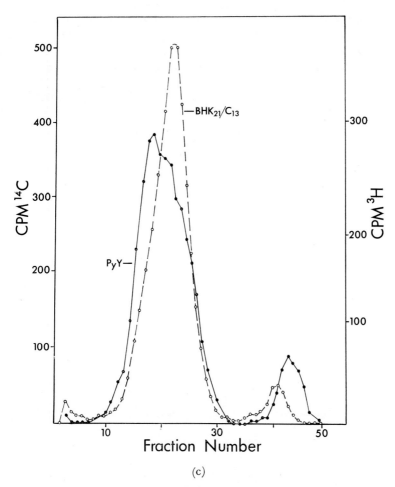

(c)

For legend see p. 275.

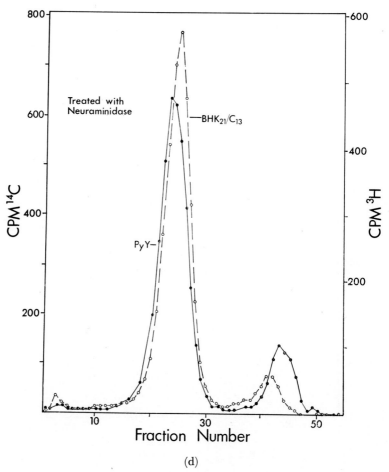

(d)

For legend see p. 275.

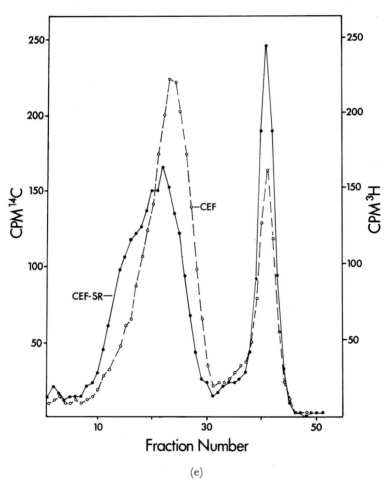

(e)

For legend see p. 275.

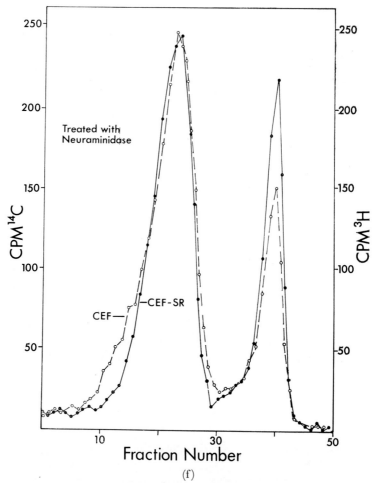

(f)

For legend see p. 275.

Since fucose residues are terminal in the small polysaccharides derived from glycoproteins we believe that there is little degradation of these moieties during digestion with pronase. Further it has been shown that there are very few amino acids associated with the fraction (Buck et al., 1970). We know little else except that from molecular weight estimations by co-chromatography with known polysaccharide standards the glycoprotein component under study consists of approximately 20–25 sugar residues. Some of these residues are derived from radioactive glucosamine (Buck et al., 1970) but in the studies discussed here the fraction is followed by radioactivity that we know to be in the L-fucose component.

We have recently found that if pronase digests of mixed surface materials are treated with purified neuraminidase from either *V. cholerae* or *Cl. perfringens* the large, early-eluting material (peak A) of the transformed cell disappears into the larger, later appearing peak (B) and it would appear that the polysaccharides of control and virus-transformed cells are virtually the same in their elution patterns (Figs. 1b,d,f) (Warren *et al.*, 1972b, 1973). Similarly if *intact* cells, labeled with radioactive fucose, are treated with neuraminidase and then with trypsin and this trypsinate is co-chromatographed with a trypsinate from labeled cells that had not been exposed to neuraminidase the elution pattern seen in Fig. 2 is obtained. The early-eluting material

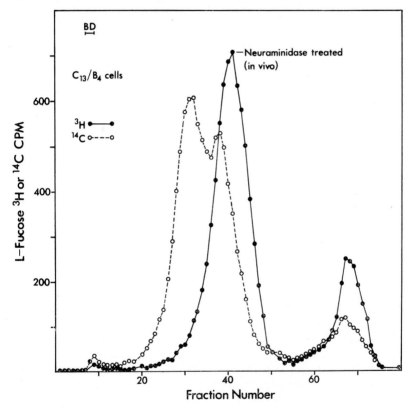

Fig. 2. Cells cultured and processed as previously described (Buck *et al.*, 1970) except that cells grown in the presence of ^{14}C-L-fucose were washed and incubated for one hour at 37°C in the presence of Eagle's minimal essential medium (MEM) without protein while cells which had been grown with ^{3}H-L-fucose were incubated in MEM containing neuraminidase from *V. cholerae*. Both cultures were then exposed to crystalline trypsin and subjected to the usual procedures.

(A) characteristic of transformed cells disappears into the second, large peak (B).

Closer examination of the elution patterns has shown that in fact there is also a small but definite shift of peak B to the right after treatment with neuraminidase suggesting the presence of some sialic acid in the components of peak B, i.e. peak B treated with neuraminidase migrates to a peak B' position.

An explanation for the presence of a large 'peak A' in transformed cells is that these cells contain a sialyl transferase capable of transferring sialic acid (NAN) from its activated form, CMP-NAN to a carbohydrate acceptor which is a component of a cell surface glycoprotein. The acceptor polysaccharide would migrate in the B or B' area before the addition of NAN. After NAN is transferred, the molecule is sufficiently enlarged so that it now would elute earlier, in the A area. A considerable amount of experimental data has now been gathered to support this hypothesis (Warren *et al.*, 1972b, 1973). At the heart of the problem is the requirement for a specific acceptor to which a hypothetical specific enzyme transfers NAN from CMP-NAN.

To demonstrate the presence of a specific transferase-acceptor system, relatively large quantities of peak A material were obtained from C_{13}/B_4 (transformed) cells labeled with ^{3}H-L-fucose. From this, sialic acid was removed enzymatically. This material largely carbohydrate in composition and migrating in the B' area, served as acceptor. We call this acceptor 'desialylated peak A'.

Our assay mixture contained buffer, CMP-NAN labeled with ^{14}C in the NAN moiety, acceptor (desialylated peak A) and crude, particulate extracts from control or virus-transformed cells as enzyme (Warren *et al.*, 1972b). After incubation blue dextran and phenol red markers were added to the mixture which was applied to a column of Sephadex G-50 (60 × 0·8 cm). Approximately 25 fractions were collected. If there has been transfer of ^{14}C-NAN, a peak of ^{14}C (NAN) should be seen immediately preceding the ^{3}H peak of acceptor (in excess). Transfer to large acceptors, (exogenous or endogenous) can be measured by a peak of ^{14}C that elutes with the blue dextran in the void volume. These patterns can be seen in Fig. 3. The peak of ^{14}C-NAN in panel 3 in which the enzyme is derived from a transformed cell (C_{13}/B_4) is considerably greater than the ^{14}C peak in panel 2 where the enzyme is from the control cell (BHK_{21}/C_{13}).

Table I also shows that transfer of NAN to desialylated peak A is considerably greater in the transformed cell than in the control. However, there is no detectable transfer to peak A material bearing its full complement of sialic acid. Peak B material can accept some NAN

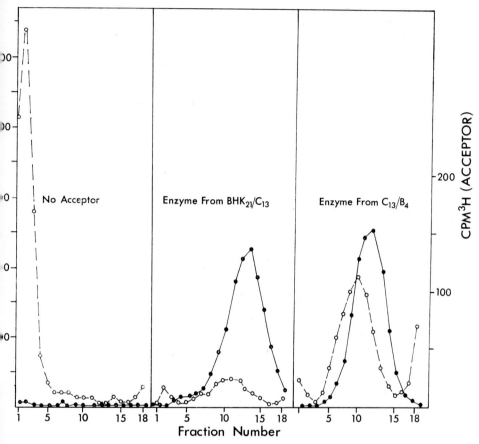

Fig. 3. Assay for sialyl transferase. The results of 3 assays. In the first, no added acceptor is present and there is no ^{3}H (acceptor) peak or ^{14}C-NAN peak in the 5th to the 15th tube region. The enzyme extract was derived from C_{13}/B_4. In the second and third assays desialylated peak A material (●—●—●) is present (^{3}H). ○—○—○ indicated transfer of ^{14}C-NAN to acceptor. 0·2 mg of particulate enzyme protein used in each assay. Incubation for 1 h. (See text and Warren et al. (1972b) for details.)

Table I
Transfer of ^{14}C-NAN to various acceptors by extracts from control and virus-transformed cells

^{14}C-NAN acceptor	Extract from BHK $_{21}/C_{13}$ cells	Extract from C_{13}/B_4 cells
	cpm/mg protein/h	
None	0	0
Non-specific (endogenous)	1690	2170
Peak B	1600	2030
Peak A	0	0
Desialylated peak A	1790	5450
Desialylated fetuin	14,800	13,500
Desialylated bovin mucin	17,700	18,000

Incubation for 1 h. (See text and (Warren et al., 1972b) for details of assay.)

but it should be noted that NAN transferase activity to endogenous acceptor ('non-specific'), peak B, desialylated fetuin or bovine mucin is the same in control and transformed cells whereas transfer to desialylated peak A is distinctly greater in the transformed cell.

To summarize our results (Warren et al., 1972b, 1973):

1. There is 2·5–11 times more NAN transferase activity in virus-transformed cells than in controls. However, there is activity in control cells. In other studies of sialyl transferases in control and transformed cells activity is lower in transformed cells (Grimes, 1970; Cumar et al., 1970; Den et al., 1971). These sialyl transferases are apparently different from the one reported here. Experiments in which extracts of control and transformed cells are mixed does not support the notion that activity is controlled by soluble activators or inhibitors.

2. Enzyme activity is sharply reduced in both BHK_{21}/C_{13} or C_{13}/B_4 cells in plateau phase of growth as compared to rapidly dividing cells (Warren et al., 1972b). This is in agreement with the observation that peak A disappears both in control and virus-transformed cells in plateau phase of growth (Buck et al., 1971a).

3. The enzyme is found in isolated surface membranes but this is probably not its only location (Warren et al., 1972b).

4. We now have data (unpublished) to show that the changes upon viral transformation seen in the surface membranes of the cells described here also take place in endoplasmic reticulum, inner and outer mitochondrial membrane and nuclear membrane. However we have

not completely ruled out some contamination of these fractions by fragments of surface membranes. Experiments are in progress to rule out the contamination.

5. Chick embryo fibroblasts transformed by the temperature-sensitive mutant of Rous sarcoma virus, T5, contain considerably more of the activity (3·5 to 4-fold) when grown at 35 °C (permissive temperature) than at 40°C at which the cells appear normal and the level is about that of control cells (Warren *et al.*, 1972b, 1973).

It would appear that we are dealing with a change in the carbohydrate component of glycoprotein that is associated with membrane of virus-transformed cells. Glycoproteins of internal membrane systems as well as the surface seem to be affected. Whether the affected carbohydrate is situated on one or more species of membrane glycoprotein or whether the protein component itself is altered is unknown. We are now attempting to find out which protein(s) of the surface and other internal membranes bear the polysaccharide of peak A. It should be pointed out that the resolving power of Sephadex G-50 is relatively poor and that peaks A, B and B' are not homogeneous. Several peaks can be resolved by DEAE-Sephadex chromatography and by high voltage paper electrophoresis. Work is progressing in identifying the components, partially separated by columns of Sephadex G-50 chromatography.

The differences uncovered in these studies between control and virus-transformed cells appears to be essentially quantitative in nature. 'Peak A material' can be observed in both control and transformed cells (Buck *et al.*, 1970) but there is far more in the transformed cells. It would seem that an enzyme activity (sialyl transferase) associated with growth, responsible for the formation of the material in the peak A area, is present in inordinately large amounts in transformed cells. This results in one (or possibly more) membrane glycoprotein containing extra sialic acid residues being present in excess. The transferase may be fairly specific and may be masked by several other sialyl transferases. Furthermore, though there are extra residues of sialic acid on the glycoproteins we are concerned with, this is only a small fraction of the total sialic acid of the cell and is not a major factor in determining the *total* sialic acid level of a cell or of its membranes. Indeed there are reports that the sialic acid content of transformed cells or its membranes is lower than that of controls (Wu *et al.*, 1969). We have found that, depending on the cell line, the sialic acid level of a virus-transformed cell may be higher, the same or lower than its normal counterpart (Hartmann *et al.*).

Further work must be done to see whether this difference between control and transformed cells can be extended to other cell systems, whether the difference disappears upon reversion and whether some sort of analogous pattern is found in human, solid tumors. The consistency with which we have found this difference, the fact that it accompanies temperature sensitivity in T5 transformed cells and its link to the growth process permits some hope that it is associated with malignancy, a disorder of growth. Our studies have focused on the cell surface where a difference has been found. This difference is consistent with modern concepts of control of cell behavior stemming from the cell surface and altered cell behavior (malignancy) deriving from changes in the cell surface which ultimately are the consequence of genetic alteration.

ACKNOWLEDGEMENTS

The invaluable assistance of Mrs Adele Gallucci, Suzanne Redmond and Miss Kerstin Malmstrom is gratefully acknowledged. This work was supported by grants from the American Cancer Society BS-16A, PRP-28 and the US Public Health Service 5 PO1 AI 0700507 and CA 12426-01.

REFERENCES

Abercrombie, M. and Heaysman, J. E. M. (1954). *Exp. Cell Res.* **6**, 293–306.
Buck, C. A., Glick, M. C. and Warren, L. (1970). *Biochemistry* **9**, 4567–4576.
Buck, C. A., Glick, M. C. and Warren, L. (1971a). *Biochemistry* **10**, 2176–2180.
Buck, C. A., Glick, H. C. and Warren, L. (1971b). *Science* **172**, 169–171.
Burger, M. M. (1968). *Nature* **219**, 499–500.
Coman, D. R. (1944). *Cancer Res.* **4**, 625–629.
Coman, D. R. (1961). *Cancer Res.* **21**, 1436–1438.
Cumar, F. A., Brady, R. O., Kolodny, E. H., McFarland, V. W. and Mora, P. T. (1970). *Proc. Natl. Acad. Sci. USA* **67**, 757–764.
Den, H., Schultz, A. M., Basu, M. and Roseman, S. (1971). *J. Biol. Chem.* **246**, 2721–2723.
Grimes, W. J. (1970). *Biochemistry* **9**, 5083–5092.
Hartmann, J. F., Buck, C. A., Defendi, V., Glick, M. C. and Warren, L. (1972). *J. Cell Physiol.* **80**, 159–165.
Inbar, M. and Sachs, L. (1969). *Nature* **223**, 710–712.
Martin, G. S. (1970). *Nature* **227**, 1021–1023.
Meezan, E., Wu, H. C., Black, P. H. and Robbins, P. W. (1969). *Biochemistry* **8**, 2518–2524.
Onodera, K. and Sheinin, R. (1970). *J. Cell Sci.* **7**, 337–355.
Sheinin, R. and Onodera, K. (1970). *Can. J. Biochem.* **48**, 851–857.
Warren, L. and Glick, M. C. (1969). *In* "Fundamental Technique in Virology" (K. Habel and N. P. Salzman, eds.) pp. 66–71. Academic Press, New York.
Warren, L., Critchley, D. and Macpherson, I. (1972a) *Nature* **235**, 275–278.
Warren, L., Fuhrer, J. P. and Buck, C. A. (1972b) *Proc. Natl. Acad. Sci. USA* **69**, 1838–1842.
Warren, L., Fuhrer, J. P. and Buck, C. A. (1973). *Fed. Proc.* **32**, 80–85.
Wu, H. C., Meezan, I., Black, P. H. and Robbins, P. W. (1969). *Biochemistry* **8**, 2509–2517.

Modification of Glycoprotein Biosynthesis in Transformed Mouse Cell Lines

PAUL W. KENT* and PETER T. MORA

Macromolecular Biology Section, Laboratory of Cell Biology
National Cancer Institute, National Institutes of Health, Maryland 20014

The biological function of oligosaccharides attached to proteins, as glycoproteins, or to lipids as in the gangliosides, still in general remains obscure. The case of blood group specificity of the ABO system however is an exception and there the terminal sugar residues have been established as playing a fundamental role as the antigenic determinants. Other cases, investigated in lesser detail, indicate that this phenomenon may have much wider implications and that oligosaccharides, especially those bound to the exterior surfaces of mammalian cell membranes have informational potential. The present study is concerned with an exploration of possible ways of altering the structures (and hence their information content) of bound oligosaccharides by intervening in their biosynthesis at the metabolite level by means of N-acylglucosamine analogues.

MONOSACCHARIDES AND GLYCOPROTEIN BIOSYNTHESIS

An accumulation of evidence indicates that, while glucosamine is rapidly transported by mammalian cells and subsequently is a precursor for intracellular biosynthesis of glycoproteins, in some circumstances it can produce interesting inhibitory effects on cell growth. Quastel and Cantero (1952) found inhibitory effects of this amino sugar and the anti-tumour activity of the compound has been frequently investigated (Becker *et al.*, 1964; Macbeth and Akpata, 1967; Voss *et al.*, 1963). Exogenous glucosamine is reported to restrict the growth of a number of experimental tumours *in vivo* (Bekesi and Winzler, 1969, 1970). *In vitro*, glucosamine inhibits the biosynthesis of protein, DNA and RNA of neoplastic tissues (Bekesi *et al.*, 1969a; Bosmann, 1971). In high

---

* Present address: Glycoprotein Research Unit, Durham University, England.

concentration (e.g. 2·5 mg/ml), glucosamine was found to inhibit its own incorporation as well as that of glycine, into macromolecular cell products in cultured retinal and chick embryonic liver cells (Richmond et al., 1968). Galactosamine and mannosamine exhibit similar though quantitatively smaller effects. It is uncertain how far these effects are specific to glucosamine and its derived intermediates, acting on enzymes and on transport mechanisms. Exposure of Walker tumour cells to 125 mM glucosamine results in degenerative changes in the nuclei and nucleoli (Molnar and Bekesi, 1972). Electron microscopic examination has shown that in other cells irreversible ultrastructural changes occur when glucosamine is present in culture medium (Molnar, 1968; Molnar and Bekesi, 1972) and changes in the viability and transplantability of Ehrlich ascites cells, as well as Sarcoma 37 and Sarcoma 80, grown with glucosamine have been investigated (Bekesi et al., 1969b). The structural defects may well arise from changes in the biosynthesis of oligosaccharide-bearing molecules (glycoprotein and glycolipids) present in intracellular as well as plasma membranes, for which glucosamine and glucosamine 6-phosphate are metabolic intermediates. Inhibition of glucosamine metabolism resulting in incomplete oligosaccharide chains in turn results in altered macromolecular and, in some cases, immunological properties. It is an important characteristic feature of both glycoprotein and glycolipid biosynthesis that the synthesis of the oligosaccharides is effected enzymically and does not depend directly on information from the genome, as does protein biosynthesis. The cell is thus provided with means for varying macromolecular function and properties in response to environmental and metabolic changes without recourse to the genetic information.

At the present time, glucosamine metabolism is known to be susceptible to at least four regulatory steps (Fig. 1). The conversion of fructose 6-phosphate into glucosamine 6-phosphate by a specific transamidase (Winterburn and Phelps, 1971) is powerfully inhibited through feedback by UDP-N-acetylglucosamine (and relieved by UTP). This reaction is also markedly dependent on glutamine concentration, decrease in which leads to diminished rates of aminosugar biosynthesis. A further important regulatory step is the subsequent N-acetylation of glucosamine 6-phosphate (Kent, 1970).

This largely irreversible step is essential to all the subsequent metabolic transformations of the aminosugar, to the extent that N-acetylglucosamine intermediates can be regarded, metabolically, as separate sugars distinct from their non-acetylated forms. A further regulatory step involves the specific transferences in which UDP-N-acetylamino sugars participate in oligosaccharide synthesis.

While considerable evidence indicates that N-acetylglucosamine can be enzymically phosphorylated by ATP by soluble mammalian enzymes (Datta, 1970), it is nevertheless utilized by cells in ways distinct from glucosamine. Injection of N-acetylglucosamine into rats showed that it is utilized much more slowly than glucosamine (Kohn et al.,

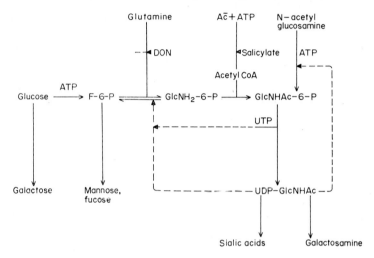

Fig. 1. Regulatory steps in intermediary metabolism of amino sugars.

1962; Richmond, 1963). Rat plasma glycoprotein obtained 3 h after administration of 1-^{14}C-N-acetylglucosamine earned only 3% of the label found in a comparable experiment with 1-^{14}C-glucosamine.

MOUSE CELL LINES

T AL/N is an epithelial-like cell line which appears in early passage in culture by 'spontaneous' transformation of a normal N AL/N cell line. SVS AL/N is an SV40 virus transformed derivative cell line from N AL/N. SVS AL/N cells are non-virus producer cells which, however, continue to carry the virus genome and exhibit overgrowth in tissue culture higher than the N AL/N or the T AL/N cells. The properties of these cell lines have been reported before (Mora et al., 1969; Smith et al., 1970; Mora and Brady, 1971), including analyses on glycolipids (Brady and Mora, 1970; Dijong et al., 1971). The rate of growth in culture (in logarithmic phase) and the size and weight (protein content) of T AL/N and SVS AL/N cells are approximately equal (cf. Mora et al., 1971). In the current studies the tumorigenic T AL/N cells were employed, rather than the normal, non-tumorigenic precursor N AL/N

Table I

Labelling of T AL/N and SVS AL/N cells by 1-¹⁴C-D-glucosamine, G-³H-L-threonine and ³H-thymidine

	Cell protein mg/per plate	³H-thymidine incorporation		¹⁴C-glucosamine incorporation		³H-threonine incorporation		$\dfrac{^{14}C\text{-glucosamine}}{^{3}H\text{-threonine}}$
		Total µCi	% input	Total µCi	Kcpm/mg protein	Total µCi	Kcpm/mg protein	Ratio of specific activities
T AL/N	1·28	1·96	(10·8)	0·485 (2·7%)	163·5	0·368 (2·0%)	76·5	2·14
SVS AL/N	3·60	2·42	(13·4)	1·4 (7·8%)	162·5	1·2 (6·7%)	92·5	1·75

cells for two reasons: first, N AL/N cells transform spontaneously in culture in early passage to T AL/N cells (Mora et al., 1969) thus only T AL/N cells can be produced in sufficient quantity for biochemical studies. Second, with T AL/N cells it is possible to study transplantability (tumorigenicity) in the syngeneic animals, including possible changes in tumorigenicity which may be altered as a consequence of the treatment with the halogenated sugars. This is not possible with the N AL/N cells.

Cell growth labelling and harvesting conditions

All cells were grown under Eagle's medium, supplemented with 10% fetal calf serum with added penicillin, streptomycin and mycostatin as described before (Mora et al., 1969) in 10-cm diameter Falcon petri dishes in a humidified incubator at 37°C containing 5% CO_2–95% O_2. Cells were seeded appropriately to achieve about 50% confluence before labelling. For the labelling experiments the old medium was removed and the cells were incubated with fresh medium (3 ml/plate) containing the halogenated sugars (the 'modifiers') but no serum. After incubation for 1 h at 37°C an equal volume of medium (with antibiotic) was added which contained 20% foetal calf serum (FCS), and 2 μCi/ml radioactive precursor. Cells were then incubated for various lengths of time (generally 16 h). When growth was continued for 6 days, fresh medium with 10% FCS and the modifiers was used to replace the old medium on the second and the fourth days and the radioactive precursor (1 μCi/ml) was added 16 h before harvesting. Harvesting was always just before the cell layer attained confluence, which corresponds to logarithmic rate of growth (Mora et al., 1969). Before harvesting the medium was carefully removed and the cell layer was washed *in situ* with cold tris-buffered saline, pH 7·4 containing when appropriate, the non-radioactive precursor in about equal concentration to that in the labelling experiment. Fresh buffered saline solution was added and the cells were removed by gentle scraping with a rubber policeman, centrifuged, and washed twice more with buffered saline.

To prepare 'acetone powder' (Kanfer, 1966) 10 volumes of chilled (-15°C) acetone was added to the washed cells. The white precipitate was collected by centrifugation (5 min 600 g), then washed once with acetone. The final precipitate was dried to a fine powder in the vacuum desiccator.

N-fluoracetyl-D-glucosamine (GlcNAcF) was synthesized (by Mr. C. G. Butchard) by the method published by Dwek et al. (1971); N-iodoacetyl-D-glucosamine (GlcNAcI) was synthesized by the method of Kent et al. (1970). Protein was determined by the method of Lowry et al. (1951).

BIOLOGICAL EFFECTS OF FLUORO SUGARS

Analogues of a variety of biochemical metabolites are available in which one or more fluorine atoms have been introduced in the course of their chemical synthesis. These have been described in detail in a recent review

(CIBA Symposium, 1972). Amongst these, a considerable number of fluoro-monosaccharides have been synthesized and characterized. As yet comparatively little information exists regarding the metabolic fate of these modified sugars. In the case of fluorinated derivatives of glucose, it would appear that there is as yet little evidence of any notable toxicity shown toward mammalian systems. Studies on 3-deoxy-3-fluoro-D-glucose (White and Taylor, 1970) have shown that this compound can be oxidized to the corresponding gluconic acid by *Pseudomonas fluorescens* without apparent development of a biochemical lesion. On the other hand, *rac*-1-deoxy-1-fluoroglycerol appears to be capable of undergoing metabolic transformations leading to fluorocitric acid and thus to lead to marked toxic effects (O'Brien and Peters 1958a, b). The role of the different stereoisomers of fluorodeoxyglycerols as inhibitors of glycerol kinase (*Candida mycoderma*) has been recently reported (Eisenthal *et al.*, 1972). Fluorine derivatives of *N*-acetyl-D-glucosamine offer particular promise in a number of ways. Firstly as close metabolic analogues, sugars such as *N*-fluoroacetyl-D-glucosamine (GlcNAcF) may provide selective enzyme inhibitors of aminosugar transformation at the sugar phosphate or nucleotide level. Secondly, the interesting NMR properties of fluorine facilitate its use as a probe. The irreversibility of *N*-acylamino sugar synthesis in mammals makes it unlikely that cleavage of GlcNAcF to fluoroacetic acid occurs. Extensive studies using the latter techniques with paramagnetic marker ions have shown that *N*-fluoroacetyl-β-D-glucosamine locates itself in the sub-site C of lysozyme in a distinctive manner. The corresponding α-anomer has different and distinguishable effects (Butchard *et al.*, 1972a; Butchard *et al.*, 1972b). The biochemical implications of the work confirm the view that fluorosugar analogues may legitimately be explored as inhibitors or modifiers of enzyme action.

EFFECTS OF *N*-FLUOROACETYL AND *N*-IODOACETYL-D-GLUCOSAMINE

On D-glucosamine-1-^{14}C, L-threonine-G-^{3}H and thymidine-^{3} incorporation by whole cells

When cell lines T AL/N and SVS AL/N were grown in logarithmic phase at pre-confluence in the presence of *N*-fluoroacetylglucosamine and *N*-iodoacetylglucosamine for about 16 h (approx. cell generation 24 h) incorporation of ^{14}C-glucosamine and ^{3}H-threonine continued. Both biochemical modifiers (in increasing concentrations up to about 1 m*M*) produced progressive *increase* in carbohydrate incorporation compared with amino acid uptake, as revealed by changes in the ^{14}C/^{3}H ratio (Fig. 2). Above 1 m*M*, the relative carbohydrate incorporations then appeared to decrease progressively, and in the case of GlcNAcI cells were killed. GlcNAcF on the other hand was markedly less toxic and both types of cells exposed up to 5 m*M* retained their normal viability and morphology. Even after extended growth (6 days) in 0·5 mM GlcNAcF appearance of cells and cell counts (growth) were comparable to the control, untreated culture.

Experiments with both T AL/N and SVS AL/N cells grown for 16 h in the presence of a fixed concentration (0·5 mM) of GlcNAcF showed that some decrease in glucosamine incorporation into whole cells had occurred: the specific activity of 0·5 mM GlcNAcF-treated T AL/N

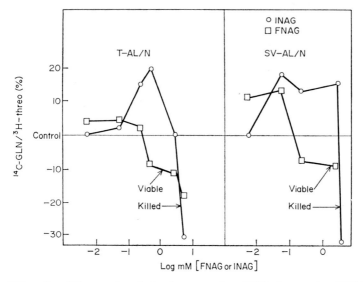

Fig. 2. Effect of modifiers on the relative incorporation of ^{14}C-glucosamine (^{14}C-GLN) and ^{3}H-threonine (^{3}H-threo). (See Table II.) ○, INAG, N-iodoacetylglucosamine (GlcNAcI); □, FNAG, N-fluoroacetylglucosamine (GlcNAcF).

cells being 0·61μCi ^{14}C/mg cell protein, compared with a value of 1·57μCi/mg cell protein for cells grown in modifier-free medium. Similarly modifier-free SVS AL/N cells had a noticeably higher rate of ^{14}C-labelling: specific activity 1·73μCi ^{14}C/mg cell protein, which was diminished to 0·49μCi ^{14}C/mg cell protein for cells grown for 16 h in the presence of 0·5 mM GlcNAcF. This modifier at 0·5 mM concentration did not appear to influence significantly protein biosynthesis as monitored by incorporation of (G-^{3}H)-threonine (Table II), and even at concentrations up to 5 mM both glucosamine and threonine continued to be utilized. On the other hand, at concentrations above 1 mM, GlcNAcI caused a sharp decline in the amount of isotopes incorporated into cells (Fig. 2).

Table II
Cells (T AL/N and SVS AL/N) grown in presence of modifier
The duration of growth was 16 h.

Cells and conditions	3H-threonine		14C-glucosamine		3H-thymidine	
	total µCi incorporated	% input	total µCi incorporated	% input	total µCi incorporated	% input
GlcNAcF						
SVS AL/N						
inhibitor-free control*	0·14	2·9	0·36	7·2	—	—
0·5 mM GlcNAcF†	1·27	—	0·42	8·3	0·34	6·8
		2·8	3·42	7·5	—	—
T AL/N						
inhibitor-free control*	0·29	5·8	0·82	16·2	—	—
0·5 mM GlcNAcF†	—	—	1·03	20·6	0·47	9·4
	2·89	6·4	6·8	15·1	—	—
GlcNAcI						
SVS AL/N						
inhibitor-free control*	0·40	7·9	0·97	19·4	—	—
0·5 mM GlcNAcI†	—	—	0·61	12·2	0·79	15·8
	4·05	9·0	9·36	20·8	—	—
T AL/N						
inhibitor-free control*	0·30	6·0	0·80	16·0	—	—
0·5 mM GlcNAcI†	—	—	0·66	13·1	0·78	15·5
	3·24	7·3	12·60	27·8	—	—

* 5 µCi of each labelled substrate in 5 ml total vol. initially.
† 45 µCi of each labelled substrate in 45 ml initially, i.e. 9 plates each of 5 µCi.

Glycolipid biosynthesis

In studies of the distribution of ^{14}C and ^{3}H in control and GlcNAcF-grown cells it was found that the glycolipid fraction (Fig. 3) of untreated T AL/N cells incorporated the sugar precursor to a larger extent (0·71% of input ^{14}C-glucosamine counts) compared with the untreated SVS AL/N cells (0·31%), in agreement with the reported absence of higher gangliosides in the latter cell line (Mora et al., 1969; Brady and Mora, 1970; Dijong et al., 1971). In the presence of GlcNAcF (0·5 mM), the glycolipid composition of both cell lines became markedly diminished (0·05% incorporation of input ^{14}C for T AL/N and 0·11% for SVS AL/N).

Glycoprotein biosynthesis

^{14}C-glucosamine counts in the cells were also distributed between low molecular weight substances (including intermediates) and high molecular weight products) and these were measured by papain digestion of cell residues after the extraction of glycolipids, followed by dialysis. With both T AL/N and SVS AL/N cell lines in the presence and absence of GlcNAcF, the proportion of ^{14}C in the diffusible fraction (low molecular weight fraction) was in the region of 30% (Table III). The non-diffusible fraction (glycoprotein fraction) in controls accounted for most of the radioactive labelling, the specific activity after 16 h growth of the SVS AL/N products (av. 0·20 µCi ^{14}C/mg protein), being again higher than in the corresponding T AL/N material (av. 0·135 µCi ^{14}C/mg protein). Similar results were obtained when the glycoprotein fraction was separated by gel chromatography (Kent and Winterbourne; also Fishman, Brady and Mora, unpublished observations). Thus it appears likely that a real difference exists in the rate, and possibly the type, of glycoprotein synthesis between untreated T AL/N and SVS AL/N cells, the values being elevated as a result of the virus-transformation. Our values now calculated as specific activities on separated fractions, confirm the suggestions that transformation by tumorigenic viruses in other cell lines is accompanied by similar glycoprotein changes (Warren et al., 1972; Buck et al., 1970, 1971; Meezan et al., 1969; Wu et al., 1969).

While the general trend of our data indicates that the virally transformed line SVS AL/N synthesizes more glycoprotein than T AL/N, it is not possible to know at this stage whether this is stimulated synthesis of existing material or whether a glycoprotein of altered structure is formed. Attempts to isolate the radioactive glycoprotein fraction by precipitation from aqueous solution by trichloroacetic acid resulted in substantial losses of radioactivity (c. 50%). Thus these products are not completely precipitable by this technique. In a number of control

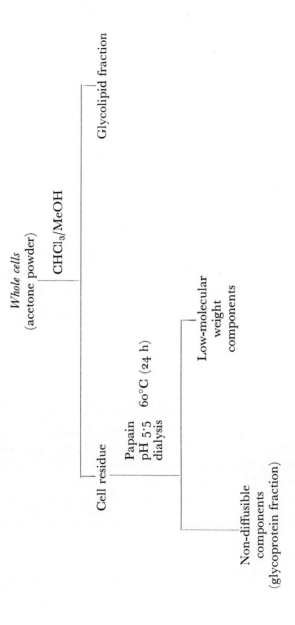

Fig. 3. Scheme for separation of labelled components from acetone-dried cells. (Conditions of Mora *et al.*, 1969.)

Table III
Effect of modifiers on the labelling of the glycolipid and glycoprotein fractions

Cells and conditions	Length of culti- vation	Acetone powder Prot. mg (1)	Glycolipid fractions μCi incorporated (total)	Papain digest prior to dialysis μCi (total)	Papain digest after dialysis μCi (total)	Protein mg (total) (2)	High molecular weight fraction Sp. act. using (1)	Sp. act. using (2)
SVS AL/N								
No modifier	1 day	66.5	0.14(0.073)*	15.86	10.857	—	0.163(0.24)*	—
0.5 mM GlcNAc	1 day	8.8	—	0.914	0.641	2.03	0.073	0.316
0.5 mM GlcNAcF	1 day	6.6	—	0.580	0.392	2.63	0.059	0.149
0.5 mM GlcNAcI	1 day	4.2	—	0.961	1.000	2.71	0.238	0.369
No modifier	6 days	—	—	0.778	0.503	1.96	—	0.257
0.5 mM GlcNAcF	6 days	—	—	3.203	2.118	1.03	—	2.056
TAL/N								
No modifier	1 day	129.5	0.32(0.129)*	16.87	13.03	—	0.14(0.13)*	—
0.5 mM GlcNAc	1 day	1.7	—	0.694	0.408	1.90	0.240	0.215
0.5 mM GlcNAcF	1 day	3.9	—	0.320	0.203	2.52	0.052	0.081
0.5 mM GlcNAcI	1 day	1.6	—	0.699	0.465	1.22	0.291	0.381
No modifier	6 days	—	—	0.760	0.467	1.55	—	0.301
0.5 mM GlcNAcF	6 days	—	—	4.819	3.104	1.82	—	1.706

* Results of duplicate experiments in parenthesis.

experiments, the effect of N-acetylglucosamine upon the ^{14}C-glucosamine labelling of these cells was included for comparison. While under comparable conditions this sugar also modifies the labelling pattern of the cells, it appears to be less effective than the halogen analogues.

Continued presence of the modifier

It is of interest that cells grown in the presence of GlcNAcF for longer periods (6 days) resulted in several-fold increase in specific activity of ^{14}C labelling of the glycoprotein fraction (Table III). The specific activity of the SVS AL/N fraction was thus 2·1 compared to 0·26 for modifier-free control. Values for T AL/N were: 1·7 μCi/mg protein for glycoprotein fraction from cells grown in GlcNAcF, and 0·30 from modifier-free control. The iodo-analogue GlcNAcI proved too inhibitory for comparable investigations.

The effect of growth in GlcNAcF appeared to bring about changes in the biosynthesis pathway without manifest alteration of growth properties of cells. While GlcNAcI displays some of the same effects it is noticeably more toxic to cells particularly in the longer term experiments. The toxic effect of GlcNAcI is to be expected on account of its known action as an alkylating agent which influences active transport and growth in *E. coli* K12 and ML 308 (Kent *et al.*, 1970; White and Kent, 1970; White, 1970). In *E. coli* and *B. subtilis* GlcNAcF brings about restriction

Table IV

^{14}C/^{3}H ratios for cells grown in low concentrations of GlcNAcF and labelled with 1-^{14}C-glucosamine together with ^{3}H-thymidine or ^{3}H-uridine

Duration of growth, 16 h.

cells	GlcNAcF (final conc.) mM	Ratios of specific activities	
		$\dfrac{^{14}C\text{-}glucosamine}{^{3}H\text{-}thymidine}$	$\dfrac{^{14}C\text{-}glucosamine}{^{3}H\text{-}uridine}$
SVS AL/N	0·5	1·98	lost
	5×10^{-2}	2·81	0·92
	5×10^{-3}	3·11	0·71
	5×10^{-4}	3·02	0·69
	5×10^{-5}	2·22	1·0
T AL/N			
	0·5	1·63	0·71
	5×10^{-2}	1·78	0·67
	5×10^{-3}	2·02	0·52
	5×10^{-4}	2·06	0·59
	5×10^{-5}	1·86	0·58

of growth though it is unlikely to act by alkylation (Kent, unpublished).

N-fluoroacetylglucosamine appears to exert little significant effect on the ^{3}H-thymidine labelling of either cell line, over a wide range of concentration; thus the modifier does not appear to inhibit DNA synthesis. Similarly no significant effect was shown by the modifier upon ^{3}H-uridine incorporation by either cell line suggesting that RNA biosynthesis too was not altered by the modifier (Table IV). The fact that these two important energy dependent pathways appear to remain unimpaired, suggests that the modifier is not acting adversely on ATP-generating processes in the cell. Preliminary metabolic experiments with mammalian epithelial cells are in keeping with this view. In other tissues, e.g. sheep colonic epithelium *in vitro* and calf tracheal explants in organ culture, GlcNAcF was not found to produce evident toxicity as shown by Q_{O_2} measurements, though as in the case of T AL/N and SVS AL/N cells, the compound was a significant modifier of the extent of glucosamine utilization by the tissues (Kent *et al.*, 1971).

As yet no direct measurements of the effect, if any, of GlcNAcF upon active transport mechanisms in T AL/N or SVS AL/N cells have been made, but the similarity of labelling of soluble low molecular weight substances in control and treated cells leads to a tentative suggestion that transport sites are not the prime points of action of the modifiers. It has been noted that in red cells, hamster intestinal ring preparations and in rat diaphragm, GlcNAcI does not inhibit active transport of ^{14}C-methyl-α-D-glucoside (Kent *et al.*, 1970). Barnett *et al.* (1971) reported that amino sugars (glucosamine, galactosamine and mannosamine) were only weakly accumulated by hamster intestinal rings and were poor non-competitive inhibitors of the active transport of galactose by the tissue. N-acetylglucosamine was almost inactive in the latter respect. The possibility that GlcNAcF undergoes metabolic transformation cannot be excluded, especially since other fluoro sugars, e.g. 6-deoxy-6-fluoro-D-galactose are known to be ready substrates for enzymes (Kent and Wright, 1972). Further effects may arise from the possible incorporation of GlcNAcF in the structural elements of the cells. Evidence of such has been reported (Virijonandha and Baxter, 1970) in the cell wall structures of *Staph. aureus*. Detailed studies using ^{19}F-NMR techniques have been made of the binding of N-fluoroacetylglucosamine to specific sites in lysozyme (Dwek *et al.*, 1971; Butchard *et al.*, 1972a, b).

The effect of GlcNAcF treatment on cell surface properties and on the resulting changes in biological properties of the cells, including the expression of cell surface antigens such as histocompatibility antigens, will be presented separately.

CONCLUSIONS

Two mouse-cell lines T AL/N and SVS AL/N have been grown for one generation (c. 16 h) in 0·5 M to 5 mM N-fluoroacetyl-D-glucosamine without detectable toxic effects. Below 1 mM, the fluoro sugar modifier increases the incorporated $^{14}C/^{3}H$ ratio for whole cells of both lines with 1-^{14}C-D-glucosamine and G-^{3}H-L-threonine as labels. The ratio is substantially diminished at concentrations above 1 mM by comparison with modifier-free controls. Isolated glycolipid and glycoprotein fractions from cells grown for 16 h in the presence of N-fluoroacetylglucosamine (0·5 mM) have specific activities appreciably lower than control fractions. Growth of cells for 6 days under these conditions however leads to elevation of the specific activities of the glycoprotein fraction compared with controls. N-iodoacetylglucosamine exhibits similar short term effects but it is more toxic. Parallel studies showed little effect of modifiers on ^{3}H-L-threonine, ^{3}H-thymidine and ^{3}H-uridine incorporation in double-labelling experiments with ^{14}C-glucosamine. Evidence indicates that the SVS AL/N-cell line synthesizes more glycoprotein than the T AL/N cell line. The action of the halogeno sugars appears to be due at least in part to selective modification of amino sugar metabolism.

ACKNOWLEDGEMENTS

We thank Leslie Danoff, Vivian McFarland, Lorenzo Waters and D. Winterbourne for technical assistance and help with the glycoprotein determination, Dr. Roscoe Brady for advice and the Medical Research Council, the Wellcome Foundation and the Cystic Fibrosis Research Trust for support to P. W. Kent, and Mrs. T. McGahan for valuable secretarial help.

REFERENCES

Atkinson, P. H. and Summers, D. F. (1971). *J. Biol. Chem.* **246**, 5162–5175.
Barnett, J. E. G., Holman, G. D., Ralph, A. and Munday, K. A. (1971). *Biochim. Biophys. Acta* **249**, 493–497.
Becker, K., Voss, H. and Lindner, J. (1964). *Arch. Geschwulstforsch.* **22**, 297–366.
Bekesi, J. G. and Winzler, R. J. (1969). *J. Biol. Chem.* **244**, 5663–5668.
Bekesi, J. G. and Winzler, R. J. (1970). *Cancer Res.* **30**, 2905–2910.
Bekesi, J. G., Bekesi, E. and Winzler, R. J. (1969a). *J. Biol. Chem.* **244**, 3766–3772.
Bekesi, J. G., Molnar, Z. and Winzler, R. J. (1969b). *Cancer Res.* **29**, 353–359.
Bosmann, H. F. (1971). *Biochim. Biophys. Acta* **240**, 74–93.
Brady, R. O. and Mora, P. T. (1970). *Biochim. Biophys. Acta* **218**, 308–319.
Buck, C. A., Glick, M. C. and Warren, L. (1970). *Biochemistry* **9**, 4567–4576.
Buck, C. A., Glick, M. C. and Warren, L. (1971). *Biochemistry* **10**, 2176–2180.
Butchard, C. G., Dwek, R. A., Kent, P. W., Williams, R. J. P. and Xavier, A. (1972a). *Europ. J. Biochem.* **27**, 548–553.
Butchard, C. G., Dwek, R. A., Ferguson, S. J., Kent, P. W., Williams, R. J. P. and Xavier, A. V. (1972b). *FEBS Letters* **25**, 91–93.

CIBA Symposium (1972). "Carbon-Fluorine Compounds; Their Chemistry, Biochemistry and Biological Activities." Assoc. Scientific Publishers, London and Amsterdam.
Cumar, F. A., Brady, R. O., Kolodny, E. G., McFarland, V. W. and Mora, P. T. (1970), *Proc. Natl. Acad. Sci. USA* **67**, 757–764.
Datta, A. (1970). *Biochim. Biophys. Acta* **220**, 51–60.
Dijong, I., Mora, P. T. and Brady, R. O. (1971). *Biochemistry* **10**, 4039–4044.
Dwek, R. A., Kent, P. W. and Xavier, A. V. (1971). *Europ. J. Biochem.* **23**, 343–348.
Eisenthal, R., Harrison, R., Lloyd, W. J. and Taylor, N. F. (1972). *Biochem. J.* **130**, 199–205.
Hayden, G. A., Crowley, G. M. and Jamieson, G. A. (1970). *J. Biol. Chem.* **245**, 5827–5832.
Kanfer, J. N. (1966). In "Methods in Enzymology" (J. Lowenstein, ed.) Vol. XIV, p. 660. Academic Press, New York and London.
Kent, P. W. (1970). "Exposés Annuels de Chemie Médicale". Sér. **30**, pp. 97–120.
Kent, P. W. and Wright, J. (1972). *Carbohyd. Res.* **22**, 193–200.
Kent, P. W., Ackers, J. P. and White, R. J. (1970). *Biochem. J.* **118**, 73–79.
Kent, P. W., Daniel, P. F. and Gallagher, J. T. (1971). *Abs. Commun. 7th Cong. Fed. Europ. Biochem. Soc.* p. 1108. Varna.
Kohn, P., Winzler, R. J. and Hoffman, R. C. (1962). *J. Biol. Chem.* **237**, 304.
Lowry, O. H., Rosebrough, N. J., Farr, A. L. and Randall, R. J. (1951). *J. Biol. Chem.* **193**, 265.
Macbeth, R. A. and Akpata, M. (1967). *Cancer Res.* **27**, 912–916.
Meezan, E., Wu, H. C., Black, P. H. and Robbins, P. W. (1969). *Biochemistry* **8**, 2518–2524.
Molnar, Z. (1968). *Proc. Amer. Assoc. Cancer Res.* **9**, 50.
Molnar, Z., and Bekesi, J. G. (1972). *Cancer Res.* **32**, 389, 756–761.
Mora, P. T. and Brady, R. O. (1971) *Transplantation Proceedings* **3**, 1213.
Mora, P. T., Brady, R. O., Bradley, R. M. and McFarland, V. W. (1969). *Proc. Natl. Acad. Sci. USA* **63**, 1290–1296.
Mora, P. T., Cumar, F. A. and Brady, R. O. (1971). *Virology* **46**, 60–72.
O'Brien, R. D. and Peters, R. A. (1958a). *Biochem. Pharmacol.* **1**, 3–18.
O'Brien, R. D. and Peters, R. A. (1958b). *Biochim. J.* **70**, 188–195.
Onodera, K. and Sheinin, R. (1970). *J. Cell. Sci.* **7**, 337–355.
Quastel, J. H. and Cantero, A. (1952). *Nature* **171**, 252–254.
Richmond, J. E. (1963). *Biochemistry* **2**, 676–686.
Richmond, J. E., Glasser, R. M. and Todd, P. (1968). *Expl. Cell Res.* **52**, 43–58.
Robinson, G. B. (1968). *Biochem. J.* **108**, 275–280.
Smith, R. W., Morganroth, J. and Mora, P. T. (1970). *Nature* **227**, 141–145.
Takemoto, K. K., Ting, R. C. Y., Ozer, H. L. and Fabish, P. (1968). *J. Natl. Cancer Inst.* **41**, 1401–1420.
Virijonandha, J. and Baxter, R. M. (1970). *Biochim. Biophys. Acta* **201**, 495–496.
Voss, H., Becker, K. and Lindner, J. (1963). *Z. Krebsforsch.* **65**, 228–240.
Warren, L., Critchley, D. and Macpherson, I. (1972). *Nature* **235**, 275–277.
White, R. J. (1970). *Biochem. J.* **118**, 89–92.
White, F. H. and Taylor, N. F. (1970). *FEBS Letters* **4**, 268–271.
White, R. J. and Kent, P. W. (1970). *Biochem. J.* **118**, 81–89.
Winterburn, P. J. and Phelps, C. F. (1971). *Biochem. J.* **121**, 701–709; 711–720; 721–730.
Wu, H. C., Meezan, E., Black, P. H. and Robbins, P. W. (1969). *Biochemistry* **8**, 2509–2517.

Effect of Influenza Virus Infection on Lipid Metabolism of Chick Embryo Fibroblasts

H. A. BLOUGH and D. B. WEINSTEIN

Scheie Eye Institute, Division of Biochemical Virology and Membrane Research
University of Pennsylvania School of Medicine
Philadelphia, Pennsylvania 19104, USA

Previous studies in this laboratory have shown that the lipid composition of the virion is dependent upon three major factors: first, the primary amino acid sequence and hence the tertiary structure of virus envelope polypeptides (Tiffany and Blough, 1969), second, host cell biosynthetic and/or catabolic pathways (Blough and Lawson, 1968) and third, environmental factors (Blough and Tiffany, 1969). Whereas lipids make up 23% by weight of the influenza virion (Blough and Merlie, 1970), there have been no studies, to date, on both synthesis and turnover of lipids following infection of chick fibroblasts with enveloped viruses. We shall direct our attention to the effect of influenza virus on the metabolism of lipids in chick embryo fibroblast cells and its relationship to envelope and hence membrane biogenesis.

MATERIALS AND METHODS

Cells and media. Primary chick embryo fibroblasts (CEF) were grown in Puck's medium with 10% fetal calf serum. When confluent, cells were washed and 'starved' for 18 h in a modified Eagle's medium (MEM) containing 1% fatty-acid poor bovine serum albumin. Uptake studies using ^{3}H-amino acids or uridine revealed that 'starving' the cells had no appreciable effect on RNA or protein synthesis.

Virus adsorption. The A_0/WSN strain of influenza virus was used throughout. Three to five pfu or 20 pfu per cell of influenza virus was adsorbed at $+4°C$ or 20°C for 45 min; controls were 'mock-infected' with dilutions of sterile allantoic fluid.

Isotopic techniques. Monolayers were pulsed at $+37°C$ for 30 min at various times post-infection with a double label containing 1 μCi/ml of ^{3}H-glycerol and 0·5 μCi of ^{14}C-acetate (in the 'starve medium'); cells were washed and 'chased' at various time intervals with MEM containing 1% BSA*. Monolayers were washed with cold medium, scraped off the plastic flasks, sedimented by low speed centrifugation, and frozen at $-70°C$ overnight.

Extraction of lipids. Lipids were extracted from the cells with $CHCl_3$-CH_3OH (2 : 1, v/v). Neutral lipids were separated by unidimensional thin layer chromatography on 36 cm plates of silica gel G using a two-solvent system; phospholipids were separated by two-dimensional thin layer chromatography. Spots were localized by exposure to iodine vapor, scraped off the glass, and one-half the sample quantitated by charring or by analyzing for inorganic phosphorus (Blough and Merlie, 1970). The other half of the sample was mixed with a toluene-based scintillation fluid and counted in an Intertechnique spectrometer. All values were corrected for overlap of channels and the efficiency calculated using known standards.

RESULTS

Infection with 3-5 pfu of influenza virus produced a significant change in the synthesis of lipids by 7 h post-infection (Fig. 1) at which

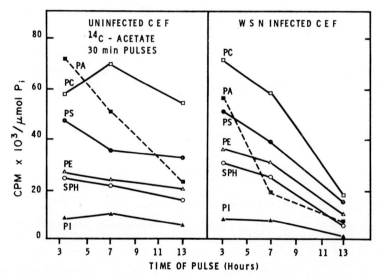

Fig. 1. Time of inhibition of lipid synthesis in chick fibroblasts infected with 3–5 pfu/cell of influenza virus (A_0/WSN).

time the synthesis, of phosphatidic acid, sphingomyelin, phosphatidylinositol and phosphatidylcholine were depressed. At 13 h all phospholipids were inhibited (range 56–94%).

*For the explanation of abbreviations see end of this chapter.

If however the multiplicity of infection is increased to 20 pfu/cell a 16–40% depression of all lipids was seen as early as four h post-infection (Fig. 2). By pulsing at various times post-infection we were able to follow both the synthesis and turnover of the glycerol backbone and fatty acyl

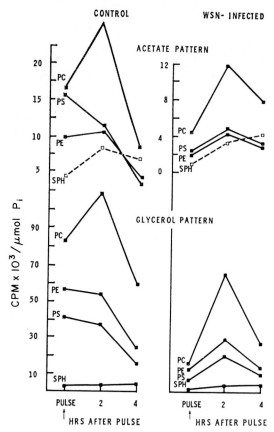

Fig. 2. Effect of high multiplicity of infection on suppression of phospholipids in CEF (pulsed at 4 h post-infection).

chains. Despite a depression of phospholipid synthesis, synthesis and turnover of neutral lipids appear relatively 'normal' even at 12 h post-infection; an outstanding difference was noted in the sterol and triglyceride patterns (Fig. 3). Thus, our results suggest an uncoupling of synthesis and turnover of phospholipids and neutral lipids.

Using short pulse-chase studies, changes of phospholipid synthesis were examined in cells infected with Von Magnus virus (3rd passage, $\log_{10}$ $EID_{50}/\log_{10}$ HA = 3·02); this virus lacks a portion of its genome

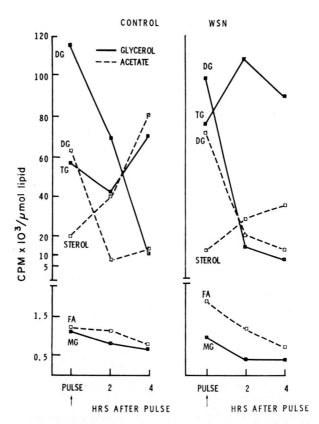

Fig. 3. Synthesis and turnover of neutral lipids in CEF and effect of influenza virus (20 pfu/cell); cells pulsed at 8 h post-infection.

(Pons and Hirst, 1969). A stimulation of phosphatidylcholine and phosphatidylethanolamine was seen at 6 h post-infection (Fig. 4); this is in striking contrast to what occurs in uninfected CEF or CEF infected with standard virus (Figs. 1 and 2).

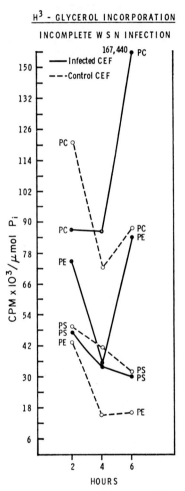

Fig. 4. Effect of Von Magnus virus on phospholipid synthesis in CEF; cells pulsed at 2, 4, or 6 h (no chase).

DISCUSSION

The results of these studies show that the 'shut-off' of host cell lipid metabolism is dependent in part upon the multiplicity of infection and is at variance with the earlier work of Kates et al. (1961) who suggested that newly synthesized lipid is not incorporated into the virus. From this study it can be shown that *de novo* synthesis of lipid continues for 4–7 h post-infection in chick cells. These results are also different from those noted in chick fibroblasts infected with an alphavirus (Waite and Pfefferkorn, 1970) where lipid synthesis is inhibited at 2–3 h when ^{3}H-choline is used as a precursor of phospholipid synthesis. The results found in Von Magnus virus infected cells agree with our earlier compositional data on incomplete influenza virus (Blough and Merlie, 1970). More important, perhaps, is the finding that an uncoupling of synthesis (and turnover) of phospholipids and neutral lipids occurs in virus-infected cells; while this may reflect differences in the number or size of precursor pools, our isotopic techniques suggest that in chick fibroblasts there may be a pool of preformed lipid which is used for membrane biogenesis (hence envelope biogenesis) during the course of virus infection.

ACKNOWLEDGEMENTS

This work was supported, in part, by the Commission on Influenza, Armed Forces Epidemiological Board, Department of the Army by a contract from the United States Army Research and Development Command (contract no. DADA 17-67C-7128).

REFERENCES

Blough, H. A. and Lawson, D. E. M. (1968). *Virology* **36**, 286–292.
Blough, H. A. and Merlie, J. P. (1970). *Virology* **40**, 685–692.
Blough, H. A. and Tiffany, J. M. (1969). *Proc. Natl. Acad. Sci. U.S.* **62**, 242–257.
Kates, M., Allison, A. C., Tyrrell, D. A. J. and James, A. T. (1961). *Biochim. Biophys. Acta* **52**, 455–466.
Pons, M. and Hirst, G. K. (1969). *Virology* **38**, 68–72.
Tiffany, J. M. and Blough, H. A. (1969). *Science* **163**, 573–574.
Waite, M. R. F. and Pfefferkorn, E. R. (1970). *J. Virol.* **6**, 637–643.

Abbreviations used in figures and text:—BSA, bovine serum albumin; PA, phosphatidic acid; PC, phosphatidylcholine; PE, phosphatidyl ethanolamine; SPH, sphingomyelin; PS, phosphatidyl serine; PI, phosphatidylinositol; MG, monoglycerides; FA, free fatty acids; DG, diglycerides; TG, triglycerides.

IV
Specific Synthetic Functions

Recent Studies on the Biosynthesis of Collagen

DARWIN J. PROCKOP, PETER DEHM, BJØRN R. OLSEN,
RICHARD A. BERG, MICHAEL E. GRANT*, JOUNI UITTO
and KARI I. KIVIRIKKO†

*Department of Biochemistry, The Rutgers Medical School
New Brunswick, New Jersey, 08903, USA*

A number of major discoveries have recently been made concerning the biosynthesis of collagen and in the present chapter a few highlights of these discoveries will be reviewed briefly. (For a more complete review, see Grant and Prockop, 1972.)

It is now clear that collagen biosynthesis occurs in several relatively discrete stages (Fig. 1). At the ribosomal stage, the three polypeptide chains of collagen are synthesized. From the size of the polysomes which synthesize collagen it appears that the synthesis of each chain requires a separate monocistronic mRNA (Lazarides and Lukens, 1971a). As is now well known, the polypeptide chains coded for by the mRNA's do not contain any hydroxyproline or hydroxylysine and they are correspondingly rich in proline and lysine. The hydroxyproline and hydroxylysine found in collagen is synthesized by the hydroxylation of proline and lysine after these amino acids have been incorporated into peptide linkages. The recent controversy concerning the question of whether the hydroxylation of proline can begin while nascent chains are still being assembled on ribosomes (see Grant and Prockop, 1972) has been largely resolved in that most laboratories now agree that some hydroxylation is initiated at the ribosomal stage (Lazarides, Lukens and Infante, 1971; Uitto and Prockop, 1973). A good part of the apparent discrepancy in previous studies (Rosenbloom and Prockop, 1969; Lane et al., 1971) is now explained by the fact that collagen has an

---

* Present address: Department of Medical Biochemistry, University of Manchester, Manchester, England.
† Department of Medical Biochemistry, University of Oulu, Box 191, SF-90100, Oulu, 10, Finland.

unusually long synthesis time and that the rate at which amino acids are incorporated into peptide linkage is only about one-third of the rate observed in *E. coli* or in reticulocytes (Vuust and Piez, 1972). It should be noted however, that if the prolyl and lysyl hydroxylases are temporarily inhibited, all of the hydroxylations can occur after the polypeptide chains are completed and released from ribosomes (Juva *et al.*, 1966; Rosenbloom and Prockop, 1969; Lazarides and Lukens, 1971b). Completion of the

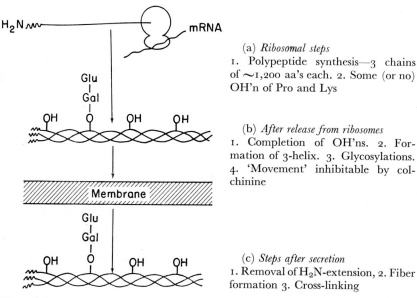

(a) *Ribosomal steps*
1. Polypeptide synthesis—3 chains of ~1,200 aa's each. 2. Some (or no) OH'n of Pro and Lys

(b) *After release from ribosomes*
1. Completion of OH'ns. 2. Formation of 3-helix. 3. Glycosylations. 4. 'Movement' inhibitable by colchinine

(c) *Steps after secretion*
1. Removal of H_2N-extension, 2. Fiber formation 3. Cross-linking

Fig. 1. Scheme summarizing the biosynthesis of collagen. For more detailed schemes, see Grant and Prockop (1972). Reproduced with permission from Prockop *et al.* (1972).

hydroxylation is not essential for the chains to be released and while the hydroxylases are inhibited by chelating agents such as α,α'-dipyridyl or by lack of oxygen, the cells accumulate a completely unhydroxylated protein precursor of collagen which we have called 'protocollagen'. It might also be noted that many connective tissues are relatively anaerobic and may be subjected to intermittent ischemia. In such tissues protocollagen may be a 'normal' intermediate in collagen synthesis. For example, it was demonstrated that newly-synthesized collagen extracted directly from 4-day old chick embryos is about 50% underhydroxylated in that its hydroxyproline content can be doubled by incubating the protein with prolyl hydroxylase (Juva and Prockop, 1965).

Recently it has been possible to isolate from fibroblasts incubated with α,α′-dipyridyl sufficient quantities of protocollagen for amino acid analysis (Jimenez et al., 1973). With this material it has been demonstrated that the protocollagen extracted from cells at 15°C is largely triple-helical and it contains the same NH_2-terminal extension as the precursor form of collagen known as 'procollagen' (see below).

The enzyme which synthesizes the hydroxyproline in collagen has been known variously as collagen proline hydroxylase, peptidyl proline hydroxylase, protocollagen proline hydroxylase and prolyl hydroxylase. The prolyl hydroxylase in chick embryos (Halme et al., 1970; Pänkäläinen et al., 1970) and in new-born rat skin (Rhoads and Udenfriend, 1970) has been well characterized, and a new affinity column procedure (Berg and Prockop, 1973) has made it possible to isolate milligram quantities of pure enzyme in short periods of time. The enzyme requires O_2, Fe^{2+}, a reducing substance such as ascorbate, and α-ketoglutarate which is stoichiometrically decarboxylated to succinate during the synthesis of hydroxyproline (for review, see Grant and Prockop, 1972). The active enzyme isolated from chick embryos has been shown to be a tetramer of about 230,000 Daltons consisting of two separate kinds of inactive monomers of about 62,000 Daltons each (Berg and Prockop, 1973). An inactive form of prolyl hydroxylase has recently been found in tissue culture cells and it has been shown that the inactive form is converted to active enzyme by addition of lactate to the tissue culture system (McGee et al., 1971). There is still some question however as to whether this inactive form of the enzyme can be regarded as a true 'proenzyme', since the specificity of the lactate effect has not been rigorously demonstrated and the data do not as yet exclude the possibility that the inactive form of the enzyme simply consists of monomers or dimers of the tetramer form.

It has been known for some time that large peptides containing sequences of -X-Pro-Gly- are better substrates for the prolyl hydroxylase than short peptides of the same structure and conformation (for reviews, see Grant and Prockop, 1972; Kivirikko, 1972). Since the size of the peptide required for optimal interaction with the enzyme is greater than any single dimension of the enzyme (Olsen et al., 1973), we have suggested that the enzyme may move laterally or 'roll' along large substrates such as protocollagen. The 'rolling hypothesis' has not been adequately tested as yet but it has been shown that there is a marked asymmetry in the hydroxylation of the peptide $(Pro-Pro-Gly)_5$ in that the fourth triplet from the NH_2-terminal end of the peptide is hydroxylated more readily than any other (Kivirikko et al., 1971). This result could be explained by a mechanism in which the initial binding of the enzyme follows a Gaussian distribution around the middle triplet and the enzyme then moves laterally toward the COOH-terminal end of the peptide as it sequentially hydroxylates prolyl residues. However, further studies will be necessary to prove this mechanism and to determine how it might apply to the hydroxylation of protocollagen in which about a hundred prolyl residues are hydroxylated before the substrate is completely converted to product.

The enzyme which synthesizes the hydroxylysine in collagen is similar to prolyl hydroxylase in that it requires the same co-factors or co-substrates (Prockop *et al.*, 1966; Hausmann, 1967; Hurych and Nordwig, 1967; Miller, 1971; Kivirikko and Prockop, 1972; Popenoe and Aronson, 1972). Also, there is a stoichiometric decarboxylation of α-ketoglutarate to succinate during the synthesis of hydroxylysine (Kivirikko *et al.*, 1972). The enzyme is however more difficult to isolate than the prolyl hydroxylase and although it has been purified about 600-fold (Kivirikko and Prockop, 1972), it is not yet obtained in pure form.

The UDP-galactosyl transferase and the UDP-glucosyl transferase which synthesize the galactosyl-hydroxylysine and the glucosylgalactosyl-hydroxylysine in collagen have been shown to require Mn^{2+} and they appear to be associated with membrane fractions of cells (Bosmann and Eylar, 1968a, b; Spiro and Spiro, 1971a, b). However the enzymes have not yet been well characterized.

In considering the hydroxylation of proline and lysine and the glycosylations of hydroxylysine during collagen synthesis, it is probably of importance to note that these reactions occur after all the information in messenger-RNA is translated and therefore they are 'post-m-RNA' reactions. As such, the reactions are not controlled by a template and experimental evidence summarized elsewhere (Grant and Prockop, 1972; Prockop *et al.*, 1973) indicates that we are faced with an important biological problem in trying to understand how these reactions are controlled in tissues synthesizing collagen so that each molecule is modified to about the same extent before it is secreted into the extracellular millieu. We have suggested that cells synthesizing collagen contain some sort of 'barrier' which prevents the secretion of molecules which have not reached an appropriate state of completeness in terms of their content of hydroxyproline, hydroxylysine and glycosylated hydroxylysine. The postulated barrier may consist of a membrane which collagen cannot pass through until it contains a minimal number of hydroxyl groups. In this type of scheme the hydroxyl groups might alter the solubility properties of the molecule or they might be required to form short-lived intermediates involved in active transport of the collagen across the membrane. Another possibility is that the postulated barrier is an enzyme such as the prolyl hydroxylase. The prolyl hydroxylase has been shown to have an extremely high affinity for protocollagen-like peptides ($K_{aff} > 10^{10}$ M^{-1}) but the affinity decreases after the protocollagen is partially hydroxylated (Juva and Prockop, 1969). On this basis one might imagine that the enzyme prevents the secretion of incompletely hydroxylated molecules by retaining them in enzyme-substrate complexes until they reach a critical level of hydroxylation.

Much of the recent excitement concerning the biosynthesis of collagen has concerned the discovery that the first polypeptide chains synthesized are longer than the α-chains found in interstitial collagen because they contain additional extensions at the NH_2-terminal ends of the chains (Layman et al., 1971; Lenaers et al., 1971; Stark et al., 1971; Bornstein et al., 1972; Dehm et al., 1972; Uitto et al., 1972; Burgeson et al., 1972; Tsai and Green, 1972). It has been suggested that the NH_2-terminal extensions might facilitate the formation of the triple-helical structure of collagen but there are as yet no published data to substantiate this possibility and our own experimental attempts have so far been negative. (Unpublished experiments carried out in collaboration with Drs. J. Engel and H. Weidner, Basel.) There is far more convincing evidence that the NH_2-terminal extensions function to facilitate transport of the collagen molecule during initial stages of its biosynthesis by keeping the molecule more soluble under physiological conditions (Layman et al., 1971; Stark et al., 1971; Dehm et al., 1972). Precise studies on the solubility of the molecule containing the NH_2-terminal extensions have been difficult to carry out, but the data indicate it is soluble under conditions where native collagen precipitates into cross-striated fibers. On this basis it seems likely that a principal function of the NH_2-terminal extensions is to keep the molecule from precipitating into fibers during the various stages of its intracellular assembly and until it reaches the appropriate extracellular site for fiber formation. There is still some disagreement as to whether the NH_2-terminal extensions are always cleaved off after the molecule is secreted and there is some evidence suggesting that cleavage might occur intracellularly in calvaria (Ehrlich and Bornstein, 1972). However, experiments carried out with tendons from rapidly growing chick embryos indicate that the cleavage readily occurs in the intact tendon but essentially no cleavage is observed during collagen synthesis by cells isolated from the same tissue by digesting away the extracellular matrix (Jimenez et al., 1971). The results indicate that although the matrix-free cells synthesize and secrete collagen at a rapid rate, the cells themselves do not contain the peptidase activity which removes the NH_2-terminal extensions. Because the NH_2-terminal extensions appear to facilitate movement of the collagen during its synthesis, it was originally suggested that the precursor molecule be called a 'transport form' of collagen (Layman et al., 1971). Most investigators in the field have employed the term 'procollagen', even though this term may introduce confusion over earlier use of the same term for a different type of collagen (Orekhovich and Shpikiter, 1958) and it may confuse the current use of the term 'procollagenase' for inactive precursor forms of collagenase.

Because the study of the 'transport form' or 'procollagen' has been actively pursued in a number of laboratories during the past two years or so, there is now extensive information concerning it. The published data agree that all three polypeptide chains of the molecule have an NH_2-terminal extension and that the molecules containing the extensions can form triple-helical structures. The data from several laboratories also agree that the NH_2-terminal extension on the $\alpha 1$ chain contains cysteine or cystine. In addition, the published reports agree in the conclusion that the amino acid composition of the NH_2-terminal extensions is different from that of the rest of the molecule. Several areas of disagreement remain however. One concerns the size of the NH_2-terminal extension. The only estimate of molecular weight by ultracentrifugation has been carried out with collagen from the skin of dermatosparactic cattle (Lenaers et al., 1971), since to date it has been impossible to obtain sufficient amounts of the precursor form from other sources. The data from the dermatosparactic collagen indicate that the pro-$\alpha 1$ chain has a molecular weight of 105,000 versus 95,000 for the $\alpha 1$ chain itself and they indicate that the molecular weight for the pro-$\alpha 2$ chain is 101,000 versus a molecular weight for the $\alpha 2$ chain of 95,000. Estimates of the size of precursor chains from other sources have only been carried out by gel filtration or polyacrylamide gel electrophoresis in denaturing solvents such as sodium dodecyl sulfate. The data suggest molecular weights ranging from about 115,000 for the precursor chains from membranous bone (Bornstein et al., 1972) to 125,000 for the precursor chains from embryonic tendon (Jimenez et al., 1971). More recent estimates of the molecular weight of the tendon precursor based on amino acid analysis suggest that each chain is about 110,000 (Uitto et al., 1972). The discrepancies in published molecular weights for the pro-$\alpha 1$ and pro-$\alpha 2$ chains may simply reflect differences in the accuracy of the various techniques employed and therefore be only of trivial interest. There are however, two other possibilities which for the moment have not yet been ruled out. One is that the NH_2-terminal extensions in the precursor forms from different tissues may be different. It is generally assumed that the NH_2-extensions on the pro-α chains from normal skin, tendon, and membranous bone and from the skin of cattle with dermatosparaxis are the same but this may not be correct. The potential problem here is illustrated by the fact that by electron microscopy of the segment-long-spacing aggregates it appears that the NH_2-terminal extensions in the procollagen from chick cartilage have an additional positively-staining band not seen in the NH_2-terminal extensions of procollagen from the tendons of chick embryos or in published photographs of procollagen from dermatosparactic cattle (Olsen et al., in preparation). A second possible explanation for the discrepancies in the published data on pro-α chains is that the procedures used to isolate the polypeptide chains may produce a loss of part of the NH_2-terminal extensions, and therefore yield a smaller polypeptide chain than the one which is present in the tissue. For example, we have recently observed that the NH_2-terminal extensions in the skin collagen from sheep with a disease resembling dermatosparaxis (see below) is easily split off if the collagen extracted with 0·05 M acetic acid is allowed to remain in 0·05 M acetic acid at 4°C for more than 24 h (Fjølstad et al., in preparation). This observation as well as others (see Uitto et al., 1972) suggests that in order to compare the data from different laboratories one must be certain that adequate precautions are taken to prevent partial degradation of the molecules during their purification.

An illustration of the problems facing us in this area is the question of whether the pro-α2 chains contain cystine or cysteine, amino acids not present in collagen. The original reports on dermatosparactic collagen indicated that the pro-α1 chain contained 6–8 residues of half-cystine and pro-α2 contained only 1 (Lenaers et al., 1971). Radioactive labeling of membranous bone (Bornstein et al., 1972) and tissue cultures of fibroblasts (Tsai and Green, 1972) suggested that the pro-α1 contained cystine whereas the pro-α2 did not. In contrast, radioactive labeling of the pro-α2 in matrix-free cells from embryonic chick tendon indicated that both chains contained about the same amount of cystine (Uitto et al., 1972). These discrepancies can be explained by the fact that the pro-α2 chain in chick tendon is different from the pro-α2 chain found in the other tissues studied. Alternatively, the discrepancies can be explained by partial loss of a cystine-containing region of the pro-α2 chain in some of the published experiments. The question of whether both the pro-α1 and the pro-α2 chains contain cystine is of considerable interest, since if they do, it becomes easier to accept the suggestion that disulfide bonds among the three chains promote the formation of the triple-helix during collagen synthesis.

The study of the 'transport form' or 'procollagen' has been greatly stimulated and advanced by the discovery in Belgium of the genetic disease in cattle known as dermatosparaxis (Lenaers et al., 1971). We recently have been able to establish a collaborative series of investigations on a similar disease described in sheep in Norway by Drs. Helle, Ness, Fjølstad and Svenkerud. In the affected lambs the skin begins to tear off on the first day of life and the sheep die of secondary infections within the first few days after birth (Helle and Ness, 1972). By electron microscopy the fiber structure of the collagen in the skin is greatly distorted. Chemically, the most striking feature is that by polyacrylamide electrophoresis in sodium dodecyl sulfate only two types of collagen polypeptides are found in extracts of the skin (Fjølstad et al., in preparation). The preliminary data suggest that these two types of polypeptides are pro-$\alpha 1$ and pro-$\alpha 2$ chains and that extraction of the skin with salt, acetic acid or urea does not yield any significant amount of $\alpha 1$ or $\alpha 2$ chains.

At the moment it is not clear whether the Norwegian sheep disease is identical with the Belgian cattle disease, or whether the sheep disease is a more severe defect. Over twenty diseased lambs have been studied, (Helle and Ness, 1972) and only one has survived for as long as 3 weeks. In contrast, the diseased calves appear to survive for 3 months or more (Lapière, personal communication). Also, a considerable fraction of the collagen in the skin from the diseased calves consists of α chains whereas, as indicated above, there is no evidence of α chains in the skin of the sheep. The presence of α chains in the skin of dermatosparactic calves is as yet unexplained, since assays for the enzyme which converts procollagen to collagen (Lapière et al., 1972) indicated that there is no demonstrable enzymic activity.

ACKNOWLEDGEMENTS

This work was supported in part by grant AM-16,516 from the United States Public Health Service.

REFERENCES

Berg, R. A. and Prockop, D. J. (1973). *J. Biol. Chem.* **248**, 1175–1182.
Bornstein, P., von der Mark, K., Wyke, A. W., Ehrlich, H. P. and Monson, J. M. (1972). *J. Biol. Chem.* **247**, 2808–2813.
Bosmann, H. B. and Eylar, E. H. (1968a). *Biochem. Biophys. Res. Commun.* **30**, 89–94.
Bosmann, H. B. and Eylar, E. H. (1968b). *Biochem. Biophys. Res. Commun.* **33**, 340–346.
Burgeson, R. E., Wyke, A. W. and Fessler, J. H. (1972). *Biochem. Biophys. Res. Commun.* **48**, 892–897.
Dehm, P., Jimenez, S. A., Olsen, B. R. and Prockop, D. J. (1972). *Proc. Natl. Acad. Sci. USA* **69**, 60–64.
Ehrlich, H. P. and Bornstein, P. (1972). *Nature (New Biol.)* **238**, 257–260.
Grant, M. E. and Prockop, D. J. (1972). *N. Engl. J. Med.* **286**, 194–199; 242–249; 291–300.
Halme, J., Kivirikko, K. I. and Simons, K. (1970). *Biochim. Biophys. Acta* **198**, 460–470.

Hausmann, E. (1967). *Biochim. Biophys. Acta* **133**, 591–593.
Helle, O. and Ness, O. (1972). *Acta Vet. Scand.* (In Press.)
Hurych, J. and Nordwig, A. (1967). *Biochim. Biophys. Acta* **140**, 168–170.
Jimenez, S. A., Dehm, P. and Prockop, D. J. (1971). *FEBS Letters* **17**, 245–248.
Jimenez, S. A., Dehm, P., Olsen, B. R. and Prockop, D. J. (1973). *J. Biol. Chem.* **248**, 720–729.
Juva, K. and Prockop, D. J. (1965). *In* "Biochimie et Physiologie du Tissu Conjonctif" (Ph. Comte, ed.) pp. 417–432. Imprimé en 1966 par Société Ormeco et Imprimerie du Sud-Est à Lyon, Lyon.
Juva, K. and Prockop, D. J. (1969). *J. Biol. Chem.* **244**, 6486–6492.
Juva, K., Prockop, D. J., Cooper, G. W. and Lash, J. (1966). *Science* **152**, 92–94.
Kivirikko, K. I. (1973). *Proceedings of the Second European Symposium on Connective Tissue Research*. (In Press.)
Kivirikko, K. I. and Prockop, D. J. (1972). *Arch. Biochem. Biophys.* **258**, 366–379.
Kivirikko, K. I., Kishida, Y., Sakakibara, S. and Prockop, D. J. (1972). *Biochim. Biophys. Acta* **271**, 347–356.
Kivirikko, K. I., Shudo, K., Sakakibara, S. and Prockop, D. J. (1972). *Biochemistry* **11**, 122–129.
Kivirikko, K. I., Suga, K., Kishida, Y., Sakakibara, S. and Prockop, D. J. (1971). *Biochem. Biophys. Res. Commun.* **45**, 1591–1596.
Lane, J., Rosenbloom, J. and Prockop, D. J. (1971). *Nature (New Biol.)* **232**, 191–192.
Lapière, C. M., Lenaers, A. and Kohn, L. D. (1971). *Proc. Natl. Acad. Sci. USA* **68**, 3054–3058.
Layman, D. L., McGoodwin, E. B. and Martin, G. R. (1971). *Proc. Natl. Acad. Sci. USA* **68**, 454–458.
Lazarides, E. and Lukens, L. N. (1971a). *Nature (New Biol.)* **232**, 37–40.
Lazarides, E. and Lukens, L. N. (1971b). *Science* **173**, 723–725.
Lazarides, E. and Lukens, L. N. and Infante, A. A. (1971). *J. Molec. Biol.* **58**, 831–846.
Lenaers, A., Ansay, M. Nusgens, B. V. and Lapière, C. M. (1971). *Eur. J. Biochem.* **23**, 533–543.
McGee, J. O'D., Langness, U. and Udenfriend, S. (1971). *Proc. Natl. Acad. Sci. USA* **68**, 1585–1589.
Miller, R. L. (1971). *Arch. Biochem. Biophys.* **147**, 339–342.
Olsen, B. R., Berg, R. A., Kivirikko, K. I. and Prockop, D. J. (1973). *Eur. J. Biochem.* **35**, 135–145.
Orekhovich, V. N. and Shpikiter, V. O. (1958). *In* "Recent Advances in Gelatin and Glue Research" (G. Stainby, ed.) p. 87. Pergamon Press, New York.
Pänkäläinen, M. Aro, H., Simons, K. and Kivirikko, K. I. (1970). *Biochim. Biophys. Acta* **221**, 559–565.
Popenoe, E. A. and Aronson, R. B. (1972). *Arch. Biochem. Biophys.* **258**, 380–386.
Prockop, D. J., Weinstein, E. and Mulveny, T. (1966). *Biochem. Biophys. Res. Commun.* **22**, 124–128.
Prockop, D. J., Dehm, P., Olsen, B. R., Berg, R. A., Jimenez, S. A. and Kivirikko, K. I. (1972). *In* "Inflammation: Mechanisms and Control" (I. H. Lepow and P. A. Ward, eds.). Academic Press, New York and London.
Rhoads, R. E. and Udenfriend, S. (1970). *Arch. Biochem. Biophys.* **139**, 329–339.
Rosenbloom, J. and Prockop, D. J. (1969). *In* "Regeneration and Repair: The Scientific Basis for Surgical Practice" (J. E. Dunphy and W. Van Winkle, Jr., eds.) pp. 117–135. McGraw-Hill Book Company, New York.

Spiro, M. J. and Spiro, R. G. (1971a). *J. Biol. Chem.* **246,** 4910–4918.
Spiro, R. G. and Spiro, M. J. (1971b). *J. Biol. Chem.* **246,** 4899–4909.
Stark, M., Lenaers, A., Lapière, C. M. and Kühn, K. (1971). *FEBS Letters* **18,** 225–227.
Tsai, R. L. and Green, H. (1972). *Nature (New Biol.)* **237,** 171–173.
Uitto, J. and Prockop, D. J. (1973). *Fed. Proc.* **32,** 650.
Uitto, J., Jimenez, S. A., Dehm, P. and Prockop, D. J. (1972). *Biochim. Biophys. Acta* **278,** 190–205.
Vuust, J. and Piez, K. A. (1972). *J. Biol. Chem.* **247,** 856–862.

DISCUSSION

Dr. Bard:

Dr. Prockop has, in this most interesting summary, omitted one of the intriguing questions in the collagen story: Why a given collagen fibril has such a constant diameter (typically 1000–1500 Å) over great lengths There is in the field the feeling that if long rod-like molecules self-assemble, they will naturally form fibrils This is not so: if one solves the differential equations governing aggregate growth, it is clear that cigar-shaped tactoids rather than fibrils will form. In the case of collagen, they would be expected to have an axial ratio of about 45:1 Such tactoids can indeed be formed by precipitating acetic acid-soluble calf skin collagen under such abnormal conditions as in the presence of DNA or after trypsin treatment when tactoids with a polarized fine structure are obtained. If collagen is precipitated in the cold in acetate buffer, pH 4·2, then very large tactoids (with diameters of 2 or 3 microns) with a symmetric fine structure are seen.

These facts show that there is an additional control in the collagen system beyond those considered by Dr. Prockop. I think that little is known about it at the sub-molecular level but as this is the control responsible for the shape of the connective tissue structural sub-unit—the collagen fibril—one must not underestimate its biological significance.

The Intracellular Translocation and Secretion of Collagen

PAUL BORNSTEIN and H. PAUL EHRLICH
*Departments of Biochemistry and Medicine, University of Washington
Seattle, Washington 98195, USA*

MORPHOLOGICAL STUDIES

Since collagen is deposited and functions extracellularly the cell must provide a means for the transit of newly synthesized protein through the cytoplasm and across the cell membrane. The subcellular mechanisms which effect this transport are of considerable intrinsic interest; their elucidation may also lead to information regarding the regulation of collagen synthesis and secretion during morphogenesis, tissue repair and regeneration. Furthermore, the ability to manipulate collagen secretion may prove useful in clinical circumstances such as the repair of tendon or nerve damage and the control of neoplastic growth.

Early studies of collagen secretion generally supported the view, now largely discarded, that ecdysis or shedding of cytoplasm (apocrine secretion) led to the initial formation of extracellular collagen fibers (see review by Ross, 1968). Many workers considered the tissue-specific and highly ordered organization of the extracellular matrix to be a result of a mandatory cellular activity (Porter, 1964). In addition, electron microscopic observations were thought to demonstrate the direct participation of fibroblasts in fiber formation.

The observation that purified soluble collagen could precipitate under physiological conditions to form fibers with a native banding pattern (Gross et al., 1955) raised the possibility that fibrillogenesis might also occur at some distance from the cell surface. Evidence that the secretion of collagen by chondrocytes of the regenerating amphibian limb resembled the merocrine pattern established for epithelial glandular cells was obtained by Revel and Hay (1963). These workers utilized light and EM autoradiography and observed a sequential labeling of the endoplasmic reticulum, Golgi complex, Golgi-derived vacuoles and extracellular space. Secretion was presumed to occur by fusion of the vacuole and cell membranes. In these studies the incorporation of radioactive proline was thought to occur largely into collagen, but a contribution by newly synthesized proteoglycan was not excluded. A similar sequence of events was suggested by Goldberg and Green (1964) for the secretory process in an established line of mouse fibroblasts in culture, although the participation of the Golgi complex in the transcellular movement of collagen was not well resolved. A merocrine pattern of secretion of protein-bound tritiated proline was also observed by Ross and Benditt (1962) in light-microscopic studies of the healing guinea pig wound.

In more extensive EM studies of the same system Ross and Benditt (1965) suggested that at least a fraction of newly synthesized collagen may be secreted by direct intermittent communication of vesicles, derived from the cisternae of the rough endoplasmic reticulum, with the extracellular space. EM autoradiographic observations of chondrocytes in the regenerating newt limb (Salpeter, 1968) also pointed to a pathway of secretion which bypassed the Golgi. In the latter studies the ground cytoplasm, possibly including vesicles smaller than 300 Å, was thought to represent the cellular compartment which fed directly into the extracellular matrix.

The presence of a fibrillar material in vesicles in embryonic chick corneal epithelium was demonstrated by Trelstad (1971). In some instances the material was shown to have a cross-striated appearance resembling that of extracellular collagen. Such structures had in fact been observed previously by a number of other workers and were considered to be secretory vacuoles containing collagen (see Trelstad, 1971). Alternate explanations, including the possibilities that the vacuoles contain phagocytic material or that they represent sectioning artifacts resulting from an invagination of extracellular fibers, were considered highly unlikely in the case of the corneal epithelium.

SECRETION OF OTHER EXTRACELLULAR PROTEINS

Since the secretion of collagen may have certain features in common with that of other extracellular proteins, it is pertinent to summarize the steps which have been identified in the production of digestive enzymes, polypeptide hormones, and immunoglobulins.

Exocrine glands

In both the pancreas and the parotid gland digestive enzymes are synthesized in the rough endoplasmic reticulum, pass through the cisternae of the reticulum to the Golgi apparatus, and are subsequently packaged as zymogen granules which serve as the immediate source of the extracellular product (Jamieson and Palade, 1967a, b; Castle et al., 1972).

The initial step in this sequence is the vectorial transport of newly synthesized protein to the cisternal space of the rough endoplasmic reticulum. Thereafter, secretory proteins reach the transitional elements of the endoplasmic reticulum at the periphery of the Golgi complex. The latter step requires neither continued protein synthesis nor metabolic energy (Jamieson and Palade, 1968). The next step in the sequence, the transport from transitional elements to condensing vacuoles, requires energy (presumably ATP) generated by oxidative phosphorylation.

Fusion of the zymogen granule membrane with the membrane of the acinar lumen has been demonstrated ultrastructurally in the exocrine pancreas and in the parotid gland (Palade, 1959; Amsterdam et al., 1969). Cyclic AMP is likely to be involved in this process in the salivary gland since the nucleotide induces secretion of amylase in tissue slices (Babad et al., 1967). Cyclic AMP may in fact play a role in a number of processes which involve membrane fission and fusion (Rasmussen, 1970).

Endocrine glands

Studies of the secretory processes in alpha and beta cells of the islets of Langerhans reveal many parallels with the sequence of events in the exocrine pancreas (Bauer et al., 1966; Howell et al., 1969; Gomez-Acebo et al., 1968). The role of the Golgi complex in the formation of the beta granule may however include the function of translocation of the granule through the cytoplasm of the cell.

Lacy et al. (1968) have implicated microtubules in the movement of beta granules to the surface of the cell membrane. Colchicine at a concentration of 10^{-6} M significantly reduced the secretion of insulin in response to glucose in isolated rat pancreatic islets. Vincristine, a mitotic spindle inhibitor, and deuterium oxide, a microtubule stabilizer, also suppressed glucose-induced insulin secretion in rat pancreatic fragments (Malaisse-Lagae et al., 1971).

The secretion of thyroid hormone is stimulated by TSH which is thought to activate an adenyl cyclase leading to an elevation of cyclic AMP levels and a stimulation of endocytosis of colloid (Zor et al, 1969; Pastan and Wollman, 1967). Both colchicine and vinblastine were found to block the TSH-stimulated release of ^{131}I from previously ^{131}I-labeled mouse thyroid glands in vitro (Williams and Wolff, 1970). ^{3}H-colchicine was found to bind to a soluble 6S protein of bovine thyroid with properties consistent with those of a microtubule subunit. Since colchicine also blocked the stimulation of ^{131}I release by added dibutyryl cyclic AMP and did not affect adenyl cyclase or 3':5'-cyclic nucleotide phosphodiesterase activities, Williams and Wolff (1970) suggested that antimitotic agents interfere with microtubule-mediated endocytosis of colloid.

Cytochalasin B at concentrations of 0·5–1 μg/ml also inhibited the TSH- or cyclic AMP-stimulated release of non-protein ^{131}I from prelabeled mouse thyroid glands (Williams and Wolff, 1971). At higher concentrations 19S iodoprotein, presumably thyroglobulin, was released indicating cellular damage.

The mode of synthesis, intracellular transport and secretion of hormones by the anterior pituitary resemble the mechanisms elucidated for similar steps in the exocrine pancreas (Farquhar, 1971). In addition, secretory granules can be disposed of by fusion with preexisting lysosomes, a process termed crinophagy.

Plasma cells

In mouse plasmacytomas synthesis of immunoglobulins occurs in the rough endoplasmic reticulum and the secreted protein passes to smooth membrane components of the cell (Golgi complex) prior to secretion (Zagury et al., 1970; Choi et al., 1971). Carbohydrates are added in the Golgi complex but N-acetyl glucosamine and mannose may be incorporated into nascent polypeptide chains on polysomes (Uhr and Schenkein, 1970; Choi et al., 1971). It has not been possible to identify subsequent steps in the secretory pathway but cytochemical studies favor transport in membrane-bound vesicles rather than discharge in soluble form through the cell sap (Zagury et al., 1970).

ROLE OF HYDROXYLATION OF PEPTIDYL PROLINE AND LYSINE

The synthesis of collagen includes the hydroxylation of a significant fraction of prolyl and lysyl residues after these amino acids have been incorporated into peptide linkage (see Grant and Prockop, 1972 for a review). The enzymes which catalyze these hydroxylations (collagen proline and lysine hydroxylases) utilize molecular oxygen and require Fe^{++}, α-ketoglutarate and a reducing agent, possibly ascorbic acid, as cofactors (Hutton et al., 1967; Kivirikko et al., 1972).

It was observed that interference with the hydroxylation of collagen *in vitro* resulted in an accumulation of underhydroxylated collagen (protocollagen) by the system. Such inhibition could be produced by incubation in an anaerobic environment (Peterkofsky and Udenfriend, 1963; Prockop and Juva, 1965), by chelation of Fe^{++} with α,α'dipyridyl (Kivirikko and Prockop, 1967b; Bhatnagar et al., 1968), by use of scorbutic guinea pig granulomas (Gottlieb et al., 1966), or by incorporation of proline analogs into collagen polypeptides (Takeuchi and Prockop, 1969; Takeuchi et al., 1969; Rosenbloom and Prockop, 1970, 1971). In several instances it was shown by autoradiography that the underhydroxylated collagen was largely retained intracellularly. Reversal of the inhibition of hydroxylation led, in the case of tissues incubated in a N_2 atmosphere or in the presence of a chelator of Fe^{++}, to the hydroxylation and secretion of the previously synthesized protein (Juva et al., 1966; Bhatnagar et al., 1967; Cooper and Prockop, 1968).

Working with 3T6 fibroblasts Margolis and Lukens (1971) showed that the chelation of Fe^{++} with α,α'-dipyridyl reduced the rate of secretion of underhydroxylated collagen to about 25–30% that of normal collagen, but did not abolish it. The secretion of collagen synthesized prior to administration of the drug was not affected suggesting that α,α'-dipyridyl does not generally inhibit the cell's secretory machinery. Experiments with inhibitors of collagen proline and lysine hydroxylases therefore generally led to the conclusion that hydroxylation of collagen was a necessary condition for its secretion.

Studies with inhibitors of hydroxylation were supported by the incorporation of proline analogues into the newly synthesized collagen of embryonic chick cartilage (Takeuchi and Prockop, 1969; Takeuchi et al., 1969; Rosenbloom and Prockop, 1970). Since azetidine-2-carboxylic acid, cis-4-fluoroproline and 3,4-dehydroproline not only substituted for proline but also, in some manner, adversely affected the hydroxylation of those prolyl and lysyl residues actually incorporated into peptide linkage, a precise chemical requirement for secretion was difficult to establish. The incorporation of cis-4-hydroxyproline differed, however, in that although the analog was substituted extensively for proline in embryonic chick cartilage, prolyl and lysyl residues which were incorporated were normally hydroxylated (Rosenbloom and Prockop, 1971). The failure to secrete collagen, synthesized in the presence of cis-4-hydroxyproline, at a normal rate therefore suggested that a certain minimal content of trans-4-hydroxyproline is required for the secretion of collagen and that the hydroxyl groups provided by cis-hydroxyproline do not fulfill these requirements.

That a normal content of trans-4-hydroxyproline may be necessary but not sufficient for secretion of collagen is suggested by experiments utilizing the incorporation of trans-4, 5-dehydrolysine into embryonic chick cartilage collagen (Christner and Rosenbloom, 1971). This analog of lysine reduced the incorporation of lysine, resulting in a corresponding reduction in hydroxylysine content, but did not significantly affect the hydroxylation of lysyl and prolyl residues actually incorporated. Again dehydrolysine-containing collagen was not secreted from cells at a normal rate (Christner and Rosenbloom, 1971).

In some circumstances underhydroxylated collagen has been identified extracellularly. Thus mouse L-929 fibroblasts in the proliferative growth phase secrete into the medium an underhydroxylated collagen which is a substrate for collagen proline hydroxylase (Gribble et al., 1969). Apparently during this phase of growth the specific activity of the hydroxylating enzyme is low. A similar observation has been made by Bates et al. (1972) in 3T6 mouse fibroblast cultures. These workers were also able to show that ascorbic acid deficiency increased the degree of underhydroxylation of secreted collagen in such cultures in both the logarithmic and stationary phases of growth. Uncertainty exists, however, regarding the size of underhydroxylated chains which are capable of leaving the cell. The data of Ramaley and Rosenbloom (1971) indicate that at least part of the collagen extruded by 3T6 cells in the presence of chelating agents represents partially degraded chains.

ROLE OF GLYCOSYLATION

Collagen is a glycoprotein containing galactosyl and glucosylgalactosyl residues linked by o-β-glycosidic bonds to hydroxyl groups of hydroxylysine (Spiro, 1969). The hypothesis that sugar moieties are necessary substituents of proteins designed for export from the cell (Eylar, 1965), although open to question (Winterburn and Phelps, 1972) has been used to explain in part the inhibition of secretion observed experimentally when hydroxylation of lysine is impaired. Since sugars are attached to the hydroxy groups of hydroxylysine, failure to hydroxylate lysine would necessarily interfere with subsequent glycosylation.

The recent identification of a heritable disorder of connective tissue (possibly a form of the Ehlers-Danlos syndrome) in which the hydroxylysine content of skin collagen is markedly reduced may provide a clue to the question of the obligatory presence of glycosylated hydroxylysine in extracellular collagen (Pinnell *et al.*, 1972). The disorder, which is characterized by scoliosis, recurrent dislocations, and hyper-extensible skin and joints is inherited in a recessive manner. The hydroxylysine content of skin collagen of two affected patients was only 5–10% that of normal. Presumably the clinical manifestations of the disorder result from the lack of hydroxylysyl residues which are required for the formation of stable intermolecular crosslinks and for the structural integrity of collagen fibers (see review by Bornstein, 1970).

There was no ultrastructural evidence for a disturbance in the secretory pattern of fibroblasts from patients with hydroxylysine-deficient collagen disease (Pinnell *et al.*, 1972). If cells from these patients could be shown to secrete collagen at a normal rate the requirement for glycosylated hydroxylysine as a 'sugar tag' for secretion would be open to serious question.

The subcellular distribution of the enzymes involved in the intracellular modification of collagen polypeptides (peptidyl hydroxylases and glycosyltransferases) may provide a clue to the pathway for the intracellular translocation of collagen. Unfortunately subcellular fractionation frequently results in dissociation of enzymes from subcellular structures. Although a substantial fraction of glucosyl- and galactosyl transferase activity is present in a rapidly sedimenting fraction of kidney cortex homogenate, activity was also found in various other particulate fractions and in the high speed supernatant (Spiro and Spiro, 1971). Similarly, both collagen proline and lysine hydroxylases have been found to be soluble enzymes (Kivirikko and Prockop, 1967b; Miller, 1971) but their association with cytoplasmic organelles cannot be excluded. The development of specific peroxidase- or ferritin-labeled antibodies to these enzymes and the localization of resulting antigen-antibody complexes by electron microscopy may prove useful in charting the course of the collagen molecule through the cell.

PARTICIPATION OF MICROTUBULES

Recently several laboratories have shown that a higher molecular weight form of collagen, procollagen, exists and may function as a biosynthetic precursor and as a transport form of the protein (Layman et al., 1971; Bellamy and Bornstein, 1971; Dehm et al., 1972). A neutral pH enzyme, procollagen peptidase, is capable of converting procollagen to collagen (Lapière et al., 1971; Bornstein et al., 1972a). Although it is not known whether conversion of procollagen to collagen occurs intra or extracellularly (see below), the intracellular presence of procollagen offers a means of following the protein during at least a part of its course.

Currently the most reliable assay for procollagen involves quantitation of its constituent pro-$\alpha 1$ chain by chromatography on CM-cellulose with urea-containing buffers (Bellamy and Bornstein, 1971; Bornstein et al., 1972b). Using this technique Ehrlich and Bornstein (1972a, b) investigated the possibility that microtubules might participate in the intracellular translocation of procollagen. In these experiments the effects of compounds or agents known to influence microtubular function were evaluated by their addition to rat or chick embryonic cranial bones which were actively synthesizing procollagen and collagen in organ culture. If microtubules are involved in the transport of procollagen at some point in the secretory pathway prior to the conversion of procollagen to collagen an accumulation of precursor (relative to product) might be expected. Such an effect is observed in Fig. 1 in which the ratio of the pro-$\alpha 1/\alpha 1$ chain, reflecting the proportion of procollagen and collagen, respectively, was reversed by addition of colchicine to rat cranial bones in culture. The activity of colchicine was dose-dependent; concentrations as low as 1×10^{-6} M were effective.

In addition to colchicine, other agents which inhibited microtubule function either by abnormal stabilization (deuterium oxide, high

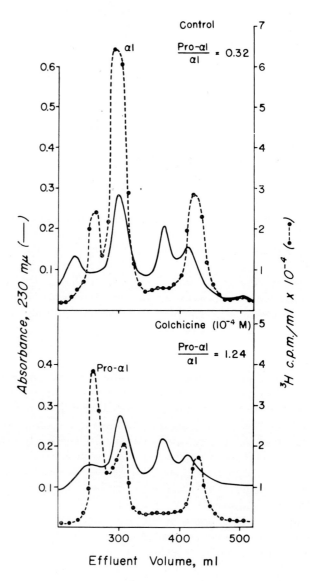

Fig. 1. CM-cellulose chromatography of procollagen and collagen from control and colchicine-treated rat cranial bones. Bones were preincubated for 20 min and pulsed with ^{3}H-proline for 45 min in the absence or presence of 10^{-4} M colchicine. The procedures for tissue extraction and CM-cellulose chromatography have been described (Bornstein et al., 1972b).

hydrostatic pressure) or by precipitation of microtubule subunits (vinblastine) also retarded conversion of procollagen to collagen (Table I). An exception was the fungistatic antibiotic griseofulvin, which reversibly disrupts the mitotic spindle of annelid oocytes (Malawista et al., 1968) and retards aggregation of melanin granules in frog skin melanocytes (Malawista, 1971). Griseofulvin (10^{-4} M) was ineffective in preventing the conversion of procollagen to collagen (Table I); similarly, no effect was noted on thyroid secretion at a concentration of 10^{-5} M (Williams and Wolff, 1970).

In order to implicate microtubules directly in procollagen transport it is necessary to exclude other possible effects of microtubule-inhibitory agents. Colchicine at a concentration of 10^{-4} M did not inhibit non-collagen protein synthesis, as measured by incorporation of radioactive tryptophan, and none of the agents listed in Table I inhibited procollagen peptidase activity in a direct assay (Ehrlich and Bornstein, 1972b). Similarly, compounds such as colchicine and vinblastine did not retard the transport of procollagen by interfering with the hydroxylation of prolyl residues in underhydroxylated procollagen since both the hydroxyproline contents and molecular weights of pro-α1 chains synthesized in the presence of these agents were normal (Ehrlich and Bornstein, 1972b). Nevertheless, caution must be used in the interpretation of these results. In leukocytes colchicine is known to have several metabolic effects including inhibition of glucose-6-phosphate and 6-phosphogluconate dehydrogenases leading to an inhibition of the hexose monophosphate pathway (DeChatelet et al., 1971). Conceivably, metabolic effects in fibroblasts unrelated to microtubules may disproportionately affect the synthesis of proteins destined for secretion (such as collagen) with a smaller effect on intracellular protein synthesis.

The effects described for colchicine and vinblastine have generally been confirmed by Diegelmann and Peterkofsky (1972) working with chick embryo cranial bone 3T3 cells, and by Dehm and Prockop (1972) with chick embryo tendon fibroblasts. However, in the former study the relative rate of collagen synthesis was unaffected by colchicine or vinblastine indicating a concordant reduction in both collagen and non-collagen protein synthesis. The use of 3T3 and chick embryo cells in culture in the two studies permitted a clearcut demonstration that microtubule-inhibitory drugs are responsible for an impairment of secretion of collagen into the medium

ENERGY-DEPENDENCE OF INTRACELLULAR TRANSPORT

The processes which affect intracellular movement are likely to be highly dependent on a continuous source of energy in the form of ATP (Rasmussen,

Table I
Modification of conversion of procollagen to collagen by microtubule-inhibitory agents*

Experiment	Labeling period (min)	Pro-α1/α1 Control	Pro-α1/α1 Exp.	Exp./Control	Procollagen + collagen synthesis % of control
Colchicine (10^{-4} M)	18	1·26	1·96	1·56	76
Colchicine (10^{-4} M)	30	0·26	1·80	6·92	71
Colchicine (10^{-4} M)	45	0·32	1·24	3·87	50
Colchicine (3×10^{-6} M)	30	0·26	1·47	5·64	91
Vinblastine (5×10^{-5} M)	30	0·54	2·07	3·81	43
Vinblastine (5×10^{-5} M)	45	0·32	1·29	4·08	60
Vinblastine (5×10^{-5} M)	10†	0·24	0·78	3·25	62
Deuterium oxide (50%)	30	0·32	0·43	1·34	22
Deuterium oxide (100%)	18	0·98	2·34	2·38	11
Deuterium oxide (100%)	10†	0·20	1·64	8·50	29
Hydrostatic pressure 10,000 psi	10‡	0·78	4·43	5·69	78
Griseofulvin (10^{-4} M)	10†	0·20	0·14	1·73	95

* Cranial bones were preincubated for 20 min with the test agent and then incubated in the presence of ^{3}H-proline for the indicated time. The proportions of pro-α1 and α1 were determined by CM-cellulose chromatography. Data obtained in part from Ehrlich and Bornstein (1972b).

† 10-min labeling period was followed by 20-min chase with excess ^{1}H-proline.

‡ A 20-min preincubation was omitted. The experimental conditions were in effect only during a 20-min chase period in the presence of excess ^{1}H-proline.

1970). Thus dinitrophenol, which uncouples oxidative phosphorylation, was shown to interfere with the translocation of insulin in the rat pancreatic β cell (Howell, 1972). Similarly Ehrlich and Bornstein (1972b) have shown that another uncoupler of oxidative phosphorylation, carbonyl cyanide m-chlorophenylhydrazone, markedly increased the proportion of procollagen to collagen in cranial bone cultures. Like microtubule-inhibitory agents, this compound had no direct inhibitory effect on procollagen peptidase or on peptidyl proline hydroxylation. Interference with oxidative phosphorylation did inhibit total protein synthesis but accumulation of procollagen cannot be ascribed to such an effect since it is known that conversion of procollagen to collagen readily occurs in the presence of high concentrations of cycloheximide (Bornstein et al., 1972a).

ATP is likely to be required in the fusion and fission of membranes which accompanies the transport of vesicle-bound proteins from one cell compartment to another (Jamieson and Palade, 1968; Howell, 1972). An additional role for ATP in the secretory process may include the formation of $3',5'$-cyclic AMP which in turn may be involved in activation of protein kinases leading to phosphorylation of elements of the cytoskeleton-vesicle complex (Rasmussen, 1970). The administration of dibutyryl cyclic AMP to cranial bone cultures (Ehrlich and Bornstein, 1972b) and to chick embryo tendon cells (Dehm and Prockop, 1972) resulted in a modest acceleration of the conversion of procollagen to collagen in bone and in a moderate stimulation of procollagen secretion by tendon cells. Conceivably, these effects would be more prominent in cells depleted of ATP.

THE SECRETORY PROCESS

Effect of cytochalasin B

The possibility that procollagen is transported in vesicles to the surface of the cell, either directly from the rough endoplasmic reticulum or via the Golgi complex, suggested that the final step in the extrusion of the secretory product may involve fusion of the vesicle and cell membranes in an exocytotic process. Fusion of the unit membrane of the secretory granule with the plasmalemma has been clearly demonstrated by electron microscopy in the exocrine pancreas, parotid and anterior pituitary glands (Palade, 1959; Castle et al., 1972; Farquhar, 1971) and is a likely event in secretory processes in the adrenal medulla (Kirschner et al., 1967) and in the β cell of the pancreatic islet (Howell and Lacy, 1971).

In order to study the extrusion of the procollagen molecule through the cell membrane, Ehrlich and Bornstein (1972b) examined the effects of cytochalasin B on procollagen synthesis and conversion in cranial bone cultures. Cytochalasin B reversibly inhibits a variety of processes including cytokinesis, cytoplasmic streaming and alterations in cell shape. These effects have been attributed to interference by cytochalasin B with a contractile microfilament system underlying the plasma membrane (Wessells et al., 1971). More recently it has become apparent that the drug is a potent competitive inhibitor of transmembrane transport of sugars, aminosugars, and possibly purines (Kletzien et al., 1972; Estensen and Plagemann, 1972). It therefore seems likely that many of the effects of cytochalasin B can be attributed to binding of the drug to the cell membrane.

Despite the uncertainty regarding its mode of action, cytochalasin B has been shown to inhibit the release of thyroxine from mouse thyroid gland organ cultures (Williams and Wolff, 1971) and to retard the prostaglandin E_2-stimulated secretion of growth hormone by bovine pituitary slices (Schofield, 1971). In contrast, cytochalasin B enhanced the glucose-induced secretion of insulin by isolated rat islets, an effect which was attributed to alteration of a microfilamentous web which might represent an obstacle to exocytosis (Orci et al., 1972).

When embryonic rat cranial bones were cultured in the presence of cytochalasin B conversion of procollagen to collagen was unimpaired (Table II). Indeed at the higher concentration of 5 µg/ml an acceleration of conversion may have occurred. However, total procollagen plus collagen synthesis, measured as chromatographically identifiable chains,

Table II

Effects of cytochalasin B on cranial bone cultures*

Experiment	Pro-$\alpha 1/\alpha 1$†	Collagen synthesis % control††	Non-collagen protein synthesis % control*†
Control**	0·63	100	100
Cytochalasin B, 2·5 µg/ml	0·48	53	—
Cytochalasin B, 5 µg/ml	0·33	55	130

* Data obtained from Ehrlich and Bornstein (1972b).

** Cultures performed in the same concentration of dimethyl sulfoxide, 0·05%, used in incubations with cytochalasin B.

† Cranial bones were preincubated for 20 min with the test agent and then incubated in the presence of ^{3}H-proline for 30 min. The proportions of pro-$\alpha 1$ and $\alpha 1$ were determined by CM-cellulose chromatography.

†† Represents procollagen and collagen synthesis assayed by CM-cellulose chromatography.

*† Cranial bones were preincubated for 20 min with the test agent and then incubated in the presence of ^{3}H-tryptophan for 20 min. Non-collagen protein was precipitated with 5% trichloroacetic acid.

was reduced by approximately 50% in the absence of an inhibition of general protein synthesis (Table II). As with microtubule-inhibitory agents, the hydroxyproline content and molecular weight of chains synthesized in the presence of cytochalasin B were normal (Ehrlich and Bornstein, 1972b). In studies reported by Diegelmann and Peterkofsky (1972) cytochalasin B, at a concentration of 10 µg/ml, did not inhibit collagen secretion or reduce collagen synthesis in excess of the inhibition resulting from the presence of 1% dimethyl sulfoxide in control cultures. Possibly the inhibition produced by dimethyl sulfoxide obscured the effect of cytochalasin B.

The finding that cytochalasin B selectively inhibits collagen synthesis without retarding the conversion of procollagen to collagen suggests an intracellular site of action for procollagen peptidase in embryonic bone. If conversion occurred extracellularly one might have expected an accumulation of procollagen relative to collagen in the presence of cytochalasin B. However, until the precise mode of action of cytochalasin B is defined, results such as those listed in Table II must be interpreted with caution. At present alternative explanations for the reduced yield of collagen chain, such as an accelerated degradation of the newly synthesized protein, cannot be excluded.

Site of conversion of procollagen to collagen

The observation by several laboratories (Layman et al., 1971; Jimenez et al., 1971; Church et al., 1971) that fibroblasts in culture secrete a procollagen-like protein into the medium indicates that in these circumstances conversion to collagen is not a requirement for secretion. Similarly cattle with dermatosparaxis, a genetic disorder resulting from a deficiency of procollagen peptidase (Lapière et al., 1971) are capable of secreting and utilizing procollagen, or derivatives thereof, in the formation of extracellular collagen fibers (Lenaers et al., 1971).

The demonstration of procollagen extracellularly together with pulse-chase experiments which indicate that in tendons (Jimenez et al., 1971) and in cranial bone (Bellamy and Bornstein, 1971) conversion of procollagen to collagen occurs relatively rapidly, has led to the assumption that conversion is an extracellular step (Jimenez et al., 1971). Although this is a reasonable interpretation of much of the data currently available, the possibility that the secretory pattern of fibroblasts in culture differs from that of cells in tissues must be considered. Certainly enzymes used in the dissociation of cells from tendons (Dehm and Prockop, 1971) may alter cell surface proteins and affect secretory patterns when studies are performed shortly after isolation of cells. A somewhat analogous situation exists with respect to insulin secretion.

Glucose-stimulated rat islets of Langerhans secrete approximately 30% of the total insulin material in the form of proinsulin into the culture medium; this proportion can be increased to 50% by addition of dibutyryl cyclic AMP, theophylline or tolbutamide in the presence of glucose (Tanese et al., 1970). In contrast, the proportion of proinsulin in human serum is usually considerably lower, although the relative concentration of the precursor may increase in the fasting state and in obese individuals (Rubenstein and Steiner, 1971).

The possibility exists that conversion of procollagen to collagen occurs intracellularly in one tissue and extracellularly in another. Thus in the corneal epithelium, which elaborates a highly ordered matrix, it may be advantageous for cells to secrete collagen molecules or even small molecular aggregates. Such aggregates would be expected to accrete rapidly onto existing fibrils permitting the cell to exert a very precise control over collagen fibrogenesis. In contrast in mammalian skin, where a more disordered fiber pattern exists, procollagen may be secreted and converted to collagen either prior to polymerization or on the fiber. At one extreme, in basement membranes, procollagen may escape conversion and function as a structural element extracellularly. Conceivably, during morphogenesis a given tissue may vary in the degree to which intra and extracellular conversion occurs and such modulation may provide flexibility in the elaboration of the extracellular matrix.

SUMMARY

Fig. 2 summarizes alternate schemes for the intracellular translocation and secretion of procollagen. Procollagen is apparently synthesized on membrane-bound ribosomes in the rough endoplasmic reticulum (RER) (1). Thereafter the protein may pass through the transitional elements of the RER to the vesicles and cisternae of the Golgi complex(2-4) where glycosylation of peptidyl hydroxylysine may occur. Alternatively, vesicles may bud off directly from the RER, bypassing the Golgi (1a). Finally, procollagen may gain access to the exterior of the cell by direct intermittent communication of the cisternae of the RER with the extracellular space (1b).

In the first two instances microtubular elements may be responsible for the transcellular movement of procollagen from its site of synthesis to a more peripheral location in the cell (1a, 5) since microtubule-inhibitory agents retard both the conversion of procollagen to collagen and the secretion of collagen. It is highly probable that procollagen is transported within membrane-bound vesicles although the existence of the protein in the soluble cell sap has not been excluded. Micro-

INTRACELLULAR TRANSLOCATION OF COLLAGEN   335

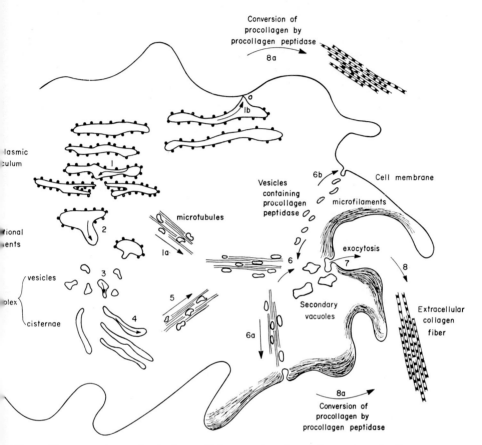

Fig. 2. Alternate schemes for the intracellular translocation and secretion of collagen. The following pathways are considered. Procollagen, synthesized in the rough endoplasmic reticulum (1) is transferred via the transitional elements (2) to the Golgi complex (3, 4). Thereafter it is transported in vesicles by microtubular elements (5) and is transferred to the extracellular space either directly (6a) or via secondary vacuoles formed by fusion with vesicles containing procollagen peptidase (6, 7). In the former case conversion to collagen by procollagen peptidase would occur extracellularly (8a). Alternatively procollagen may bypass the Golgi (1a) and then follow either one of the above two sequences. Finally, procollagen may gain access to the extracellular space by direct intermittent communication of the cisternae of the rough endoplasmic reticulum (1b). In this instance conversion to collagen by procollagen peptidase (8a) would also occur extracellularly. See text for additional details.

tubules may also play a role in movement of elements of the RER within the cell.

Extrusion of procollagen may occur in one of at least two ways. Vesicles containing procollagen may fuse with the cell membrane (1b). Procollagen is then converted extracellularly by procollagen peptidase (8a) which itself may reach the matrix via vesicles (6b). Alternatively, fusion of vesicles containing procollagen with those containing enzyme may occur intracellularly (6) leading to the formation of secondary vacuoles, the contents of which are extruded by an exocytotic process (7). This last step may be sensitive to cytochalasin B. In either scheme polymerization of collagen to form fibers occurs extracellularly (8). Conceivably, one or more of these pathways are utilized by cells in different tissues or by the same cell population depending on the cellular environment and the rate of collagen synthesis.

ACKNOWLEDGEMENTS

We thank Dr. Russell Ross for helpful discussions and for his critical review of the manuscript. Paul Bornstein is the recipient of Research Career Development Award K4-AM-42582 from the United States Public Health Service. Paul Ehrlich is a postdoctoral fellow supported by NIH Training Grant AM 1000.

REFERENCES

Amsterdam, A., Ohad, I. and Schramm, M. (1969). *J. Cell Biol.* **41**, 753–773.
Babad, H., Ben-Zvi, R., Bdolah, A. and Schramm, M. (1967). *Europ. J. Biochem.* **1**, 96–101.
Bates, C. J., Prynne, C. J. and Levene, C. I. (1972). *Biochim. Biophys. Acta* **263**, 397–405.
Bauer, G. E., Lindahl, Jr., A. W., Dixit, P. K., Lester, G. and Lazarow, A. (1966). *J. Cell Biol.* **28**, 413–421.
Bellamy, G. and Bornstein, P. (1971). *Proc. Natl. Acad. Sci. USA* **68**, 1138–1142.
Bhatnagar, R. S., Kivirikko, K. I. and Prockop, D. J. (1968). *Biochim. Biophys. Acta* **154**, 196–207.
Bhatnagar, R. S., Prockop, D. J. and Rosenbloom, J. (1967). *Science* **158**, 492–494.
Bornstein, P. (1970). *Am. J. Med.* **49**, 429–435.
Bornstein, P., Ehrlich, H. P. and Wyke, A. W. (1972a). *Science* **175**, 544–546.
Bornstein, P., von der Mark, K., Wyke, A. W., Ehrlich, H. P. and Monson, J. M. (1972b). *J. Biol. Chem.* **247**, 2808–2813.
Castle, J. D., Jamieson, J. D. and Palade, G. E. (1972). *J. Cell Biol.* **53**, 290–311.
Choi, Y. S., Knopf, P. M. and Lennox, E. S. (1971). *Biochemistry* **10**, 668–679.
Christner, P. J. and Rosenbloom, J. (1971). *J. Biol. Chem.* **246**, 7551–7556.
Church, R. L., Pfeiffer, S. E. and Tanzer, M. L. (1971). *Proc. Natl. Acad. Sci. USA* **68**, 2638–2642.
Cooper, G. W. and Prockop, D. J. (1968). *J. Cell Biol.* **38**, 523–537.
DeChatelet, C. R., Cooper, M. R. and McCall, C. E. (1971). *Infection and Immunity* **3**, 66–72.

Dehm, P., Jimenez, S. A., Olsen, B. R. and Prockop, D. J. (1972). *Proc. Natl. Acad. Sci. USA* **69,** 60–64.
Dehm, P. and Prockop D. J. (1971). *Biochim. Biophys. Acta* **240,** 358–369.
Dehm, P. and Prockop, D. J. (1972). *Biochim. Biophys. Acta* **264,** 375–382.
Diegelmann, R. F. and Peterkofsky, B. (1972). *Proc. Natl. Acad. Sci. USA* **69,** 892–896.
Ehrlich, H. P. and Bornstein, P. (1972a). *Fed. Proc.* **31,** 479.
Ehrlich, H. P. and Bornstein, P. (1972b). *Nature* **238,** 257–260.
Estensen, R. D. and Plagemann, P. G. W. (1972). *Proc. Natl. Acad. Sci. USA* **69,** 1430–1434.
Eylar, E. H. (1965). *J. Theor. Biol.* **10,** 89–113.
Farquhar, M. (1971). *In* "Subcellular Organization and Function in Endocrine Tissues", Mem. Soc. Endocrinology, No. 19, pp. 79–122. Cambridge University Press, London.
Goldberg, B. and Green, H. (1964). *J. Cell Biol.* **22,** 227–258.
Gomez-Acebo, J., Parilla, R. and R-Candela, J. L. (1968). *J. Cell Biol.* **36,** 33–44.
Gottlieb, A. A., Kaplan, A. and Udenfriend, S. (1966). *J. Biol. Chem.* **241,** 1551–1555.
Grant, M. E. and Prockop, D. J. (1972). *New Engl. J. Med.* **286,** 242–249 and 291–300.
Gribble, T. J., Comstock, J. P. and Udenfriend, S. (1969). *Arch. Biochem. Biophys.* **129,** 308–316.
Gross, J., Highberger, J. H. and Schmitt, F. O. (1955). *Proc. Natl. Acad. Sci. USA* **41** 1–7
Howell, S. L. (1972). *Nature (New Biol.)* **235,** 85–86.
Howell, S. L., Kostianovsky, M. and Lacy, P. E. (1969). *J. Cell Biol.* **42** 695–705.
Howell, S. L. and Lacy, P. E. (1971). *In* Mem. Soc. Endocrinol. No. 19, pp. 469–478. Cambridge University Press, London.
Hutton, J. J., Jr., Tappel, A. L. and Udenfriend, S. (1967). *Arch. Biochem. Biophys.* **118,** 231–240.
Jamieson, J. D. and Palade, G. E. (1967a). *J. Cell Biol.* **34,** 577–596.
Jamieson, J. D. and Palade, G. E. (1967b). *J. Cell Biol.* **34,** 597–615.
Jamieson, J. D. and Palade, G. E. (1968). *J. Cell Biol.* **39,** 589–603.
Jimenez, S. A., Dehm, P. and Prockop, D. J. (1971). *FEBS Letters* **17,** 245–248.
Juva, K., Prockop, D. J., Cooper, G. W. and Lash, J. W. (1966). *Science* **152,** 92–94.
Kirschner, N., Sage, H. J. and Smith, W. J. (1967). *Mol. Pharmacol.* **3,** 254–265.
Kivirikko, K. I. and Prockop, D. J. (1967a). *Biochem. J.* **102,** 432–442.
Kivirikko, K. I. and Prockop, D. J. (1967b). *Arch. Biochem. Biophys.* **118,** 611–618.
Kivirikko, K. I., Shudo, K., Sakakibara, S. and Prockop, D. J. (1972). *Biochemistry* **11,** 122–129
Kletzien, R. F., Perdue, J. F. and Springer, A. (1972). *J. Biol. Chem.* **247,** 2964–2966.
Lacy, P. E., Howell, S. L., Young, D. A. and Fink, C. J. (1968). *Nature* **219,** 1177–1179.
Lapière, C. M., Lenaers, A. and Kohn, L. D. (1971). *Proc. Natl. Acad. Sci. USA* **68,** 3054–3058.
Layman, D. L., McGoodwin, E. B. and Martin, G. R. (1971). *Proc. Natl. Acad. Sci. USA* **68,** 454–458.
Lenaers, A., Ansay, M., Nusgens, B. V. and Lapière, C. M. (1971). *Eur. J. Biochem.* **23,** 533–543.
Malaisse-Lagae, F., Greider, M. H., Malaisse, W. J. and Lacy, P. E. (1971). *J. Cell Biol.* **49,** 530–535.

Malawista, S. E. (1971). *J. Cell Biol.* **49,** 848–855.
Malawista, S. E. Sato, H. and Bensch, K. G. (1968). *Science* **160,** 770–771.
Margolis, R. L. and Lukens, L. N. (1971). *Arch. Biochem. Biophys.* **147,** 612–618.
Miller, R. L. (1971). *Arch. Biochem. Biophys.* **147,** 339–342.
Orci, L., Gabbay K. H. and Malaisse, W. J. (1972). *Science* **175,** 1128–1130.
Palade, G. E. (1959). *In* "Subcellular Particles" (T. Hayashi, ed.) pp. 64–83. Ronald Press Co., New York.
Pastan, I. and Wollman, S. H. (1967). *J. Cell Biol.* **35,** 262–266.
Peterkofsky, B. and Udenfriend, S. (1963). *J. Biol. Chem.* **238,** 3966–3977.
Pinnell, S. R., Krane, S. M. Kenzora, J. E. and Glimcher, M. J. (1972). *New Engl. J. Med.* **286,** 1013–1020.
Porter, K. R. (1964). *Biophys. J.* **4,** 167–196.
Prockop, D. J. and Juva, K. (1965). *Proc. Natl. Acad. Sci. USA* **53,** 661–668.
Ramaley, P. B. and Rosenbloom, J. (1971). *FEBS Letters* **15,** 59–64.
Rasmussen, H. (1970). *Science* **170,** 404–412.
Revel, J.-P. and Hay E. D. (1963). *Z. Zellforsch. Mikrosk. Anat.* **61,** 110–144.
Rosenbloom J. and Prockop, D. J. (1970). *J. Biol. Chem.* **245,** 3361–3368.
Rosenbloom, J. and Prockop, D. J. (1971). *J. Biol. Chem.* **246,** 1549–1555.
Ross, R. (1968). *In* "Treatise on Collagen" (B. S. Gould, ed.) Vol. 2A, pp. 1–82. Academic Press, London and New York.
Ross, R. and Benditt, E. P. (1962) *J. Cell Biol.* **15** 99–108.
Ross, R. and Benditt E. P (1965). *J. Cell Biol.* **27,** 83–106.
Rubenstein, A. H. and Steiner D. F. (1971). *Ann. Rev. Med.* **22,** 1–18.
Salpeter, M. M. (1968). *J. Morphol.* **124,** 387–422.
Schofield, J. G. (1971). *Nature (New Biol.)* **234,** 215–216.
Spiro, R. G. (1969). *J. Biol. Chem.* **244,** 602–612.
Spiro, R. G. and Spiro, M. (1971). *J. Biol. Chem.* **246,** 4919–4925.
Takeuchi, T. and Prockop, D. J. (1969). *Biochim. Biophys. Acta* **175,** 142–155.
Takeuchi, T., Rosenbloom, J. and Prockop, D. J. (1969). *Biochim. Biophys. Acta* **175,** 156–164.
Tanese, T., Lazarus, N. R., Devrim, S. and Recant, L. (1970). *J. Clin. Invest.* **49,** 1394–1404.
Trelstad, R. L. (1971). *J. Cell Biol.* **48,** 689–694.
Trelstad, R. L. and Coulombre, A. J (1971). *J. Cell Biol.* **50,** 840–858.
Uhr, J. W. and Schenkein, I. (1970). *Proc. Natl. Acad. Sci. USA* **66,** 952–958.
Wessells, N. K., Spooner, B. S., Ash, J. F., Bradley, M. O., Luduena, M. A., Taylor, E. L. A., Wrenn, J. T. and Yamada, K. M. (1971). *Science* **171,** 135–143.
Williams, J. A. and Wolff, J. (1970). *Proc. Natl. Acad. Sci. USA* **67,** 1901–1908.
Williams, J. A. and Wolff, J. (1971). *Biochem. Biophys. Res. Commun.* **44,** 422–425.
Winterburn, P. J. and Phelps, C. F. (1972). *Nature* **236,** 147–151.
Zagury, D., Uhr, J. W., Jamieson, J. D. and Palade, G. E. (1970). *J. Cell Biol.* **46,** 52–63.
Zor, N., Kaneko, T., Lowe, I. P., Bloom, G. and Field, J. B. (1969). *J. Biol. Chem.* **244,** 5189–5195.

On the Nature of the Polypeptide Precursors of Collagen

GEORGE R. MARTIN, PETER H. BYERS
and BARBARA D. SMITH
*Laboratory of Biochemistry, National Institute of Dental Research
NIH, Bethesda, Md. 20014 USA*

Recent studies (Layman *et al.*, 1971; Müller *et al.*, 1971; Bellamy and Bornstein, 1971; Lenaers *et al.*, 1971; Dehm *et al.*, 1972) demonstrate that collagen is formed from a biosynthetic precursor called procollagen. Two components, pro-α1 and pro-α2, analogous to but larger than the α chains of collagen, due to an additional peptide at the amino terminal end, have been isolated from the denatured protein (Lenaers *et al.*, 1971; Bornstein *et al.*, 1972a). It has been suggested that the additional sequences align the chains and allow their spontaneous assembly into procollagen molecules (Speakman, 1971), although evidence consistent with an enzyme catalyzed reaction has been presented (Müller *et al.*, this book p. 349). Procollagen is more soluble than collagen (Layman *et al.*, 1971; Lenaers *et al.*, 1971) and may also be important in the transport of collagen from the cell to the site of fiber formation. Subsequent conversion to collagen is enzymatic and involves the removal of the amino terminal extensions (Lapière *et al.*, 1971; Bornstein *et al.*, 1972b).

It has been suggested that there are even earlier and larger precursors from which the pro-α chains arise. The procollagen fraction from fibroblasts in culture, when denatured, produces components 2 to 4 times the size of pro-α chains (Layman *et al.*, 1971; Ramaley and Rosenbloom, 1971; Church *et al.*, 1971). To establish the relation of these precursors to pro-α chains, we have isolated procollagen from the most recently synthesized collagenous fraction of skin and calvaria. Their behavior on ion exchange and molecular sieve chromatography has been compared to the fibroblast and some other precursor molecules. Although some heterogeneity was observed, the results indicate that the procollagens are composed of pro-α1 and pro-α2 chains or their derivatives.

MATERIALS AND METHODS

Preparation of procollagen

The calvaria from 100 newborn rats were pulse labeled for 20 min with ^{3}H-proline (Müller et al., 1971; Bellamy and Bornstein, 1971; Vuust and Piez, 1972). To inhibit the conversion of procollagen to collagen and to dissolve these proteins, the calvaria were extracted with cold 0·1 M acetic acid for 24 h, homogenized and centrifuged. The supernatant fluid was used as a source of labeled procollagen

Since large quantities of collagen are extracted from rat skin with 0·1 M acetic acid, we have used 0·15 M NaCl, 0·05 M tris-HCl pH7·4 plus 0·02 M EDTA as a solvent. Neutral salt solutions are known to extract the most recently synthesized collagen which is only a small proportion of that extracted by acetic acid (Jackson and Bentley, 1960; Gross, 1969). EDTA was added because it has been shown to inhibit the conversion of procollagen to collagen by procollagen peptidase (Lapière et al., 1971). Tsai and Green (1972) have used a similar solvent to prepare intracellular precursors of collagen.

Chromatographic procedures

We have used chromatographic methods to isolate procollagen from the extracts of skin and calvaria. The extracts were adjusted to contain 2 M urea, 0·05 M tris-HCl, pH 7·4. These solutions were centrifuged to remove particulate material and applied to a DEAE-cellulose column equilibrated with the same solvent (Müller et al., this book p. 349). A linear gradient to 0·3 M NaCl over a total volume of 600 ml was used for elution.

CM-cellulose chromatography (Piez et al., 1963) and molecular sieving (Piez, 1968) of denatured proteins were carried out as previously described. Radioactivity in the various samples was measured in a liquid scintillation spectrometer by standard procedures.

RESULTS

Preparation and properties of procollagen from rat skin and calvaria

The acetic acid extracts of pulse labeled calvaria were found to contain a large portion of collagen in precursor form as judged by chromatography on CM-cellulose (similar to Fig. 5 on p. 355). Procollagen was separated from collagen and other proteins in the extract by DEAE-cellulose chromatography. Under the conditions used two radioactive peaks were eluted by the salt gradient (Fig. 1). The first peak contained the majority of the procollagen in the extract as judged by the recovery of pro-α1 after CM-cellulose chromatography. The second peak also yielded pro-α chains when subsequently chromatographed on CM-cellulose, however, other unidentified components were also present. It is possible that this material represents a minor procollagen component, since we have also observed precursor-rich material prepared from the skins of dermatosparactic cattle to elute in this position. Tentatively we have labeled this peak (Fig. 1), dermatosparactic collagen.

In addition to ion exchange chromatography, we fractionated the components of the major radioactive peak on a molecular sieve under denaturing conditions. Collagen extracted from rat skins with acetic

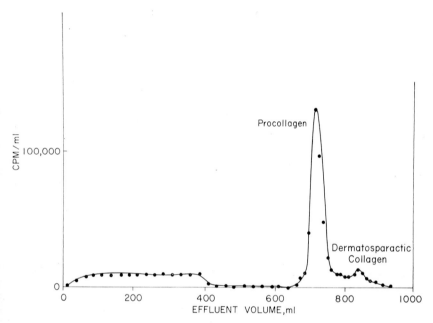

Fig. 1. DEAE-cellulose chromatography of an acid extract of pulse labeled calvaria. The column was equilibrated with 0·05M tris-HCl, pH 7·4, 2M urea. Protein was eluted with a linear salt gradient.

acid was added to the labeled sample to mark the positions at which α chains, β-components and larger components emerged from the column. The labeled sample separated into two fractions estimated to have molecular weights of 120,000 and 220,000 (Fig. 2). The smaller material corresponds in size to pro-α chains. The nature of the larger peak will be considered below.

Initial studies on extracts of rat skin using the techniques (i.e. acid extraction) described for calvaria indicated that label was first incorporated into components with the chromatographic characteristic of

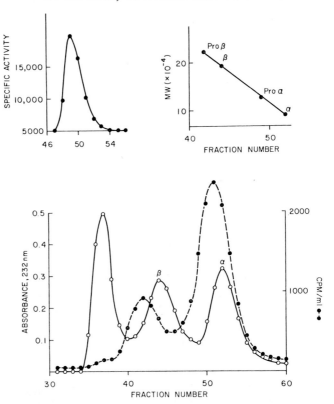

Fig. 2. Molecular sieve chromatography of the procollagen peak isolated by DEAE-cellulose chromatography (Fig. 1). A column of 8% agarose equilibrated with $1M$ $CaCl_2$, $0.05M$ tris-HCl, pH 7.4 was used. Bottom: (O———O), the absorbance of carrier collagen; (●----●), radioactivity from procollagen. Upper left: the specific activity across the pro-α, α-region of the chromatogram. Upper right: the estimated size of the labeled peaks.

pro-α chains but after two hours almost all the label was associated with α1 and α2 (Fig. 3).

Pro-α1 was a negligible constituent as judged by its absorbance in these acid extracts. However, extracts of skin, using the neutral salt-EDTA solvent, were found to be markedly enriched in procollagen, as judged by the ratio of pro-α1 to α1. Procollagen was separated from the collagen in the extract by DEAE-cellulose chromatography. Chromatography of extracts of skins from rats given a labeled amino acid 10 min prior to killing showed one major radioactive peak that eluted exactly where procollagen from calvaria emerged. The pro-

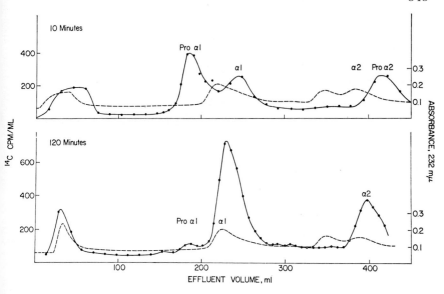

Fig. 3. CM-cellulose chromatography of skins from rats killed 10 min (above) and 120 min (below) after the injection of labeled amino acids. ●——●, ^{14}C; ----, absorbance.

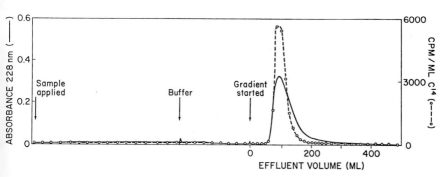

Fig. 4. Rechromatography on DEAE-cellulose of procollagen from rat skins. Conditions are those listed in Fig. 1.

collagen fraction from several chromatograms was pooled and rechromatographed under the same conditions (Fig. 4). This preparation contained pro-α1 and α1 in the ratio of 4 to 1. The amino acid compositions of pro-α1 and pro-α2 are consistent with those already published (Lenaers et al., 1971; Bornstein et al., 1972a; Ehrlich and Bornstein, 1972). In comparison to α-chains, they contain reduced levels of glycine and hydroxyproline, more aspartic and glutamic acid, and six residues of cysteine in pro-α1 and two in pro-α2. Using the procedures outlined here, approximately 10–20 mg of purified procollagen can be isolated from the skins of 100 rats.

DISCUSSION

Nature of procollagen components

In the brief time that has elapsed since the initial report on procollagen, similar or identical proteins have been observed in a variety of systems. Some of these systems and the sizes of precursor components obtained from the denatured proteins are listed in Table I. The range of molecular

Table I
Components of denatured procollagens

Fibroblast Media
>95,000 (a)
>200,000 (b)
5–600,000 (c)
125,000   250,000 (d)

Organ Culture (Pulse Labeled)
pro-α1 — 120,000 (e)   pro-α2 — 110,000 (f)
115,000 (g)

Protocollagen
95,000 (h)
130,000 (i)

Dermatosparaxis
pro-α1   104,000 (j)
pro-α2   100,000 (j)

Enzymatically Dissociated Cells
125,000 (k)
3–400,000 (k)

(a) Layman et al., 1971; (b) Ramaley and Rosenbloom, 1971; (c) Church et al., 1971; (d) Smith et al., 1972; (e) Bornstein et al., 1972a; (f) Ehrlich and Bornstein, 1972; (g) Vuust and Piez, 1972; (h) Müller et al., 1971; (i) Tsai and Green, 1972; (j) Lenaers et al., 1971; (k) Dehm et al., 1972.

weights reported for pro-α chains (100,000–130,000) may reflect real differences in molecular weight or may be the results of special problems encountered in estimating their size. The mobilities of collagen α chains and globular proteins of equal molecular weight differ markedly during molecular sieving (Piez, 1968), and electrophoresis in sodium dodecyl sulfate (Furthmayr and Timpl, 1971). Pro-α chains appear to have both collagenous and globular portions so that there are no appropriate proteins to use as calibration standards.

Some of the molecular weight disparities observed among procollagen components from different systems may result from limited proteolysis occurring in the tissue or during extraction. We observe some α chains in our chromatographically purified procollagen. These α chains probably occur in 'hybrid' molecules (molecules containing both pro-α and α chains) which are formed by the cleavage of the amino terminal extension of pro-α chains resulting in a degree of molecular heterogeneity that does not appear to alter the chromatographic properties of the native molecule. Multiple forms of the pro-α1 chain have been observed in extracts of dermatosparaxic cattle skin (Lenaers et al., 1971) as well as in extracts of pulse labeled calvaria (Vuust and Piez, 1972). These observations could be explained if the conversion of procollagen to collagen were the result of sequential removal of peptidyl material from the amino terminal end of the molecule rather than a single step. Alternatively, other proteases may have caused some degradation.

We cannot assume that the procollagen isolated from any system is chemically homogeneous. The usual extract of pulse labeled tissue probably consists or precursor molecules that vary in the extent of hydroxylation and glycosylation. Such extracts could contain pro-α chains not yet assembled into molecules to give still additional heterogeneity. The procollagens prepared from dermatosparactic cattle skin (Lenaers et al., 1971) or cell culture media (Layman et al., 1971) are primarily extracellular, and should be fully hydroxylated and glycosylated but may be heterogeneous due to cross linking and proteolysis.

In addition to pro-α chains, larger components have been obtained following the denaturation of procollagen (Table I). The procollagen secreted by fibroblasts in culture and isolated by DEAE-cellulose chromatography contained in addition to pro-α chains a 250,000 molecular weight component that was identified as a dimer of pro-α1 chains linked by disulfide bonds (Smith et al., 1972). Pro-α1 contains several cysteine residues (Lenaers et al., Dehm et al., 1972; Bornstein et al., 1972) whereas pro-α2 appears to have at most two (Tsai and Green, 1972; Ehrlich and Bornstein, 1972a). These compositional differences may account for the preferential incorporation of pro-α1 into disulfide linked dimers. A component of similar size is observed in chromatographically purified procollagen from pulse labeled calvaria (Fig. 2). Similar results have been obtained by Burgeson et al. (1972). Dehm et al. (1972) observed a larger aggregate (300–400,000) in their isolated cell system and suggested it contained three pro-α chains linked by disulfides. We have isolated a component from fibroblast cultures that is larger than the dimer, however, it contained less hydroxyproline than is found in collagen or procollagen and this was liberated as pro-α chains after reduction with mercaptoethanol in 8 M urea. We presume that this aggregate contained pro-α chains linked to noncollagenous material by disulfide bonds.

In general, the cysteine in proteins functions either catalytically in enzymes or structurally, as cystine, to stabilize inter and intrachain interactions. While the function of the cysteine containing regions of pro-α chains has not been elucidated, it is possible that disulfide bond formation may precede helix formation and hold the chains in register. The incorporation of pro-α chains into disulfide linkage with non-collagenous material could also be a normal step in the biosynthesis of procollagen. Furthermore, it is possible that collagen in the junctions between tissues could be linked covalently to noncollagenous structural proteins via disulfides.

SUMMARY

Procollagen was prepared from the calvaria and skin of rats. Purification steps included DEAE-cellulose chromatography which removed collagen and other contaminating proteins. The purified protein was largely composed of pro-α chains although a larger component judged to be a dimer of the pro-α chains was observed. Identification of the pro-α chains was based on pulse-chase experiments and the composition of the fractions.

The larger components observed after denaturing procollagen are apparently due to disulfide bond formation between pro-α chains and to pro-α chains linked to other proteins by disulfide bonds. Considerable heterogeneity results from limited proteolysis by procollagen specific proteases and other proteases.

REFERENCES

Bellamy, G. and Bornstein, P. (1971). *Proc. Natl. Acad. Sci. USA* **68**, 1138–1142.
Bornstein, P., von der Mark, K., Wyke, A. W., Ehrlich, H. P. and Monson, J. M. (1972a). *J. Biol. Chem.* **247** 2808–2813.
Bornstein, P., Ehrlich, P. H. and Wyke, A. W. (1972b). *Science* **175**, 544–546.
Burgeson, R. E., Wyke, A. W. and Fessler, J. H. (1972). *Biochem. Biophys. Res. Commun.* **48**, 892–897.
Church, R. L., Pfeiffer, S. E. and Tanzer, M. L. (1971). *Proc. Natl. Acad. Sci. USA* **69**, 2638–2642.
Dehm, P., Jimenez, S. A., Olsen, B. R. and Prockop, D. J. (1972). *Proc. Natl. Acad. Sci. USA* **69**, 60–64.

Ehrlich, H. P. and Bornstein, P. (1972). *Biochem. Biophys. Res. Commun.* **46**, 1750–1756.
Furthmayr, H. and Timpl, R. (1971). *Anal. Biochem.* **41**, 510–516.
Gross, J. (1969). In "Ageing of Connective and Skeletal Tissue" (A. Engel and T. Larsson, eds.). Nordiska Bokhandelns Förlag, Stockholm.
Jackson, D. S and Bentley, J. P (1960). *J Biophys Biochem. Cytol.* **7**, 37–42.
Lapière, C. M., Lenaers, A and Kohn L. D. (1971). *Proc. Natl. Acad. Sci. USA* **68**, 3054–3058.
Layman, D. L., McGoodwin, E. B. and Martin, G. R. (1971). *Proc. Natl. Acad. Sci. USA* **68**, 454–458.
Lenaers, A., Ansay, M., Nusgens, B. V. and Lapière, C. M. (1971). *Eur. J. Biochem.* **23**, 533–543.
Müller, P. K., McGoodwin, E. B. and Martin, G. R. (1971). *Biochem. Biophys. Res. Commun.* **44**, 110–117.
O'Hara, P. J., Read, W. K., Romane, W. M. and Bridges, C. H. (1970). *Lab. Invest.* **23**, 307–314.
Piez, K. A. (1968). *Anal. Biochem.* **26**, 305–312.
Piez, K. A., Eigner, E. A. and Lewis, M. S. (1963). *Biochemistry* **2**, 58–66.
Ramaley, P. B. and Rosenbloom, J. (1971). *FEBS Letters* **15**, 59–64.
Smith, B. D., Byers, P. H. and Martin, G. R. (1972). *Proc. Natl. Acad. Sci. USA* **69**, 3260–3262.
Speakman, P. (1971). *Nature* **229**, 241–243.
Tsai, R. L. and Green, H. (1972). *Nature (New Biol.)* **237**, 171–173.
Vuust, J. and Piez, K. A. (1972). *J. Biol. Chem.* **247**, 856–862.

Intracellular Forms of Collagen— Underhydroxylated Collagen

Peter K. Müller, Ermona B. McGoodwin
and George R. Martin

*Laboratory of Biochemistry, NIDR, NIH, Bethesda, Md. 20014, USA
and Max-Planck-Institut für Eiweiss- und Lederforschung, Munich, Germany*

Several years ago it was demonstrated that anaerobiosis and iron chelators prevented the formation of hydroxyproline and hydroxylysine (Peterkofsky and Udenfriend, 1963; Chvapil, 1968; Rosenbloom and Prockop, 1969). The formation of the polypeptide chains of collagen was not inhibited, however, and lysine and proline-rich polypeptides accumulated (Gottlieb et al., 1965). This material was termed protocollagen, since it was thought to be a biosynthetic intermediate in collagen synthesis (Juva and Prockop, 1965). Most of the evidence available now, indicates that the hydroxylation of prolyl and lysyl residue occurs in nascent chains attached to polyribosomes and that protocollagen is an abnormal form, an underhydroxylated collagen (Lazarides et al., 1971).

Hydroxyproline and hydroxylysine are relatively unique to collagen. Hydroxylysine serves as a site of attachment for carbohydrate and is also used in crosslinking. The role of hydroxyproline is obscure, but could be structural since it accounts for 10% of the molecule. Alternatively, since underhydroxylated collagen is not secreted normally (Bhatnagar et al., 1967) hydroxylation could alter either the solubility of collagen or its transport. The structure and properties of underhydroxylated collagen are not well characterized. Previously, we have reported that underhydroxylated collagen is composed largely of precursor molecules, the pro-α chains (Müller et al., 1971). In this chapter, we report further details on the preparation, macromolecular structure and properties of underhydroxylated collagen.

MATERIALS AND METHODS

Preparation of underhydroxylated collagen

The calvaria from 17-day old chick embryos were incubated in Dulbecco-Vogt media lacking glycine and supplemented with penicillin, ascorbic acid and β-aminopropionitrile-HCl, as previously described (Müller et al., 1971).

α, α'-Dipyridyl (1·4 mmole) was added to the media to inhibit the hydroxylation of proline and lysine residues. In general, 20 calvaria were preincubated in 20 ml of media at 37°C for 20 min and this was replaced with 20 ml of fresh media containing ^{14}C-proline (40 μc) and ^{14}C-glycine (100 μc) or proline-3, 4-^{3}H(250 μc). At the end of the incubation period, the labeled media and calvaria were separated and dialyzed in tubing against a large volume of 0·5% acetic acid. After 2 days the calvaria were homogenized in the retentate and extracted overnight. Insoluble material was removed from the calvaria homogenate by ultracentrifugation. Initial experiments utilizing proline-3,4-^{3}H-labeled calvaria and peptidyl proline hydroxylase from rat skin to estimate underhydroxylated collagen by the tritium release assay of Hutton et al. (1966) indicated that essentially all the labeled, collagenous protein was recovered in the final supernatant fluid.

Labeled collagen was prepared from calvaria in the same manner except that the α, α'-dipyridyl was omitted from the media.

CHROMATOGRAPHIC PROCEDURES

Carboxymethyl-(CM-) cellulose chromatography of denatured collagens

CM-cellulose chromatography as described by Piez et al. (1963) was used to separate the components of collagen after denaturation. Minor modifications in buffer composition have been described earlier (Müller et al., 1971). Purified collagen (20–25 mg) obtained from the skins of lathyritic rats was added to labeled samples prior to chromatography.

CNBr cleavage and chromatography of the resulting peptides

α1 chains (90–100 mg) from chick skin collagen were added to labeled α1 and pro-α1 from underhydroxylated collagen. CNBr digestion followed the procedures described by Miller et al. (1969). Peptides were separated on CM-cellulose or phosphocellulose as described by Epstein et al. (1971).

CM-cellulose chromatography of native collagens

Purified rat skin collagen (5 mg) was added to solutions of underhydroxylated collagen and dialyzed against 0·02 M potassium acetate, 1 M urea, pH 5·4. These samples were applied to a 2·5 × 8 cm column of CM-cellulose, equilibrated with the same buffer and maintained at 8°C. Elution was effected using a linear salt gradient to 0·5 M NaCl over a total volume of 600 ml.

Collagen, procollagen and their underhydroxylated forms cochromatograph under these conditions (Fig. 1). However, if denatured by heating to 50°C prior to chromatography, these proteins do not bind to the column but elute

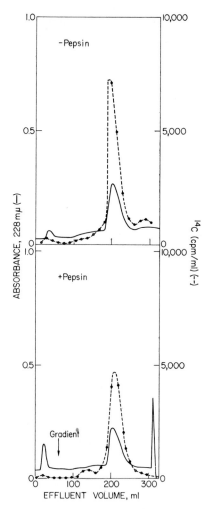

Fig. 1. CM-cellulose chromatography of native collagens. Purified rat skin collagen (5 mg) added to solutions of radioactive underhydroxylated collagen and incubated with and without pepsin (0·1 mg/ml) at 15°C in 0·5M acetic acid for 6 h. Subsequently, pepsin was inactivated and the samples were dialyzed against 0·02M potassium acetate, 1M urea at pH 5·4. The samples were applied to a 2·5 × 8 cm column of CM-cellulose equilibrated with this same buffer at 8°C. A linear salt gradient to 0·5M NaCl in a total volume of 600 ml was used for elution.

at the front. Since only the native macromolecules bind to the column, we have used this technique to distinguish between native and denatured collagens.

DEAE-cellulose chromatography of native collagens

Solutions of labeled collagen or underhydroxylated collagen were dialyzed against 1 M urea, 0.01 M tris-HCl, pH 7.4 at 4°C and applied to a DEAE-cellulose column (2.5 × 8 cm) maintained at 8°C and equilibrated with the same solvent. A linear NaCl gradient from 0–0.5 M over a total volume of 600 ml was used for elution (Fig. 2).

Three major radioactive peaks were observed when extracts of α, α'-dipyridyl treated calvaria were chromatographed in this system (Fig. 2). Peak B was found to contain α chains and peak C pro-α and some α chains when subsequently denatured and chromatographed on CM-cellulose (Fig. 3).

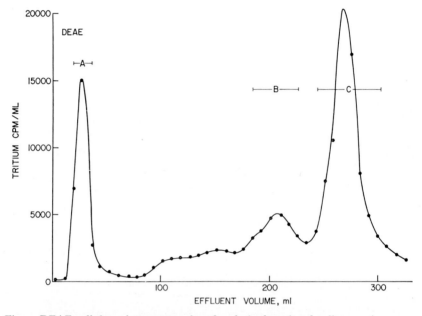

Fig. 2. DEAE-cellulose chromatography of underhydroxylated collagens. An extract containing underhydroxylated collagen was dialyzed against 0·01M tris-HCl, ph 7·4, 1M urea and applied to a 2·5 × 8 cm column of DEAE-cellulose equilibrated with the same solvent and maintained at 8°C. A linear NaCl gradient from 0 to 0·5M over a total volume of 600 ml was used for elution.

Similar chromatographic properties are observed with the proteins extracted from pulse labeled calvaria in the absence of α, α'-dipyridyl. Collagen is found in the peak B region and procollagen in peak C. We have used this procedure to separate procollagen from collagen and other proteins.

RESULTS

Characterization of underhydroxylated collagen

Previous studies have shown that chelating agents such as α,α'-dipyridyl prevent the hydroxylation of prolyl and lysyl residues in cartilage

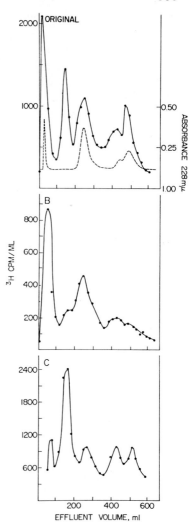

Fig. 3. CM-cellulose chromatography of aliquots from peaks B and C from Fig. 2 and the original extract under denaturing conditions (45°C). Purified rat skin collagen (20 mg) was added to the samples prior to chromatography and the absorbance of the carrier collagen was monitored at 228 mµ (-----). The top pattern shows the distribution of ³H in the original extract (●———●); the middle pattern was obtained with peak B from Fig. 2; and the bottom pattern was observed with peak C from Fig. 2. The chromatographic conditions are essentially those described by Piez et al. (1963).

and causes the accumulation of an underhydroxylated collagen (Chvapil, 1968; Lukens, 1966, 1970; Kivirikko and Prockop, 1967). In preliminary experiments we have found that α, α'-dipyridyl has a similar effect on calvaria. No hydroxyproline was synthesized from proline incorporated into peptide linkage and underhydroxylated collagen was detected in acetic acid extracts of the tissue using the tritium release assay of Hutton et al. (1966).

Comparison of underhydroxylated with hydroxylated collagen is complicated by the presence of precursor molecules in both preparations.

(Müller et al., 1971; Bellamy and Bornstein, 1971). The amount of precursor present in each preparation can be estimated by CM-cellulose chromatography of the denatured preparation since pro-α1 separates from α1 in this system. CM-cellulose chromatograms of extracts of calvaria with ^{3}H-proline in the presence and absence of α, α'-dipyridyl are shown in Fig. 4. Collagen prepared from the skins of lathyritic rats was added to each sample to serve as carrier and an internal

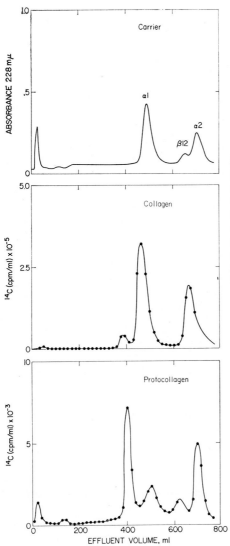

Fig. 4. CM-cellulose chromatography of denatured labeled collagen and underhydroxylated collagen from calvaria labeled for 6 h in the presence (bottom) and absence (centre) of α,α'-dipyridyl. Carrier collagen (20 mg) was added to each sample prior to denaturation (top). The chromatographic conditions are essentially those described by Piez et al. (1963).

standard. In addition to the α1 and α2 chains obtained from this membranous bone in approximately a 2:1 ratio, another peak elutes prior to α1. This peak is a much greater proportion of the total radioactivity in the underhydroxylated preparation than in the uninhibited system. We, and others have suggested that this material is a precursor of the α1 chain, namely pro-α1. Conversion of procollagen to collagen can be demonstrated by pulse-chase techniques in calvaria continually exposed to α, α'-dipyridyl (Fig. 5).

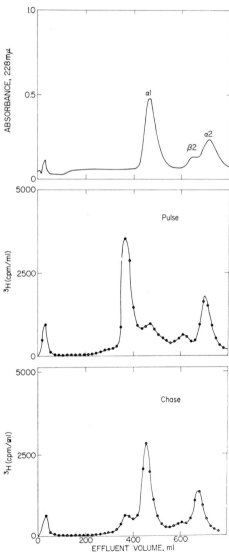

Fig. 5. CM-cellulose chromatography of denatured underhydroxylated collagen, after a 3 h pulse with ^{3}H-proline (middle) and a subsequent chase without ^{3}H-proline (bottom). Other conditions are those referred to in Fig. 4.

Identification of the components observed in preparation of underhydroxylated collagen has been based in large part on their chromatographic characteristics on CM-cellulose and molecular sieves and activity as substrates for peptidyl proline hydroxylase (Müller et al., 1971).

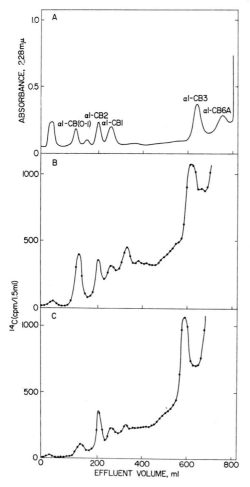

Fig. 6. Phosphocellulose chromatography of CNBr digests of underhydroxylated pro-α1 (C). Unlabeled α1 from chick skin (A) was added to each sample prior to digestion with CNBr. CNBr digestion was according to Miller et al. (1969) and chromatography according to Epstein et al. (1971).

To obtain more direct identification of the peaks and observe possible effects of hydroxylation at the peptide level, we have carried out the limited cleavage of underhydroxylated pro-α1 and α1 with CNBr. The radioactive profile of the peptides from underhydroxylated pro-α1 and α1 as separated on phosphocellulose is shown in Fig. 6. A close correspondence exists between this preparation and the peptides produced

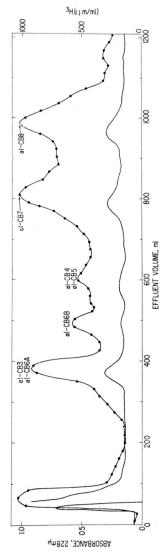

Fig. 7. CM-cellulose chromatography of a CNBr digest of underhydroxylated pro-α1 (●———●) and chick skin collagen α1 (———). CNBr digestion was according to Miller et al. (1969) and chromatography according to Epstein et al. (1971).

from authentic α1 except that the precursor has an additional peak eluting after α1–CB1. Similar CNBr digests of these materials were also chromatographed on CM-cellulose and were not found to be different (Fig. 7, only underhydroxylated pro-α1 is shown).

Molecular structure of underhydroxylated collagen

We have used indirect methods to establish whether the underhydroxylated collagen that forms in the present of α, α'-dipyridyl has a native collagen structure or is randomly coiled chains. As described in above, native collagen binds to CM-cellulose at 8°C in 1 M

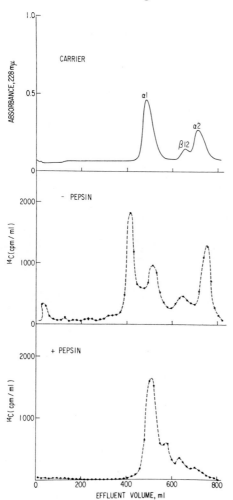

Fig. 8. CM-cellulose chromatography of underhydroxylated collagen treated with (bottom) and without (middle) pepsin as indicated in Fig. 1. Residual pepsin was inactivated prior to denaturation as previously described (Müller et al., 1971). Chromatographic conditions were those used previously.

urea, 0·02 M acetate, pH 5·4 and is eluted as a single peak by the salt gradient. Under these same conditions, denatured collagen does not bind and emerges at the front. Underhydroxylated collagen (Fig. 1) cochromatographs with native collagen molecules, indicating its native triple helical structure.

It is well known that native collagen except for the terminal nonhelical portions, resists digestion by pepsin while the denatured protein is rapidly degraded to small peptides. Another aliquot of the calvaria extract to which collagen had been added was incubated with pepsin at 15°C and then chromatographed on the cold CM-cellulose column. It was observed that the collagen in the preparation was largely resistant to digestion by pepsin and that two-thirds of the protocollagen survived the incubation (Fig. 1). Chromatography of aliquots of these same preparations on CM-cellulose at 45°C under the conditions necessary to resolve components of the denatured proteins indicated that incubation with pepsin markedly reduced the recovery of radioactivity in the $\alpha 2$ region (Fig. 8). In addition, pepsin treatment caused a loss of material from the position where the precursor of $\alpha 1$ had eluted. When the underhydroxylated collagen and carrier were heated at 50°C prior to incubation with pepsin, both were digested and no collagen or underhydroxylated collagen was found (not shown). These studies indicate that underhydroxylated collagen chromatographs like native rather than denatured collagen. When exposed to pepsin, it has the properties expected for the native molecule. The changes noted in the recovery of components indicate that some differences in sensitivity to pepsin exists between the two materials.

Dissociation of pro-α chain synthesis from formation of the macromolecule

Previous studies with embryonic chick tibia have indicated that underhydroxylated collagen is retained by the cells (Bhatnagar *et al.*, 1967). In contrast, rapidly dividing cells secrete this material (Gribble *et al.*, 1969). To establish the pattern in calvaria we have measured the quantities of underhydroxylated collagen in tissue and media as a function of time in culture (Fig. 9). During the first several hours essentially all of the underhydroxylated collagen is associated with the tissue and after this time it accumulates in the media. The amount of material synthesized was constant during this period. If the media were changed at 6 h and isotope added, then the underhydroxylated collagen formed in the subsequent six hours was found in the tissue not the media (not shown). To determine if the underhydroxylated collagen extracted from the tissue had the same macromolecular structure as that found in the media, we exposed portions of the media and cell extract to pepsin at 15°C for 6 h. In addition, another aliquot of the cell extract was heated at 45°C for 20 minutes and then incubated at 15°C. Sixty per cent of the underhydroxylated collagen extracted from the calvaria survived incubation with pepsin whereas only 18% of the material in the media was not degraded. Denaturation prior to incubation with pepsin caused the loss of all substrate activity. These results indicate that a large proportion of the underhydroxylated collagen

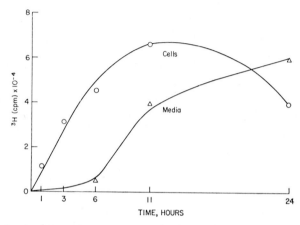

Fig. 9. Levels of underhydroxylated collagen in media or tissue extracts (cells) of samples incubated for various times with ^{3}H-3,4-proline and α,α'-dipyridyl. Rat skin peptidyl proline hydroxylase was used to measure underhydroxylated collagen in the assay devised by Hutton et al. (1966).

extracted from the calvaria had a native type structure whereas the material in the media behaved differently from native molecules. Characterization (not shown) of the media protein on CM-cellulose and molecular sieve indicated that the proteins had a mean size similar to that from the cell but chromatographed as $\alpha 1$ and $\alpha 2$ chains rather than precursor chains. The peaks of material were broader than those observed with the cell protein indicating some heterogeneity of charge and size.

DISCUSSION

Conversion of procollagen to collagen

While it has been observed in some systems that the properties of underhydroxylated collagen are anomalous (Kivirikko and Prockop, 1967; Prockop and Juva, 1965), our calvaria preparation was found to behave similarly to collagen during extraction and purification. Differences in properties noted here can be attributed to the larger proportion of precursor molecules present in the extracts of α, α'-dipyridyl treated tissues rather than the degree of hydroxylation. Indeed, the behavior of procollagen on DEAE-cellulose and of pro-α chains on CM-cellulose appears to be independent of the degree of hydroxylation of either proline or lysine. The peptides obtained from underhydroxylated pro-α1 and α1 following CNBr cleavage resembled that obtained with authentic chick skin collagen α1, except that the underhydroxylated pro-α1 produced an additional peak that was not observed in digests of α1. It is possible that this peptide represents the amino terminal extension of the pro-α1 chain. However, since the preparation was labeled solely with proline, separated proline-poor peptides may not have been noticed (Stark et al., 1971).

The conversion of procollagen to collagen is markedly slowed in the α, α'-dipyridyl treated calvaria but does occur as demonstrated by appropriate pulse-chase experiments. Since procollagen contains amino terminal peptidyl extensions on each α chain which can be cleaved with pepsin it has been suggested that a specific protease effects the conversion to collagen (Lapière et al., 1971; Bornstein et al., 1972). It is possible that α, α'-dipyridyl slows the conversion of procollagen to collagen by inhibiting this specific enzyme (procollagen protease). However, other substances, e.g. vinblastine and colchicine (Diegelmann and Peterkofsky, 1972; Dehm and Prockop, 1972; Ehrlich and Bornstein, 1972) which inhibit conversion appear to do so by maintaining precursor material in an intracellular pool, suggesting that α, α'-dipyridyl may inhibit conversion by similarly inhibiting secretion.

Since collagen is largely insoluble in physiological solutions, it has been suggested that procollagen is the intracellular form and that conversion to collagen occurs outside the cell. Evidence in accord with this suggestion has been presented. Since it is likely that the same cell synthesizes procollagen and procollagen protease some mechanism must exist to prevent premature reaction and precipitation of collagen within the cell.

Function of hydroxyproline

Little is known of the function of hydroxyproline in collagen. Previously, it was suggested that the hydroxyl group of hydroxyproline was an additional site for hydrogen bonding and that this stabilized the helix. Subsequent studies have shown that the stability of the collagen molecule is proportional to the sum of hydroxyproline plus proline (Piez, 1967) and that the two amino acids are equivalent in conferring stability. It is thought that the pyrrolidine ring stabilizes the structure by inhibiting free rotation around the peptide bond. In these past studies, collagens from different species varying in hydroxyproline and proline content were studied. We have sought to compare the properties of collagens from the same tissue which differ only in the hydroxylation of proline and lysine. So far it has not been possible to isolate underhydroxylated collagen free of collagen so that the comparison of properties has been made by indirect methods. The studies outlined here indicate that the underhydroxylated preparation does form a native, collagen-like structure. However, this preparation was not identical to collagen in stability. Incubation with pepsin caused the loss of a third of the underhydroxylated native preparation without a noticeable effect on collagen. Incubation with pepsin also markedly reduced the recovery of $\alpha 2$. Thus it is possible that hydroxyproline stabilizes the structure of certain regions of the collagen molecule more than proline as indicated by the increased resistance to pepsin digestion in this material

Possible enzymatic step in macromolecular assembly

The renaturation of the collagen molecule from isolated α chains occurs slowly and with low efficiency (Piez, 1967). This observation prompted Speakman (1971) to propose the existence of precursors of collagen with regions whose specific function was to align the chains and ensure rapid and efficient molecular assembly, spontaneously following the completion of the pro-α chains. We have carried out preliminary studies on the renaturation of heat denatured procollagen and have not found it to proceed faster than with collagen (unpublished observations). Thus, there is no evidence yet for the spontaneous assembly of pro-α chains.

The results reported here suggest that synthesis of the pro-α chains can be uncoupled from macromolecular assembly. In particular, calvaria in tissue culture maintain a constant rate of pro-α chain synthesis but stop making procollagen after a few hours. That this relates to depletion of the media can be demonstrated by changing the tissue to fresh media which again permits molecular-assembly. If we can uncouple synthesis from assembly, it seems plausible that an energy dependent or enzymatic reaction is involved. One possible mechanism would involve the enzyme-catalyzed formation of disulfide bonds linking the pro-α chains and their subsequent cleavage. In particular, we (Smith et al., in press) and others (Dehm et al., 1972) have noted the occurrence of polymers of the pro-α chains linked by disulfide bonds after denaturation of procollagen preparations. Disulfide bond formation could stabilize weak but specific associations between the amino terminal ends of the pro-α chains. Since pro-α1 contains several cysteine residues, it may be that an enzyme catalyzing disulfide exchange reactions is required to align the appropriate half-cystines. An enzyme, catalyzing disulfide reactions has been already identified and is thought to play a role in producing the native structure of ribonuclease and other proteins (Goldberger et al., 1967).

SUMMARY

Underhydroxylated collagen from calvaria incubated with α, α'-dipyridyl has been isolated and characterized. A large proportion of the protein was found to have a native, triple helical structure and to be composed of pro-α1 and pro-α2 chains. Some differences from collagen in the susceptibility of the underhydroxylated protein to enzymatic digestion were noted and suggest that hydroxyproline stabilizes certain regions of the molecule.

Under certain conditions chains were synthesized but not assembled into a native macromolecule. It is possible that the formation of the native structure is not spontaneous but involves an energy dependent or enzymatic step.

REFERENCES

Bellamy, G. and Bornstein, P. (1971). *Proc. Natl. Acad. Sci. USA* **68**, 1138–1142.
Bhatnagar, R. S., Prockop, D. J. and Rosenbloom, J. (1967). *Science* **158**, 492–494.
Bornstein, P., Ehrlich, H. P. and Wyke, A. W. (1972). *Science* **175**, 544–546.
Chvapil, M. (1968). *Int. Rev. Connect. Tissue Res.* **4**, 67–196.
Dehm, P. and Prockop, D. J. (1972). *Biochim. Biophys. Acta* **264**, 375–382.
Dehm, P., Jimenez, A. S. and Prockop, D. J. (1972). *Proc. Natl. Acad. Sci. USA* **69**, 60–64.
Diegelmann, R. F. and Peterkofsky, B. (1972). *Proc. Natl. Acad. Sci. USA* **69**, 892–896.
Ehrlich, H. P. and Bornstein, P. (1972). *Fed. Proc.* **31**, 479.
Epstein, E. H., Jr., Scott, R. D., Miller, E. J., and Piez, K. A. (1971). *J. Biol. Chem.* **246**, 1718–1724.
Goldberger, R. F., Epstein, C. H., Jr. and Anfinsen, C. B. (1964). *J. Biol. Chem.* **239**, 1406–1410.
Gottlieb, A. A., Peterkofsky, B. and Udenfriend, S. (1965). *J. Biol. Chem.* **240**, 3099–3103.
Gribble, T. J., Comstock, J. P. and Udenfriend, S. (1969). *Arch. Biochem.* **129**, 308–316.
Hutton, J. J., Tappel, A. L. and Udenfriend, S. (1966). *Anal. Biochem.* **16**, 384–394.
Juva, K. and Prockop, D. J. (1965). *Abstr. Amer. Chem. Soc.* 150th Meeting, Atlantic City, p. 11c.
Kivirikko, K. I. and Prockop, D. J. (1967). *Biochem. J.* **102**, 432–442.
Lapière, C., Lenaers, A. and Kohn, L. D. (1971). *Proc. Natl. Acad. Sci. USA* **68**, 3054–3058.
Lazarides, E. L., Lukens, L. N. and Infante, A. A. (1971) *J. Mol. Biol.* **58**, 831–846.

Lukens, L. N. (1966). *Proc. Natl. Acad. Sci. USA* **55**, 1235–1243.
Lukens, L. N. (1970). *J. Biol. Chem.* **245**, 453–461.
Miller, E. J., Lane, J. M. and Piez, K. A. (1969). *Biochemistry* **8**, 30–39.
Müller, P. K., McGoodwin, E. B. and Martin, G. R. (1971). *Biochem. Biophys. Res. Commun.* **44**, 110–117.
Peterkofsky, B. and Udenfriend, S. (1963). *J. Biol. Chem.* **238**, 3966–3977.
Prockop, D. J. and Juva, K. (1965). *Proc. Natl. Acad. Sci. USA* **53**, 661–668.
Piez, K. A. (1967). In "Treatise on Collagen" (G. N. Ramachandran, ed.) Vol. 1, pp. 207–252. Academic Press, New York.
Piez, K. A., Eigner, E. A. and Lewis, M. S. (1963). *Biochemistry* **2**, 58–66.
Rosenbloom, J. and Prockop, D. J. (1969). In "Repair and Regeneration. The Scientific Basis of Surgical Practice" (J. E. Dunphy and W. Van Winkle, Jr., eds.) pp. 117–135. McGraw-Hill Book Company, New York.
Smith, B. D., Byers, P. H. and Martin, G. R. *Proc. Natl. Acad. Sci. USA*. (In Press.)
Speakman, P. T. (1971). *Nature* **229**, 241–243.
Stark, M., Lenaers, A., Lapière, C. and Kühn, K. (1971). *FEBS Letters* **18**, 225–227.

Approach to the Hydroxylation of Collagenous Proline

Josef Hurych, Pavel Hobza*, Jiřina Rencová
and Rudolf Zahradník*

*Institute of Hygiene and Epidemiology, Centre of Industrial Hygiene
and Occupational Diseases, 100 42 Prague, Czechoslovakia*

Dedicated to the memory of Jan Rosmus

Collagen hydroxyproline is formed by enzyme catalysis of selected peptide-bound proline residues. Purified preparations of the enzyme in question have an absolute requirement for Fe^{2+}, α-ketoglutarate and atmospheric oxygen. This hydroxylase belongs to a new type of oxygenase: one atom of oxygen is incorporated into the substrate and the other into the succinate formed by decarboxylation of α-ketoglutarate (Cardinale et al., 1971). The absorption spectrum of highly purified proline hydroxylase corresponds to a simple protein. It does not contain the components of the α-ketoglutarate-dehydrogenase complex (thiamine pyrophosphate, flavin, pyridine nucleotide, Rhoads and Udenfriend, 1970; Pänkäläinen and Kivirikko, 1971).

It is not yet clear, in which phase and in which way ferrous ions participate in the hydroxylation. If addition of Fe^{2+} to the hydroxylating systems is omitted the activity of hydroxylase isolated from different sources is influenced in a different degree: in enzyme from chick embryos the activity decreases by 85%, in enzyme from new-born rat skin by 60% and in that from granulation tissue by 100% (Hutton et al., 1967). It ought to be mentioned that these experiments were not carried out with pure enzyme preparations. Experiments with p-chloromercuribenzoate and N-ethylmaleimide tend to show that the sulfhydryl groups also play a role in the hydroxylation of collagen proline. The active SH-group or groups probably are near the site for binding of α-ketoglutarate (Popenoe et al, 1969). The disulfide-bridges have still

---

* J. Heyrovský Institute of Physical Chemistry and Electrochemistry, Prague, Czechoslovakia.

a further function. They link the enzyme subunits so that the actual molecule of proline hydroxylase forms a dimer. In determination of the molecular weight of the enzyme no identical results were achieved (Rhoads and Udenfriend, 1970; Pänkäläinen et al., 1970).

A further cofactor required for proline hydroxylation is ascorbic acid, whose role is not yet quite clear. It does not seem to take part in electron transfer. The hydroxylation does not occur in the presence of dehydroascorbic acid (Rhoads and Udenfriend, 1970).

In an earlier paper (Zahradník et al., 1971) we investigated the electron distribution in proline, lysine and peptidyl proline. The excess of electrons on the hydroxylable carbon atoms suggest an electrophilic mechanism of the hydroxylation.

In this report the correlation of the known facts of collagen proline hydroxylation with quantum chemical calculations are presented and the probable mechanism of this reaction is proposed. EHT (Hoffmann, 1963) type of calculation was performed by standard method. The ionization potentials for Fe orbitals were taken from Zerner and Gouterman (1966) (IP(4s) = -7.9 eV, IP(3d) = -8.7 eV).

ACTIVATION OF MOLECULAR OXYGEN

There is not as yet experimental proof of the structure of the oxygen complex, which is able to hydroxylate collagen proline. We introduced the following model: Fe^{2+} is bound to the sulfhydryl group (S^-, SH) of the enzyme. The side chain of an amino acid residue is simulated by the methyl group. From different types of approach of molecular oxygen to Fe^{2+} as the most favourable the perpendicular model with equilibrium distance 2.7 Å (Fig. 1) was found. Let us add that interaction between non-ionized sulfhydryl group and Fe^{2+} also leads to similar results. A characteristic feature of this complex is a considerable transfer of electrons from Fe^{2+} ($\delta = +3.36$) to oxygen ($\delta = -0.99$) (Fig. 1). This clarifies why the catalytic effectiveness is limited to Fe^{2+}. It is not probable that under physiologic conditions Fe^{3+} should change to higher valency.

POSSIBILITIES OF INTERACTION OF HYDROXYLATING COMPLEX

The isolated removal of the one oxygen atom in the direction of the O-Fe connecting line from the complex, is connected with a strong decrease of electron density on Fe. From the chemical viewpoint it is extremely improbable especially if we consider that the reaction occurs under physiological conditions.

In the stable complex (Fig. 1) there is a considerable electron excess on the oxygen atoms and therefore the acid-base equilibrium is reached very quickly; it is possible, of course, that the OOH particle or hydrogen peroxide may be formed. Formation of the OOH particle seems more probable because the addition of catalase (Kivirikko and Prockop,

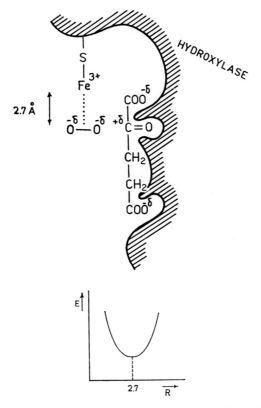

Fig. 1. Formation of a complex between oxygen and iron bound to the enzyme. Activation of molecular oxygen takes place.

1967) into the hydroxylating system is not connected with a decrease of hydroxylation and, even, a certain increase was observed.

From the quantum chemical characteristic of this interaction it may be assumed that migration of the OOH particle is not a process demanding energy. OOH is either in an anionic form or as a radical.

It should be now decided, whether the OOH particle will first react with the peptidyl proline or with the α-ketoglutarate. With reference to the conclusions of the previous paper (Zahradník et al., 1971), only the electrophilic attack of proline by the hydroxylating agent should take place. In view of the possibility that there may apply the nucleophilic attack of oxygen from OOH (or OOH$^-$) on carbon of the carbonyl group in α-ketoglutarate (carbon charge amounts to $+1\cdot08$) we believe the reaction with α-ketoglutarate to be the first step. The plausibility of this suggestion is supported by the experiment carried out by Lindblad et al. (1969) who pointed out that *tert*-butyl-hydroperoxide reacts with α-ketoglutarate acid under the formation of succinic acid. In view of the plausibility of this mechanism we did not examine the direct interaction of the OOH particles with proline.

INTERACTION OF THE OOH PARTICLE WITH α-KETOGLUTARATE

Approach of the OOH particle to the α-ketoglutarate is connected with no stabilization energy. If however we fix the OOH particle and approach only one oxygen atom by prolonging the O-O bond towards carbon of the carbonyl group, we obtain an energy minimum (Fig. 2).

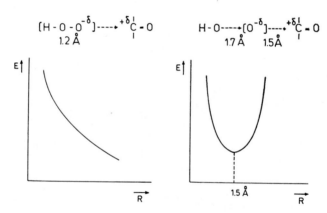

Fig. 2. Potential energy curves for two different interactions between the OOH particle and the carbonyl group of α-ketoglutarate. Left, approach of the whole particle; right, approach of the one oxygen atom only.

We cannot remove the OH residue of the considered particle completely because OH would split off in anionic form. With regard to the fact that the hydroxylable positions in proline (Zahradník et al., 1971) possess a negative charge, only repulsion would result. Should on the other hand a hydroxylable position of proline appear in the vicinity of protonized oxygen, hydroxylation could occur (Fig. 3). Radicals are mostly electrophilic agents. It should be feasible to investigate in this connection energies of the mentioned complex and proline. But these calculations exceed the possibilities of available computers.

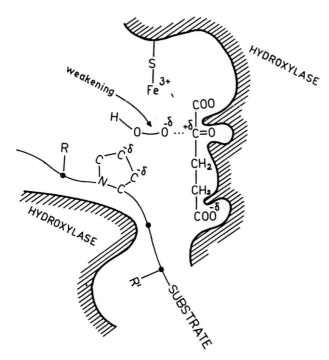

Fig. 3. Formation of the hydroxylating complex.

DISCUSSION

With regard to the mechanism we assume that Fe^{2+} is oxidized to Fe^{3+}. In favor of this hypothesis is the comparison of spectra of particular cofactors with and without the enzyme (Fig. 4). Whereas an aqueous solution of α-ketoglutarate with ferrous ions has no absorption between 250–450 nm, there occurs an increase in the absorption in this region after addition of the purified hydroxylase. The shape of the absorption curve resembles qualitatively that of ferric ions with α-ketoglutarate in water so that a change in the valency of iron may be expected. On the other hand the addition of albumin to the aqueous solution of ferrous ions with α-ketoglutarate does not change the spectrum in any way.

Electron absorption spectra support the hypothesis on the action of ascorbic acid. After addition of ascorbic acid to the whole hydroxylating system the spectrum resembles in the region between 30–20 kK (330–500 nm) the spectrum of α-keloglutarate and ferrous ions in water (see Fig. 4). In the region between 40–30 kK the absorption band of ascorbic acid interferes.

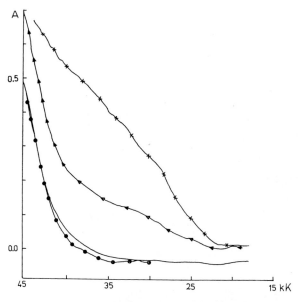

Fig. 4. Electron absorption spectra of proline hydroxylase (PCH) with particular cofactors. Spectra were measured with a spectrophotometer Unicam SP 700 in air atmosphere at 25°C 40 min after addition of individual components in 10 mm cell. Concentration of the enzyme was 0·5 mg protein/ml, tris buffer pH 7·8, Fe^{2+} 0·1 mM, α-ketoglutarate 0·5 mM. ●——●, PCH + α-ketoglutarate | PCH; ———, H_2O + α-ketoglutarate + Fe^{2+} | H_2O; ×———×, PCH + α-ketoglutarate + Fe^{2+} | PCH + α-ketoglutarate; ◀———◀ H_2O + α-ketoglutarate + Fe^{3+} | H_2O + α-ketoglutarate.

It remains to be explained how the whole system is regenerated. Experiments *in vitro* with dehydroascorbate (Rhoads and Udenfriend, 1970) disclosed that the ascorbate does not function as an electron carrier in a cyclic manner. The ascorbate which is probably present in considerable excess could reduce Fe^{3+} itself without being a part of a more extensive redox system.

The hydroxylation of γ-butyrobetaine to carnitine seems to be similar to proline hydroxylation. Both enzymes require the same cofactors. Lindstedt and Lindstedt (1970) assume that a transient hydroperoxide of the substrate is formed, the anionic form of which possibly attacks the α-ketoglutarate. The formed complex is decomposed under the formation of the hydroxylated substrate, succinate and CO_2. An analogous mechanism is considered by Cardinale *et al.* (1971) in proline hydroxylation. In the mechanism proposed by us based on quantum chemical calculations we succeeded in coordinating the theoretical data with the experimental results only under the assumption that in the first step the α-ketoglutarate is attacked and that proline hydroxylation occurs as the next step. With respect to the comparable values of electron densities in hydroxylable positions in γ-butyrobetaine and in proline (-0.23 and -0.29 respectively) we consider our model to be suitable also for the hydroxylation of γ-butyrobetaine.

Let us summarize the individual steps of proline hydroxylation as follows:

1. Oriented fixation of α-ketoglutarate and Fe^{2+} ions on surface of hydroxylase.
2. Activation of molecular oxygen by forming a complex with Fe^{2+} bound through the sulfhydryl group of the enzyme or through another element electronegative towards carbon (Fig. 1).
3. Monoprotonation of oxygen and migration of the OOH particle in anionic form (possibly radical).
4. Nucleophilic addition of OOH particle to carbon of the carbonyl group in α-ketoglutarate under simultaneous weakening of O—O bond (Fig 3).
5. Interaction of mentioned system with peptidyl proline; formation of hydroxyproline by substituting the H atom by the OH group and decarboxylation of α-ketoglutarate.

SUMMARY

Known experimental data on peptidyl proline hydroxylation and quantum chemical (EHT) model calculations were applied for the explanation of the hydroxylation mechanism. Intermediate steps consist in the activation of molecular oxygen by Fe^{2+} formation of an OOH particle (radical or anion), addition of this particle to the carbonyl group in α-ketoglutarate and interaction of the complex formed with peptidyl proline. During the hydroxylation the oxidation of $Fe^{2+} \rightarrow Fe^{3+}$ is assumed. Absorption spectra of the hydroxylase with Fe^{2+} support this hypothesis.

REFERENCES

Cardinale, G. J., Rhoads, R. E. and Udenfriend, S. (1971). *Biochem. Biophys. Res. Commun.* **43**, 537–543.
Hoffmann, R. (1963). *J. Chem. Phys.* **39**, 1397–1412.

Hutton, J. J., Tappel, A. L. and Udenfriend, S. (1967). *Arch. Biochem. Biophys.* **118**, 231–240.
Kivirikko, K. I. and Prockop, D. J. (1967). *J. Biol. Chem.* **242**, 4007–4012.
Lindblad, B., Lindstedt, G., Tofft, M. and Lindstedt, S. (1969). *J. Am. Chem. Soc.* **91**, 4604–4606.
Lindstedt, G. and Lindstedt, S. (1970). *J. Biol. Chem.* **245**, 4178–4186.
Pänkäläinen, M. and Kivirikko, K. I. (1971). *Biochim. Biophys. Acta* **229**, 504–508.
Pänkäläinen, M., Aro, H., Simons, K. and Kivirikko, K. I. (1970). *Biochim. Biophys. Acta* **221**, 559–565.
Popenoe, E. A., Aronson, R. B. and Van Slyke, D. D. (1969). *Arch. Biochem. Biophys.* **133**, 286–292.
Rhoads, R. E. and Udenfriend, S. (1970). *Arch. Biochem. Biophys.* **139**, 329–339.
Zahradník, R., Hobza, P. and Hurych, J. (1971). *Biochim. Biophys. Acta* **251**, 314–319.
Zerner, M. and Gouterman, M. (1966). *Theoret. chim. Acta* **4**, 44–63.

Collagen Proline Hydroxylase Activity and Anaerobic Metabolism

U. LANGNESS and S. UDENFRIEND[*]

Medical University Clinic II, D 2300 Kiel, Metzstr. 53-57, Germany

Biosynthesis of collagen can be envisaged as a two step process involving polypeptide chain formation and hydroxylation of certain peptidyl-proline residues to hydroxyproline residues by collagen proline hydroxylase. This enzyme was shown to be increased prior to appearance of new collagen in the wounds as was demonstrated by Mussini et al. (1967). It was, therefore, suggested to be the regulating factor in collagen synthesis. Gribble et al. (1969) demonstrated a direct correlation between hydroxylase activity and hydroxyproline formation.

Investigating the growth cycle of fibroblasts in tissue culture Gribble et al. (1969) also found that newly synthesized active hydroxylase did not appear until near the end of the log phase and then increased up to 10-fold. Our own investigations corroborated these results and, in addition, revealed that cultured fibroblasts, mainly in early log phase, produce an inactive precursor of the active collagen proline hydroxylase (Langness et al., 1971).

The mechanism of activation of the apparent 'proenzyme' is not yet known. Gribble et al. (1969) first demonstrated that the increase of hydroxylase activity is related to cell density. Since lactate accumulates during the stationary phase of cell growth it was suggested to be involved in the regulation of enzyme activity. In fact, specific activation of enzyme independent of new protein synthesis could be shown by adding lactate to cultured fibroblasts (Table I). *In vitro* studies indicated that lactate does not affect proline hydroxylase directly and that the intact cell is required for the expression of this effect. Pyruvate, succinate, citrate, α-ketoglutarate, fumarate and glucose had no effect on enzyme activation *in vivo* or *in vitro*.

The following studies were done to investigate further the influence

[*] Roche Institute of Molecular Biology, Nutley, N. J. 07110, USA.

Table I
Effect of cell concentration and lactate administration on collagen proline hydroxylase activity

Experiment	Enzyme activity milliunits/mg protein
Cell concentration	
control	19·6
cells concentrated 4×	53·5
Lactate administration	
control	6·0
cells + lactate	51·4

of anaerobic metabolism on activation of collagen proline hydroxylase as it is a general pathophysiological observation that necrobiotic processes which are accompanied by increased anaerobic metabolism can initiate abnormal collagen formation.

METHODS

Procedures for growing fibroblasts and preparation of cell sonicates have been described in detail elsewhere (Gribble et al., 1969). The collagen proline hydroxylase activity of cell sonicates was measured by the tritium release assay of Hutton et al. (1966). Lactic acid was determined by the enzymatic method of Rosenberg and Rush (1966) or, in the case oxamic acid was applied, by the method of Barker and Summerson (1941).

RESULTS

Galactose instead of glucose in the medium

Fig. 1 shows enzyme activity during the growth cycle of L 929 mouse fibroblasts in relation to cell number of two parallel lines growing in different medium. One line grew in Eagle's Minimum Essential Medium, the other in the same medium except that it contained galactose instead of glucose. Significant differences in cell number could not be found up to 144 h growth; then 25–30% less cells were counted in galactose containing medium. Most evident was the difference in specific enzyme activity as convincingly shown in the figure. As also demonstrated in the figure lactate determination, absolutely specific by the enzymatic method, revealed an enormous difference in lactate concentration between both media. Lactate content in galactose medium did not increase at all, it even decreased compared to the value when cell growth started.

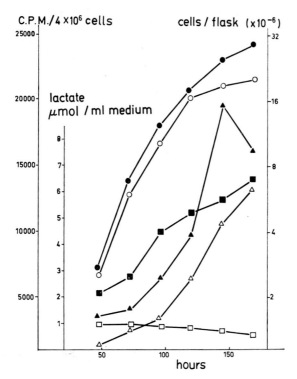

Fig. 1. Growth cycle of L 929 mouse fibroblasts in monolayer cultures. ●, cell number in flasks containing MEM medium; ▲, collagen proline hydroxylase activity in these cells; ■, lactate concentration in medium; ○, cell number in flasks fed with MEM medium containing galactose instead of glucose; △, corresponding enzyme activity; □, lactate concentration in the galactose containing medium.

Oxamic acid effect

In a second series of experiments (Table II) oxamic acid was added to the medium. In the first line oxamic acid at a concentration of 10 μmoles/ml medium was added at the time of inoculation and again when the medium was changed. In a second line oxamic acid was added to the medium after cells had grown over a period of 144 h and remained for another 20 h. The third parallel line was grown without oxamic acid. Oxamic acid had no influence on cell growth, but large

Table II
Effect of oxamic acid on cell growth, on collagen proline hydroxylase activity, and on lactate concentration

	Regular growing L 929 mouse fibroblasts; 164 h	Oxamic acid, 10 μmoles/ml medium on cells for 164 h	Oxamic acid, 10 μmoles/ml medium on cells over 20 h after 144 h regular growth
Cells/flask inoculation 1·6 × 10⁶/flask	15·5 × 10⁶	16·8 × 10⁶	15·7 × 10⁶
Enzyme activity cpm per 4 × 10⁶ cells	60,440	30,826	29,000
Lactate concentration in medium; μmoles/ml	6·9	3·21	3·17

differences of 40–50% were found in collagen proline hydroxylase activity. There was no difference whether oxamic acid had been on the cells over a period of 6 days or only 20 h. Lactate determinations revealed differences in lactate concentration in media corresponding to differences in enzyme activity.

DISCUSSION

The data presented convincingly confirm the influence of anaerobic metabolism. Substitution of glucose by galactose in the medium reduced enzyme activation significantly and is explained by the differences of glucose and galactose in anaerobic metabolism. Glucose is required for synthesis of pyruvate and lactate. The latter are formed to a far lesser extent from galactose and in a separate pathway. This was corroborated by the significantly different lactate concentration in glucose and galactose containing medium.

The effect of oxamic acid can be interpreted in the same way. Novoa et al. (1959) first showed that oxamic acid, which is structurally related to pyruvic acid had a specific inhibitory effect on anaerobic glycolysis. Oxamate inhibited lactic dehydrogenase in beef heart as was shown by Glaid et al. (1955), but also in rabbit skeletal muscle and Ehrlich ascites tumor as was demonstrated by Papaconstantinou and Colowick

(1961). At the same time pyruvate accumulated. In addition, these authors showed that oxamic acid does not compete with pyruvate or related substances in other enzyme systems. The interaction of oxamate with NADH takes place forming the inactive LDH-NADH-oxamate complex. Taking into consideration this specific effect, the experiment presented is explainable. Oxamic acid was expected to reduce lactate concentration as, indeed, could be demonstrated by determination in the media.

Addition of lactate to sonicates of early log phase cells had no activation effect on enzyme. The intact cell was required. This demonstrates that it is not lactate itself which activates 'proenzyme'. Possibly ATP or some other intermediate formed on lactate metabolism is responsible for converting 'proenzyme' to active collagen proline hydroxylase.

REFERENCES

Barker, S. B. and Summerson, W. H. (1941). *J. Biol. Chem.* **138**, 535–541.
Glaid, A. J., Hakala, M. T. and Schwert, G. W. (1955). *Abstr. Meeting Am. Chem. Soc.*, p. 18c.
Gribble, T. J., Comstock, J. P. and Udenfriend, S. (1969). *Arch. Biochem. Biophys.* **129**, 308–316.
Hutton, J. J., Tappel, A. L. and Udenfriend, S. (1966). *Anal Biochem.* **16**, 384–394.
Langness, U., McGee, J. and Udenfriend, S. (1971). *Fed. Proc.* **30**, 1196.
Mussini, E., Hutton, J. J. and Udenfriend, S. (1967). *Science* **157**, 927–929.
Novoa, W. B., Winer, A. D., Glaid, A. J. and Schwert, G. W. (1959). *J. Biol. Chem.* **234**, 1143–1151.
Papaconstantinou, J. and Colowick, S. P. (1961). *J. Biol. Chem.* **236**, 278–284.
Rosenberg, J. C. and Rush, B. F. (1966). *Clin. Chem.* **12**, 299–307.

Procollagen and Procollagen Peptidase in Skin as a Functional System

Ch. M. LAPIÈRE, A. LENAERS and G. PIERARD
*Service de Dermatologie, Hôpital de Bavière
Université de Liège, 4000 Liège, Belgium*

As observed for several enzymes, fibrinogen and one hormone, all polypeptides performing a function in the extracellular fluid, the operating compound is activated after its synthesis by limited proteolysis and removal of a portion of its polypeptide chain. Collagen is another example of such a process: this structural macromolecule is synthesized in the form of a precursor (Layman *et al.*, 1971; Bellamy and Bornstein 1971; Lenaers *et al.*, 1971; Dehm *et al.*, 1972) bearing an extension at the N-terminal extremity of its 3 α chains (Stark *et al.*, 1971; Dehm *et al.*, 1972). This additional peptide is removed by the proteolytic activity of a specific enzyme system, procollagen peptidase (Lapière *et al.*, 1971).

The three constitutive polypeptides of the collagen molecules, in skin two α1 chains and one α2, are synthesized inside specific cells and organized into a triple helix. The extracellular properties of the collagen depend on chemical modifications of some amino acid side chains. This occurs before the release of the molecules from the cell through the action of intracellular enzymes. The tertiary structure of collagen is unstable under physiological conditions and under these conditions the collagen rapidly forms polymers, stabilizing it. The collagen molecules are operational after association into well defined polymers deposited in the extracellular space at specific locations distant from the site of synthesis. The degradation of collagen depends on the activity of a collagenase, an enzyme operating in the extracellular space, secreted as a precursor and activated after its release.

Difficulties in understanding the process of fibrogenesis *in vivo* result from the various properties of native collagen isolated from the extracellular fibrils. Such problems could be overcome by considering the properties of procollagen and taking into account its delayed conversion to collagen by procollagen peptidase (PCP).

(a) The N-terminal extension of procollagen polypeptides could be involved in the post-synthesis organization of the tertiary structure of the collagen molecule and/or the association of the α chains in the correct proportion as well as the stabilization of their structural organization.

(b) The polypeptide extension could reduce the rate of polymerization of procollagen and represents the appendage allowing for the secretion of the molecule from the cells and/or their transport in a soluble form.

(c) The N-terminal extension could be further involved in the extracellular polymerization process by hiding some reactive amino acid side chains and preventing their interaction with reacting side chains on neighbouring molecules before the polymers are correctly arranged.

(d) Either by steric hindrance or due to its amino acid composition (the presence of cystine), the additional peptide might prevent the activity of collagenase during the diffusion of the collagen precursor, during the process of fibril formation or even later.

(e) The additional peptide might represent a temporary connecting linkage between collagen and other connective tissue compounds also synthesized in the fibroblasts and associated with the fibrils in the extracellular space.

Although few of these functions are supported by direct evidence, all of them are possible when taking into account what is known about procollagen peptidase in normal skin and procollagen in dermatosparaxis, a heritable disorder of the connective tissues in the calf. This defect results from the absence of procollagen peptidase and provides a biological model mainly suitable for investigating the relevance of collagen precursor to the process of fibrogenesis.

NATURE AND PROPERTIES OF PROCOLLAGEN

The extended collagen molecules extracted from dermatosparactic skin (Lenaers *et al.*, 1971) are similar if not identical with collagen precursors isolated from cultures of tissues or cells (Layman *et al.*, 1971; Bellamy and Bornstein, 1971; Dehm *et al.*, 1972). The labelled products collected from calf fibroblast cultures when eluted from a CM-cellulose column occur at the same position as the precursor polypeptides from dermatosparactic skin collagen (unpublished results). The reaction products isolated from an *in vitro* protein synthesizing system using chick embryo polysomes contains labeled polypeptides similar to those of dermatosparactic calf skin collagen (Kerwar *et al.*, 1972). Immunesera directed specifically against the additional peptide of dermatosparactic collagen react with normal calf or human connective tissues (Timpl *et al.*, 1973), suggesting also close similarity between dermatosparactic collagen and collagen precursor in various connective tissues. The extended polypeptides of dermatosparactic collagen, the precursor collected from culture and the *in vitro* products from the cell free system are converted into the extracellular types of α chains by procollagen peptidase.

The overall amino acid composition of the peptide extension of procollagen is not compatible with the classical repeating sequence of collagen in which glycine occupies every third position. It bears many similarities with that of an acidic glycoprotein or the polypeptide core of the proteoglycans. It is rich in diacidic amino acids and leucine; it contains half cystine residues and at least one tryptophane residue (unpublished information). The additional peptide of α1 seems to be twice as large as that of α2, as determined by analytical ultracentrifugation or sodium dodecyl sulfate acrylamide gel electrophoresis. Recent evidence (Furthmayr *et al.*, 1972) suggests a molecular weight of $\pm 20,000$ for the isolated additional peptide of α1. This is in agreement with previous measurements if one accepts a non-collagen type of structure and uses values for globular proteins as standards. Since, as seen on SLS in the electron microscope, the N-terminal extension of procollagen represents only 8% of the total length of the collagen molecule (± 250 Å), the additional peptides could not be fully extended, at least when observed on purified collagen. Its size should be small enough to fit in the hole zone of the native type fibrils. Even though it should not overlap with the preceding molecule in the polymer, the capacity of procollagen to form fibrils is impaired *in vitro* and *in vivo*.

Collagen extracted from dermatosparactic skin in neutral salt solution is composed of a mixture of normal molecules and procollagen in a ratio of 1:1 as observed by CM-cellulose chromatography of formaldehyde crosslinked collagen (Lapière and Hanset, 1972). At neutral pH, in NaCl solution (ionic strength ranging from 0·15–0·6) dermatosparatic purified collagen forms a gel at 37°C less opaque than normal collagen and more slowly, particularly at high ionic strength. The reduced opacity seems to be related to the formation of abnormal polymers (thin filaments) since the proportion of polymerizable collagen is similar in both normal and dermatosparatic samples. The rate of polymer formation of the two types of collagen is similarly affected by the pH, the optimal lying between 7·2 and 7·8 In the dermatosparactic calf skin, the collagen framework made in part of collagen precursor, is also disorganized and lacks its classical mechanical properties. The skin is fragile, the cylindrical fibers are absent and replaced by thin filaments associated in ribbons twisted along their longitudinal axis

(O'Hara et al., 1970; Simar and Betz, 1971; Bailey and Lapière, 1973). These abnormal polymers do not form normal bundles due to the absence of parallel packing. In tangential section, as seen in the scanning electron microscope, the abnormal fibrils are bundled together like balls of yarn, some thin filaments joining adjacent masses (Lapière and Hanset, 1971).

The fibrillar collagen of the dermatosparactic skin or the fibrils reconstituted from purified dermatosparactic skin collagen do not form the reducible intermolecular crosslinks although the precursor (allysine) and the intramolecular crosslinks (aldol condensation products) are present in normal and even increased amounts (Bailey and Lapière, 1973). A different type of intermolecular linkage should however exist since collagen in dermatosparactic skin is less extractable than that of normal skin (Lapière and Hanset, 1971). The proportion of collagen extracted from normal and dermatosparactic skin in salt solution at neutral pH is not significantly different while the acid extractable collagen from dermatosparactic skin is strikingly reduced and even more when mercaptoethanol ($0.05\ M$) is added to the extractants. Since the presence of mercaptoethanol does not modify the extractibility of collagen in normal skin the resistance to extraction of the collagen in dermatosparactic skin is probably not mediated by disulfide bridges.

Immunological studies further suggests that the additional peptide is involved in the process of fibrogenesis since immunofluorescence studies using an antibody specific for the polypeptide extension react with the connective tissue fibrils of various organs (Timpl et al., 1972). This observation supports the presence in the extracellular space of either collagen precursor or its two parts, collagen and the additional peptide.

PROCOLLAGEN PEPTIDASE IN SKIN

Procollagen peptidase (PCP) is the enzyme which converts collagen precursors into collagen. This proteolytic activity is present in all normal connective tissues and absent or reduced in dermatosparaxis (Lapière et al., 1971). Since the peptidase activity is specifically modified in this genetic disease, it represents in all probability a specific enzyme system. Its activity on procollagen is indeed more specific than that performed by proteases such as trypsin, chymotrypsin, papain, tissue cathepsins etc. All these enzymes degrade completely the non-helical extension of native collagen in solution while procollagen peptidase does not remove the telopeptides but specifically cleaves their extensions.

In various animal species (rat, guinea pig, calf, human) and many types of connective tissues (skin, tendon, bone, cartilage, blood vessels, lung, cornea) PCP has been detected in significant amounts. The enzyme activity is present in the connective tissue-rich structures and not associated with the endothelium or epithelium. White blood cells, blood serum or plasma, capillary endothelium, and isolated epidermis do not display a significant amount of PCP activity. In cell cultures, fibroblast produces this enzyme while Hela cells, BHK cells, epidermis and lymphocytes do not.

Procollagen peptidase activity is related to collagen metabolism. The amount of enzyme activity in the skin of growing rats is related to the rate of collagen accretion. In human skin, PCP activity is increased in diseases associated with stimulated fibroblastic activity (Pierard et al., 1972).

Procollagen peptidase activity does not accumulate in the cells responsible of its synthesis. In normal calf fibroblasts, most of the enzyme is present in the culture fluid. In connective tissues such as skin or tendon, PCP can be collected in large proportions without extensive disintegration of the tissue (unpublished results). The mode of action of procollagen peptidase is also compatible with its extracellular activity. This enzyme can excise the additional peptide from procollagen in solution as well as from polymeric collagen. The immunological demonstration of the additional peptide in the fibrillar framework also supports this hypothesis.

DISCUSSION

In dermatosparaxis, fibrogenesis in skin occurs to a significant extent using the precursor of collagen . Polymerization seems to proceed normally up to the formation of filaments made of the parallel arrangement of a few molecules in cross section some type of units of polymeric collagen. These long linear polymers do not however associate with similar adjacent units, perhaps due to the presence of the additional peptide. In such polymers, the classical intermolecular crosslinks are prevented from forming but links are established between adjacent molecules or polymer units through the additional peptides. These observations indicate that the additional peptides might have a function in fibrogenesis even in normal tissues. The enzyme responsible for its removal seems to operate in the extracellular fluid.

Besides other possible functions, the procollagen—procollagen peptidase system might operate in the organization of the collagen fibers in the tissues according to the following hypothetical steps. Procollagen would be secreted from the fibroblasts and allowed to diffuse to sites distant from the secreting cells. The filaments made of precursors would organize in the extracellular space and connect with similar elements, using perhaps the additional peptide as a recognition site for attachment. These similar units could bind loosely together in such a way that mechanical force could place them in a suitable position. After the action of procollagen peptidase, the progressive removal of the additional peptide would allow intermolecular bonds to form, eventually establishing the definitive fibrous framework capable of insuring the mechanical properties of the connective tissue.

Perhaps, the abnormal genome responsible of dermatosparaxis will provide a unique tool to understand the relationship between the synthesis of biopolymers and their mechanical properties.

REFERENCES

Bailey, A. J. and Lapière, Ch. M. (1973). *Eur. J. Biochem.* **34,** 91–96.
Bellamy, G. and Bornstein, P. (1971). *Proc. Natl. Acad. Sci. USA* **68,** 1138–1142.
Dehm, P., Jimenez, S. A., Olsen, B. R. and Prockop, D. J. (1972). *Proc. Natl. Acad. Sci. USA.* **69,** 60–64.
Furthmayr, H., Timpl, R., Stark, M., Lapière, Ch. M. and Kühn, K. (1972). *FEBS Letters* **28,** 247–250.
Kerwar, S. S., Kohn, L. D., Lapière, Ch. M. and Weissbach, H. (1972). *Proc. Natl. Acad. Sci. USA* **69,** 2727–2731.
Lapière, Ch. M. and Hanset, R. (1971). *Bull. Acad. R. Med. Belg.* **11,** 747–785.
Lapière, Ch. M., Lenaers, A. and Kohn, L. (1971). *Proc. Natl. Acad. Sci. USA* **68,** 3054–3058.
Layman, D. L., McGoodwin, E. B. and Martin, G. R. (1971). *Proc. Natl. Acad. Sci. USA* **68,** 454–458.
Lenaers, A., Ansay, M., Nusgens, B. and Lapière, Ch. M. (1971). *Eur. J. Biochem.* **23,** 533–543.
O'Hara, P. J., Read, W. K., Romane, W. M. and Bridges, C. H. (1970). *Lab. Invest.* **23,** 307–314.
Pierard, G., Lenaers, A. and Lapière, Ch. M. (1972). Joint Meeting, The Society for Investigative Dermatology, Inc. and European Society for Dermatological Research, Amsterdam, 25 May, Abstract, *J. Invest. Derm.* **58,** 264.
Simar, L. J. and Betz, E. H. (1971). *Hoppe-Seylers Z. Physiol. Chem.* **352,** 13.
Stark, M., Lenaers, A., Lapière, Ch. M. and Kühn, K. (1971). *FEBS Letters* **18,** 225–227.
Timpl, R., Wick, G., Furthmayr, H., Lapière, Ch. M. and Kühn, K. (1973). *Eur. J. Biochem.* **32,** 584–591.

The Role of Hydroxylysine in the Stabilization of the Collagen Fibre

ALLEN J. BAILEY and SIMON P. ROBINS

*Agricultural Research Council, Meat Research Institute
and Department of Animal Husbandry, University of Bristol
Langford, Bristol, England*

Hydroxylysine is an amino acid unique to collagen. Hence, it is reasonable to assume that it has an important role to play in some aspects of the molecular structure, organization, mechanical stability, metabolism or the immunological properties of collagen fibre. Although discovered in 1938 by van Slyke it was some years later before a role was identified for this residue.

CARBOHYDRATE ATTACHMENTS

The first demonstration of a specific role came from the studies of Butler and Cunningham (1966) who isolated a hydroxylysine residue with galactose and glucose attached. Spiro (1967) demonstrated that the structure was hylys-gal-glu and Bosman and Eylar (1966) have demonstrated the presence and specificity of the glucose and galactose transferases.

The existence of hylys-gal-glu and of hylys-gal have now been demonstrated in a number of different tissues (Spiro, 1970). It is interesting to note that the invertebrate, basement membrane and cartilage collagens, in addition to containing a higher hydroxylysine content, have a much higher proportion of these residues carbohydrated (80%). In basement membrane hydroxylysine-linked carbohydrate occurs in the form of disaccharide units, whilst in fibrillar collagens a substantial percentage, up to 50%, of the hydroxylysine linked units occurs as single galactose units. However, the role of the these residues is still far from clear and at the present time one can only speculate as to their possible role in the organization and stabilization of the collagen fibre.

Piez et al. (1970) have shown that one such hydroxylysine residue in rat skin is located at the N-terminal end of α1-CB5, i.e. at one end of the overlap region. Morgan et al. (1970) have therefore suggested that a large carbohydrate unit may be important in locating adjacent molecules during fibrogenesis. However, it remains to be seen where the other similar residues are located. Corneal collagen is interesting in this respect since it possesses a larger number of sugars than might be expected as quarter-stagger locating points yet still forms normal banded fibres, the excess units apparently not disturbing the alignment. An additional possibility is that the sugar residue may in some way be involved in the cross-linking reactions (Bailey et al., 1970) which will be discussed in more detail in the next section.

A number of workers have noted an apparent correlation of the hexose content and fibril size (e.g. Spiro, 1970). Since almost all the hexose present in collagen can be accounted for by that glycosidically linked to hydroxylysine, one of the indirect roles of hydroxylysine might be to control fibre size. Thus, the carbohydrate units are spaced farthest apart in the fibrillar collagens possessing a high degree of organization under the electron microscope, and they are closest together in the amorphous basement membranes. Corneal collagen has an intermediate density of hydroxylysine-linked units and consists of very fine fibrils. However, it must be remembered that the basement membranes possess an additional non-collagenous peptide attached to the collagen and this may well also interfere with the fibre organization.

It has also been suggested that hydroxylysine may be necessary for extrusion from the cell on the basis of the general hypothesis that for export of proteins from cells carbohydration must occur (Eylar, 1965). Winterburn and Phelps (1972) have refuted this suggestion and quote a number of extracellular proteins that are devoid of sugars, and they propose that the sugars may determine the extracellular fate of the protein molecule. Furthermore, Pinnell et al. (1972) have recently reported a heritable connective tissue disorder in which the skin collagen contains no hydroxylysine, and hence no sugars, yet it is still extruded from the fibroblast to form fibres. It is unlikely therefore that hydroxylysine is essential for extrusion from the cell.

Thus, although hydroxylysine is required for the subsequent addition of sugar residues the functional role of these units requires to be established.

BIOSYNTHESIS OF INTERMOLECULAR CROSSLINKS

A more clearly defined function of hydroxylysine is its involvement in intermolecular crosslinks. Three major intermolecular crosslinks have been shown to be present in young collagen. The precursors of the crosslinks are lysine and hydroxylysine-derived aldehydes, allysine and hydroxyallysine respectively and both reside in the N-terminal telopeptide and probably in the C-terminal telopeptide (Stark et al., 1971). All the crosslinks involve hydroxylysine either by reaction of the ε-NH_2 or of the aldehyde formed after oxidative deamination (Fig. 1).

(1) dehydro-OH-LNL results from the condensation of one residue of allysine with ε-NH_2 of hydroxylysine from an adjacent molecule (Bailey and Peach, 1968).

$$HOOC\diagdown_{CH-(CH_2)_3}\diagup^{H_2N} CH=N-CH_2-\underset{\underset{OH}{|}}{CH}-(CH_2)_2-CH\diagup^{COOH}_{\diagdown NH_2}$$

(2) dehydro-diOH-LNL is formed from a residue of hydroxyallysine with the ε-NH_2 of hydroxylysine from an adjacent molecule (Davis and Bailey, 1971; Mechanic et al., 1971).

$$HOOC\diagdown_{CH-(CH_2)_2-}\diagup^{H_2N}\underset{\underset{OH}{|}}{CH}-CH=N-CH_2-\underset{\underset{OH}{|}}{CH}-(CH_2)_2-CH\diagup^{COOH}_{\diagdown NH_2}$$

This aldimine bond spontaneously undergoes a rearrangement *in vivo* to the 5-keto form, thus accounting for the unusual stability of this crosslink (Robins and Bailey, 1973).

(3) Fr. C. The structure of this crosslink has not yet been completely elucidated but it appears to be derived from the condensation of the aldehyde moiety of the intramolecular aldol with hydroxylysine and some other residue, from an adjacent molecule. The structure of this compound has now been shown to be histidinohydroxymersodesmosine, formed by the condensation of hydroxylysine and histidine with the allysine aldol (Tanzer et al., 1973). However, this compound is believed to be an artefact of the reduction procedure (Robins and Bailey, 1973).

The proportion of these three crosslinks, all Schiff bases, varies in different tissues.

(i) Skins contains two of the Schiff bases, dehydro-hydroxylysinonorleucine (dehydro-OH-LNL) and the unknown component designated Fr. C.

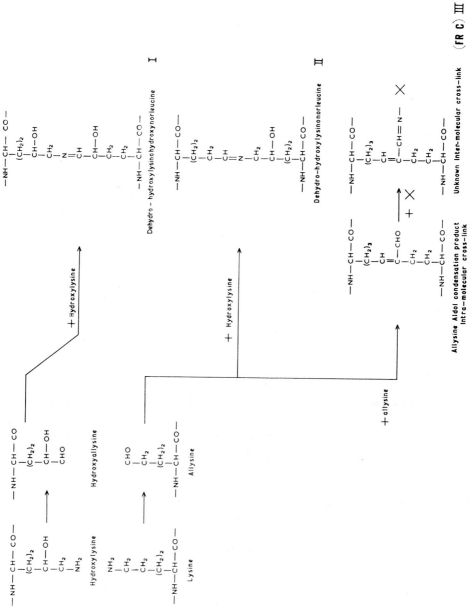

Fig. 1. Biosynthetic pathway of the formation of the Schiff base crosslinks present in skin (II and III), tendon (I, II

(ii) Tendon contains three Schiff bases, dehydro-hydroxylysino-hydroxynorleucine (dehydro-di-OH-LNL), dehydro-OH-LNL and Fr. C.

(iii) Bone and cartilage contain one major Schiff base dehydro-di-OH-LNL.

Amino acid sequence analysis of the N-terminal telopeptide in various tissues reveals that only lysine is present in the skin telopeptide, but this residue is 30–40% hydroxylated in tendon, whilst in bone and cartilage up to 60–70% hydroxylation occurs (Barnes et al., 1971a). Enzymic oxidative deamination of these residues leads to allysine and hydroxyallysine, and the variation in the extent of hydroxylation in different tissues may therefore act as a control on the type of crosslink present. The nature of the crosslink is of considerable importance since we have shown that the bond involving hydroxyallysine is thermally and chemically more stable than the other two Schiff base crosslinks. Furthermore, dehydro-di-OH-LNL is the major crosslink of the very insoluble collagens, bone and cartilage, although the exact relationship between solubility and crosslinks has not yet been clearly established.

Further support for the importance of hydroxylysine in the stabilization of the molecule has recently been demonstrated in the studies on vitamin C deficient collagen. Levene et al. (1972a) demonstrated that collagen synthesized by 3T6 mouse fibroblasts in the absence of vit.C was more salt soluble than collagen from vit.C supplemented cultures. They found little change in the intramolecular crosslinking, the increased solubility being due to a change in the intermolecular crosslinks. Since these crosslinks are derived from hydroxylysine, the hydroxylation of the lysine residues specifically involved in crosslinking must have been inhibited in vit.C deficient cultures. Confirmation of this was achieved by analysis of the reducible crosslinks when a decrease in di-OH-LNL and OH-LNL was noted (Levene et al., 1972b). A concomitant increase in LNL occurred due to condensation of allysine with a lysine residue present in the reactive position which would normally have been hydroxylated. Furthermore, analysis of the aggregated procollagen in the culture media indicated a decreased stability when underhydroxylated (Bates et al., 1972).

Location of the crosslinks

The formation of both allysine and hydroxyallysine occurs in the N- and C-terminal telopeptide regions. It is almost certain therefore that the crosslinks OH-LNL and di-OH-LNL are located at the same position, the difference lying in the extent of hydroxylation of the telopeptide lysine. The actual location has not yet been established but the precise alignment of the tropocollagen molecules in a quarter-stagger fashion limits the possible locations at which these aldehyde groups can crosslink to adjacent molecules.

As the order of the CNBr peptides in rat skin collagen has been established, it is possible to predict the probable peptides involved in the intermolecular crosslinking. Recently Kang (1972) reported that CNBr cleavage of reduced collagen and chromatographic separation of the peptides revealed the presence of a peptide linking $\alpha 1$-CB1 and $\alpha 1$-CB6. Since the exact position of the crosslink within $\alpha 1$-CB6 is not known, this does not confirm the quarter-stagger or overlap hypotheses but is strongly suggestive of its correctness since these peptides occur at the N- and C-terminal ends respectively. Analysis of the CNBr peptides of cartilage collagen may prove more informative since this collagen contains three identical α chains (Miller, 1971) and a single type of crosslink.

Changes in the crosslinks with age

The previous results demonstrating the presence of reducible crosslinks in collagen were all carried out on young tissues. It is generally agreed that with increasing age the collagen fibre steadily increases in stability to external influences, e.g. thermal denaturation, swelling, solubility and enzymes. All these changes could be accounted for by a gradual increase in the number of covalent crosslinkages between the peptide chains as originally proposed by Verzar (see review, 1964). Because of the labile nature of most types of the reducible intermolecular bonds an increase in their concentration with age would not account for the observed decrease in solubility: for this the presence of bonds of higher thermal stability would be required. Analysis of collagen from skin, tendon and cartilage revealed that the proportion of these reducible crosslinks at first increased with age from birth reaching a maximum during the rapid growth in infancy. The proportion of the crosslinks then decreased until at maturity they were virtually absent (Fig. 2). The crosslinks thereafter remain at a relatively low level, no further changes being observed during senescence.

Since the formation of these reducible crosslinks involved a chemical reaction of the carbonyl function of the allysine and hydroxyallysine residues with another reactive species, and in view of the low metabolic activity of collagen it was conceivable that the process was time dependent rather than following the physiological age. However, comparison of the rates of decrease of the reducible crosslinks showed that the process was virtually complete at 4–5 years for the bovine tissues, and 17–20 years for the human tissues, i.e. at physiological maturity (Bailey

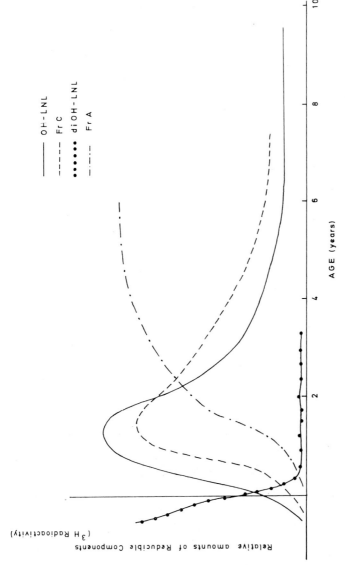

Fig. 2. Variation in the proportion of the various reducible components present in skin collagen (bovine) with increasing age.

and Shimokomaki, 1971). The crosslinking process therefore follows the physiological rather than the temporal age of the animals.

These results clearly demonstrate that the aldehyde-derived crosslinks do indeed change with age and are probably intermediate crosslinks as originally proposed. The maximum content of reducible crosslinks reflects the point at which the maximum rate of growth occurs, i.e. when the maximum amount of newly synthesized collagen is present; as the growth rate subsequently slows down the proportion of reducible crosslinks decreases. It might have been expected that the number of crosslinks would remain constant after reaching the maximum value but they decrease until at true maturity when growth stops, i.e. the closing of the epiphyseal plates, only a small proportion of these crosslinks are present. The decrease must therefore be due to some change in the reducible crosslinks since it is clear from the increased stability of collagen that there cannot be a decrease in the number of crosslinks. We therefore proposed that as the animal matures the labile Schiff bases are replaced by more thermally stable non-reducible crosslinks, the Schiff bases being regarded as intermediates.

The mechanism by which these crosslinks are stabilized is not at present known. The most obvious possibility is reduction *in vivo* of the aldimine bond. This would produce a stable non-reducible crosslink. However, analysis of old non-borohydride reduced collagen for this form of the crosslink failed to reveal detectable quantities of any of the three major Schiff bases (Bailey and Peach, 1971; Robins *et al.*, 1973). Furthermore, addition of radioactively labelled crosslinks of known specific activity to a hydrolysate of 5 gm of old (15 years) bovine skin failed to reveal any significant reduction in the specific activity when re-isolated, again indicating the absence of reduction *in vivo*. As a further test, skin was reduced with borodeuteride and a single molecular ion was recorded, there being no evidence of an additional ion of (M-1) due to reduction *in vivo* with hydrogen rather than the heavier isotope deuterium. Although we could find no evidence for reduction *in vivo*, Mechanic *et al.* (1972) have reported that 25–50% of the Schiff base dehydro-di-OH-LNL is reduced *in vivo*. These results are based on the borodeuteride technique in the mass spectrometer but this extensive proportion of the reduced crosslink in native collagen was not confirmed by isolation of the crosslink.

Parallel with the decrease in the reducible crosslinks it was found that two of the minor components in young tissue, designated Fr. A, increased with age. This was apparent not only in the different tissues examined, but also in different species of animal and in man. These components have been isolated and characterized. The first of the two peaks (Fr. A1) was shown to be a mixture of the condensation products of lysine with mannose and glucose (Robins and Bailey, 1972). The components comprising Fr. A2 are anhydro-derivatives of these complexes, and are therefore solely artefacts of the acid hydrolysis procedure. Analogous derivatives involving hydroxylysine have also been isolated and characterized but these are present in much smaller quantities. Obviously these components do not constitute crosslinkages between collagen molecules nor do they participate in the binding of collagen to glycoproteins or proteoglycans since the sugar reducing end-group is no longer available.

The immediate reaction is that these lysine-glycosylamines are artefacts produced during the preparation of the tissue. However, the presence of such glycosylamine under physiological conditions has previously been demonstrated for normal human haemoglobin (Bookchin and Gallop, 1968) and the possibility that these compounds also exist in collagen cannot be dismissed outright. The observed increase in the amount of these glycosylamines might indicate a systematic increase in the binding of collagen to free hexose or the reducing end groups of a polysaccharide, a fact which would have considerable influence on the organization of the fibrils due to the change in the charge profile.

Embryonic skin collagen

In contrast to collagen from the skin of young chicks or young mammals, the major reducible crosslink of embryonic skin was found to be identical to the crosslink previously observed in bone, dehydro-di-OH-LNL (Fig. 2) (Bailey and Robins, 1972). The presence of this crosslink indicates that hydroxylysine occurs in the N-terminal telopeptide of embryonic collagen. All the amino acid sequence studies carried out on this region of the collagen from young chick, rat, calf or man, demonstrated the absence of hydroxylysine. Confirmation that hydroxylysine occurs in the telopeptide of embryonic skin has been provided by Barnes et al. (1971b). Miller et al. (1971) also recently suggested the presence of a different collagen possibly of the $(\alpha 1)_3$ type in addition to the normal $(\alpha 1)_2 \alpha 2$ type, in new-born human skin. This highly hydroxylated embryonic collagen must be resorbed; or at least diluted out, during the rapid post-natal growth phase by the normal $(\alpha 1)_2 \alpha 2$ type. It is interesting in this respect to note that unlike other soft connective tissue collagen, the collagen of the uterus, non-gravid and immediately post-partum is similar to embryonic collagen; dehydro-di-OH-LNL being the major crosslink.

CONCLUSION

In summary, hydroxylysine is necessary for the addition of the small amounts of carbohydrates linked to collagen, but the role of these sugar units is not yet clear. The importance of hydroxylysine involvement in the crosslink is more clearly established. The extent to which the lysine residue in the N-terminal telopeptide is hydroxylated, and hence the extent to which a higher proportion of the more stable crosslink dehydro-di-OH-LNL is produced, is clearly of great importance in differentiating between tissues. However, these presently identified crosslinks are only the intermediates and the role of the hydroxylysine residues in the more stable crosslinks has not yet been clarified. This may well prove to be a more important aspect of hydroxylysine in the stabilization of the collagen fibre.

REFERENCES

Bailey, A. J. and Peach, C. M. (1968). *Biochem. Biophys. Res. Communs.* **33**, 812–819.
Bailey, A. J. and Peach, C. M. (1971). *Biochem. J.* **121**, 257–259.
Bailey, A. J. and Robins, S. P. (1972). *FEBS Letters* **21**, 330–334.
Bailey, A. J. and Shimokomaki, M. (1971). *FEBS Letters* **16**, 86–88.
Bailey, A. J., Peach, C. M. and Fowler, L. J. (1970). *Biochem. J.* **117**, 819–831.
Bates, C. J., Bailey, A. J. and Levene, C. I. (1972). *Biochim. Biophys. Acta* **278**, 372–390.
Barnes, M. J., Constable, B. and Kodicek, E. (1971a). *Biochem. J.* **125**, 433–437.
Barnes, M. J., Constable, B., Morton, L. F. and Kodicek, E. (1971b). *Biochem. J.* **125**, 925–928.
Bookchin, R. M. and Gallop, P. M. (1968). *Biochem. Biophys. Res. Commun.* **32**, 86–93.
Bosmann, H. B. and Eylar, E. H. (1968). *Biochem. Biophys. Res. Commun.* **33**, 340–346.
Butler, W. T. and Cunningham, L. W. (1966). *J. Biol. Chem.* **241**, 3882–3888.
Davis, N. R. and Bailey, A. J. (1971). *Biochem. Biophys. Res. Commun.* **45**, 1416–1422.
Eylar, E. H. (1965). *J. Theor. Biol.* **10**, 89–113.
Kang, A. H. (1972). *Biochemistry*, **11**, 1828–1835.
Levene, C. I., Shoshan, S. and Bates, C. J. (1972a). *Biochim. Biophys. Acta* **257**, 384–388.
Levene, C. I., Bates, C. J. and Bailey, A. J. (1972b). *Biochim. Biophys. Acta* **263**, 574–584.
Mechanic, G., Gallop, P. M and Tanzer, M. L. (1971). *Biochem. Biophys. Res. Commun.* **45**, 644–653.
Miller, E. J. (1971). *Biochem. Biophys. Res. Commun.* **45**, 444–451.
Miller, E. J., Epstein, E. H. and Piez, K. A. (1971). *Biochem. Biophys. Res. Commun.* **42**, 1024–1029.
Morgan, P. H., Jacobs, H. G., Segrest, J. P. and Cunningham, L. W. (1970). *J. Biol. Chem.* **245**, 5042–5048.
Piez, K. A., Miller, E. J., Lane, J. M. and Butler, W. T. (1970). *In* "Chemistry and Molecular Biology of the Intercellular Matrix" (E. A. Balazs, ed.), Vol. 1, p. 117. Academic Press New York.

Pinnell, S. R., Krane, S. M., Kenzora, J. E. and Glimcher, M. J. (1972). *N. Engl. J. Med.* **286**, 1013–1020.
Robins, S. P. and Bailey, A. J. (1972). *Biochem. Biophys. Res. Commun.* **48**, 76–84.
Robins, S. P. and Bailey, A. J. (1973). *FEBS Letters.* (In Press.)
Robins, S. P., Shimokomaki, M. and Bailey, A. J. (1973). *Biochem. J.* **131**, 771–780
Spiro, R. (1967). *J. Biol. Chem.* **242**, 4813–4823.
Spiro, R. (1970). *Annu. Rev. Biochem.* **39**, 599–630.
Stark, M., Rauterberg, J. and Kühn, K. (1971). *FEBS Letters*, **13**, 101–104.
Tanzer, M. L., Housley, T., Bernbi, L., Fairweather, R., Franzblau, C. and Gallop, P. M. (1973). *J. Biol. Chem.* **248**, 393–402.
Van Slyke, D., Hiller, A., Dillon, R. T. and MacFadyan, D. (1938). *Proc. Soc. Exp. Biol. Med.* **38**, 548–554.
Verzar, F. (1964). *Intern. Rev. Connect. Tissue Res.* **2**, 243–300.
Winterburn, P. J. and Phelps, C. F. (1972). *Nature* **236**, 147–151.

Ascorbic Acid and Collagen Synthesis

C. I. LEVENE and C. J. BATES

Dunn Nutritional Laboratory, Medical Research Council and University of Cambridge, Cambridge, England

The clinical consequences of fibrosis may be beneficial to the host, e.g. the healing of a fractured bone or of a surgical wound, or they may be harmful, e.g. mitral stenosis following rheumatic fever or cirrhotic fibrosis of the liver. Furthermore, fibrous tissue, once laid down, tends, with few exceptions, to remain permanently. Experimental osteolathyrism offered the possibility of inhibiting the crosslinking of collagen and elastin whose functions as tensile elements depend on their intact crosslinking. A number of experimentally induced fibrotic conditions were therefore investigated to see whether treatment with lathyrogenic compounds would halt the fibrotic process. The most severe and progressive of these conditions was pulmonary silicosis in the rat, induced by injecting quartz intratracheally. Silicotic nodules of fibrosis appeared in the lungs whose collagen content rapidly increased within

Table I

Effect of BAPN treatment on the collagen content of the lungs of silicotic rats

	Right lungs (mg)	Left lungs (mg)
Normal controls	21·28 ± 1·4	11·95 ± 1·13
Silicotic controls	76·10 ± 18·1	21·86 ± 3·15
Silicotics treated with BAPN	34·73 ± 1·8	16·75 ± 1·60

eight weeks; however, if these animals were subsequently treated with daily injections of β-aminopropionitrile—the lathyrus factor (BAPN), which had been shown to inhibit the formation of crosslinks in collagen—the total collagen content of the silicotic lungs was diminished (Table I) despite the fact that the BAPN-treated silicotic rats had gained more body weight during the eight weeks than the untreated silicotic controls (Levene et al., 1968).

It thus appeared that the *in vivo* inhibition of fibrosis, at least in the rat, was feasible and we therefore embarked on a study of the fibroblast in a controlled model system, to shed some light on the mechanism of collagen synthesis, and the role, if any, of the glycosaminoglycans (GAGs) in this process.

METHODS AND MATERIALS

The fibroblast selected was a mouse line, 3T6, which had been isolated by Green. The decision to study a cell line rather than using the whole

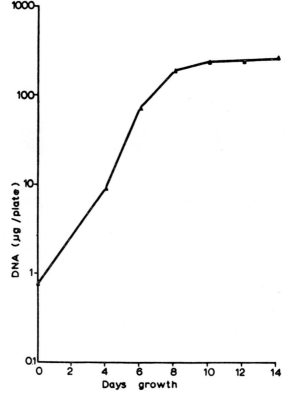

Fig. 1. Growth of 3T6 cultures measured by DNA accumulation.

animal, was based on a desire to control as many variables as possible; the selection of 3T6 was based on its having been well documented and its known ability to synthesize collagen (Green and Todaro, 1967).

The cells were plated at 25,000/60-mm Falcon petri dish and fed every 2 days with 4 ml of Dulbecco and Vogt's modification of Eagle's MEM, supplemented with 10% foetal calf serum and equilibrated with 5% CO_2 for a total of 14 days (Levene and Bates, 1970). The cells grew logarithmically for 8–10 days and then remained in stationary phase for the next 4–6 days (Fig. 1) when they had reached $12-15 \times 10^6$ cells/dish. Electron microscopic studies of the cell layer showed them to have reached a thickness of 8–10 cells.

RESULTS

Preliminary studies during stationary phase showed that collagen accumulated with time in the cell layer (Table II) and that some,

Table II
Cumulative hydroxyproline values for cell layer and growth medium in the presence of ascorbic acid

Culture age, days	Growth medium hydroxyproline, $\mu g/10^6$ cells	Cell layer hydroxyproline, $\mu g/10^6$ cells
8	0·80	0·24
10	2·47	0·49
12	3·85	1·61
14	5·56	2·02

at least, was present as fibrils as seen in the electron microscope, exhibiting their familiar periodicity (Fig. 2). It was also clear that much more hydroxyproline-containing material was released into the medium confirming the results of previous workers (Macek *et al.*, 1967; Schafer *et al.*, 1967; Aleo, 1969; Priest and Davies, 1969).

Most of the cell-layer collagen was found to be insoluble in acetic acid (Table III), suggesting the possibility that crosslinks had been formed in the cell layer collagen.

Similarly, GAGs were found to be present both in the cell layer and in the medium. They were partially characterized chemically

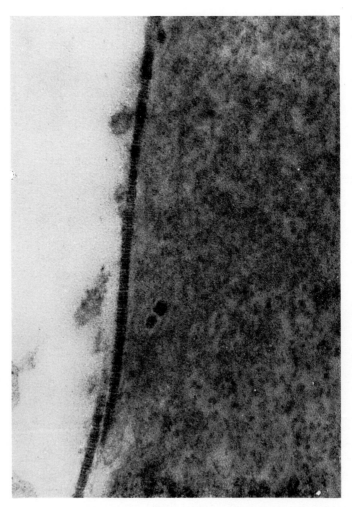

Fig. 2. Electron micrograph of collagen fibrils lying at edge of 3T6 fibroblast; stained with phosphotungstic acid, × 100,000.

Table III
Solubility of cell layer collagen in a 14-day culture grown in the presence of ascorbic acid

Collagen fraction	Total hydroxyproline		Radioactive hydroxyproline	
	µg/plate	% of total	dpm/plate	% of total
Neutral salt-soluble	0·30	15	7700	8
Acid-soluble	0·52	27	22,700	25
Insoluble	1·17	59	60,800	67

as hyaluronic acid, chondroitin sulphate and heparan sulphate (Bates and Levene, 1970). Their role in the formation of collagen was, and remains, unknown.

These cells behaved in culture in a predictable and quantifiable manner, vis-à-vis cell growth and the synthesis of their major macromolecules, collagen and the GAGs.

Our main interest is in collagen; hitherto, at least five diseases of the collagen molecule are known—scurvy, lathyrism (Levene and Gross, 1959) and Ehlers-Danlos syndrome (Mechanic, 1972), the latter two affecting crosslink formation, dermatosparaxis in calves (Lenaers et al., 1971) where procollagen is not converted into tropocollagen, and recently, a human collagen found to be deficient in hydroxylysine (Pinnell et al., 1972). Undoubtedly many more pathological conditions of the collagen molecule remain to be discovered. In this paper, discussion will be confined to the effects of ascorbic acid and of its deficiency on collagen formation in our 3T6 system.

EFFECTS OF ASCORBIC ACID

Ascorbic acid, at 50 µg/ml concentration in the culture medium, produced only a slight increase in cell proliferation (5–10%).

It had previously been shown (Antonowicz and Kodicek, 1968) that there is a fall in the macromolecular-bound galactosamine in the healing tendonectomy wound of the scorbutic guinea-pig, a finding which we subsequently confirmed (Bates et al., 1969). We tested the effect of ascorbate deficiency in 3T6 cultures and found no effect and must therefore consider the possibility that the galactosamine effect in the whole animal may not be a primary result of ascorbate deficiency.

The main effect of ascorbate deficiency which we observed in 3T6 cultures was a diminution in the amount of hydroxyproline in the cell layer (Table IV); there was incidently a much larger amount of hydroxy-

Table IV

Distribution of ^{3}H-hydroxyproline in the cell layer and growth medium during four days growth in stationary phase in the presence and absence of ascorbic acid

Fraction	— ascorbic acid		— ascorbic acid	
	dpm per µg DNA	% of total	dpm µg DNA	% of total
DNA content per plate	234 µg		236 µg	
Cell layer	700	14·6	237	7·4
Growth medium-non-diffusible	2910	60·4	2094	65·6
Growth medium-diffusible	1205	25·0	865	27·0

proline liberated into the medium of which the major part was in macromolecular form and the minor part in dialysable form. We interpret the presence of dialysable hydroxyproline in the medium as indicative of degradation of collagen and hence of the production by 3T6 of a collagenolytic enzyme.

We wished to understand the reason for the greatly diminished deposition of hydroxyproline in the cell layer—was this indicative of a smaller amount of fully hydroxylated collagen or of an unchanged amount of under-hydroxylated collagen? The solution was helped by the availability of a collagenase which was practically free of proteolytic activity and which specifically degraded collagen in our system irrespective of its degree of hydroxylation. This bacterial collagenase after partial purification by the method of Peterkofsky and Diegelman (1971), had the following properties:

(1) It failed to liberate any diffusible C^{14} from a C^{14} trytophan-labelled protein preparation.

(2) Proline/hydroxyproline radioactivity ratios of 1·0 were obtained from the collagenase-susceptible fraction from optimally hydroxylated 3T6 collagen even when less than 10% of the protein-bound proline was present in collagen.

(3) Proline/hydroxyproline radioactivity ratios were unaffected by N-ethyl maleimide up to 25 mM concentration.

(4) Two different commercial sources of bacterial collagenase gave identical results.

By the use of ^{3}H-proline, ^{3}H-lysine and collagenase, it became evident that ascorbate deficiency had no effect on the synthesis of general proteins or of collagenase-sensitive material but that it did shift some collagenase-sensitive material from the cell layer into the medium (Table V). Examination of the effect of ascorbic acid concen-

Table V
Effect of ascorbic acid on total protein synthesis, and distribution of collagen between cell layer and growth medium, in late logarithmic phase

Assay	Ratio: $\dfrac{+ \text{ascorbate}}{- \text{ascorbate}}$
Total proline incorporation per 10^6 cells	1·01
Total lysine incorporation per 10^6 cells	0·98
Proline in collagen per 10^6 cells	0·99
Lysine in collagen per 10^6 cells	0·91
% collagen in cell layer	
(a) Proline label	1·39
(b) Lysine label	1·39

tration on proline hydroxylation showed that below 10 µg ascorbate/ml, collagen in both the cell layer and the medium was sub-optimally hydroxylated and that the degree of hydroxylation was dosage-dependent; above 10 µg/ml, optimal hydroxylation as indicated by a proline/hydroxyproline ratio of 1, was achieved in both cell layer and medium collagen (Fig. 3).

We observed, like van Robertson's group who had studied human fibroblast cultures (Schafer *et al.*, 1967) that ascorbate-deficient cell layers were more fragile to handle than ascorbate-supplemented cell layers; we therefore compared their behaviour to salt and acetic acid extraction by three independent methods and found that the collagen in ascorbate deficiency was more soluble than in the ascorbate-treated cell layers (Table VI) (Levene *et al.*, 1972b). This finding supported the view that ascorbic acid might be involved in the formation of the hydroxylysine-derived intermolecular collagen crosslinks found by Bailey's group (Bailey, 1968). In order to see whether intramolecular crosslinks were also affected we examined TCA-ethanol-purified collagen from the cell layer by disc-gel electrophoresis. These intramolecular crosslinks are believed to be formed via lysine and so, theoretically, there should be no effect due to ascorbate deficiency as, indeed, we found (Levene *et al.*, 1972b) although the α chains in the

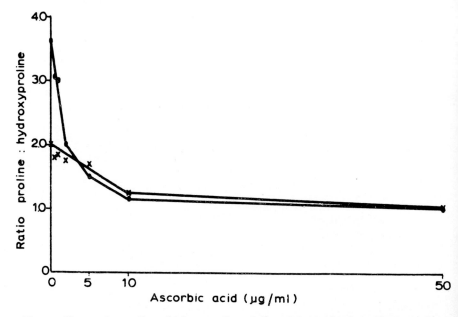

Fig. 3. Change in proline: hydroxyproline radioactivity ratios in collagen—with increasing ascorbic acid concentration. ×———×, cell layer; ●———●, growth medium.

ascorbate-deficient cultures appeared underhydroxylated demonstrating the 'scorbutic' effect.

Our earlier extraction studies had suggested that 3T6 were able to synthesize crosslinks; in order to obtain direct evidence we examined, in collaboration with Dr. Allen J. Bailey, the collagen in cell layers at various ages, following sodium borohydride reduction and found that the cells were able to produce the reduced aldehydic products from hydroxylysine (dihydroxynorleucine), lysine (hydroxynorleucine) and the three reduced Schiff-base crosslinks—hydroxylysino-hydroxynorleucine (reduced Schiff base of hydroxylysine and hydroxylysine aldehyde), hydroxylysinonorleucine (reduced Schiff base of hydroxylysine and lysine aldehyde) and lysinonorleucine (reduced Schiff base of lysine aldehyde and lysine). It was not possible to measure the amount of crosslinks quantitatively but our impression was that they increased

Table VI

Effect of ascorbic acid on the extractibility of cell layer collagen in stationary phase

	+ Ascorbic acid				− Ascorbic acid			
	Neutral salt soluble collagen (%)	Acid soluble collagen (%)	Insoluble collagen (%)	Ratio of insoluble to salt soluble collagen	Neutral salt soluble collagen (%)	Acid soluble collagen (%)	Insoluble collagen (%)	Ratio of insoluble to salt soluble collagen
Total hydroxyproline chemically estimated	16·8	21·4	61·8	3·7	69·6	24·9	5·5	0·1
³H-hydroxyproline	16·1	3·0	80·9	5·0	31·7	4·2	64·1	2·0
Total collagenase-digestible material	30·2	6·5	63·3	2·1	59·4	5·2	35·4	0·6

with age (Levene et al., 1972a). Further evidence that the crosslinks were derived from the oxidative deamination of lysine and hydroxylysine via aldehydic intermediates was obtained by testing the effect of various lathyrogenic compounds which are known to inhibit aldehyde formation; the results indicated a considerable drop in their formation. We were also able to confirm the previous observation by Martin's group (Layman et al., 1972) that lysyl oxidase, the enzyme responsible for the oxidative deamination of lysine to aldehyde, was released into the medium by 3T6 (Cuthbert and Levene, unpublished data). On examining the effect of ascorbate deficiency on the synthesis of the previously mentioned crosslinks, it appeared that the hydroxylysine-derived crosslinks, hydroxylysino-hydroxynorleucine and hydroxylysinonorleucine, were diminished whilst there was a corresponding and presumably related increase in the lysine-derived crosslink, lysinonorleucine (Fig. 4). These findings support the role of ascorbate in the formation of hydroxylysine-mediated crosslinks.

We subsequently examined the effect of ascorbate deficiency on the hydroxylation of ^{3}H lysine in collagen both in the cell layer and in the medium and were surprised to find that, unlike proline hydroxylation which was extremely susceptible to ascorbate deficiency, the hydroxylation of lysine was hardly affected at all (Table VII). Following alkali hydrolysis, it was evident that the glycosylation of hydroxylysine was unaffected and the small amount of underhydroxylation which was found, was confined to the free, non-glycosylated hydroxylysine residues.

Table VII

Effect of ascorbic acid on the hydroxylation of proline and lysine, and on the non-glycosylated hydroxylysine content of cell layer and growth medium collagens in late logarithmic phase

Assay	Ratio: $\dfrac{+ \text{ ascorbate}}{- \text{ ascorbate}}$
Hydroxylation of proline	
(a) cell layer	2·6
(b) growth medium	6·0
Hydroxylation of lysine	
(a) cell layer	1·1
(b) growth medium	1·4
Free (non-glycosylated) hydroxylysine	
(a) cell layer	1·9
(b) growth medium	1·9

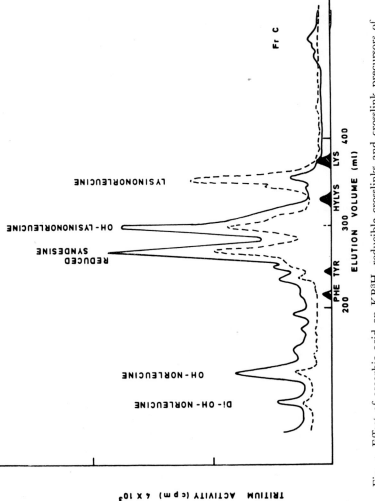

Fig. 4. Effect of ascorbic acid on KB³H₄-reducible crosslinks and crosslink precursors of collagen in the cell layer. -----, minus ascorbic acid; ———, plus ascorbic acid.

We presume that these residues would have provided the source for hydroxylysine-derived crosslinks, explaining the effects of ascorbate deficiency on crosslink formation. We hope to explore further the different susceptibilities of proline and lysine residues to hydroxylation when deprived of ascorbic acid; one possible explanation may be the different K_m values of protocollagen prolyl hydroxylase and protocollagen lysyl hydroxylase for ascorbic acid, as shown by Kivirikko and Prockop (1972).

In summary then, ascorbic acid deficiency, in our system, had virtually no effect on cell growth, general protein synthesis or GAG synthesis; protocollagen synthesis was unaffected but hydroxylation of proline was severely curtailed. The secretion of collagen into the medium despite underhydroxylation, was not reduced. Collagen deposition in the cell layer was diminished; crosslinking of collagen was also diminished. We have also found a consistent effect of ascorbic acid deficiency on the soluble collagen precursor found in the medium (Bates et al., 1972).

Finally, we would like to comment on the difficulty of defining differentiation in terms of collagen synthesis in the fibroblast. Prior to the advent of a specific collagenase, it had been possible to quantitate collagen only by measuring the hydroxyproline content. However, with the use of collagenase it was also possible to measure total collagenase-sensitive material of various degrees of hydroxylation; during log phase, fully hydroxylated collagen was never formed even in the presence of ascorbic acid. We were also able to measure the activity of the hydroxylating enzyme, protocollagen proline hydroxylase in the cell layer during most of the 14-day growth period (Aleo and Levene, unpublished data). When these data were plotted on a single graph on a per 10^6 cell basis (Fig. 5) it was evident that net hydroxyproline synthesis is the result of two processes which change with time in opposite directions. In contrast to the aphorism 'a cell divides or produces its export molecule', we find that protocollagen synthesis is maximal during log phase. Protocollagen proline hydroxylase is, however, minimally active at this time and reaches maximal activity during stationary phase. The production of hydroxyproline is thus the net result of these two processes changing in opposite directions and whereas, early, on, enzyme provides the limiting factor, at later times substrate becomes limiting. These observations confirm the studies of Udenfriend who used cultured L929 fibroblasts (Gribble et al., 1969).

Clearly the dynamics of collagen synthesis are complex and difficult to apply to the definition of differentiation in the fibroblast, since in log phase this would imply protocollagen formation, whilst in stationary phase it would imply the hydroxylation of protocollagen.

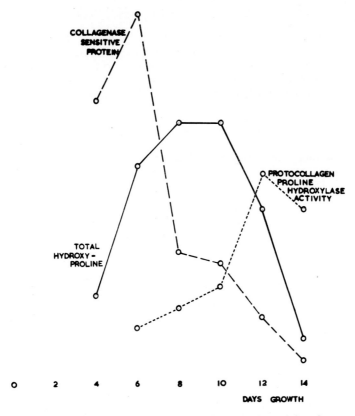

Fig. 5. Effect of culture age on collagen polypeptide synthesis, activity of protocollagen proline hydroxylase, and net formation of hydroxyproline.

ACKNOWLEDGEMENTS

We would like to thank Dr. D. C. Barker of the Dunn Nutritional Laboratory, Medical Research Council and University of Cambridge, for the electron microscopy, Miss Celia J. Prynne for devoted assistance, the editors of *Biochimica et Biophysica Acta* for permission to reproduce Figs. 1, 3 and 4 and Tables IV and VI, and the editors of the *Journal of Cell Science* for permission to reproduce Tables II and III.

REFERENCES

Aleo, J. J. (1969). *Proc. Soc. exp. Biol. Med.* **130**, 451–454.
Aleo, J. J. and Levene, C. I. (Unpublished data).
Antonowicz, I. and Kodicek, E. (1968). *Biochem. J.* **110**, 609–616.
Bailey, A. J. (1968). In "Comprehensive Biochemistry", Vol. 26B, p. 297. Elsevier, New York.

Bates, C. J., Bailey, A. J. Prynne, C. J. and Levene, C. I. (1972). *Biochim. Biophys. Acta* **278**, 372–390.
Bates, C. J., Levene, C. I. and Kodicek, E. (1969). *Biochem. J.* **113**, 783–790.
Bates, C. J. and Levene, C. I. (1970). *Biochim. Biophys. Acta* **237**, 214–226.
Cuthbert, J. and Levene, C. I. (Unpublished data).
Green, H. and Todaro, G. (1967). *Ann. Rev. Microbiol.* **21**, 573–600.
Gribble, T. J., Comstock, J. P. and Udenfriend, S. (1969). *Arch. Biochem. Biophys.* **129**, 308–316.
Kivirikko, K. I. and Prockop, D. J. (1972). *Biochim. Biophys. Acta* **258**, 366–379.
Layman, D. L., Narayanan, A. S. and Martin, G. R. (1972). *Arch. Biochem. Biophys.* **149**, 97–101.
Lenaers, A., Ansay, M., Nusgens, B. V. and Lapière, C. M. (1971). *Eur. J. Biochem.* **23**, 533–543.
Levene, C. I. and Bates, C. J. (1970). *J. Cell Sci.* **7**, 671–682.
Levene, C. I., Bates, C. J. and Bailey, A. J. (1972a). *Biochim. Biophys. Acta* **263**, 574–584.
Levene, C. I., Bye, I. and Saffiotti, U. (1968). *Brit. J. Exp. Path.* **49**, 152–159.
Levene, C. I. and Gross, J. (1959). *J. exp. Med.* **110**, 771–790.
Levene, C. I., Shoshan, S. and Bates, C. J. (1972b). *Biochim. Biophys. Acta* **257**, 384–388.
Macek, M., Hurych, J. and Chvapil, M. (1967). *Cytologia* **32**, 426–443.
Mechanic, G. (1972) *Biochem. Biophys. Res. Commun.* **47**, 267–272.
Peterkofsky, B. and Diegelmann, R. (1971). *Biochemistry* **10**, 988–994.
Pinnell, S. R., Krane, S. M., Kenzora, J. E. and Glimcher, M. J. (1972). *N. Engl. J. Med.* **286**, 1013–1020.
Priest, R. E. and Davies, L. M. (1969). *Lab. Invest.* **21**, 138–142.
Schafer, I. A., Silverman, L., Sullivan, J. C. and Robertson, W. van B. (1967). *J. Cell Biol.* **34**, 83–95.

DISCUSSION

Jonathan Bard:

I am not sure that we can have an ascorbate-minus culture as there is ascorbic acid in the serum. Bio-Cult (our suppliers in foetal calf serum) inform us that they estimate there to be about 20 μg/ml so that there is about 2 μg/ml in normal medium. Otherwise I was pleased to see that 3T6 cells respond to vitamin C in a similar way as early subcultures of human embryonic lung fibroblasts and interested to hear what was really going on.

C. I. Levene:

We have measured the ascorbic acid content of several batches of Flow Laboratories' foetal calf serum and have found less than 2 μg ascorbic acid per ml of serum. Moreover, we find an increased level of hydroxylation in cultures to which 0·5 to 1·0 μg ascorbic acid per ml medium has been added.

Biosynthesis of Collagen in the Axial Organs of Young Chick Embryos

SUZANNE BAZIN and GEORGES STRUDEL*

Service de Pathologie Expérimentale, Institut Pasteur, 92380 Garches, France

For many years we have been studying the influence exercised by the spinal cord and the notochord in chick vertebral cartilage differentiation. It is well known that the extirpation of the spinal cord and of the notochord prevents vertebral cartilage differentiation (Strudel, 1953). According to recent findings (Strudel, 1971) the extirpation of the axial organs not only results in failure of vertebral chondrocytes to differentiate but the metachromatic extracellular material, existing prior to cartilage differentiation as a dense network in the acellular area around the axial organs, fails also to differentiate. In another chapter in this book (p. 93), Strudel demonstrates that the periaxial extracellular material is, in some way, involved in the mechanisms of vertebral cartilage differentiation. It is therefore of a great importance to learn more about the origin, the chemical composition and the role of this periaxial extracellular material. Electron microscopic studies (O'Connell and Low, 1970; Strudel, 1971) showed that this material is composed of flocculent masses of electron dense granules and of microfilaments (Figs. 1 and 2). These components are embedded in an electrontranslucent ground substance. At the present time, the results of the chemical investigations of different workers indicate that the periaxial material of 3-day old embryos contains hyaluronic acid, chondroitin 4 and 6 sulfates and sialoglycoproteins. One supposes that microfilaments could be immature collagen fibrils. Our purpose is to investigate whether the axial organs of $2-3\frac{1}{2}$-day old embryos contain a collagenous material and whether this collagenous material is identical with the extracellular periaxial microfilaments. In order

* Laboratoire d'Embryologie Expérimentale, Collège de France, 49bis, avenue de la Belle Gabrielle, 94130 Nogent-sur-Marne, France.

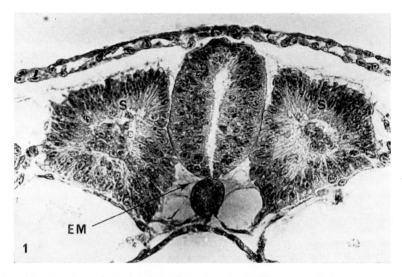

Fig. 1. Section through the somites of a 45 h embryo. The extracellular space around the notochord, the neural tube and the somites (S) contain extracellular (EM) metachromatic material (×370).

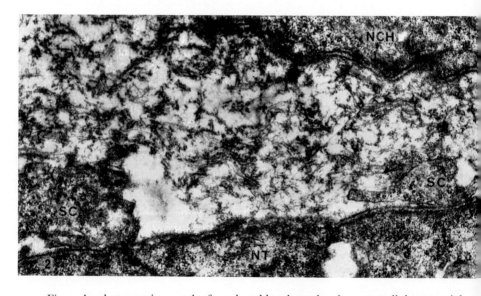

Fig. 2. An electron micrograph of a 3-day old embryo showing extracellular material located between the neural tube (NT) and the notochord (NCH), both limited by a basal lamina. The extracellular material shows microgranules and microfilaments embedded in an electron-translucent ground substance. Two emigrated sclerotome cells (SC) have already reached the axial organs (×18300).

to try to solve this problem, we undertook three types of experiments: 1, a chemical analysis, 2, a histochemical analysis and 3, a histoautoradiographic analysis. All the experiments have been performed with white Leghorn eggs. The axial organs consist of the spinal cord, the notochord and the two adjoining rows of somites. The techniques utilized in light and electron microscopy and in organ culture are widely described elsewhere (Strudel, 1971).

CHEMICAL ANALYSIS

The axial organs from $2-3\frac{1}{2}$-day old embryos were defatted and digested with pronase at pH 7·2 at 4°C for 24 h. The solubilized material was dialyzed against diluted acetic acid. During the dialysis a precipitate was formed. This precipitate and its supernatant as the residue of pronase digestion were hydrolyzed in 6N hydrochloric acid at 105°C for 24 h. Hydroxyproline was isolated from the hydrolysates by microchromatography according to a modified technique by Moore et al. (1958) and quantitatively determined by the method of Stegemann and Stalder (1967).

Hydroxyproline was detected in the three fractions. It was much more abundant in the material solubilized by the pronase digestion, especially in the precipitate obtained after dialysis. We found 36 μg of hydroxyproline for 100 mg of dried axial tissues. This hydroxyproline was incorporated in non-dialysable peptides. This indicates that the axial organs contain a collagenous material. If the hydroxylation of proline reaches its normal level the amount of the detected hydroxyproline would correspond 270 μg of collagen, approximately 0·27% of the dry weight.

The next problem which arose, was to determine the localization of this collagenous material in the axial organs. In order to solve this we undertook assays with histochemical and histoautoradiographic techniques.

HISTOCHEMICAL ANALYSIS

The extracellular periaxial material can be stained for light microscopy by different procedures but we could not obtain a specific staining of the microfilaments. The specific connective tissue stain of Mallory and Weigert gave no positive results. Even the polychrome technique of Gaussen, which gives a clear specific staining of the collagen fibrils within the embryonic chick corneal stroma is unable to stain specifically the microfilaments in the periaxial extracellular material. We assume that the staining affinity of the microfilaments is probably masked by the glycoproteins with which they are associated.

Ultrastructural studies showed that the periaxial microfilaments varied in length and diameter according to the developmental stage of the embryos. It was not possible to detect a real periodic striation. The action of different enzymes has been studied by means of light as well as by electron microscopy. With purified protease free collagenase (kindly provided by Dr. Kühn, Munich, Germany) the periaxial microfilaments of $2\frac{1}{2}$-day old embryos were hydrolyzed and digested. If the same material is treated with purified testicular hyaluronidase (purchased from Leo, Sweden) the microfilaments are preserved but the electron-dense granules and the electron-translucent ground substance disappear. Since the chemical analysis demonstrated the presence of a collagenous material and since protease-free collagenase hydrolyzes specifically the extracellular microfilaments and the testicular hyaluronidase does not, one can conclude that the collagenous material of the axial organs exists as microfilaments.

HISTOAUTORADIOGRAPHIC ANALYSIS

In an attempt to determine the sites of biosynthesis of early collagen fibrils we analyzed the uptake of tritiated proline. We injected 0·05 ml of an aqueous solution of ^{3}H-L-proline (Sp. activity 28 C/mM, Saclay, France) containing 25 or 50 µc of radioactivity, into embryos of 48 and 72 h of incubation. The embryos were fixed after 3, 5 and 8 h of incorporation of the radioactive precursor. The slides were dipped in Ilford K 5 emulsion, stored for exposure for 15 days and developed with Microdol from Kodak. The uptake of proline and the qualitative appreciation of the labeling were obtained by the histological examination of the radiographs by light microscopy.

After 3 h of incorporation the silver grains are uniformly distributed in all the tissues with a slight predominance at the internal side of the epithelia along the basal lamina. After 5 h, one observes a more intense uptake and labeling of the sclerotome cells, the spinal cord and the notochord. Only few silver grains can be observed in the extracellular area around the notochord and the spinal cord.

Eight hours after the injection (Fig. 3) a partial diminution of the silver grains in the sclerotome, the spinal cord and the notochord can be observed. But an important labeling is noticed in the acellular area. Especially the three dimensional network around the notochord shows a considerable concentration of silver grains in the sites where the microfilaments are the most numerous.

Sections of embryos having incorporated the isotopic precursor for 8 h were incubated for 1 or 2 h at 38°C with purified protease-free collagenase. The radiographs show the disappearance of most of the silver grains located in the acellular area around the axial organs (Fig. 4). The radioactive extracellular collagenous proteins were

BIOSYNTHESIS OF COLLAGEN IN CHICK EMBRYOS 415

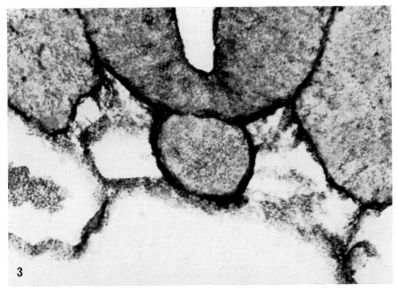

Fig. 3. Light histoautoradiograph showing ^{3}H-proline incorporation in a 56 h embryo. The embryo was fixed after 8 h of incorporation. The silver grains over the extracellular material are shown ($\times 560$).

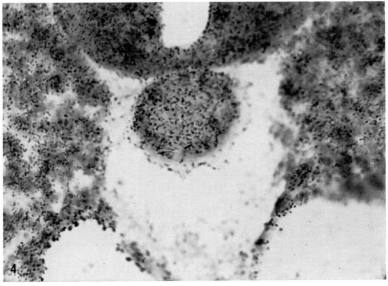

Fig. 4. Light micrograph of a section from the same embryo as in Fig. 3, treated with purified collagenase for 1 h at 38°C. One observes the disappearance of most of the silver grains located in the extracellular area, whereas they subsist within the cells ($\times 560$).

digested. The remaining radioactivity represents the non-collagenous intracellular proteins. This result confirms that the periaxial extracellular microfilaments of 2–3-day old embryos are chiefly collagen fibrils.

The present results indicate that the axial organs of young chick embryos:

(a) Contain a non-dialysable peptide fraction of high molecular weight containing hydroxyproline. This peptide fraction can be considered as collagenous material.

(b) The microfilaments of the periaxial extracellular material can be selectively digested by purified collagenase.

(c) The silver grains located in the periaxial area where the microfilaments are the most abundant after 8 h of incorporation of ^{3}H-proline, disappear by the action of collagenase whereas they subsist within the cells.

In conclusion these data allow us to claim that the periaxial extracellular material of chick embryos of $2-3\frac{1}{2}$ days' incubation contains collagen.

ACKNOWLEDGEMENTS

We wish to express our thanks to Miss N. Briquelet, Mr J. C. Allain and Mr G. Gateau for their excellent technical assistance.

REFERENCES

Moore, S., Spackman, D. and Stein, W. H. (1958). *Anal. Chem.* **30**, 1158–1163.
O'Connell, J. J. and Low, F. N. (1970). *Anat. Rec.* **167**, 425–438.
Stegemann, H. and Stalder, K. (1967). *Clin. Chim. Acta* **18**, 267–273.
Strudel, G. (1953). *Ann. Sc. Nat. Zool.* **11/XV**, 251–329.
Strudel, G. (1971). *C.R. Acad. Sci. (Paris)* **272**, 673–676.

Collagen Degradation *in vivo* and *in vitro*

D. S. JACKSON

Department of Medical Biochemistry, Medical School
University of Manchester, England

Although, in general, collagen has a relatively low metabolic turnover, there are a number of situations in which insoluble collagen is degraded. There is, for example, a rapid loss of collagen during the post-partum involution of the uterus and during the latter stages of the carrageenan granuloma. There is also evidence that as much as one third of newly synthesised collagen is degraded before it can be incorporated into insoluble fibres (see Kivirikko, 1971, for review). *In vitro* collagen seemed to be degraded only by bacterial collagenase. Proteolytic enzymes appeared to be ineffective—no specific collagenase could be demonstrated. The problem was apparently solved by the discovery of collagenases first by Gross and Lapière (1962) in the resorbing tadpole tail.

ACTION OF COLLAGENASE

Following the discovery of tadpole tail collagenase, a similar enzyme was found in a large number of tissues, both normal and pathological. All the enzymes seemed to act in the same way, i.e. by cleaving the triple helical structure across all three peptide chains yielding two unequal segments. With these findings it has generally been considered that the *in vivo* degradation of collagen is most likely initiated by collagenase. However, it has been pointed out (Gross, 1970) that the evidence for mammalian collagenase having a primary role in collagen degradation is circumstantial.

A few reports provide evidence that there may be a number of circumstances in which collagenase is not the prime factor.

1. The method of assay used by almost all workers involves the use of either reconstituted tropocollagen or tropocollagen in solution, as the substrate. *In vivo* the bulk of the collagen digested is in the form of insoluble collagen fibres (polymeric collagen).

2. The enzyme found in human granulocytes will degrade only tropocollagen in solution, and not in the reconstituted form (Lazarus et al., 1968).

3. Collagenase from rheumatoid synovium will not attack human synovial polymeric collagen (Leibovich and Weiss, 1971a). This suggests that in some circumstances some other enzymic attack must occur before collagenase can effectively degrade polymeric collagen.

ACTION OF PROTEOLYTIC ENZYMES

Proteolytic enzymes are ineffective against the helical part of the tropocollagen molecules, but can attack non-helical segments at the N- and C-terminal ends (Rubin et al., 1963; Leibovich and Weiss, 1970). These regions are also the site of formation of the covalent crosslinks which stabilise the fibre structure. Proteolytic enzymes such as pepsin can attack these regions in polymeric collagen resulting in depolymerisation which yields tropocollagen molecules from which the N- and C-terminal ends have been removed (Steven, 1966a, b, c). In vivo, the most likely proteolytic enzymes are the lysosomal cathepsins already implicated in the degradation of proteoglycans. Electron microscope studies have also clearly shown the presence of collagenous particles within the lysosomal vacuoles of phagocytic cells (Parakkal, 1969a, b) and within fibroblasts of the carrageenan granuloma (Perez-Tamayo, 1970).

It had previously been suggested that during the resorption phase of the carrageenan granuloma, the first phase of collagen degradation was the depolymerisation of polymeric collagen. This was inferred from the marked increase in neutral salt extractable collagen during the resorptive phase (Jackson, 1957). This was supported by a similar effect during the resorption of the post-partum uterus (Woessner, 1963).

DEPOLYMERISATION BY CATHEPSIN

The effect of lysosomal enzymes on polymeric collagen has recently been studied in this department. The details of the techniques involved, both chemical and with the electron microscope, have been described in the paper by Milsom et al. (1973). The results may be summarised as follows:

(a) Polymeric collagen from bovine and human tendon was depolymerised by lysosomal cathepsins derived from human and rat liver. The extent of depolymerisation was 100% in the case of bovine tendon and 20–40% in the case of human tendon. The pH optimum for depolymerising activity was close to pH 4.

(b) When examined by electron microscopy the process of depolymerisation was as follows: first a number of cleavages in the fibre structure were seen. The ends then formed showed evidence of 'unwinding' of exposed protofibrils. Secondly, the fibres were broken up into small segments in which the characteristic collagen striations were visible and in which the unwinding effect was marked. These segments were cleaved further into a mass of fine protofibrils.

This process is identical to that described in electron microscope studies of collagen degradation *in vivo*. Simultaneous fibre fragmentation and unwinding with the subsequent release of fine filaments was described in the carrageenan granuloma and in immune-damaged cornea (Perez-Tamayo, 1970; Mohos and Wagner, 1969). Collagen segments similar to those described above were found in the involuting uterus and in macrophages in the hair follicles during loss of hair (Parakkal, 1969a, b).

(c) Almost 100% of the solubilised collagen was undialysable and behaved like tropocollagen: for example, it could be reconstituted into fibrils at neutral pH.

(d) Depolymerase activity could not be detected biochemically. However, numerous short segments and free ends with protofibrillar unwinding could be detected in the electron microscope.

It was concluded that rat and human liver cells contain an enzyme probably lysosomal in origin which is capable of depolymerising human and bovine polymeric collagen to tropocollagen molecules. Inhibition studies suggested that the enzyme was probably not cathepsin D or E. Recent studies have implicated Cathepsin B1 (Etherington, 1972) at pH 7.

CONCLUSIONS

It is suggested that, acting in high concentration and under locally favourable pH conditions on or near the cell membrane, the depolymerising enzyme(s) cleave the fibrils into short segments. After phagocytosis these segments will be digested further within secondary lysosomes to protofibrillar particles or tropocollagen molecules. These would then be further acted upon by specific collagenases which may themselves have been activated from zymogen (Vaes, 1972; Harper *et al.*, 1971) by a lysosomal protease.

In this we follow the mechanism proposed for the digestion of proteoglycans (Dingle, 1969).

The products of collagenase digestion could be further degraded by the peptidases described by Strauch which can attack peptides having collagen specific sequences of amino acids (Strauch and Vencelj, 1967; Harper, 1970).

It should be emphasised that this mechanism may be only one of several which may be effective in the degradation of collagen in various situations. An outline of the mechanism is shown in Fig. 1.

```
        Polymeric collagen ─────────────→ short segments
                              EXTRACELLULAR
    endocytosis                    collagenase
   ─────────────→ tropocollagen ─────────────→ oligopeptides
    cathepsin B1

    peptidases
    (collagen-specific)
   ─────────────→ small peptides and amino acids
        cathepsins
```

Fig. 1. Outline of a proposed mechanism for collagen degradation.

REFERENCES

Dingle, J. T. (1969). *In* "Lysosomes in Biology and Pathology" (J. T. Dingle and D. H. Fell, eds.) Vol. 2, pp. 421–436. North Holland, Amsterdam.
Etherington, D. J. (1972). *Biochem. J.* **127**, 685–692.
Gross, J. and Lapière, C. M. (1962). *Proc. Natl Acad. Sci. USA* **48**, 1014–1022.
Gross, J. (1970). *In* "Chemistry and Biology of the Intercellular Matrix" (E. A. Balazs, ed.) Vol. 2, pp. 1623–1636. Academic Press, London.
Harper, E. (1970). *In* "Chemistry and Molecular Biology of the Intercellular Matrix" (E. A. Balazs, ed.) Vol. 3, pp. 1653–1661. Academic Press, London.
Harper, E., Glock, K. and Gross, J. (1971). *Fed. Proc.* **30**, 831.
Jackson, D. S. (1957). *Biochem. J.* **65**, 277–284.
Kivirikko, K. (1971). *Int. Rew. Connect. Tissue Res.* **5**, 93–163.
Lazarus, G. S., Daniels, J. R., Brown, R. S., Bladen, H. A. and Fullmer, H. M. (1968). *J. Clin. Invest.* **47**, 2622–2629.
Leibovich, S. J. and Weiss, J. (1970). *Biochim. Biophys. Acta* **214**, 445–454.
Leibovich, S. J. and Weiss, J. (1971). *Biochim. Biophys. Acta* **251**, 109–118.
Milsom, D. W., Steven, F. S., Hunter, J. A. A., Thomas, H. and Jackson, D. S. (1973). *Conn. Tiss. Res.* **1**, 251–265.
Mohos, S. C. and Wagner, B. M. (1969). *Arch. Pathol.* **88**, 3–20.
Parakkal, P. F. (1969a). *J. Ultrastruct. Res.* **29**, 210–217.
Parakkal, P. F. (1969b). *J. Cell Biol.* **41**, 345–354.
Perez-Tamayo, R. (1970). *Lab. Invest.* **22**, 142–159.
Rubin, A. L., Drake, M. P., Davison, P. F., Pfahl, D., Speakman, P. T. and Schmitt, F. O. (1965). *Biochemistry* **5**, 301–312.
Steven, F. S. (1966a). *Biochim. Biophys. Acta* **130**, 190–195.
Steven, F. S. (1966b). *Biochim. Biophys. Acta* **130**, 196–201.
Steven, F. S. (1966c). *Biochim. Biophys. Acta* **130**, 202–217.
Strauch, L. and Vencelj, H. (1967). *Hoppe-Seyler's Z. physiol. Chem.* **348**, 465–468.
Vaes, G. (1972). *Biochem. J.* **126**, 275–289.
Woessner, J. F. (1962). *Biochem. J.* **83**, 304–314.

Intracellular Localisation of Connective Tissue Polyanions

J. E. SCOTT and J. DORLING

*M.R.C. Rheumatism Unit, Canadian Red Cross Memorial Hospital
Taplow, Maidenhead, Berkshire, England*

There are many reasons for wanting to visualise and localise intracellular polyanionic carbohydrates. Valid techniques could be used in investigations of normal and abnormal biosynthesis and breakdown, with all the implied consequences in physiology and pathology. Totally specific and reliable histochemical techniques for the acidic glycosaminoglycans are not yet available, but we have reached a stage in the development of two separate approaches where the application to intracellular material gives interesting information. One approach based on periodate oxidation of the uronic acid moiety (Scott and Dorling, 1969) requires a separate discussion, and this paper will deal with a method based on the critical electrolyte concentration (CEC) for the identification of polyanions in the light and electron microscopes.

Basic dyes, particularly toluidine blue, have long been used to visualise acid glycosaminoglycans, but one immediately encounters a major snag, in the present context, viz. that intracellular nucleic acid usually has far greater affinity for dyestuffs than do other polyanions, due to the strong short range binding of nucleic acid bases to the dye chromophores, in structures similar to that in Fig. 1. Toluidine blue, which is very similar in size and shape to the acridine shown in Fig. 1, has been especially used to demonstrate RNA. Clearly, such dyes *cannot* be used to locate intracellular glycosaminoglycan. However, by using appropriately shaped and sized molecules, the nucleic acid interaction can be minimised. The phthalocyanin chromophore, for example, exactly fits the Crick-Watson base pair (Fig. 2) and can intercalate and bind beautifully to nucleic acids, but if bulky substituents are added, as are the S-methylene tetramethyl isothiouronium

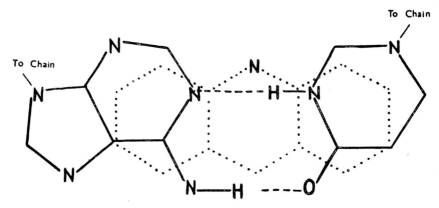

Fig. 1. Arrangement of an acridine molecule across a hydrogen bonded base pair of DNA as proposed by Lerman (1961). Dotted lines represent the acridine.

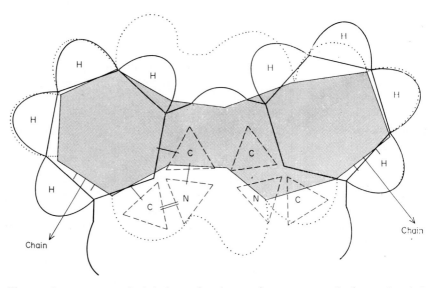

Fig. 2. Arrangement of phthalocyanin chromophore across a hydrogen-bonded base (GC) pair (Scott, 1972). The outlines are drawn directly from Courtauld atomic models.

sidechains in Alcian blue, the molecule can no longer intercalate (Scott, 1972). Binding to nucleic acid is then quite weak, and Alcian blue behaves like a coloured cetylpyridinium molecule, displaceable from polycarboxylates and polyphosphates at lower salt concentrations (CECs) than from polyester sulphates (Table I). For this reason, it is

Table I

Staining of spots containing polyanions on filter paper in Alcian blue–pH 5·8–MgCl$_2$

Type anionic group	Polyanion	Molarity of MgCl$_2$									
		0·0	0·025	0·05	0·1	0·2	0·3	0·45	0·65	0·8	1·0
—COO⁻	Hyaluronate	++	++	++	∓	−	−	−	−	−	−
	Alginate	∓	+	++	++	−	−	−	−	−	−
=PO$_4^-$	RNA	∓	++	++	++	∓	−	−	−	−	−
	Polyphosphate	∓	+	+	+++	∓	−	−	−	−	−
	DNA	∓	+	++	+++	∓	−	−	−	−	−
—COO⁻ and —OSO$_3^-$	CSA	++	++	++	++	++	++	++	−	−	−
	Heparin	∓	++	++	++	++	++	++	∓	−	−
—OSO$_3^-$	Keratan sulphate	++	++	++	+++	+++	+++	+++	+++	+++	+

Symbols denote strength of staining. +++, very strong; ++, strong; +, moderate; ∓, weak; −, none. (Taken from Scott and Dorling, 1965).

quite well suited to the intracellular demonstration of non-nucleotide polyanions.

The accumulation of sulphated polysaccharide in cultured fibroblasts from Hurler's syndrome and other mucopolysaccharidoses reported by Danes et al. (1970) seemed a particularly attractive application, and in collaboration with them normal (Fig. 3a) and abnormal (Fig. 3b and c) fibroblasts were stained with Alcian blue in various concentrations of magnesium chloride. The results, initially at least, were as hoped, with abnormal cells showing basophilic accumulations of presumably sulphated material. Fig. 4 shows a single fibroblast, with the nucleus showing as empty 'space', and cytoplasmic granules present where in the normal fibroblast (Fig 3a) there is nothing at all. Later results on the same and other cell lines were not so consistent and probably no single explanation is sufficient. Accumulations of basophilic material were sometimes seen *outside* the cells, which were empty.

To get some idea of what factors might be important, unfixed postmortem liver from a Hurler's syndrome was examined, and aggregates of heavily stained material were observed lying *on* the section (Fig. 5a), reminiscent of the extracellular localisation sometimes seen in the tissue cultures. A similar result was reported by Haust and Landing (1961) using toluidine blue. They and others had recognised that some method of fixation was essential to retain all the polyanion in its original position. We therefore tried the accepted methods, based on formalin or glutaraldehyde, with or without cetylpyridinium chloride, but without much success. However, good results were obtained with cyanuric chloride in acetone or DMF (dimethylformamide). This polyfunctional compound presumably reacts with some -OH groups on the polysaccharide and subsequently with tissue bound nucleophilic groups, thus 'anchoring' the polysaccharide to the tissue.

The CEC of the stain was between 0·6 and 0·9, which compares favourably with that expected (e.g. the CEC of CPC complexes of Hurler's polysaccharide goes up to $1·2M$ NaCl (Prodi and Prodi, 1965) although in the *absence* of fixatives, no stainable material was present above $0·3$–$0·4M$ $MgCl_2$. This may be relevant to the observation that the fibroblast material was likewise unobservable at higher than $0·4M$ $MgCl_2$, which would imply a very low molecular weight and/or degree of sulphation.

Observations on the intracellular acid glycosaminoglycan during breakdown must therefore be performed on the freshest possible material, with quick and adequate fixation, probably involving covalent bonds between the substrate and the tissue. Otherwise the great

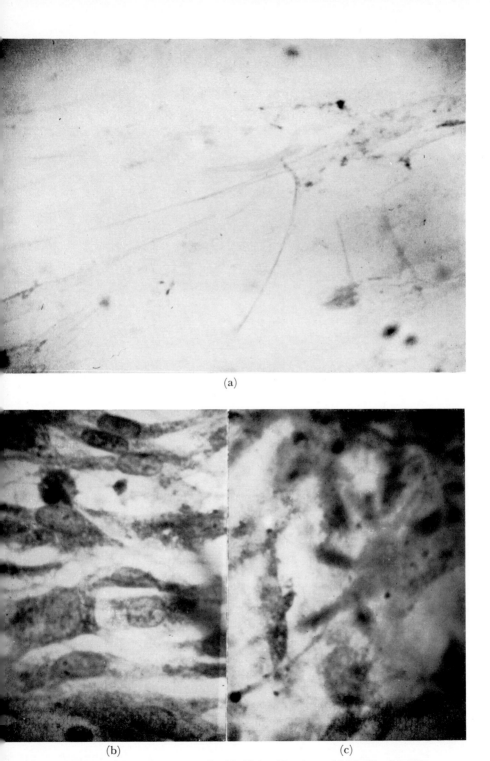

Fig. 3. (a) Normal fibroblasts, stained with Alcian blue in $0 \cdot 2M$ $MgCl_2$, (b) Fibroblasts from explants taken from a Hurler's Syndrome, stained in $0 \cdot 2M$ $MgCl_2$ and (c) in $0 \cdot 3M$ $MgCl_2$.

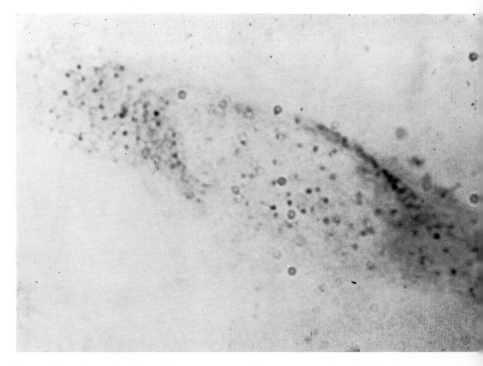

Fig. 4. A single fibroblast ($\times 1250$), stained with Alcian blue in $0\cdot 3M$ MgCl$_2$, from a Hurler's syndrome explant.

mobility of the polyanion in aqueous media may give erroneous impressions. For instance, one cannot tell whether the heavy staining of some nuclei (Fig. 5b) is due to absorption of diffusible polyanion, although this seems very likely.

During biosynthesis accumulations of polyanion similar to that in mucopolysaccharidotic tissues seems unlikely to occur. Observations will probably have to be made on a few molecules at the most and this implies the use of an electron microscope. Alcian blue and related phthalocyanins possess a chelated metal atom which confers electron density. Fig. 6a and b shows the use of the CEC approach in electron microscopy. The dramatic decrease in electron density of the nucleus on going from $0\cdot 2M$ to $0\cdot 3M$ MgCl$_2$ parallels the change seen in light microscopy. Unfortunately, the Cu atom in Alcian blue does not have sufficiently high electron density for the best contrast, but we have now synthesised new analogues containing Pt, and we hope for improved results.

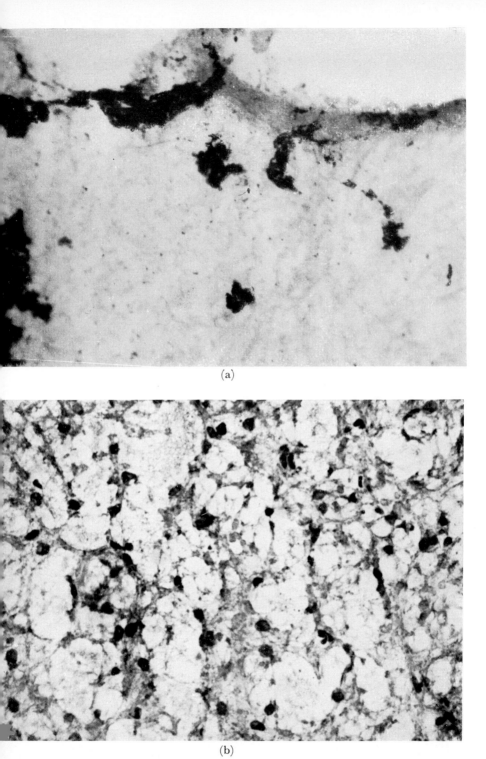

Fig. 5. (a) Post mortem liver, fresh frozen unfixed section from a Hurler's syndrome, stained with Alcian blue in 0·4M MgCl$_2$, (b) A similar section, fixed in 1% w/v cyanuric chloride in dimethyl formamide for 1 h, and stained as in (a).

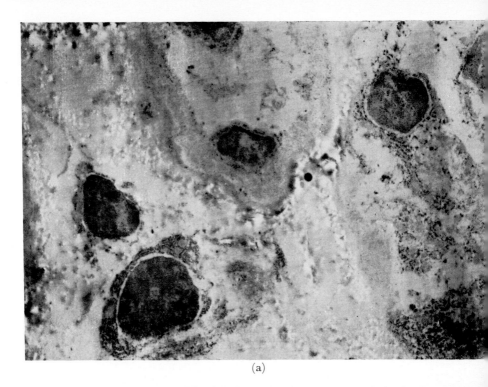

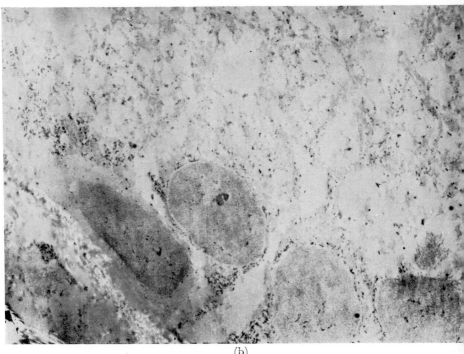

Fig. 6. Electron micrographs of bovine nasal septal soft tissue cells, stained with Alcian blue in (a) 0·2M $MgCl_2$, (b) 0·3M $MgCl_2$. Fresh frozen tissue (*not* treated with osmium etc.)

ACKNOWLEDGEMENTS

Our thanks are due to Dr. Paul Whiteman for providing tissues from mucopolysaccharidotic patients.

REFERENCES

Danes, B. S., Scott, J. E. and Bearn, A. G. (1970). *J. exp. Med.* **132**, 765–774.
Haust, M. D. and Landing, B. H. (1961). *J. Histochem. Cytochem.* **9**, 79–86.
Lerman, L. S. (1961). *J. Mol. Biol.* **3**, 18–30.
Prodi, G. and Prodi, M. P. (1965). *Ital. J. Biochem.* **14**, 321–327.
Scott, J. E. (1972). *Histochemie* **32**, 191–212.
Scott, J. E. and Dorling, J. (1965). *Histochemie.* **5**, 221–233.
Scott, J. E. and Dorling, J. (1969). *Histochemie.* **19**, 295–301.

Biosynthesis of the L-Iduronic Acid Unit of Heparin

ULF LINDAHL, MAGNUS HÖÖK, ANDERS MALMSTRÖM*
and LARS-ÅKE FRANSSON*

*The Institute of Medical Chemistry, University of Uppsala
Uppsala, Sweden*

The biosynthesis of sulfated glycosaminoglycans has been studied extensively during the last ten years, and is now characterized in considerable detail (Rodén, 1971; Stoolmiller and Dorfman, 1969). The polysaccharides are formed by stepwise transfer of monosaccharide units from the appropriate nucleotide sugars to the nonreducing termini of nascent chains. Substitution with sulfate groups occurs subsequent to polymerization and requires the presence of 3'-phosphoadenylylsulfate.

Two uronic acids, D-glucuronic acid and L-iduronic acid, have been identified as constituents of glycosaminoglycuronans. The role of UDP(uridine diphosphate)-glucuronic acid in the formation of glycosaminoglucuronans has been firmly established (Rodén, 1971; Stoolmiller and Dorfman, 1969). The formation of L-iduronic acid units has, on the other hand, remained unclear. Although C-5 epimerization of UDP-glucuronic acid has been demonstrated in mammalian tissues (Jacobson and Davidson, 1962; Fransson, 1970) the resulting UDP-iduronic acid has not been isolated and no evidence has yet been obtained to indicate the involvement of this nucleotide sugar in the biosynthesis of glycosaminoglycans.

The aim of the present study has been to elucidate the biosynthesis of the iduronic acid component of heparin (Cifonelli and Dorfman, 1962; Perlin and Sanderson, 1970; Lindahl and Axelsson, 1971). Evidence will be shown indicating that the formation of this unit does not require UDP-iduronic acid, but involves epimerization, on the polymer level, of D-glucuronic acid residues previously incorporated

* Department of Physiological Chemistry 2, University of Lund, Lund, Sweden.

into the polysaccharide chain (cf. the biosynthesis of alginic acid; Larsen and Haug, 1971; Haug and Larsen, 1971). Particular attention has been given to the relationship between this epimerization process and sulfation of the polysaccharide.

EFFECT OF SULFATION ON THE URONIC ACID COMPOSITION OF MICROSOMAL GLYCOSAMINOGLYCAN

The incubation conditions employed (Lindahl et al., 1972; Höök et al., 1973) were essentially based on previous studies by Silbert (1963, 1967). A microsomal fraction from neoplastic murine mast cells (Furth et al., 1957) were incubated with UDP-^{14}C-glucuronic acid and unlabeled UDP-N-acetylglucosamine, either in the presence or in the absence of 3′-phosphoadenylylsulfate, as indicated in Table I. Electro-

Table I

Incubations of mastocytoma microsomal fraction with nucleotide sugars and 3′-phosphoadenylylsulfate

Preparation	Incubation*	
	0–60 min	60–120 min
A	UDP-^{14}C-GlcUA + UDP-GlcNAc	—
B	UDP-^{14}C-GlcUA + UDP-GlcNAc + PAPS	—
C	UDP-^{14}C-GlcUA + UDP-GlcNAc	UDP-GlcUA + PAPS

* The incubation conditions were similar to those described in a previous study (Lindahl et al., 1972). Incubations A and B were interrupted after 60 min at 37°C. In incubation C the incorporation of ^{14}C-radioactivity was terminated, by the addition of excess amounts of unlabeled UDP-glucuronic acid, simultaneously with the addition of 3′-phosphoadenylylsulfate. Labeled polysaccharides were isolated along with carrier heparin, by gel chromatography (Sephadex G-50) after digestion of the boiled incubation mixtures with papain (Lindahl et al., 1972). (PAPS, 3′-phosphoadenylylsulfate; GlcUA, D-glucuronic acid; GlcNAc, 2-acetamido-2-deoxy-D-glucose.)

phoresis of the resulting polysaccharide preparations showed that the material formed in the absence of 3′-phosphoadenylylsulfate (preparation A) was nonsulfated, in contrast to the sulfated preparations B and C, which contained components migrating similar to commercial heparin (Höök et al., 1973). Analysis of the isolated polysaccharides showed that the nonsulfated preparation A contained ^{14}C-glucuronic acid as the only labeled uronic acid component detectable (Fig. 1A), whereas the sulfated preparation B yielded ^{14}C-iduronic acid amounting to approximately one third of the total ^{14}C-uronic acid (Fig. 1B). These findings suggest that the formation of iduronic acid depends on the sulfation of the polymer.

Previous results of Silbert (1967) have shown that sulfation *in vitro* of microsomal polysaccharide may occur during polymerization as well as subsequent to the formation of the complete polysaccharide chain.

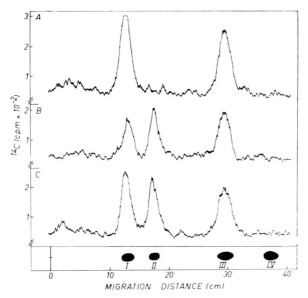

Fig. 1. Paper chromatography (ethyl acetate-acetic acid-water, 3:1:1) of ^{14}C-labeled uronic acids and lactones liberated from nonsulfated (preparation A); and sulfated (preparations B and C, respectively; see Table I) microsomal polysaccharides. The standards shown below the tracings are: (I) D-glucuronic acid; (II) L-iduronic acid; (III) D-glucuronolactone; (IV) L-iduronolactone. The degree of lactonization of iduronic acid was variable, and generally lower than that of glucuronic acid.

The relation between chain elongation and incorporation of iduronic acid residues could therefore be studied by pulse-chase incubations, in which sulfation was initiated after completed ^{14}C-labeling (Table I). Analysis of the resulting polysaccharide (preparation C) clearly showed that sulfation of preformed polysaccharide yielded the same proportion of iduronic acid (Fig. 1C; Table II) as did sulfation in conjunction with polymerization (Fig. 1B; Table II). Since the preformed polysaccharide lacked iduronic acid (Fig. 1A), it is concluded that iduronic acid was formed by C-5 epimerization of glucuronic acid residues, on the polymer level.

Table II

Uronic acid composition of ^{14}C-labeled poly- and oligosaccharides

Preparation	Corresponding polymer structure*	$^{14}C\text{-}IdUA / (^{14}C\text{-}IdUA + {}^{14}C\text{-}GlcUA)$** (%)
A	Nonsulfated polysaccharide	<5
B	Sulfated polysaccharide	36
C	Sulfated polysaccharide	31
Nonsulfated disaccharide	—UA—GlcNSO$_3^-$—	9
Monosulfated disaccharide	—UA—GlcNSO$_3^-$— with OSO$_3^-$ or —UA—GlcNO$_3^-$— with OSO$_3^-$	47
Disulfated disaccharide	—UA—GlcNO$_3^-$— with OSO$_3^-$ OSO$_3^-$	>95
Low-sulfated hexasaccharide	—UA—(GlcNAc—UA)$_2$—GlcNSO$_3^-$— varying amounts of ester sulfate	10
High-sulfated hexasaccharide		80

* The isolated oligosaccharides contain one sulfate residue less than the corresponding polymeric structures, as the N-sulfate groups were lost during the deamination reaction.

** Determined by liquid scintillation spectroscopy on eluted uronic acids and lactones, after separation by paper chromatography (Fig. 1). Samples were prepared for analysis by acid hydrolysis in combination with nitrous acid deamination, as described previously (Lindahl et al., 1972); by this degradation method about 70% of the uronic acid content of heparin (or of N-acetylheparin) is recovered as monosaccharide. Analysis of disaccharides did not involve the entire degradation procedure, but merely the last hydro-

CHARACTERIZATION OF OLIGOSACCHARIDES DERIVED FROM SULFATED MICROSOMAL GLYCOSAMINOGLYCAN

The relationship between sulfation and uronic acid epimerization was studied in more detail with regard to the location of sulfate substituents required for the epimerization of a particular glucuronic acid unit. Such information was obtained by structural characterization of oligosaccharides isolated after deaminative cleavage of sulfated micro-

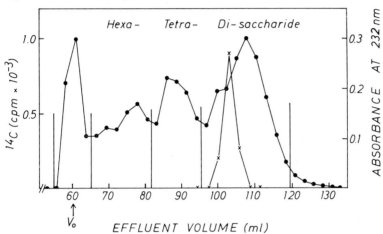

Fig. 2. Gel chromatography of sulfated microsomal polysaccharide (preparation C) after treatment with nitrous acid (Lagunoff and Warren, 1962). The sample (6×10^5 cpm) was applied to a column (1×193 cm) of Sephadex G-25, and eluted with 0.2 M NaCl at a rate of 4–5 ml/h. Fractions of about 2.5 ml were collected, analyzed for radioactivity (●——●), and combined as indicated by the vertical lines. A nondeaminated sample of preparation C emerged quantitatively at the void volume of the column. The pooled fractions were desalted by passage through a a column of Sephadex G-15. Superimposed is a chromatogram of the unsaturated disaccharide △Di-OS (2-acetamido-2-deoxy-3-O-(β-O-gluco-4-enepyranosyluronic acid)-D-galactose) (Yamagata et al., 1968), which was quantitated by its absorbance at 232 nm (x——x).

somal polysaccharide (preparation C). The oligosaccharides are listed in Table II, along with the probable structures of the corresponding polysaccharide segments (Lindahl and Axelsson, 1971; Helting and Lindahl, 1971; Cifonelli and King, 1972; Lindahl et al., 1973).

Preparation C was degraded by treatment with nitrous acid* (Lagunoff and Warren, 1962), and the degradation products were fractionated by gel chromatography on Sephadex G-25 (Fig. 2). The

* In this procedure glucosamine residues having either unsubstituted or sulfated amino groups are attacked, with concomitant cleavage of the corresponding glucosaminidic linkage; N-acetylated glucosamine units are resistant towards deamination. Glucosamine residues susceptible to deamination are converted to 2,5-anhydromannose units.

disaccharide and hexasaccharide fractions (the latter fraction containing, in addition, larger oligosaccharides) were separated according to the degree of sulfation by paper electrophoresis at pH 1·7. The disaccharide fraction thus yielded three components, migrating like nonsulfated, monosulfated and disulfated uronosylanhydromannose (Lindahl and Axelsson, 1971), respectively (Fig. 3), whereas the hexasaccharide fraction showed several incompletely separated components. The most anionic hexasaccharide subfraction migrated faster than the disulfated uronosylanhydromannose disaccharide standard.

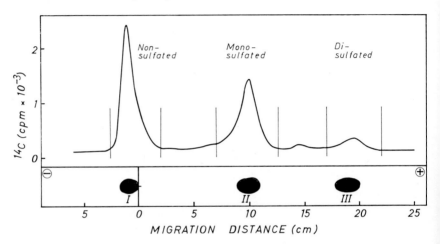

Fig. 3. Paper electrophoresis at pH 1·7 of the disaccharide fraction obtained by the gel chromatography illustrated in Fig. 2. The standards shown below the tracing are: (I) D-glucuronic acid; (II) monosulfated uronosylanhydromannose (Lindahl and Axelsson, 1971); (III) disulfated uronosylanhydromannose.

The uronic acid composition of the labeled oligosaccharide fractions varied with the sulfate contents (Table II). Within the hexasaccharide fraction the ratio of ^{14}C-iduronic acid to total ^{14}C-uronic acid thus varied from 0·1 in the low-sulfated to 0·8 in the high-sulfated species. It is obvious that the high-sulfated hexasaccharide fraction must contain ^{14}C-iduronic acid residues linked both at C-1 and at C-4 to N-acetylglucosamine units. Referring to the model trisaccharide structure shown in Fig. 4, this finding implies that neither of the N-sulfate groups on GlcN I and GlcN II would seem to be essential for the epimerization of the interjacent uronic acid residue (UA). The monosulfated disaccharide fraction showed a ratio of ^{14}C-iduronic to total ^{14}C-uronic acid of about 0·5, whereas the disulfated disaccharide appeared to contain ^{14}C-iduronic acid as the only labeled uronic acid component. These results clearly suggest that the sulfate ester groups on C-2 and C-6 of the UA and GlcN II residues, respectively, may be of significance for the epimerization of the UA unit. However, epimerization can occur also in the absence of these particular substituents, as the nonsulfated disaccharide fraction was consistently found to contain small amounts of ^{14}C-iduronic acid. Provided that these results are not due to desulfation at the polymer or oligosaccharide stage, it is concluded that the connection between sulfation and uronic

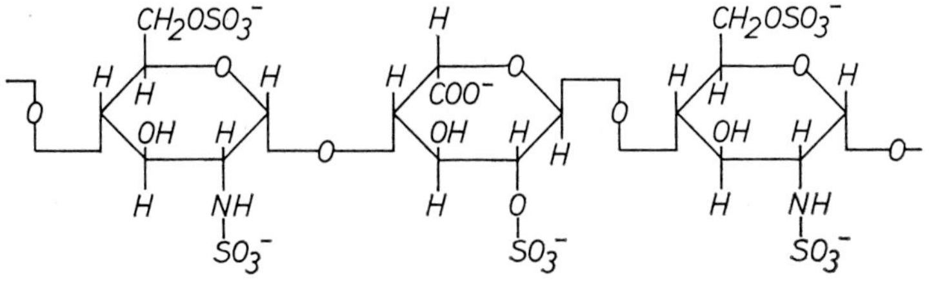

Fig. 4. Probable structure of a fully sulfated heparin trisaccharide sequence, including one iduronic acid and two glucosamine residues.

acid epimerization may also involve the sulfate group at C-6 of the glucosamine residue linked to C-4 of the epimerizing uronic acid (GlcN I in Fig. 4). Effects of sulfate substituents more remote from this uronic acid cannot be excluded.

ACKNOWLEDGEMENTS

This work was supported by grants from the Swedish Medical Research Council (13P-3431; 13X-139; 13X-2309), the Swedish Cancer Society (53), Konung Gustaf V:s 80-årsfond and the Faculty of Medicine, University of Uppsala.

REFERENCES

Cifonelli, J. A. and Dorfman, A. (1962). *Biochem. Biophys. Res. Commun.* **7**, 41–45.
Cifonelli, J. A. and King, J. (1972). *Carbohyd. Res.* **21**, 173–186.
Fransson, L.-Å. (1970). In "The Chemistry and Molecular Biology of the Intercellular Matrix" (E. A. Balazs, ed.) Vol. 2, pp. 823–842. Academic Press, New York.
Furth, J., Hagen, P. and Hirsch, E. I. (1957). *Proc. Soc. Exp. Biol. Med.* **95**, 824–828.
Haug, A., and Larsen, B. (1971). *Carbohyd. Res.* **17**, 297–308.
Helting, T., and Lindahl, U. (1971). *J. Biol. Chem.* **246**, 5442–5447.
Höök, M., Lindahl, U., Malmström, A. and Fransson, L.-Å. (1973). (In Press.)

Jacobson, B. and Davidson, E. A. (1962). *J. Biol. Chem.* **237**, 638–642.
Lagunoff, D. and Warren, G. (1962). *Arch. Biochem. Biophys.* **99**, 396–400.
Larsen, B. and Haug, A. (1971). *Carbohyd. Res.* **17**, 287–296.
Lindahl, U. and Axelsson, O. (1971). *J. Biol. Chem.* **246**, 74–82.
Lindahl, U., Bäckström, G. and Jansson, L. (1973). (In Press.)
Lindahl, U., Bäckström, G., Malmström, A. and Fransson, L.-Å. (1972). *Biochem. Biophys. Res. Commun.* **46**, 985–991.
Perlin, A. S. and Sanderson, G. R. (1970). *Carbohyd. Res.* **12**, 183–192.
Rodén, L. (1971). *In* "Metabolic Conjugation and Metabolic Hydrolysis" (W. H. Fishman, ed.) Vol. 2, pp. 345–442. Academic Press, New York.
Silbert, J. E. (1963). *J. Biol. Chem.* **238**, 3542–3546.
Silbert, J. E. (1967). *J. Biol. Chem.* **242**, 5146–5152.
Stoolmiller, A. C. and Dorfman, A. (1969). *In* "Comprehensive Biochemistry" (M. Florkin and E. H. Stotz, eds.) Vol. 17, pp. 241–275. Elsevier, Amsterdam.
Yamagata, T., Saito, H., Habuchi, O. and Suzuki, S. (1968). *J. Biol. Chem.* **243**, 1523–1535.

Biosynthesis of Dermatan Sulfate in Fibroblasts

Lars-Åke Fransson Anders Malmström,
Ulf Lindahl* and Magnus Höök*
*Department of Physiological Chemistry 2
University of Lund, Lund, Sweden*

It is generally accepted that dermatan sulfate (DS) has a copolymeric structure composed of sequences of L-iduronic acid-containing disaccharide periods interspersed with sequences of D-glucuronic acid-containing periods as shown in Fig. 1.

Previous studies on the biosyntheses of DS have primarily been concerned with the formation of its unusual hexuronic acid moiety L-iduronic acid. The finding of Rodén and Dorfman (1958) that glucose-6-^{14}C was incorporated into the carboxyl group of L-iduronic acid led to the proposal of two alternative pathways for the formation of this hexuronic acid (Fig. 1). Route 1, which proposes formation of UDP-iduronic acid from UDP-glucose via UDP-idose, has never received experimental support. Route 2, however, depicting a direct epimerization at C-5 of UDP-glucuronic acid to form UDP-iduronic acid was demonstrated in rabbit skin extracts by Jacobson and Davidson (1962). The significance of the UDP-glucuronic acid-5'-epimerase in the biosynthesis of DS has been obscured by the fact that the involvement of UDP-iduronic acid in the polymerization process has remained hypothetical. In recent years it has become apparent that C-5 epimerizations of hexuronic acid moieties can take place on the polymer level. Thus, in the biosynthesis of bacterial alginate C-5 epimerization of D-mannuronic acid to L-guluronic acid occurs in the intact polysaccharide (Haug and Larsen, 1971). Furthermore, during biosynthesis of heparin by a microsomal fraction from mouse mastocytoma L-iduronic acid is formed by epimerization at C-5 of D-glucuronic acid residues previously incorporated into the chain (Lindahl et al., 1971).

* Institute of Medical Chemistry, University of Uppsala, Uppsala, Sweden.

For DS biosynthesis it is conceivable that a chondroitin-like segment is transformed into a dermatan-like segment as depicted in route 3 in Fig. 1. Thus, it is possible that the L-iduronic acid in DS may have been formed by epimerization of D-glucuronic acid residues at the microsomal stage of the synthetic process.

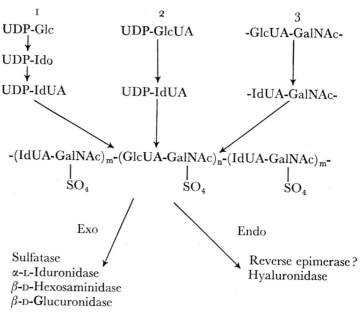

Fig. 1. Current theories concerning biosynthesis and degradation of dermatan sulfate.

The present report demonstrates interconversion on the polymer level between L-iduronic and D-glucuronic acid residues during the later stage of DS biosynthesis in cultured skin fibroblasts. A major part of the D-glucuronic acid residues of secreted DS have arisen by epimerization of L-iduronic acid residues in the intact polymer.

EXPERIMENTAL AND RESULTS

The fibroblast cultures were established from skin specimens of human fetuses. The cells were grown as monolayers in Roux's flasks in Earle's Minimal Essential Medium supplemented with 10% calf serum. After 5–10 transfers the cells were grown in sulfate-free medium until confluence was achieved. At this stage each culture contained approximately 10–15 million cells in 50 ml of medium. Cell cultures in stationary phase were then given 250 μCi of $Na_2^{35}SO_4$ and after 3 days of incubation media and cells were collected separately. The medium was dialyzed against 0.1 M ammonium sulfate followed by water. The retentate was lyophilized, 0·7 mg of carrier DS was added to each sample and the mixture was digested with papain (Antonopoulos et al., 1964). The cells were washed twice with saline and similarly digested with papain. Finally, glycosaminoglycans were isolated from the various digests by precipitation with cetyl pyridinium chloride (CPC) followed by precipitation with ethanol (Olsson et al., 1968). Approximately 60% of the total incorporated radioactivity was found in the extracellular products.

The isolated radioactive products were subjected to various microcolumn analyses to assess molecular size polydispersity, degree of sulfation and, in particular, the ratio of L-iduronic acid to D-glucuronic acid (Fransson et al., 1970). Barium acetate-ethanol cellulose chromatography, which primarily separates according to uronic acid composition gave the results shown in Fig. 2. It is seen that the radioactive

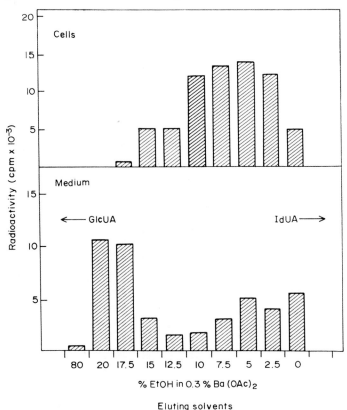

Fig. 2. Barium acetate-ethanol profiles of radioactive dermatan sulfate from cells and medium after incubation with radiosulfate.

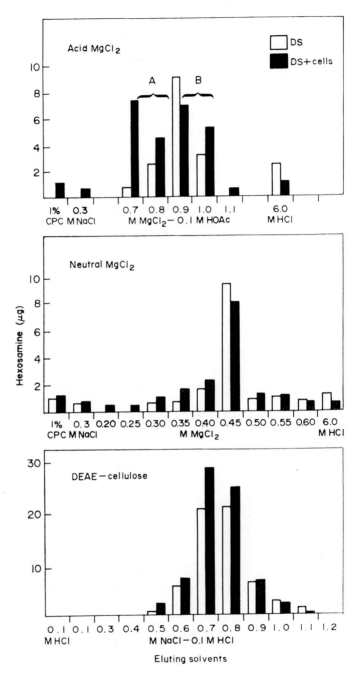

Fig. 3. Various microcolumn analyses of dermatan sulfate before (□) and after (■) incubation with fibroblasts in culture.

DS isolated from the cells was considerably more iduronic acid-rich than that excreted into the medium. It might be argued that fibroblasts secrete selectively those DS molecules that have the highest content of D-glucuronic acid. However, in view of the fact that almost half of the secreted DS seemed to contain more glucuronic acid than any of the intracellular molecules, this explanation appears unlikely.

In order to test the possibility that the cells synthesized an iduronic acid-rich DS, which was subsequently transformed to a glucuronic acid-rich DS, cells were grown in the presence of 5 mg of DS per culture. After 5 days of incubation the polysaccharide was reisolated from the medium as described above and analyzed for uronic acid composition (Dische, 1947; Brown, 1946), molecular size polydispersity and degree of sulfation. As shown in Fig. 3 acid $MgCl_2$ profiles (Fransson et al., 1970) revealed that DS, which had been in contact with fibroblasts contained more glucuronic acid than the starting material. The neutral $MgCl_2$ and acid DEAE profiles (Fransson et al., 1970) indicated that no substantial depolymerization or desulfation had taken place. In a large scale experiment DS which had been in culture with fibroblasts was fractionated into two fractions, A and B in Fig. 3 (upper graph). Analyses of these two fractions, which are shown Table I yielded a

Table I

Analyses of dermatan sulfate fractions A and B isolated after incubation of dermatan sulfate with fibroblasts in culture

Fraction	HexN	UA (Carbazole)	Carbazole / Orcinol	Δ O.D. 232 nm after digestion with chondroitinase	
				-AC	-ABC
A	19·4	10·8	0·50	0·050	0·424
B	20·7	10·6	0·38	0·031	0·518

higher carbazole to orcinol ratio for fraction A than for fraction B, indicating that fraction A contained as much as 15% glucuronic acid of total uronic acid. Similarly, digestion of fractions A and B with chondroitinase-AC (Saito et al., 1968) yielded a proportionally larger increase in ultraviolet absorption for fraction A than for fraction B; the values correspond to a glucuronic acid content of approximately 12%.

The results presented thus far appear to indicate that DS added to fibroblast cultures either promotes the synthesis of relatively glucuronic acid-rich DS (Fig. 4a) or is transformed by an epimerization reaction

(a) $^{35}SO_4 = \xrightarrow[\text{DERMATAN-SO}_4]{(+)} \text{DERMATAN-}^{35}SO_4$ (GlcUA-rich)

(b) $^{35}SO_4 = \rightarrow \text{DERMATAN-}^{35}SO_4 \rightarrow \text{DERMATAN-}^{35}SO_4$
         (IdUA-rich)   (GlcUA-rich)
                ↑
         DERMATAN-SO$_4$
         (IdUA-rich)

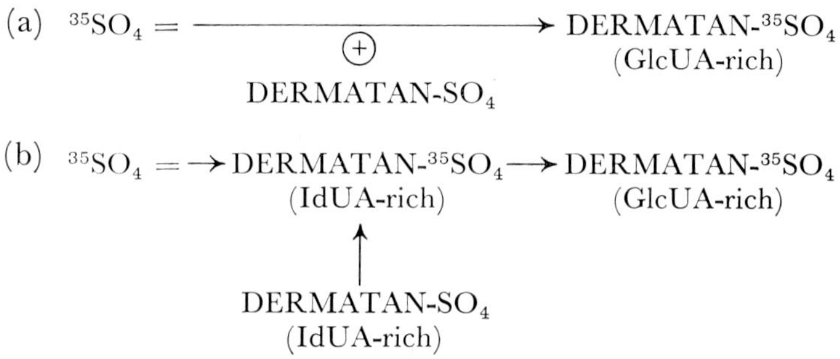

Fig. 4. Possible pathways for the synthesis of 'glucuronic acid-rich' dermatan sulfate.

to glucuronic acid-rich DS (Fig. 4b). These postulates were tested by the following method. Cells were grown for 3 days in the presence of (a) Na$_2$^{35}SO$_4$ alone and (b) both Na$_2$^{35}SO$_4$ and non-radioactive DS. If DS stimulates synthesis as in pathway a (Fig. 4) more radioactivity would be incorporated into glucuronic acid-rich DS. However, if pathway b is operating the dilution of radioactivity in the iduronic acid-rich DS pool by the exogenous material would be expected to reduce the amount of radioactivity incorporated into glucuronic acid-rich DS. The results of such experiments are shown in Fig. 5. It is apparent that in the presence of exogenous DS the secreted radioactive products are considerably more iduronic acid-rich (middle graph in Fig. 5) than when Na$_2$^{35}SO$_4$ was administered alone (upper graph). In fact the profile of the secreted material in the former experiment (middle graph) closely resembles that of the intracellular product (Fig. 2, upper graph). When the distribution of hexosamine was recorded in the same experiment (lower graph in Fig. 5) it can be seen that a marked transformation of the nonradioactive, exogenous DS had indeed occurred.

Attempts were also made to explore whether this epimerization reaction was taking place intracellularly or extracellularly. In these experiments the DS substrate had been exhaustively digested with testicular hyaluronidase, β-glucuronidase and chondroitinase-AC (Fransson and Malmström, 1971) to obtain an extremely iduronic acid-rich product. When this material was hydrolyzed and subjected to ion-exchange chromatography (Fransson et al., 1968) the result shown

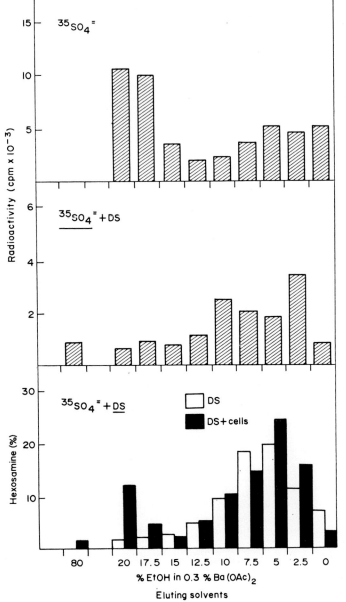

Fig. 5. Barium acetate-ethanol profiles of dermatan sulfate from cell medium after incubation with radiosulfate, or radiosulfate + dermatan sulfate. In the upper two graphs the radioactivity is plotted, while in the lower graph the amount of hexosamine is recorded.

in Fig. 6 was obtained. It seen (upper left) that the preparation still contained a few glucuronic acid residues, presumably originating from the linkage region to protein which is not totally removed by this enzymic treatment. Fibroblasts were then grown for 8 days in the

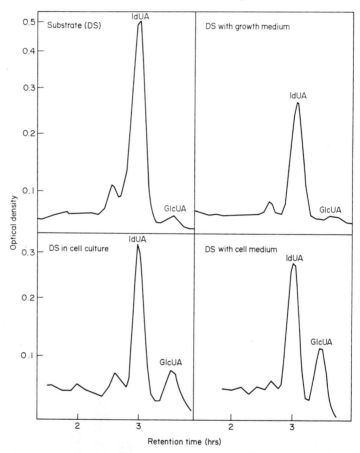

Fig. 6. Ion-exchange chromatography of hydrolysates of a dermatan sulfate substrate (upper left), the substrate after incubation with cells in culture (lower left), the substrate after incubation with growth medium alone (upper right) and the substrate incubated with the cell medium (lower right).

presence of this DS preparation. The polysaccharide was re-isolated by gel chromatography on Sephadex G-50 after papain digestion of the medium (Antonopoulos *et al.*, 1964), finally hydrolyzed and subjected to ion-exchange chromatography. As shown in the lower left graph (Fig. 6) DS which had been in contact with cells contained considerably more glucuronic acid than the substrate itself. However, DS, which had been incubated with the growth medium alone, was unaffected (upper right), whereas in DS incubated with the cell medium (medium in which cells had grown for 3 days) an increase in the amount glucuronic acid was observed (lower right). It was estimated that the glucuronic acid content had increased to approximately 20% of total uronic acid. It should be added that when chondroitin 4-sulfate was incubated in this system no formation of iduronic acid was observed. On the other hand, chemically desulfated DS (Kantor and Schubert, 1957) was slightly transformed, the amount of glucuronic acid formed, however, was approximately one-tenth of that obtained with DS. Finally, preliminary experiments have indicated that a microsomal fraction from fibroblasts does not catalyze the formation of polymer-bound glucuronic acid.

DISCUSSION

The present data indicate that fibroblasts in culture synthesize and secrete a product which contains very little, if any, D-glucuronic acid. Some of the L-iduronic acid moieties of this polymer are subsequently transformed to D-glucuronic acid residues *without* cleavage of the DS chains. This epimerization reaction seems to take place extracellularly and is presumably catalyzed by an enzyme derived from the fibroblasts.

The results presented here raise a number of interesting questions, notably (*a*) is the 'epimerase' solely the result of cell damage or is it actively secreted from the fibroblasts? (*b*) do fibroblasts synthesize a DS composed exclusively of L-iduronosyl-*N*-acetyl galactosamine repeating units? In view of the results obtained it would appear as if fibroblasts synthesize and secrete an iduronic acid-rich DS. This molecule is subsequently epimerized at specific sites to produce the final product. This reaction may induce conformational changes necessary to give the molecule its final shape. This step may also play a role in degradation of DS, since it would provide an extracellular pathway based on endoglycosidase action (Fig. 1) as an alternative to the intracellular pathway of exoglycosidases (Matalon *et al.*, 1971).

ACKNOWLEDGEMENTS

Our research was supported by grants from the Swedish Medical Research Council (B72-13X-2309-05A; B70-13X-139-06B), the Swedish Cancer Society (53-B71-04XC), Gustaf V:s 80-års fond, and the Medical Faculty, University of Lund.

REFERENCES

Antonopoulos, C. A., Gardell, S., Szirmai, J. A. and DeTyssonsk, E. R. (1964). *Biochim. Biophys. Acta* **83,** 1–19.
Brown, A. H. (1946). *Arch. Biochem. (Biophys.)* **11,** 269–275.
Dische, Z. (1947). *J. Biol. Chem.* **167,** 189–195.
Fransson, L.-Å. and Malmström, A. (1971). *Eur. J. Biochem.* **18,** 422–430.
Fransson, L.-Å., Rodén, L. and Spach, M. L. (1968). *Anal. Biochem.* **23,** 317–330.
Fransson, L.Å., Anseth, A., Antonopoulos, C. A. and Gardell, S. (1970). *Carbohyd. Res.* **15,** 73–89.
Haug, A. and Larsen, B. (1971). *Carbohyd. Res.* **17,** 297–304.
Jacobson, B. and Davidson, E. A. (1962). *J. Biol. Chem.* **237,** 638–643.
Kantor, T. G. and Schubert, M. (1957). *J. Am. Chem. Soc.* **79,** 152–154.
Lindahl, U., Bäckström, G., Malmström, A. and Fransson, L.-Å. (1972). *Biochem. Biophys. Res. Comm.* **46,** 985–991.
Matalon, R., Cifonelli, J. A. and Dorfman, A. (1971). *Biochem. Biophys. Res. Comm.* **42,** 340–345.
Olsson, I., Gardell, S. and Thunell, S. (1968). *Biochim. Biophys. Acta* **165,** 309–323.
Rodén, L. and Dorfman, A. (1958). *J. Biol. Chem.* **233,** 1030–1033.
Saito, H., Yamagata, T. and Suzuki, S. (1968). *J. Biol. Chem.* **243,** 1536–1542.

Genetic Defects in the Degradation of Glycosaminoglycans: the Mucopolysaccharidoses

ALBERT DORFMAN, REUBEN MATALON, JERRY N. THOMPSON
and J. ANTHONY CIFONELLI

*Departments of Pediatrics and Biochemistry
Joseph P. Kennedy, Jr., Mental Retardation Research Center
and the La Rabida-University of Chicago Institute
Pritzker School of Medicine
University of Chicago, Chicago, Illinois, 60637, USA*

The mucopolysaccharidoses are a group of heritable diseases characterized by the intracellular deposition and urinary excretion of glycosaminoglycans. The clinical description, pathology, and early history of the study of their etiology have been reviewed by Dorfman and Matalon (1972), Spranger (1972) and McKusick (1966).

Elucidation of the metabolic defects in these diseases started with the demonstration by Brante (1952) that livers of patients with Hurler's disease contain increased quantities of glycosaminoglycans. Dorfman and Lorincz (1957) discovered that dermatan sulfate and heparan sulfate are excreted in large amounts in the urine of patients with the Hurler syndrome. Subsequently, studies in a number of laboratories confirmed and extended these observations (Dorfman and Matalon, 1972). Although Hunter's and Hurler's syndromes were distinguished on genetic grounds, the clearcut correlation of chemical findings and and clinical symptoms first became apparent after the observations by Harris (1961) and Sanfilippo et al. (1962) that the excessive urinary excretion of only heparan sulfate was characteristic of a distinctive clinical entity. The classification of the mucopolysaccharidoses by McKusick et al. (1965) served to clarify the concept that mucopolysaccharidoses are a group of clinically distinct diseases.

That Hurler's syndrome represents a lysosomal disease due to faulty degradation of glycosaminoglycans was first suggested by Van Hoof and Hers (1964) on the basis of electron microscopic studies. Study of the etiology of this group of diseases was advanced by the discovery by Danes and Bearn (1966) that fibroblasts cultured from the skin of Hurler and Hunter patients demonstrated large numbers of metachromatic granules. Independently, Matalon and Dorfman (1966) found that such cultured fibroblasts accumulated large amounts of dermatan sulfate.

Fratantoni et al. (1968a) concluded that accumulation of glycosaminoglycans results from faulty degradation. These investigators (1968b) also observed that growth of Hunter or Hurler fibroblasts in the presence of normal fibroblasts results in the disappearance of metachromasia and the correction of the metabolic defect as measured by the uptake or disappearance of $^{35}SO_4$-containing macromolecules. Correction was also achieved by addition of medium exposed to normal cells. Subsequent studies showed that fibroblasts derived from most of the specific syndromes were deficient in a factor produced by fibroblasts of all other syndromes. Addition of the appropriate factor resulted in correction of the metabolic defect. In the case of Scheie's and Hurler's syndromes mutual correction was not observed (Wiesmann and Neufeld, 1970). Neufeld and Cantz (1971) discovered that fibroblast supernatants of certain patients with Sanfilippo's disease cross-correct each other, and therefore are enzymically distinct, each presumably lacking a different factor required for normal mucopolysaccharide metabolism. Neufeld and Cantz (1971) also indicated that fibroblasts from 'I-cell' disease require both the Hunter and Hurler factors for correction.

Corrective factors have been found in urine and Barton and Neufeld (1971) have extensively purified the Hurler factor. Sanfilippo A factor has been purified by Kresse and Neufeld (1972) and the Hunter factor has been partially purified by Cantz et al. (1970).

DEGRADATION OF GLYCOSAMINOGLYCANS

The existence of a large number of glycosidases in mammalian tissues has long been known. The recent appreciation of their localization to lysosomes and their role in the pathogenesis of glycosphingolipidoses has stimulated a more detailed study of their properties. In general, such glycosidases are specific with respect to the aglycone portion of the glycoside and the anomeric linkage. However, exceptions to this concept exist, e.g. β-N-acetylhexosaminidase appears to act on both N-acetylglucosamine and N-acetylgalactosamine glycosides.

In contrast α-N-acetylglucosaminidase is specific for the glucosamine configuration (Weissmann and Hinrichsen, 1969). Different isozymes may have different specificities. An example of this will be discussed below.

Our approach to the study of the etiology of the mucopolysaccharidoses has been a consideration of the structural features of the glycosaminoglycans.

Table I indicates the established glycoside linkages of the known glycosaminoglycans. Those listed seem reasonably well-established although certain structural features of the keratan sulfates remain to be elucidated. Table II indicates the known sulfate linkages. The specificity of sulfatases for specific linkages and the sequence of action of glycosidases and sulfatases is not clear. The studies of Tudball and Davidson (1968) indicated that partial degradation of chondroitin sulfate precedes sulfatase action.

The studies of Knecht et al. (1967) showed that the glycosaminoglycans excreted in urine and deposited in tissues of patients with the Hurler syndrome are partially degraded. In the case of dermatan sulfate such degradation could be accounted for by the action of hyaluronidase. The findings of N-acetylgalactosamine-SO_4 nonreducing terminal groups in dermatan sulfate fragments isolated from the urine of a Hurler parient by Fransson et al. (unpublished results) indicates further degradation by β-glucuronidase.

The fragments of heparan sulfate that were isolated by Knecht et al. (1967) could have arisen by the action of an endoglycosidase though no such enzyme acting on heparin or heparan sulfate has been discovered. The finding by Matalon and Dorfman (1968) that the molecular weight of dermatan sulfate isolated from Hurler fibroblasts was comparable to that prepared from skin suggested that fibroblasts are devoid of hyaluronidase activity. Assays of fibroblast extracts for hyaluronidase by the sensitive viscosity reduction method failed to reveal hyaluronidase (Matalon and Dorfman, unpublished results). When labeled chondroitin sulfate was subjected to digestion by extracts of normal fibroblasts, no oligosaccharides could be detected by gel chromatography. A small peak which behaved as a monosaccharide was observed. These results led to the conclusion that degradation of dermatan sulfate might proceed by two different pathways. In hyaluronidase-containing tissues such as liver, degradation occurs by way of hyaluronidase followed by the serial action of exoglycosidases, while in fibroblasts, degradation is confined to the action of exoglycosidases.

On the basis of these conclusions a search has been made for possible defects in degradation in the mucopolysaccharidoses.

Table I

Glycoside linkages of glycosaminoglycans

	β-xylose	β-galactose	β-N-acetyl-glucosamine	β-N-acetyl-galactosamine	α-N-acetyl-glucosamine	β-glucuronic acid	α-iduronic acid
Hyaluronic acid			X			X	
Chondroitin 4/6-SO₄	X	X		X		X	
Dermatan sulfate	X	X		X		X	X
Heparan sulfate	X	X			X	X	X
Heparin	X	X			X	X	X
Keratan sulfate I*		X	X				
Keratan sulfate II†		X	X				

* Linkage to protein is by way of asparagine.
† Linkage to protein is probably by way of glycoside of N-acetylgalactosamine to serine and threonine. Other linkages may be present.

Table II
Sulfate linkages of glycosaminoglycans

	Hexosamine-O-SO_4	Iduronic acid-O-SO_4	Galactose-O-SO_4	Sulfamide
Chondroitin 4/6-sulfate*	X			
Dermatan sulfate	X	X		
Heparan sulfate	X	X		X
Heparin	X	X		X
Keratan sulfate I	X		X	
Keratan sulfate II	X		X	

* Additional O-SO_4 may be present in some chondroitin sulfates.

DEFECTS IN DEGRADATION IN MUCOPOLYSACCHARIDOSES

Examination of the linkages in Tables I and II indicates that dermatan sulfate and heparan sulfate, the glycosaminoglycans deposited and excreted in Hurler, Hunter and Scheie's disease, share α-L-iduronosyl and probably iduronic acid sulfate linkages. Search for evidence of α-L-iduronidase was initially hampered by the lack of an adequate substrate. In an attempt to demonstrate α-L-iduronidase, desulfated dermatan sulfate was incubated with testicular hyaluronidase, β-glucuronidase, and tissue extracts (Matalon et al., 1971). The hyaluronidase and β-glucuronidase were added in an attempt to bare a maximal number of nonreducing terminal iduronic acid residues. On paper chromatography the release of free L-idurone was demonstrated by extracts of fibroblasts, leukocytes, and liver as well as by preparations obtained from normal urine. Striking was the limited release of idurone by extracts of fibroblasts cultured from Hurler patients. In order to obtain a substance which did not require the addition of other

Table III

α-L-Iduronidase activity of liver, cultured fibroblast extracts and urine protein

	Extract	Phenol $\mu moles/mg\ prot./24\ h$
Fibroblasts*	Normal	0·260
	Hunter	0·310
	Sanfilippo	0·180
	'I-Cell'	0·080
	Hurler	0·020
Liver†	Normal	0·042
	Sanfilippo	0·067
	'I-Cell'	0·033
	Hurler	0·002
Urine†	Normal	0·137
	Hurler	0·007
Amniotic cells	Normal	0·120

* Assays performed with 300 mcg of α-phenyliduronide per assay.
† Assays performed with 60 mcg of phenyliduronide per assay.

enzymes, a disaccharide, α-L-iduronosyl-anhydromannose, was prepared by the action of nitrous acid on desulfated heparin (Cifonelli *et al.*, 1971; Dorfman *et al.*, 1972). Utilizing this substrate, α-L-iduronidase activity was again demonstrated in extracts of normal fibroblasts, liver, and leukocytes, and urine but was found to be deficient in preparations derived from Hurler patients. Preparations derived from Hunter and Sanfilippo patients showed iduronidase activity comparable to preparations from normal individuals.

Subsequently, Weissmann and Santiago (1972) demonstrated the presence of α-L-iduronidase in rat liver utilizing a synthetic substrate, α-phenyl-L-iduronide. This substrate which was made available to us through the courtesy of Dr. Bernard Weissmann of the University of Illinois permitted more quantitative studies. Utilizing this compound, the absence of α-L-iduronidase in extracts of Hurler fibroblasts was once again demonstrated (Matalon and Dorfman, 1972). The data in Table III summarizes these results. The low iduronidase activity in 'I-Cell' disease is similar to the diminished activity of a number of lysosomal enzymes in fibroblasts derived from such patients. Mixing experiments indicated that the lack of iduronidase activity in Hurler's fibroblasts was not due to the presence of inhibitors.

Studies by Bach *et al.* (1972) have also shown a markedly diminished α-L-iduronidase activity in Hurler's fibroblasts. They have found an absence of a α-L-iduronidase activity in extracts of Scheie's fibroblasts, a finding that has been confirmed in this laboratory. Purified Hurler factor isolated from normal urine was found to exhibit α-L-iduronidase activity.

These results indicate that an absence of α-L-iduronidase represents the defect in Hurler's and Scheie's syndrome. Presumably these two genetic diseases are allelic mutations.

The defect in Hunter's disease remains unknown but studies are now underway utilizing a similar analysis which led to the successful unraveling of the defect in Hurler's syndrome.

Sanfilippo's disease is characterized primarily by the excretion of heparan sulfate. Reference to Table I indicates that at least two linkages exist in heparan sulfate and heparin which do not occur in the other connective tissue glycosaminoglycans. These are α-N-acetylglucosaminosyl and sulfamate linkages. The earlier studies of Neufeld and Cantz (1971) indicated that two distinct Sanfilippo's syndromes exist which have been designated Sanfilippo A and Sanfilippo B syndromes. Kresse and Neufeld (1972) have concluded that the Sanfilippo A factor is a heparan sulfate sulfatase. However, definitive proof is still lacking regarding the nature of the defect in this disease.

Studies in this laboratory summarized in Table IV showed the presence of α-N-acetylglucosaminidase in a variety of fibroblast extracts. However, O'Brien (1972) has recently demonstrated that α-N-acetylglucosaminidase is absent from extracts of fibroblasts of patients with Sanfilippo B disease. The data in Table V confirm these findings. Some of the fibroblasts studied were from the same patients as those studied by O'Brien.

Table IV

α-N-Acetylglucosaminidase activity in tissue extracts

Extract		Activity μmoles nitrophenol/mg prot./h
Fibroblast	Normal	0·038
	Sanfilippo A	0·045
	Hurler	0·040
	Hunter	0·034
	'I-Cell'	0·014
	Cystic Fibrosis	0·030
	Fabry	0·020
Liver	Normal	0·020
	Sanfilippo A	0·050
	GM_1 Gangliosidosis	0·070
	Hurler	0·036

Table V

α-N-Acetylglucosaminidase activity in fibroblast extracts

Extract	Activity μmoles phenol/mg prot./24 h
Normal	0·340
Sanfilippo A*	0·234
Sanfilippo A*	0·161
Sanfilippo A	0·150
Sanfilippo A	0·071
Sanfilippo B	N.D.†
Sanfilippo B	N.D.
Sanfilippo B*	N.D.

* Cases previously studied by O'Brien and typed by Dr. Elizabeth Neufeld.
† Not detectable.

An additional defect of glycosaminoglycan metabolism involving the absence of β-glucuronidase has been recently discovered by Hall et al. (personal communication). Stumpf and Austin (1972) in a preliminary report have observed a diminution of arylsulfatase B in tissues of a patient with Maroteaux-Lamy disease.

The results summarized indicate that like other storage diseases, the mucopolysaccharidoses are a group of heritable diseases characterized by specific defects in lysosomal degradative enzymes. Fig. 1 illustrates the defects that have been elucidated to date.

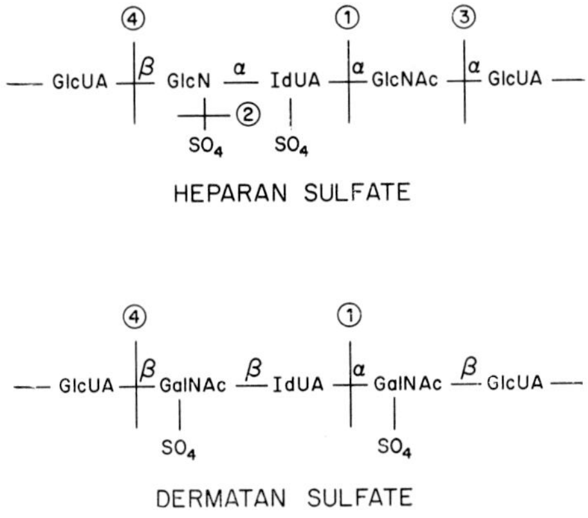

$$\text{---GlcUA} \xrightarrow{\beta}^{④} \text{GlcN} \xrightarrow{\alpha} \text{IdUA} \xrightarrow{\alpha}^{①} \text{GlcNAc} \xrightarrow{\alpha}^{③} \text{GlcUA---}$$
$$\quad\quad\quad\quad\quad +②\;|$$
$$\quad\quad\quad\quad SO_4\quad SO_4$$

HEPARAN SULFATE

$$\text{---GlcUA} \xrightarrow{\beta}^{④} \text{GalNAc} \xrightarrow{\beta} \text{IdUA} \xrightarrow{\alpha}^{①} \text{GalNAc} \xrightarrow{\beta} \text{GlcUA---}$$
$$\quad\quad\quad\quad |\quad\quad\quad\quad\quad |$$
$$\quad\quad\quad\quad SO_4\quad\quad\quad\quad SO_4$$

DERMATAN SULFATE

Fig. 1. The metabolic defects in the mucopolysaccharidoses. 1, Hurler disease; Scheie disease; 2, Sanfilippo A disease (?); 3, Sanfilippo B disease; 4, β-glucuronidase deficiency.

INTERRELATIONSHIPS OF STORAGE DISEASES

Glycoproteins and glycosphingolipids share a number of glycoside linkages with the glycosaminoglycans. If the same glycosidases are involved in the degradation of these three classes of compounds, it might be anticipated that storage of partially degraded products of each class should occur in the absence of a specific glycosidase. Whereas it is beyond the scope of this chapter to discuss all of the structural interrelationships of carbohydrate-containing macromolecules, one example that has been recently investigated is of considerable interest. It is now well-established that Tay-Sachs's disease is characterized by an absence of β-N-acetylhexosaminidase A (Okada and O'Brien, 1969) while in Sandhoff-Jatzkewitz's syndrome there is an absence of both N-acetylhexosaminidases A and B (Sandhoff et al., 1968). The nature of the interrelationships of the two isozymes is not yet clear. Recent studies indicate that the two hexosaminidases are immunologically cross-reactive (Srivastava and Beutler, 1972). Considerable evidence indicates the β-N-acetylhexosaminidases are concerned with the degradation of the glycosaminoglycans yet neither Tay-Sachs's disease nor Sandhoff-Jatzkewitz's disease show characteristics of mucopolysaccharidoses either clinically or chemically. Strecker and Montreuil (1971) found in the urine of a patient with Sandhoff-Jatzkewitz syndrome an increased amount of an oligosaccharide fraction composed primarily of N-acetylglucosamine and mannose but no increased levels of uronic acid-containing glycosaminoglycans.

In order to investigate this anomaly further, a labeled oligosaccharide was prepared biosynthetically by the following reaction by Dr. Allen C. Stoolmiller in this laboratory:

$$\text{UDP-}^{14}\text{C-GalNAc} + (\text{GlcUA-GalNAc})_3 \rightarrow$$
$$\underset{|}{}$$
$$\text{SO}_4$$

$$^{14}\text{C-GalNAc-(GlcUA-GalNAc)}_3$$
$$\underset{|}{}$$
$$\text{SO}_4$$

The enzyme utilized was prepared from 13-day old chick epiphyses by the method of Telser et al. (1965), and the product was isolated by chromatography on a Sephadex G-25 column. The labeled heptasaccharide substrate was incubated in an acetate buffer containing 0·15 M NaCl, pH 4·5, with fibroblast extracts. The reaction products were separated on a G-25 Sephadex column with 0·05 N NaCl in 15% ethanol as eluant. Extracts were also assayed for β-N-acetylhexosaminidase activity utilizing 4-methylumbelliferyl-β-D-N-acetylglucosaminide as substrate. The heptasaccharide was readily cleaved by extracts of normal fibroblasts to yield free N-acetylgalactosamine. The reaction is approximately linear up to 72 h and shows a pH optimum of 4·5. The extent of cleavage is exceptionally high since approximately 25% of the substrate is hydrolyzed in 24 h by 1 mg of protein of crude extract.

Table VI summarizes the results of an experiment comparing the activity of a series of fibroblast extracts. Of note is the complete absence of activity of extracts of fibroblasts derived from patients with Tay-Sachs's and Sandhoff-Jatzkewitz's disease and the almost complete absence of activity in heat-inactivated normal extract. The marked diminution of activity in fibroblasts of an 'I-Cell' patient is in keeping with the previous finding of diminished lysosomal enzyme activities in this disease. Mixing of extracts showed no evidence of the presence

Table VI
Reaction of skin fibroblast culture
β-hexosaminidases with heptasaccharide

Enzyme source	Release of ^{14}C-GalNAc (cpm)/mg prot./24 h
Normal-1	5135
Normal-2	5129
Hurler	5453
Hunter	4652
Sanfilippo	3577
'I-Cell'	473
Tay-Sachs	0
Sandhoff	0
Normal-2 (heat-inactivated)*	370

* Enzyme extract was heated in a water bath at 50 °C for 2 h.

Table VII
Activity of β-hexosaminidase in cultured skin fibroblast using the 4-methylumbelliferone-β-D-N-acetylglucosamine substrate

Enzyme source	μmoles of 4-methylumbelliferone /mg prot./h
Normal-2	1.52
Normal-2 (heat inactivated)	0.77
Tay-Sachs	0.71
Sandhoff	0.04
Hurler	2.10
Hunter	5.15
Fetal calf serum (BBL)	0.01

of inhibitors or multiple factors in either Tay-Sachs's and Sandhoff-Jatzkewitz's disease. Table VII summarizes the results obtained with some of the same extracts when assayed by the 4-methyl-β-D-N-acetylglucosaminide substrate. The data are consistent with previously published results showing the absence of all activity in Sandhoff-Jatzkewitz extracts and diminished activity in Tay-Sachs and heat-inactivated normal extracts. The high levels of activity in Hurler and Hunter disease have been previously reported.

These studies indicate that hexosaminidase A is required for the degradation of this heptasaccharide substrate. The results are particularly striking when compared with the fact that Strecker and Montreuil (1971) found a marked difference in the urinary oligosaccharide pattern of Tay-Sachs's and Sandhoff-Jatzkewitz's diseases. Their results indicate that the N-acetylhexosaminidase B is important in degradation of glycoproteins. Perhaps more interesting is the suggestion that a pathway of glycosaminoglycan degradation exists which circumvents the obligatory participation of the β-N-acetylhexosaminidases. Further studies along these lines are in progress.

SUMMARY AND CONCLUSIONS

The mucopolysaccharidoses are a group of storage diseases which result from deficient activity of specific hydrolases concerned with the degradation of glycosaminoglycans. The specific deficiency in Hurler's and Scheie's disease is α-L-iduronidase while that in Sanfilippo B disease is α-N-acetylglucosaminidase. A rare form of mucopolysaccharidosis involves a deficiency in β-glucuronidase activity. A heparan sulfate sulfatase has been implicated in Sanfilippo A disease which aryl sulfatase B has been suggested as a deficiency in Maroteaux-Lamy disease.

The pathways of degradation and the specificity of hydrolases involved in degradation of carbohydrate-containing macromolecules are still not completely understood.

POSTSCRIPT

Since presentation of this paper the definition of the enzymic defect in Sanfilippo A disease has been further clarified. Matalon and Dorfman (1973) have shown the release of $^{35}SO_4$ from $^{35}SO_4$-heparin specifically labeled in the N-SO_4 group by extracts of normal, Hurler, Hunter and Sanfilippo B fibroblasts but almost no release by extracts of Sanfilippo A fibroblasts. That the defect in Hunter's disease is due to an absence of iduronic acid sulfate sulfatase has been postulated by the isolation of a terminal disulfated disaccharide, containing sulfated iduronic acid sulfate, following chondroitinase ABC digestion of dermatan sulfate isolated from Hunter fibroblasts. This disaccharide was not found when dermatan sulfate isolated from Hurler's fibroblasts was digested with chondroitinase (Sjöberg et al., unpublished results). Similar conclusions based on the activity of the Hunter factor have been reached by Bach et al., 1973.

ACKNOWLEDGEMENTS

This work was supported by USPHS Grants AM-05996, HD-04583, and AM-05589, and a grant from the Illinois and Chicago Heart Association. Reuben Matalon is a scholar at Joseph P. Kennedy, Jr., Mental Retardation Research Center. The authors are grateful to Mrs. Minerva Deanching, Miss Angelita A. Labudiong, and Mrs. B. Kancharla for their expert technical assistance.

REFERENCES

Bach, G. S., Friedman, R., Weissmann, B. and Neufeld, E. F. (1972). *Proc. Natl. Acad. Sci. USA* **69**, 2048–2051.
Bach, G. S., Cantz, M., Okada, S. and Neufeld, E. F. (1973). *Fed Proc.* **32**, 483.
Barton, R. W. and Neufeld, E. F. (1971). *J. Biol. Chem.* **246**, 7773–7779.
Brante, G. (1952). *Scand. J. Clin. Lab. Invest.* **4**, 43–46.
Cantz, M., Chrambach, A. and Neufeld, E. F. (1970). *Biochem. Biophys. Res. Commun.* **39**, 936–942.
Cifonelli, J. A., Matalon, R. and Dorfman, A. (1971). *Fed. Proc.* **30**, 1207.
Danes, B. S. and Bearn, A. G. (1966). *J. Exp. Med.* **123**, 1–16.
Dorfman, A. and Lorincz, A. E. (1957). *Proc. Natl. Acad. Sci. USA* **43**, 443–446.
Dorfman, A. and Matalon, R. (1972). In "The Metabolic Basis of Inherited Diseases" (J. B. Stanbury, J. B. Wyngaarden and D. S. Frederickson, eds.) pp. 1218–1272. McGraw-Hill, New York.
Dorfman, A., Matalon, R., Cifonelli, J. A., Thompson, J. and Dawson, G. (1972). In "Sphingolipidoses" (B. Volk and S. Aronson, eds.) pp. 195–210. Plenum Press, New York.
Fransson, L.-Å., Sjöberg, I. and Dorfman, A. (Unpublished results).
Fratantoni, J. C., Hall, C. W. and Neufeld, E. F. (1968a). *Science* **162**, 570–572.
Fratantoni, J. C., Hall, C. W. and Neufeld, E. F. (1968b). *Proc. Natl. Acad. Sci. USA* **60**, 699–706.
Hall, C. W., Cantz, M., Neufeld, E. F., Sly, W. S., Quinton, B. A., McAlister, W. H. and Rimoin, D. L. and Rimoin, D. L. (Personal communication).
Harris, R. C. (1961). *Am. J. Dis. Child.* **102**, 741.
Knecht, J., Cifonelli, J. A. and Dorfman, A. (1967). *J. Biol. Chem.* **242**, 4652–4661.
Kresse, H. and Neufeld, E. F. (1972). *J. Biol. Chem.* **247**, 2164–2170.
Matalon, R. and Dorfman, A. (1966). *Proc. Natl. Acad. Sci. USA* **56**, 1310–1316.
Matalon, R. and Dorfman, A. (1968). *Proc. Natl. Acad. Sci. USA* **60**, 179–185.
Matalon, R. and Dorfman, A. (1972). *Biochem. Biophys. Res. Commun.* **47**, 959–964.
Matalon, R. and Dorfman, A. (1973). *Ped. Res.* **7**, 156.
Matalon, R. and Dorfman, A. (Unpublished results).
Matalon, R., Cifonelli, J. A. and Dorfman, A. (1971). *Biochem. Biophys. Res. Commun.* **42**, 340–345.
McKusick, V. A. (1966). In "Heritable Disorders of Connective Tissue," 3rd Edition, pp. 325–399. C. V. Mosby Co., St. Louis.
McKusick, V. A., Kaplan, D., Wise, D., Hanley, W. B., Sudderth, S. B., Serick, M. E. and Maumenee, A. E. (1965). *Med.* **44**, 445–483.
Neufeld, E. F. and Cantz, M. J. (1971). *Ann. N. Y. Acad. Sci.* **179**. 580–587.
O'Brien, J. S. (1972). *Proc. Natl. Acad. Sci. USA* **69**, 1720–1722.
Okada, S. and O'Brien, J. S. (1969). *Science* **165**, 698–700.

Sandhoff, K., Andreae, U. and Jatzkewitz, H. (1968). *Life Sci.* **7,** 283–288.
Sanfilippo, S. J., Podosin, R., Langer, L. O., Jr. and Good, R. A. (1962). *J. Peds.* **63,** 837–838.
Spranger, J. (1972). *Ergeb. In. Med. Kinderheilkd.* 166–265.
Srivastava, S. K. and Beutler, E. (1972). *Biochem. Biophys. Res. Commun.* **47,** 753–759.
Strecker, G. and Montreuil, J. (1971). *Clin. Chim. Acta* **33,** 395–401.
Stumpf, D. and Austin, J. H. (1972). *Proc. Am. Neurol. Assoc. (Chicago)* **6.**
Telser, A., Robinson, H. C. and Dorfman, A. (1965). *Proc. Natl. Acad. Sci. USA* **54,** 912–919.
Tudball, N. and Davidson, E. A. (1968). *Biochim. Biophys. Acta* **171,** 113–120.
Van Hoof, F. and Hers, H. G. (1964). *C. R. Acad. Sci. (Paris)* **259,** 1281–1283.
Weissmann, B. and Hinrichsen (1969). *Biochemistry* **8,** 2034—2043.
Weissmann, B. and Santiago, R. (1972). *Biochem. Biophys. Res. Commun.* **46,** 1430–1433.
Wiesmann, U. and Neufeld, E. F. (1970). *Science* **169,** 72–74.

Chemistry of Excretory Products in the Hunter Syndrome During Plasma Infusion

Lars-Åke Fransson, Ingrid Sjöberg and Gösta Blennow[*]

Department of Physiological Chemistry 2, University of Lund, Lund, Sweden

Recent work has shown that the various inborn errors of glycosaminoglycan (GAG) metabolism are due to defects in the synthesis of specific glycosidases, which are involved in the normal catabolism of GAG, notably dermatan sulfate (DS) and heparan sulfate (HS) (Dorfman, this book, p. 449). The various enzymic factors have been found in serum and urine and also partially purified from urine (Cantz et al., 1970; Barton and Neufeld, 1971; Kresse and Neufeld, 1972). In cell culture experiments these factors assist abnormal fibroblasts in degrading GAG. Furthermore, it has been reported that abnormal fibroblasts fail to accumulate GAG when grown in the presence of normal human serum (Hors-Cayla et al., 1968). This finding led DiFerrante et al. (1971) to explore the effects of plasma infusion on patients with GAG storage diseases. It was reported that such treatment of Hurler as well as Hunter patients resulted in a decreased urinary excretion of polysaccharide material, whereas the excretion of oligosaccharides was increased.

We have studied in somewhat more detail the chemical characteristics of the urinary products during plasma infusion of a Hunter patient. It will be demonstrated that the level of GAG excretion was not markedly affected. However, the excreted DS became more glucuronic acid-rich during treatment. Simultaneously, excretion of free D-glucuronic acid was increased approximately two-fold. These observations will be unified in a proposed pathway for degradation of DS initiated by the infusion of plasma.

EXPERIMENTAL AND RESULTS

A nine-year old boy afflicted with Hunter's disease (Dorfman, 1966) received during one day one litre of fresh-frozen and thawed human plasma.

[*] Department of Pediatrics, University of Lund, Lund, Sweden.

Urine was collected the days before, during, and after treatment, using thymol as preservative, concentrated and fractionated by gel chromatography. The material was chromatographed on columns of Sephadex G-50 (1·1 × 195 cm) or G-10 (1·1 × 189 cm), superfine, which were eluted with 0·5 M acetate buffer, pH 5·0, and 10% ethanol respectively. In either case polysaccharide material was recovered from the void volume fractions, whereas oligosaccharides were resolved on Sephadex G-10. The various fractions were subsequently analysed by several chromatographic techniques. In the case of the polysaccharide the molecular size, the degree of sulfation and the

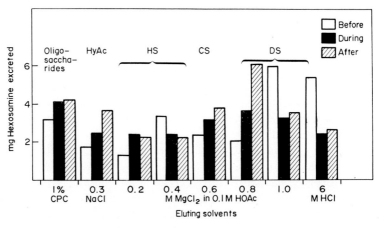

Fig. 1. CPC-cellulose microcolumn analyses of urinary GAGs excreted before, during and after plasma infusion. The elution position of standard GAGs are indicated above the appropriate fractions.

ratio of iduronic to glucuronic acid were assessed by microcolumn analyses (Fransson et al., 1970). The oligosaccharide fractions were analyzed by various ion exchange procedures (Fransson et al., 1968) as well as by specific colour reactions (Brown, 1946; Dische, 1947).

When the polysaccharide material excreted before, during and after plasma infusion was analyzed on CPC-cellulose microcolumns eluted primarily with acid $MgCl_2$ solutions the result shown in Fig. 1 was obtained. It is seen that the patient excreted DS and HS in the diseased state (open bars). During plasma infusion there appeared to be an increased excretion of relatively long oligosaccharides, which are preferentially eluted in the first two fractions. However, the most striking effect was the shift in the profile in the sense that excreted sulfated GAG were eluted in positions of chondroitin sulfate (CS) and intermediary to CS and DS.

This effect which remained for some days after treatment might be the result of depolymerization, desulfation, formation of relatively glucuronic acid-rich DS, or combinations thereof. As shown in Fig. 2 neutral $MgCl_2$ profiles as well as acid DEAE-cellulose chromatography gave no indication of extensive depolymerization or desulfation.

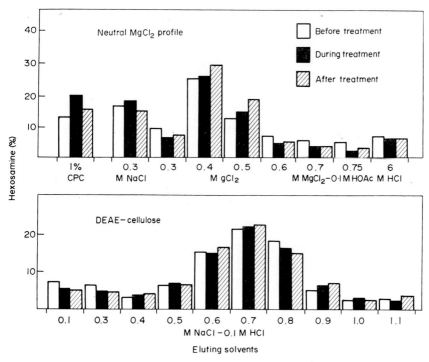

Fig. 2. Microcolumn analyses of urinary GAGs excreted before, during and after plasma infusion.

However, analyses of the uronic acid composition of the urinary GAGs (Table I) revealed that the material excreted during treatment contained a larger proportion of D-glucuronic acid. In order to corroborate these findings the urinary GAGs were also digested with chondroitinase-AC, which characteristically cleaves linkages between hexosamine and glucuronic acid (Saito et al., 1968; Fransson and Malmström, 1971). As shown in Fig. 3 gel chromatography on Sephadex G-25 (upper graphs) of chondroitinase-AC digests of urinary GAGs revealed no major differences between the GAGs excreted before and after treatment. However, when the void volume peaks were subfractionated on Sephadex G-50 (lower graphs) a larger proportion of relatively long oligosaccharides were obtained from the GAGs excreted after

Table I

Analyses of urinary polysaccharides before, during and after treatment

Day	Carbazole / Orcinol	L-IdUA / D-GlcUA
1	0.64	4
2*	0.84	3
3	0.78	

* Day of treatment.

treatment (material eluted at effluents volumes of 100–140 ml, Fig. 3) than that excreted before treatment. These results are interpreted to indicate that the DS excreted during and after treatment contained more glucuronic acid units than the material excreted before treatment.

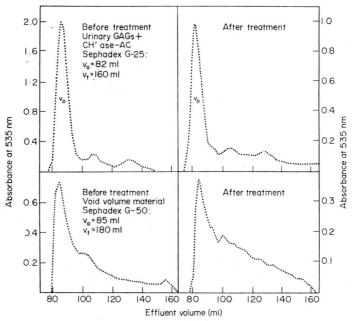

Fig. 3. Gel chromatography of chondroitinase-AC digests of urinary GAGs on columns (1.2 × 180 cm) Sephadex G-25 (upper graphs). The void volume materials from the G-25 runs were subsequently chromatographed on columns (1.2 × 230 cm) of Sephadex G-50 (lower graphs). Aliquots of the fractions were analyzed by an automated version of the carbazole-borate technique (Bitter and Muir, 1962).

EXCRETORY PRODUCTS IN THE HUNTER SYNDROME   467

Futhermore, these glucuronic acid-containing repeating units were located as single units rather than in clusters.

The low-molecular weight material excreted before, during and after treatment was fractionated by gel chromatography on Sephadex G-10. As shown in Fig. 4 four carbazole-positive peaks were regularly observed, the third of which (peak 3) was increased during plasma

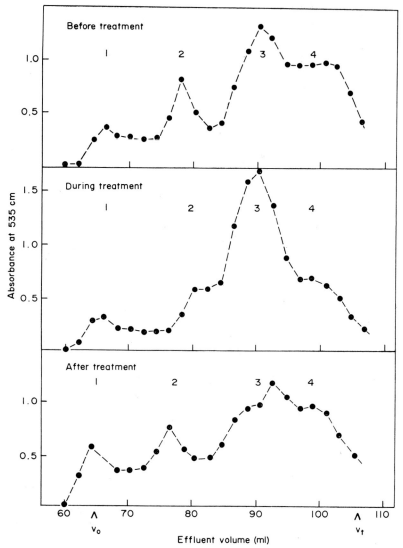

Fig. 4. Gel chromatography on Sephadex G-10 of urine samples obtained before, during and after treatment.

infusion. This material was further purified by chromatography on Dowex 1 eluted with a linear salt gradient. The chromatogram in Fig. 5 shows that peak 3 contained three components, all of which were excreted to a larger extent during treatment. Component c which was increased approximately two-fold was identified as D-glucuronic acid by its retention time on anion exchange resin and by its carbazole-to-orcinol ratio (Fig. 6).

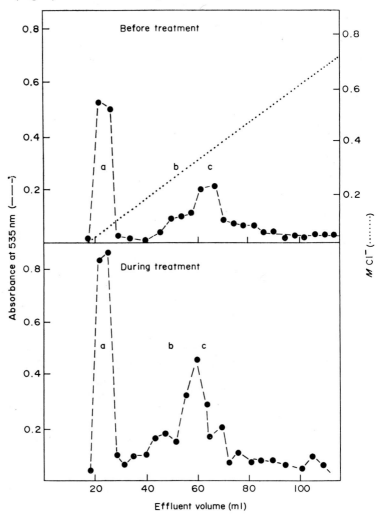

Fig. 5. Chromatography of peak 3 (Fig. 4) on a column (1·2 × 45 cm) of Dowex AG1-X8 (Cl-form) eluted with a linear LiCl gradient. Aliquots of the fractions were analyzed by the carbazole reaction (Dische, 1947).

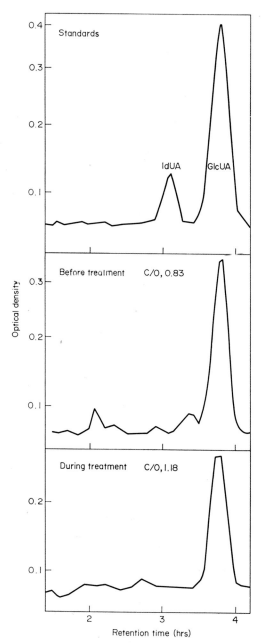

Fig. 6. Anion exchange chromatography of peak 3c (Figs. 4 and 5.) before and during treatment.

DISCUSSION

Treatment of a Hunter patient by plasma infusion gave the following results: (a) transformation of iduronic acid-rich DS to relatively glucuronic acid-rich DS, (b) increased urinary excretion of medium-sized oligosaccharides, and (c) increased excretion of D-glucuronic acid. These observations have been unified in a proposed pathway for the degradation of DS which is depicted in Fig. 7.

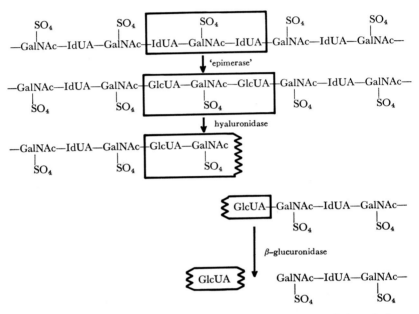

Fig. 7. Proposed scheme for degradation of DS initiated by the infusion of plasma.

It is assumed that iduronic acid-rich DS is transformed to glucuronic acid-rich DS through multiple attacks by an 'epimerase' present in the infused plasma (see also Fransson et al., this volume, p. 439). When a site composed of two adjacent glucuronic acid-containing periods is subsequently attacked by hyaluronidase nonreducing terminal glucuronic acid is formed. Finally, the action of β-glucuronidase would result in the liberation of free glucuronic acid. Medium-sized oligosaccharides would correspond to segments between two sites initially attacked by the 'epimerase'. It might be worth adding that single glucuronic acid-containing units are not attacked by hyaluronidase (Fransson and Malmström, 1971). Consequently, epimerization of one single iduronic acid residue to a glucuronic acid residue would not be sufficient to initiate the pathway outlined in Fig. 7. Accordingly, the DS excreted during treatment was found to contain primarily single glucuronic acid units. The present results are thus interpreted in view of the recent finding of DS:CS 'epimerase' in fibroblast secretions (Fransson *et al.*, this volume, p. 439). It is thus possible that plasma infusion augments an already operating alternative pathway for the degradation of DS.

ACKNOWLEDGEMENTS

Our research was supported by grants from the Swedish Medical Research Council (B70-13X-139-06C), Gustaf V:s 80-årsfond and the Medical Faculty, University of Lund.

REFERENCES

Barton, R. W. and Neufeld, E. F. (1971). *J. Biol. Chem.* **246**, 7773-7779.
Bitter, T. and Muir, H. (1962). *Anal. Biochem.* **4**, 330-335.
Brown, A. H. (1946). *Arch. Biochem. (Biophys.)* **11**, 269-275.
Cantz, M., Chrambach, A. and Neufeld, E. F. (1970). *Biochem. Biophys. Res. Comm.* **39**, 936-942.
DiFerrante, N., Nichols, B. L., Donnelly, P. V., Neri, G., Hrgovcic, R. and Berglund, R. K. (1971). *Proc. Natl. Acad. Sci. USA* **68**, 303-307.
Dische, Z. (1947). *J. Biol. Chem.* **167**, 189-195.
Dorfman, A. (1966). *In* "The Metabolic Basis of Inherited Disease" (J. B. Stanbury, J. B. Wyngaarden and D. S. Fredrickson, eds.) pp. 963-994. McGraw-Hill, New York.
Fransson, L.-Å and Malmström, A. (1971). *Eur. J. Biochem.* **18**, 422-430.
Fransson, L.-Å., Rodén, L. and Spach, M. L. (1968). *Anal. Biochem.* **23**, 317-330.
Fransson, L.-Å., Anseth, A., Antonopoulos, C. A. and Gardell, S. (1970). *Carbohyd. Res.* **15**, 73-89.
Hors-Cayla, M. C., Maroteaux, P. and de Grouchy, J. (1968). *Ann. Gene. (Paris)* **11**, 265-270.
Kresse, H. and Neufeld, E. F. (1972). *J. Biol. Chem.* **247**, 2164-2170.
Saito, H., Yamagata, T. and Suzuki, S. (1968). *J. Biol. Chem.* **243**, 1536-1542.

Aspartylglycosaminuria in vitro; Electronmicroscopic and Enzymatic Studies on Cultured Fibroblasts

PERTTI AULA, SEPPO AUTIO, VEIKKO NÄNTÖ
and MARJA-LEENA LAIPIO
*The Children's Hospital and Third Department of Pathology
University of Helsinki, Helsinki, Finland*

Aspartylglycosaminuria (AGU) is a newly discovered clinical entity, which in Finland was first described by Palo (1967) as peptiduria in eleven cases found among mentally retarded patients. Jenner and Pollitt (1967) found two similar cases in England, and they could show that the abnormal spot in paper chromatography of the urine samples of these patients was due to aspartylglycosamine. It was later discovered that the incidence of this disease was relatively high among the Finnish population (Palo and Mattsson, 1970; Autio, 1972). So far 48 cases have been detected in Finland. Autio (1972) has recently summarized the clinical picture and the hereditary nature of this disease, which constantly leads to a severe mental retardation and which is associated with some peculiar somatic signs. Family studies clearly point to an autosomal recessive mode of inheritance.

The diagnosis is made by demonstration of large amounts of a glycopeptide, aspartylglycosamine (AADG) in the urine, which is not found in the urine of healthy persons. It is suggested that AGU is a lysosomal storage disease affecting the metabolism of glycoproteins. The nature of the intracellular storage material seen in several tissues of AGU patients is not exactly known, but recent studies (Palo and Savolainen, 1972) suggest that it might contain an oligosaccharide together with AADG. Ultrastructural studies have shown that liver, kidney and brain cells contain large quantities of abnormal and enlarged lysosomes (Arstila *et al.*, 1972). The activity of an AADG cleaving hydrolase, N-aspartyl-β-glucosaminidase (AADGase) is decreased in the brain and liver tissues of AGU patients (Palo *et al.*, 1972) whereas the activity of some other closely related lysosomal hydrolases, particularly N-acetyl-β-glucosaminidase and N-acetyl-β-galactosaminidase are increased, as compared with control samples. A decreased AADGase activity was earlier reported by Pollitt and Jenner (1969) in the serum and seminal fluid of their AGU patient, but the relatives of this patient had a normal activity of this enzyme in the serum.

We have carried out ultrastructural and enzymatic studies on cultured fibroblasts from AGU patients in order to find out whether the morphological changes in the lysosomes and the enzymatic deficiences described above can be demonstrated in cell cultures. Earlier studies (unpublished results) had demonstrated an increased amount of metachromasia in cultured fibroblasts with toluidin blue staining according to Danes and Bearn (1966) indicating possible changes in the lysosomal structures.

CELL CULTURES

Fibroblast cultures were initiated from small skin biopsies of AGU patients and of healthy controls. The cells were grown in plastic bottles (Falcon Plastics); the growth medium consisted of 10% calf serum in medium F 10 (Flow Laboratories, Scotland). Cell cultures were harvested for both ultrastructural and enzymatic studies at a stationary growth phase, usually six days after the previous cell passage.

ULTRASTRUCTURAL STUDIES

Cells were scraped off from the bottles with a rubber policeman, spun down, fixed in 2% glutaraldehyde, post-fixed in osmium tetroxide, embedded in Epon, thin-sectioned and stained with uranyl acetate and lead citrate and finally studied with a Zeiss EM 9A electron microscope. Particular attention was paid to the structure of lysosomes.

The cultured fibroblasts from AGU patients contained large amounts of lysosomal bodies of various types which were also seen in control cells. Besides normal lysosomes, however, some cells of AGU cultures also contained abnormally enlarged lysosomes with electron lucent material. Morphologically the altered lysosomes were identical to the abnormal lysosomes in the liver tissues of these patients, as illustrated in Fig. 1. Most of them are limited by a single membrane, and the diameter varies from 0·5 to 5 microns. The abnormal lysosomes contain electron lucent material which ranges from fine granular to networklike material. This type of lysosomal body was only occasionally seen in control cells. The other cellular organelles, mitochondria, Golgi apparatus and nuclei showed no changes in AGU cells.

ENZYME STUDIES

So far, we have determined the activities of the specific AADG cleaving hydrolase, N-aspartyl-β-glucosaminidase (E.C. 3.5.1.–, AADGase) and arylsulphatase A in fibroblasts of one AGU patient and of three separate control individuals.

AADGase activity in fibroblasts was determined by a method used by Makino et al. (1966). Substrate for the enzyme reaction was β-aspartylglucosamine (AADG) which was isolated from the urine of an AGU patient using ion exchange and paper chromatographic techniques. N-acetylglucosamine liberated by enzymatic hydrolysis was measured photometrically with Morgan-Elson reaction. (Reissig et al., 1955). Pooled cells from three culture bottles (250 cc) were harvested mechanically and suspended in a small volume of phosphate buffer, pH 7·0, containing 0·1% Triton-X. The cells

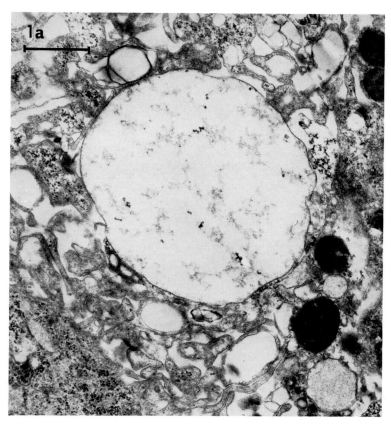

Fig. 1(a). For (b) and caption, see next page.

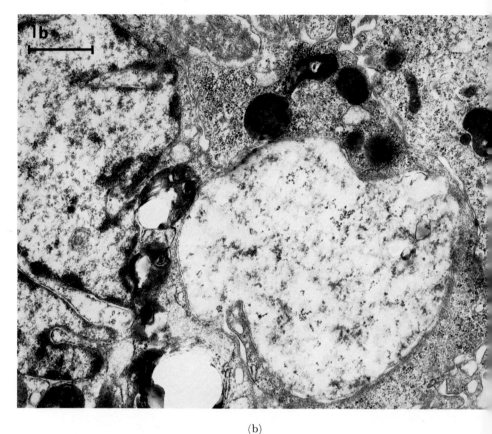

(b)

Fig. 1, a and b. Examples of abnormal lysosomes in cultured fibroblasts from AGU patients. The size of the altered lysosomes varies from one to several microns, and they contain electron lucent material ranging from granular to network-like in appearance. (Bar = one micron.)

were then disrupted by an ultrasound sonicator and the total cell homogenate was used for enzyme studies. Total protein concentration was determined by the method of Lowry *et al.* (1951). The activity of arylsulphatase A was measured (as a cell viability control) using 4-nitro-catecholsulphate as substrate according to the method of Baum *et al.* (1959).

In three separate experiments the AADGase activity of the fibroblasts from the AGU patient was found to be greatly decreased, as compared with fibroblasts from control cultures. The activity of AADGase was approximately 20 times higher in control fibroblasts than in AGU fibroblasts, as illustrated in Fig. 2. The activity of arylsulphatase A was equal both in AGU and control cell cultures.

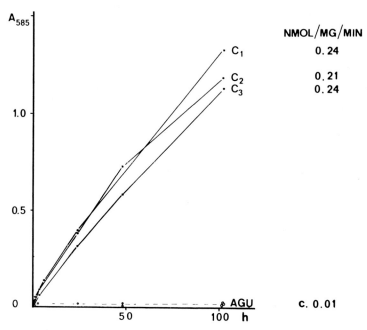

Fig. 2. Liberation of N-acetylglucosamine by AADGase from β-aspartylglucosamine in AGU and in three separate control (C_1, C_2 and C_3) fibroblasts. 0·1 ml of cell homogenate in $1/15$ M phosphate buffer, pH 7·0, containing 0·1% Triton-X and 0·05 ml of substrate solution were incubated at 37°C up to 100 h. An aliquot of 0·150 ml of the reaction mixture was used for the Morgan-Elson reaction. Enzyme activities were expressed by nmol of the liberated N-acetylaminosugar/mg protein/minute. Absorbance values in the Morgan-Elson reaction are given in the abscissa.

CONCLUSIONS

Our preliminary results on cultured fibroblasts from AGU patients indicate that the cellular pathology, i.e. the defect of AADG cleaving acid hydrolase and the morphologic changes in the lysosomes, which are known to be characteristic of this disease *in vivo*, are also expressed in cultured cells. Altered lysosomes are not pathognomonic for AGU but are seen in a large group of lysosomal storage diseases, such as various mucopolysaccharidoses and other related diseases (Hers and van Hoof, 1966).

The low activity of AADGase in cultured fibroblasts from AGU patients is an interesting difference, when compared with the results obtained by Palo et al. (1972) in liver and brain tissues. It may, of course be simply due to the insensitivity of our methods. The residual activity of the deficient enzyme is usually much higher in other genetic diseases due to a lysosomal enzyme defect. The connective tissue, on the other hand, seems to be severely affected in AGU (see Autio, 1972), a fact which is in good agreement with the absence of AADGase activity in fibroblasts. These preliminary results which are to be confirmed in later experiments offer some interesting aspects for further studies. The most evident practical consequence is the possibility for a prenatal diagnosis of this disease. As a rule, the changes in the enzyme activities present in the fibroblasts of post-natal skin biopsies are also expressed in cell cultures from prenatal cells obtained by amniotic puncture. More data are needed, however, on enzyme activities in the fibroblasts from AGU patients, and particularly from their parents, the obligate heterozygotes,* before the prenatal diagnosis can be put into practice. As mentioned earlier, Pollitt and Jenner (1969) found normal AADGase activity in the sera of the relatives of AGU patients.

Studies on cultured fibroblasts from patients with various types of genetic mucopolysaccharidoses have revealed the presence of 'correcting' factors which are specific for each type of mucopolysaccharidosis (Fratantoni et al., 1968). These factors, derived from the fibroblasts of healthy persons or even from patients with another type of mucopolysaccharidosis, or which can also be isolated from the urine of the respective individuals are capable of normalizing the metabolic defect *in vitro*, when measured by the degree of the accumulation of radioactive sulphate. These studies have clarified the genetic heterogeneity in this group of diseases (McKusick et al., 1972) and also have led to interesting therapeutic trials *in vivo* (DiFerrante et al., 1971).

Further studies with AGU fibroblasts should reveal the presence of similar factors which could correct the metabolic defect in the fibroblast cultures of AGU patients.

ACKNOWLEDGEMENTS

This research has been supported by the Sigrid Jusélius Foundation and by The National Research Council for Medical Sciences, Finland.

REFERENCES

Arstila, A. U., Palo, J., Haltia, M., Riekkinen, P. and Autio, S. (1972). *Acta Neuropathol. (Berlin)* **20**, 207–216.
Autio, S. (1972). *J. Ment. Defic. Res.* Monogr. Ser. I, 1–93.
Baum, H., Dodgson, K. S. and Spencer, B. (1959). *Clin. Chim. Acta* **4**, 453–455.
Danes, B. S. and Bearn, A. G. (1966). *J. Exp. Med.* **123**, 1–16.

* Our recent studies have demonstrated an intermediate activity of AADGase in fibroblast cultures from parents of AGU patients (Aula et al., 1973, *Clin. Genet.* in press).

DiFerrante, N., Nichols, B. L., Donnelly, P. V., Neri, G., Hrgovcic, R. and Berglund, R. K. (1971) *Proc. Natl Acad. Sci USA* **68,** 303–307.
Fratantoni, J. C., Hall, C. W. and Neufeld, E. (1968). *Science* **162,** 570–572.
Hers, H. G. and van Hoof, F. (1969). *In* "Lysosomes in Biology and Pathology" (J. T. Dingle and H. B. Fell, eds.) Vol. 2, pp. 19–40. North-Holland Publ. Comp., Amsterdam.
Jenner, F. A. and Pollitt, R. J. (1967). *Biochem. J.* **103,** 48–49.
Lowry, O. H., Rosebrough, W. J., Farr, A. L. and Randall, R. J. (1951). *J. Biol. Chem.* **193,** 265–275.
Makino, M., Kojima, T. and Yamashina, I. (1966). *Biochem. Biophys. Res. Commun.* **24,** 961–966.
McKusick, V. A., Howell, R. R., Hussels, I. E., Neufeld, E. F. and Stevenson, R. E. (1972). *Lancet* **1,** 903–996.
Palo, J. (1967). *Acta Neurol. Scand.* **43,** 573–579.
Palo, J. and Mattsson, K. (1970). *J. Ment. Defic. Res.* **14,** 168–173.
Palo, J. and Savolainen, H. (1972). *Clin. Chim. Acta* **37,** (In Press.)
Palo, J., Riekkinen, P., Arstila, A. U., Autio, S. and Kivimäki, T. (1972). *Acta Neuropathol. (Berlin).* **20,** 217–224.
Pollitt, R. J. and Jenner, F. A. (1969). *Clin. Chim. Acta* **25,** 413–416.
Reissig, J. L., Strominger, J. L. and Leloir, L. F. (1955). *J. Biol. Chem.,* **217,** 959–966.

V
Inflammation, Repair and Fibrosis

10% heat-inactivated human serum, and supplemented with L-glutamine. Penicillin and streptomycin were added at a concentration of 100 μg/ml, and the medium was changed completely every other day. When cell outgrowth was sufficient, usually at 14–21 days post-explantation, the outgrowth was dispersed with trypsin, 1.0 mg/ml, (Worthington 2× crystallized) in phosphate-buffered saline at pH 7·0. This cell suspension was used to initiate monolayer subcultures with 1·0 × 10^6 cells per serum dilution bottle. Each serum dilution bottle was nourished by 10·0 ml of the standard medium and reached confluent growth in approximately 7 days. Cells were enumerated and sized with a Coulter model B electronic cell counter (Coulter Electronics, Hialeah, Florida).

Preparation of connective tissue activating peptide (CTAP)

We previously described the extraction of CTAP using a thiol-rich neutral buffer (Castor, 1971c). More recently it has proved advantageous to carry out the initial extraction and subsequent gel permeation chromatography under acidic conditions. Human splenic tissue was homogenized at 4°C with 10 volumes of acidic buffer, 0·1 M glycine, pH 2.2, containing 0·1% β-mercaptoethanol. The homogenate was stirred with cold buffer for 4 h, and the particulate material centrifuged at 5°C at 17,000 × g for 10 min. CTAP activity was found in the supernatant portion of this initial extract. The extract was placed on a Sephadex G-25 column, and the void volume emerging from this column was applied to a Sephadex G-50 column. Both column procedures were carried out with the acidic, thiol-rich glycine buffer. A broad band of protein retarded by Sephadex G-50 was accepted for further processing. The dilute retentate from the Sephadex G-50 column was made 70% with respect to ammonium sulfate and allowed to stand at 4°C overnight. The resulting precipitate was harvested by centrifugation at 10,400 × g at 5°C for 10 min. The peptide fraction was dissolved in distilled water containing 0·0005 M cysteine and placed in dialysis at 4°C against 0·0005 M cysteine. Following 24 h of dialysis the CTAP preparation may be assayed for activity. Experience suggests that it is reasonably stable when frozen for several weeks, but tends to lose activity rapidly with lyophilization.

Assay of CTAP

The assay of CTAP was carried out by placing 1·0 × 10^6 normal synovial cells in T-15 flasks in regular media and allowing 4-6 h for the cells to attach and spread on the glass surface of the flask. At this time the serum-containing medium was removed, and 2·0 ml of 'assay medium' was introduced (Eagle's synthetic medium buffered with 0·02 M HEPES buffer at pH 7·6). The material to be tested for biological activity was introduced in a volume ranging from 0·05 to 0·3 ml. The test flasks and their appropriate vehicle controls were incubated at 37°C for 40 h. Cell protein was measured by the method of Oyama and Eagle, (1956) and the medium was analyzed for hyaluronic acid, residual glucose, and lactic acid. Hyaluronic acid was measured in terms of its uronic acid moiety by a modified carbazole procedure (Bitter and Muir, 1962) after isolation of the polymer (Castor et al., 1968). Glucose was measured by a glucose oxidase method (Gibson et al., 1964) and lactate by the Barker-Summerson technique (1941). Earlier studies demonstrated that the amount of hyaluronic acid synthesized by the incubated synovial cells was proportional to the time of incubation and to the amount of CTAP added. We consequently have arbitrarily defined a unit of CTAP in terms of the relation:

$$\text{units CTAP/ml} = \frac{A_1 - A_0}{10 \times V}$$

where A_1 is the hyaluronic acid synthesis rate (μg/mg cell protein/24 h) in the experimental flasks, and A_0 is this measurement in control flasks. The term 'V' represents the volume of test material in ml.

CHARACTERISTICS OF SYNOVIAL CELL ACTIVATION

The multiple effects of CTAP on normal synovial cells

CTAP prepared from leukocytes, spleen, lymph nodes, and fibroblast sources markedly activates connective tissue cells derived from human synovial membrane. Interestingly, CTAP from these sources

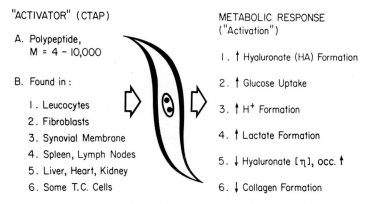

Fig. 1. A schematic representation of the synovial cell 'activation' process. (Reproduced from *J. Lab. Clin. Med.* (Castor, 1973) with the permission of the publishers.)

has very little effect on fibroblasts derived from human skin. On the other hand, skin fibroblasts have been demonstrated to be sensitive to extracts derived from human skin fibroblasts (Castor, 1971c). Fibroblasts from human retrobulbar tissue are moderately sensitive to CTAP from some of these sources (Sisson, 1971).

The multiple effects of CTAP on normal human synovial cells are outlined schematically in Fig. 1. CTAP derived from multiple sources influenced synovial connective tissue cells to produce large quantities of hyaluronic acid, which usually exhibited diminished intrinsic viscosity. Carbohydrate metabolism was promptly altered, with large increases in glucose consumption and lactate formation and a substantial increase in the lactate:pyruvate ratio. Modest activation of hyaluronate synthesis may be demonstrated in a nutrient-free saline medium, but it was clear that a higher order of hyaluronic acid synthesis was supported when simple sugars were present in the saline medium (Castor, 1972a). Glucose, galactose, mannose, fructose and glucosamine all served to support marked incremental formation of hyaluronic acid in the face of CTAP stimulation. Interestingly, when galactose was the sole hexose, the stimulation of hyaluronic acid synthesis occurred without significant increase in the rate of hexose uptake.

Data to date suggest that formation of fibrous and soluble collagen *in vitro* is depressed in the face of active CTAP stimulation of hyaluronate synthesis and glycolysis. The effects of CTAP on synovial cells apparently do not include a characteristic effect on cell growth. CTAP has demonstrated no reproducible effect in acute experiments on the cellular content of acid phosphatase or β-glucuronidase, nor did it induce the extrusion of these enzymes into the medium.

The CTAP content of various types of cells

Studies of the CTAP content of various mammalian cells demonstrated early that most rheumatoid synovial cell strains had significant amounts of CTAP activity. Rheumatoid synovial cells had approximately four times the level of CTAP activity which could be detected in normal cells. In significant measure, this may account for many of the 'abnormalities' of rheumatoid synovial cells which were observed *in vitro*. Not all cells, however, are well endowed with CTAP. For instance, the mouse LM strain seems to have virtually no CTAP activity, while, on the other hand, the human HEp-2 cell strain, derived from a laryngeal epidermoid carcinoma, possesses high levels of this active principle.

BIOLOGICAL SIGNIFICANCE OF CONNECTIVE TISSUE ACTIVATION

CTAP flux in the cotton pellet granuloma

In the studies of cotton pellet granulomas harvested from male Sprague-Dawley rats, it was possible to demonstrate that CTAP concentration in the granuloma varied with the time in the life history of the granuloma, as shown in Fig. 2. CTAP was elevated by the fourth day, reached a peak at day 7, and returned to low levels by day 14.

It was of interest that the hyaluronic acid concentration of the granuloma was also greatest on day 7, while the concentration of sulfated mucopolysaccharides tended to be highest by 14 days. The molecular weight (intrinsic viscosity) of hyaluronate in the early granuloma was significantly greater than that found at 14 days and subsequently. The elevated level of CTAP found early in the inflammatory process in the cotton pellet granuloma is in agreement with its postulated role as a

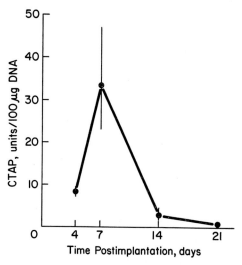

Fig. 2. CTAP flux in cotton pellet granuloma followed for 21 days. Vertical bars at the data points include the range of the measurements. (Reproduced from *J. Lab. Clin. Med.* (Castor, 1973) with the permission of the publishers.)

regulator of the transition from the exudative phase of inflammation to the reparative phase, and the concomitant peak concentration of hyaluronate in the granuloma is in agreement with the known *in vitro* actions of this substance (Castor and Horn, 1972; Castor, 1973).

CTAP in plasma and joint fluid

Studies of human plasma and joint fluid make it clear that CTAP is detectable in these biological fluids (Horn and Castor, 1972). Semiquantitative measurements of CTAP levels in plasma of patients with rheumatic diseases generally gave values that were elevated over those found in a normal human population. Approximately 40% of the CTAP in circulating blood has been detected in the plasma compartment, the remainder being divided between leukocytes and platelets. The evidence suggests that the level of CTAP activity in the plasma of rheumatic subjects is directly related to the activity of the disease process (see Fig. 3). Measurements of CTAP in joint fluids of patients with rheumatoid, traumatic, and osteoarthritic effusions revealed modest levels of activity in the fluid phase of the synovial fluid, with

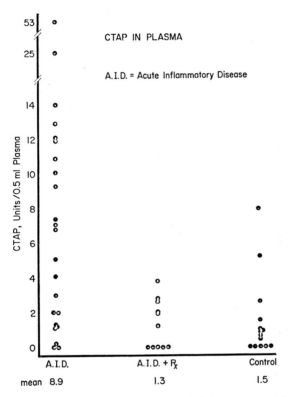

Fig. 3. Acute inflammatory disease (A.I.D.) included patients with rheumatoid arthritis, Reiter's syndrome, and psoriasis, all receiving either no medication or nonsteroidal agents. The 'treated' patients (A.I.D. + Rx) were receiving at least the equivalent of 7·5 mg prednisone per day and/or a cytotoxic agent.

very little activity in the joint fluid white cell population. The appropriate interpretation of the data from human subjects is not yet clear.

Excess CTAP in rheumatoid synovial cells

Those studies showing elevated levels of CTAP in rheumatoid fibroblasts open provocative alternatives for consideration in relation to possible fundamental defects in the connective tissue cells of rheumatoid patients. The retention time of exogenous CTAP added to synovial cells *in vitro* suggests that the half-time of this material is measured in days or hours (Castor, 1971b). It seems unlikely that the high endogenous levels found in rheumatoid synovial cells maintained *in vitro* for many weeks or months is, in fact, retained from mononuclear cells infiltrating the subsynovial stroma or from the polymorphonuclear cells of the joint fluid. If the high endogenous CTAP level found in rheumatoid cells reflects accelerated synthesis or retarded degradation of this substance, it would be important to understand the mechanism that determined this abnormal function. Careful karyologic studies of 9 normal and 9 rheumatoid cell strains failed to provide clear-cut evidence of specific chromosomal defects in rheumatoid cells (Tartof and Castor, unpublished results). While it has been fashionable to indict latent or slow viruses as possible modifiers of rheumatoid synovial cell function, no affirmative data have been forthcoming from either routine or sophisticated virologic studies, or from electron-microscopic examination of the cells (Wynne-Roberts and Castor, 1972). It seems possible that non-viral, non-nucleic acid substances might induce prolonged abnormalities of synovial cell function, bacterial endotoxin being one possible candidate for this role (Buckingham and Castor, 1972). As shown in Table I, extracts of many gram-negative organisms have the capacity to 'activate' normal human synovial cells. This biological activity seems to reside largely in the endotoxins usually associated with these organisms (Table II). The mechanism of this endotoxin activity is incompletely understood, but may depend in part on stimulation of endogenous CTAP synthesis.

MECHANISM OF CTAP ACTION

Relation to energy metabolism

Present data indicate that the energy required to support the activation process may be derived primarily from glycolysis, but that maximal activation of synovial cells requires the participation of oxidative processes (Castor, 1972a). Effective activator peptides stimulated normal human synovial cells under an atmosphere of nitrogen, but maximal stimulation was only accomplished in the presence of an oxygen-containing atmosphere. Uncoupling of oxidative phosphorylation with $4 \times 10^{-5} M$ 2,4-dinitrophenol led to inhibition of cellular activation, as did blockade of glycolysis by $2 \cdot 0 \times 10^{-3} M$ sodium fluoride.

Effects of metabolic inhibitors

Inhibition of DNA synthesis and mitosis by large concentrations (ranging up to 10 μg/ml) of cytosine arabinoside did not interrupt the

Table I

'Activation' by microorganismal extracts

Synovial cell line	Preparation assayed*	Hyaluronic acid synthesis rate†	Ratio treated/control
F.C. Fibroblasts Type: Normal (10^6 cells/assay flask)	Control	8.0	1.0
	Control	8.5	1.0
	E. coli (biliary tract infection)	60.0	7.5
	E. coli (urinary tract infection)	142.0	18.0
	Proteus mirabilis	55.0	7.0
	Pseudomonas aeruginosa type 38	11.5	1.5
	Klebsiella pneumoniae	31.0	3.8
	Corynebacterium acnes	18.0	2.2
	Neisseria gonorrhoeae	14.0	1.7
	Streptococcus (β-hemolytic)	11.5	1.5
	Staphylococcus aureus	9.5	1.1
	Diplococcus pneumoniae	11.0	1.3
	Monilia albicans	9.5	1.3
	Mycoplasma hyorhinis	8.5	1.03
	Mycoplasma hominis II	8.5	1.03

* Extracts prepared by sonication of whole organisms in a Raytheon sonicator (Raytheon Co., Lexington, Mass.). The $600 \times g$ supernate taken as the whole bacterial extract. Sonicates were prepared from an initial concentration of 0.05 g wet organisms/ml of phosphate-buffered saline.
† Micrograms hyaluronic acid/mg cell protein/24 h.
(Reproduced from the *Journal of Clinical Investigation* (Buckingham and Castor, 1972) with the permission of the publisher).

Table II
Effect of purified endotoxin on synovial fibroblast function

Endotoxin	Fibroblast* strain	Endotoxin conc. µg/ml	Endotoxin preparative method‡	Hyaluronic acid synthesis rates† (ratio, treated/control)
S. typhosa 0901	H.H.	40	TCA	7·0
S. typhosa 0901	H.H.	45	PHW	7·2
E. coli 0111:B4	H.H.	90	TCA	2·0
E. coli 0111:B4	H.H.	100	PHW	8·4
S. marcescens	F.C.	13	TCA	5·4
S. marcescens	H.H.	9	TCA	2·8

* Strains all derived from normal synovial tissue, initials refer to donor.
† Initials under preparative method indicate the following: TCA, trichloroacetic acid; PHW, phenol-water. S. typhosa and E. coli endotoxins purchased from Difco Laboratories. S. marcescens endotoxin was a gift of A. G. Johnson, Ph.D, Department of Microbiology, of Michigan Medical School.
‡ Micrograms hyaluronic acid/mg cell protein/24 h.

(Reproduced from the Journal of Clinical Investigation (Buckingham and Castor, 1972) with the permission of the publisher).

activation process. On the other hand, simultaneous administration of CTAP and agents which interfere with transcription or translation led to prompt abolition of the activation process in cultured human synovial cells (Castor, 1972a). Inhibitors of mRNA synthesis which have been effective in blocking connective tissue cell activation included actinomycin D, acridine orange, chromomycin A_3, mithramycin, and α-amanitin. Inhibitors of protein synthesis which block activation included puromycin, cycloheximide, and acetoxycycloheximide. In experiments where actinomycin D was added to cultures either simultaneously with CTAP or at varying time intervals after addition of CTAP, it was learned that inhibition of activation only occurred when the agent was added within 4 h of the activating event. In a similar vein, cycloheximide added to cultures simultaneously with CTAP, or during the first 2 h after addition of the peptide, would essentially abolish the activation process. If, however, the inhibitor of protein synthesis was added to synovial cultures 6 h after CTAP, hyperformation of hyaluronic acid was only suppressed by about 50%. If cycloheximide was added 16 h after the activating event, it had no inhibitory effect.

Effects of antirheumatic drugs

Antirheumatic drugs with the capacity to block activation of synovial cells by CTAP included hydrocortisone, phenylbutazone, indomethacin, acetylsalicylic acid, sodium salicylate, meclofenamic acid, flufenamic acid and mefenamic acid (Castor, 1972b). On the other hand, gold thiomalate, colchicine, chloroquine diphosphate, and hydroxychloroquine sulfate had negligible capacity for blocking the action of CTAP on synovial cells. The ability to inhibit activation did not reside in an effect of the drug on the CTAP molecule itself. The effective concentrations of antirheumatic drugs in the *in vitro* test situation resembled the levels achieved in clinical usage. It was clear that the presence of serum in the test system inhibited the capacity of the drug to block the activation process to a measurable extent (Fig. 4). Of considerable interest was the evidence indicating that the effective antirheumatic drugs were most efficient in blocking synovial cell activation if given simultaneously with the activating agent or within 2 h of the addition of CTAP to cultures. Most antirheumatic drugs given 16 h after the activating event had little effect on the activated cell cultures, suggesting that perhaps a major mechanism of antirheumatic drug action resides in the capacity to inhibit the transcription and translation induced by an effective activating stimulus. On the other hand, as noted below, this time-dependent effect of antirheumatic drugs may reflect an action on either

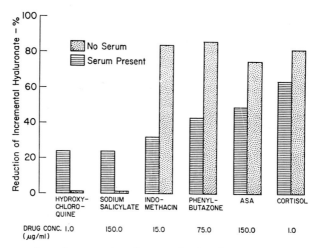

Fig. 4. Suppression of connective tissue activation by antirheumatic drugs; the effect of serum. Hydroxychloroquine and sodium salicylate show little anti-activation capacity, and serum depresses that capacity for the more potent agents. (Reproduced from *Arthritis Rheum.* (Castor, 1972b) with the permission of the publishers.)

CTAP receptor sites or those mechanisms governing intracellular cyclic nucleotide levels.

Role of cyclic adenosine 3′:5′-monophosphate

The addition of dibutyryl 3′:5′-cyclic AMP to synovial cell cultures (12·5–50·0 μg/ml) induced no measurable change in the rate of hyaluronic

Table III

Potentiation of synovial cell activation by dibutyryl cyclic AMP

Cell strain	Control	+ dibut. cAMP (25 µg/ml)	+ CTAP	+ dibut. cAMP + CTAP
		Hyaluronic acid synthesis rate µg HA/mg cell protein/24 h*		
IW-E	10·6	10·1	36·8†	70·8
RS-E	12·9	7·6	29·1†	45·2
RJ-E	3·0	6·3‡	22·4§	113·9

* Values recorded are the means of duplicate or triplicate observations.
† CTAP preparation S-230 from human spleen, 0·1 ml/assay flask.
‡ Dibutyryl cyclic AMP, 50 µg/ml.
§ CTAP preparation from human leukocytes, 0·3 ml/assay flask.

acid synthesis nor in glucose uptake or lactate formation. It was of considerable interest, however, to note that the addition of these small amounts of the acylated cyclic nucleotide markedly potentiated the effect of CTAP added simultaneously to synovial cultures, as illustrated in Table III. While lesser concentrations of dibutyryl cyclic AMP had no direct effect on hyaluronate synthesis, it soon became evident that

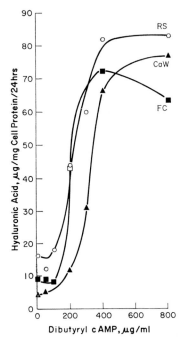

Fig. 5. Three nonrheumatoid human synovial cell strains were stimulated to form large amounts of hyaluronic acid in the presence of high levels of dibutyryl 3':5'-cyclic AMP in the medium.

concentrations in the range from 200–800 μg/ml produced major stimulation of hyaluronate synthesis. Fig. 5 illustrates dose response data derived from 3 nonrheumatoid synovial cell strains. Similar data are available for 5 other normal synovial cell strains. It is of interest that these high concentrations of the cyclic nucleotide will also stimulate hyaluronate synthesis by rheumatoid synovial cells, human skin fibroblasts, and even the 3T6 mouse fibroblast line. The higher levels of dibutyryl cyclic AMP also induce a modest accentuation of glucose uptake and lactate formation. Theophylline, $1 \cdot 0 \times 10^{-3}\ M$, potentiates the effects of both CTAP and lesser concentrations (50 μg/ml) of

Table IV
Theophylline potentiation of CTAP and dibutyryl cyclic AMP

Additives	Glucose uptake, μmoles/mg cell protein/24 h	Lactate output, μmoles/mg cell protein/24 h	Hyaluronate synthesis rate, μg/mg cell protein/24 h
None	1·98 (1·65–2·49)	4·43 (4·18–4·67)	5·8 (5·3–6·7)
Theophylline, 1·0 × 10^{-3} M	0·72 (0·63–0·81)	1·88 (1·65–2·12)	9·4 (9·2–9·5)
CTAP*	3·85 (3·85–3·85)	7·08 (6·96–7·19)	11·4 (10·7–12·1)
CTAP + theophylline, 1·0 × 10^{-3} M	2·49 (2·36–2·62)	4·67 (4·67–4·67)	19·7 (18·2–21·3)
Dibutyryl cAMP, 50 μg/ml	2·08 (1·79–2·37)	4·27 (4·27–4·27)	7·1 (6·4–7·8)
CTAP + dibutyryl cAMP, 50 μg/ml	0·68 (0·50–0·86)	3·08 (2·95–3·20)	16·1 (15·5–16·7)

* CTAP extracted from approximately 1·2 × 10^6 human leukocytes. Normal synovial cell strain AMc was employed as the target strain in this assay. The mean of duplicate or triplicate observations is recorded and the range of the data is included in parentheses.

Table V
Propranolol blockade of synovial cell activation

Additives	Glucose uptake, μmoles/mg cell protein/24 h	Lactate output, μmoles/mg cell protein/24 h	Hyaluronate synthesis rate, μg/mg cell protein/24 h
Vehicle	2·03 (1·42–2·34)	3·60 (2·88–4·80)	3·6 (3·2–3·8)
CTAP	7·91 (7·67–8·33)	14·36 (12·61–17·65)	13·3 (11·4–15·2)
Propranolol, 1·0 × 10⁻⁴ M	3·36 (2·77–3·77)	8·48 (6·42–10·09)	4·0 (3·4–4·7)
CTAP + Propranolol, 1·0 × 10⁻⁴ M	7·41 (7·14–7·54)	14·41 (12·97–16·81)	5·6 (5·1–6·4)

CTAP was extracted from human leukocytes, and normal cell strain WA, 7th passage, was used as the target strain in this assay.

dibutyryl cyclic AMP (Table IV). Since it is known that the simultaneous addition of cycloheximide and CTAP to synovial cultures blocks the activating effect of the peptide, it was of considerable interest to learn that under the same circumstances cycloheximide was unable to block the stimulatory effect of high concentrations of dibutyryl cyclic AMP. Further, hydrocortisone, which also obliterates the cellular response to simultaneously added CTAP, was without inhibitory effect on the stimulation of hyaluronate synthesis by high

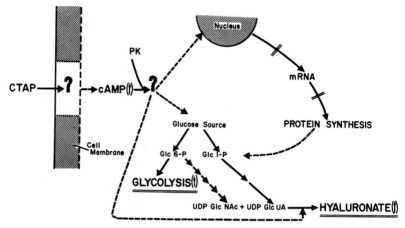

Fig. 6. A speculative schema relating events involved in activation of the human synovial cell.

concentrations of dibutyryl cyclic AMP. On the other hand, the β-adrenergic blocking agent propranolol, $1 \cdot 0 \times 10^{-4}\,M$, led to effective inhibition of incremental hyaluronate synthesis induced by simultaneously added CTAP (Table V).

Our tentative interpretation of the role of cyclic AMP in the connective tissue activation process is summarized in Fig. 6. CTAP may initiate its effects through a synovial membrane receptor site, and it may be that cyclic AMP-mediated reactions are interposed between the receptor site and the initiation of specific transcriptional events. It also seems likely that cyclic AMP modifies post-translational events, perhaps by recognized pathways that regulate the levels of phosphorylated sugar intermediates or, possibly, by modifying the activity of hyaluronate synthetase.

ACKNOWLEDGEMENTS

This investigation was supported by U.S.P.H.S. Grant AM-10728.

REFERENCES

Barker, S. B. and Summerson, W. H. (1941). *J. Biol. Chem.* **138**, 535–554.
Bitter, T. and Muir, H. M. (1962). *Anal. Biochem.* **4**, 330–334.
Buckingham, R. B. and Castor, C. W. (1972). *J. Clin. Invest.* **51**, 1186–1194.
Castor, C. W. (1962). *J. Lab. Clin. Med.* **60**, 788–798.
Castor, C. W. (1965). *J. Lab. Clin. Med.* **65**, 490–499.
Castor, C. W. (1970). *J. Lab. Clin. Med.* **75**, 798–810.
Castor, C. W. (1971a). *J. Lab. Clin. Med.* **77**, 65–75.
Castor, C. W. (1971b). *Arthritis Rheum.* **14**, 55–66.
Castor, C. W. (1971c). *Arthritis Rheum.* **14**, 41–54.
Castor, C. W. (1972a). *J. Lab. Clin. Med.* **79**, 285–301.
Castor, C. W. (1972b). *Arthritis Rheum.* **15**, 504–514.
Castor, C. W. (1973). *J. Lab. Clin. Med.* **81**, 95–104.
Castor, C. W. and Dorstewitz, E. L. (1966). *J. Lab. Clin. Med.* **68**, 300–313.
Castor, C. W. and Horn, J. R. (1972). *Arthritis Rheum.* **15**, 104–105.
Castor, C. W. and Muirden, K. D. (1964). *Lab. Invest.* **13**, 560–574.
Castor, C. W. and Prince, R. K. (1964). *Biochim. Biophys. Acta* **83**, 165–177.
Castor, C. W. and Yaron, M. (1969). *Arthritis Rheum.* **12**, 374–386.
Castor, C. W., Wright, D. and Buckingham, R. B. (1968). *Arthritis Rheum.* **11**, 652–659.
Gibson, Q. H., Swoboda, B. E. P. and Massey, V. (1964). *J. Biol. Chem.* **239**, 3927–3934.
Horn, J. R. and Castor, C. W. (1972). *Ann. Intern. Med.* **76**, 881.
Oyama, V. I. and Eagle, H. (1956). *Proc. Soc. Exp. Biol. Med.* **91**, 305–307.
Sisson, J. C. (1971). *Exp. Eye Res.* **12**, 285–292.
Tartof, D. and Castor, C. W. (Unpublished observations).
Wynne-Roberts, C. R. and Castor, C. W. (1972). *Arthritis Rheum.* **15**, 65–83.
Yaron, M. and Castor, C. W. (1969). *Arthritis Rheum.* **12**, 365–373.

Activation of Connective Tissue Enzymes in Early Wound Healing

J. RAEKALLIO

Department of Forensic Medicine, University of Turku
Turku, Finland

Healing begins at the moment of injury and there is no metabolically inert 'lag', or latent phase during the first days after wounding (Raekallio, 1960, 1965, 1970; McMinn, 1967). One of the earliest phenomena in the healing process is the activation of connective tissue enzymes.

ENZYME HISTOCHEMICAL STUDIES

When the enzymes are demonstrated histochemically, two zones can be seen around the wound soon after injury (Raekallio, 1960, 1965, 1970). In the immediate vicinity of the wound edge, a central or superficial zone, 200–500 μ in depth, shows decreasing enzyme activity in the connective tissue cells. This decrease begins 1–8 h after injury, depending on the enzyme studied (Table I). The decrease in enzyme activity

Table I
Enzyme histochemical changes in early wound healing

Enzyme activity studied	Increase in the peripheral zone (from h)	Decrease in the central zone (from h)
Monoamine oxidase	1	2
β-Glucuronidase	1	2
Non-specific esterases	1	1
Adenosine triphosphatases	1	1
Aminopeptidases	2	2
Acid phosphatases	4	4
Alkaline phosphatases	8	4
Glycosyltransferases	8	8
Succinate dehydrogenase	8	4
Cytochrome oxidase	8	4

is an early sign of imminent necrosis (Raekallio, 1960). In the irreversibly damaged central zone the first signs of necrosis, karyolysis and karyorrhexis, are histologically demonstrable in 8-h wounds, becoming more pronounced 16 h after injury.

Surrounding the central area, a 100–300 μ deep peripheral zone exhibits an increase in enzyme activity (Fig. 1). The enzymes are activated consecutively, beginning from 1–8 h after wounding (Table I).

Fig. 1. Nonspecific esterase activity in an 8-h rat skin wound. The activated fibroblasts and migrant leukocytes form a distinct peripheral zone. (Substrate: naphthol AS acetate).

The initial increase in enzyme activity appears in the local fibroblasts and in the invading leukocytes. Among the accumulating cells, neutrophilic polymorphonuclear cells are most prominent during the first 12–16 h. Thereafter, macrophages become more numerous, and from about the third day onwards, invading fibroblasts appear migrating from the nearby connective tissue (Stearns, 1940; Ross and Benditt, 1961). The immigrating cells exhibit an intense enzyme activity (Lindner, 1962) and they thus contribute to the overall activity, histochemically demonstrable in the wound periphery.

ENZYME BIOCHEMICAL STUDIES

In addition to histochemical demonstration and localization of enzymes, biochemical methods are needed to elucidate further the enzymatic response to injury. One of the problems is the origin of the wound enzymes. They could be derived from blood or from the injured connective tissue itself or from both of these sources.

If the increase in enzyme activity is largely due to blood plasma, its enzyme pattern ought to be qualitatively similar to that in the wound tissue of the same individual. To elucidate this, Raekallio and Mäkinen (1967) made an experimental study on rats. Aminopeptidases were chosen as an example group of enzymes and they were studied in blood and skin tissue before and, at various intervals, after wounding. Control blood was obtained by cutting the tail. Immediately thereafter, two square skin wounds were excised in a dorsal area of the animals. Blood and tissue samples from each animal were processed separately and the results were compared reciprocally. For technical reasons (Hess, 1963), blood aminopeptidases, like other enzymes of identical properties in plasma and in serum (McDonald et al., 1964), are assayed in serum. The modifier characteristics and the fractionation by gel filtration and by ion exchange chromatography showed that the quantitatively increased aminopeptidases in the wounded tissue differed qualitatively from the enzymes appearing in the control serum. The augmented aminopeptidases are thus hardly derived from serum in any great measure.

It has been claimed (Hou-Jensen, 1968) that the increased enzyme activity in wounds is derived solely from the invading leukocytes. Raekallio and Mäkinen (1969) have demonstrated by fractionation that there are qualitative differences between the aminopeptidases of rat blood leukocytes and of wound tissue. Further differences were noticed by studying the substrate specificity and the modifier characteristics. For example, dithiothreitol activated the wound tissue aminopeptidases by up to 500% at a concentration of approx. $10^{-4} M$, but the thiol compound (at concentrations from $10^{-6} M$ to $10^{-3} M$) had no effect on the leukocyte enzymes. According to these results, a considerable part of wound aminopeptidases is derived from other than leukocytic sources.

Further experiments have similarly shown that erythrocytes cannot be an important source of wound aminopeptidases (Raekallio and Mäkinen, 1971). According to the studies reviewed in this chapter, the initial increase in wound enzymes is not derived from serum (plasma) by any significant amount. Furthermore, wound tissue aminopeptidases differ qualitatively in many respects from the leukocytic enzymes. The increased aminopeptidase activity in wounds cannot thus be derived exclusively from the invading leukocytes either, although it is obvious that these cells, histochemically showing an intense activity of aminopeptidases and other enzymes, participate in the enzymatic response to injury. Our histochemical and biochemical results support the view that the increased enzyme activity in the peripheral zone originates not only in the invading leukocytes, but also in the local connective tissue cells. Ultrastructural studies combined with histochemistry are needed to confirm the localization of the enzymes in the various cell types. When light microscopy is used, the increased activity of several hydrolases, transferases, and oxidoreductases is histochemically demonstrable both in the invading leukocytes and in the local connective tissue cells very soon after wounding (Table I).

Several hypotheses have been put forward to explain the significance of enzymes in wounds. As an example, Schmidt (1968) suggests that the function of collagenases and other peptidases in wounds is probably two-fold, one being the removal of fragmented fibers, and the other the assuring of a sufficient amino acid pool for essential protein syntheses. Such reutilization of the products of hydrolysis was observed by Klein and Weiss (1965), who found granulation tissue capable of reincorporating amino acids digested from 'wound collagen' into newly synthesized collagen fibers.

SUMMARY

Soon after injury, two zones can be histochemically demonstrated around wounds. In the immediate vicinity of the wound edge, a superficial zone shows decreasing enzyme activity as an early sign of imminent necrosis. Surrounding this, a peripheral zone, consisting of local connective tissue cells and of invading leukocytes, exhibits an activation of enzymes. Some of the enzymes appear in this zone one hour after wounding, others after 2, 4 or 8 h.

Biochemical studies on aminopeptidases show that the enzymes in wound tissue differ qualitatively from those in blood. The data presented thus support the view that the augmented enzymes in wounds are derived, to a considerable degree, from the injured connective tissue itself.

ACKNOWLEDGEMENTS

Supported by grants from Sigrid Jusélius Foundation, Helsinki, Finland.

REFERENCES

Hess, B. (1963). "Enzymes in blood plasma", p. VII. Academic Press, New York.
Hou-Jensen, K. (1968). *J. Forensic Med.* **15**, 91–105.
Klein, L. and Weiss, P. H. (1965). *Biochem. Biophys. Res. Commun.* **21**, 311–317.
Lindner, J. (1962). *Langenbecks Arch. Chir.* **301**, 39–70.
McDonald, J. K., Reilly, T. J. and Ellis, S. (1964). *Biochem. Biophys. Res. Commun.* **16**, 135–140.
McMinn, R. M. H. (1967). *Int. Rev. Cytol.* **22**, 63–145.
Raekallio, J. (1960). *Nature* **188**, 234–235.
Raekallio, J. (1965). *Exp. mol. Pathol.* **4**, 303–310.
Raekallio, J. (1970). "Enzyme Histochemistry of Wound Healing". Gustav Fischer, Stuttgart.
Raekallio, J. and Mäkinen, P.-L. (1967). *Ann. Med. Exp. Biol. Fenn.* **45**, 224–229.
Raekallio, J. and Mäkinen, P.-L. (1969). *Experientia* **25**, 929–930.
Raekallio, J. and Mäkinen, P.-L. (1971). *Experientia* **27**, 1276–1277.
Ross, R. and Benditt, E. P. (1961). *J. Biophys. Biochem. Cytol.* **11**, 677–700.
Schmidt, A. J. (1968). "Cellular Biology of Vertebrate Regeneration and Repair" p. 203. University of Chicago Press, Chicago.
Stearns, M. L. (1940). *Am. J. Anat.* **66**, 133–176.

Local and Systemic Factors which Affect the Proliferation of Fibroblasts

I. A. SILVER

Department of Pathology, University of Bristol
Bristol, England

Very little is known of the 'trigger' which starts fibroblast proliferation in many situations and most particularly during the process of wound healing. A similar lack of knowledge obtains in respect of the conditions which determine whether a fibroblast will remain quiescent, will become an actively dividing cell or will undertake synthetic activity, either for a short or long term and will then revert to a quiescent state. Some evidence has been accumulated during the past few years on the effects of local physical environment on cell behavior; this has been carried out both *in vitro* (Glinos, 1967; Werrlein and Glinos, 1972) and *in vivo* (Silver, 1969). Much more is known with regard to the factors which act systemically although their mode of action may be equally obscure. This chapter deals with attempts to measure the local environment in which fibroblasts live and to try to correlate environmental conditions with the activity of the cells. Systemic events which radically alter the local environment will also be mentioned. While it is clear that some of the parameters measured are closely correlated with cellular function it is equally clear that others have no such obvious relationship, although they may be involved in some subtle interaction which is as yet not understood.

METHODS

Two systems have been used in these investigations; the first involved the growth of fibroblasts in transparent rabbit ear chambers and the second involved various strains of fibroblast in tissue culture.

Measurements were made with various types of microelectrodes sensitive to PO_2, PCO_2, pK, pNa and pH (Silver, 1966, 1972; Thomas, 1970; Walker, 1971). The PO_2 and PCO_2 electrodes were used in close contact with cell surfaces and at fixed distances from the cells. The pK, pNa and pH electrodes were used similarly, but could also be used for intracellular studies.

Output from the ion sensitive electrodes was recorded through an Analog Devices 311K amplifier and from the PO_2 electrodes through a Times Instrument AAA electrometer amplifier and fed to a Grass six channel polygraph.

In vivo model

Transparent ear chambers made of polymethylpentene and of a design modified from that of Wood *et al.* (1966), were inserted into the ears of half-lop rabbits (Belgian hare x full-lop). By modifying the coverslips over the ear chambers (Silver, 1969) it was possible to insert probes into the chambers and then to seal them again after measurements had been made. These measurements were made both intracellularly (where possible) and extracellularly at various points in the growing tissue. The areas investigated were (*a*) the growing edge of the granulation tissue, (*b*) the synthetic zone where fiber formation was most evident and (*c*) the established fibrosed tissue at the edge of the chamber. The environment of the cells was manipulated (*a*) by altering the local blood supply through pressure on arteries or veins and (*b*) through altering the systemic blood pressure by inducing hypovolaemia. This was carried out by bleeding the animal into a reservoir until the blood pressure was 50 mm Hg where it was maintained by subsequent withdrawals or reinjections of blood as necessary. Conditions were also changed by making the animal breathe various gas mixtures (10%, 20%, 40%, 60% and 100% O_2 in nitrogen).

Local manipulation of the wound environment was also achieved by the introduction of inert foreign bodies (silica particles, nylon particles, carbon particles) and by the introduction of infective organisms (*Pseudomonas* and *Staphylococcus* spp.).

In vitro model

Fibroblasts of human, rabbit, murine and porcine origin have been grown as primary and established cell lines for the investigation of the environment which the cells create around themselves. The effects of collagen antibodies was observed on the cells in an attempt to determine how these materials affected the cells.

Antibody experiments

Antibodies were raised in guinea pigs against soluble rabbit collagen. The serum from the guinea pigs was collected, the γ-globulin isolated and the anticollagen purified and labeled with FITC (fluorescine isothiocyanate). Antibodies against collagen were applied to the ear chamber either directly by removing the cover-slip and flooding the area after which the material was washed off with saline or by close arterial injection through the main artery feeding the ear chamber. The distribution of the fluorescent material was observed through a uv microscope.

Optical measurements

Intracellular redox states of fibroblasts were measured by the fluorescence at 450 nm excited by five and ten micron spots of uv light at 366 nm using a Zeis Ultrapak system. The fluorescence excitation from this signal was measured with a photomultiplier system designed by Chance and Legallais (1959) and indicated the degree of reduction or oxidation of the pyridine nucleotide component of the respiratory chain.

RESULTS

The normal fibroblast environment

In the 'inactive' phase of their life many fibroblasts lie sparsely distributed among fibers in a situation in which the oxygen tension gradients are relatively shallow and gradients in CO_2 are difficult to detect. Most cells had a membrane potential of between 35–60 millivolts which was maintained by a relatively high intracellular potassium concentration and a relatively low intracellular sodium concentration. The intracellular pH was around 6·95. Treatment of frozen sections of fibrous tissue with FITC labeled anti-soluble collagen showed fine fluorescent fibers on the surface of crosslinked collagen bundles which were continuous with a 'basket-work' of fluorescent fibers around individual fibroblasts. This suggested that fibers around the fibroblasts were less well protected by mucopolysaccharide than were the larger fibers in the bundles. The oxygen tension in quiescent fibrous tissue was relatively high in spite of sparse capillary perfusion, and the oxygen uptake as judged by the fall in PO_2 when the circulation was restricted, was relatively low. The oxygen tension in fibrous tissue frequently reached 50 mm Hg or more.

Fibroblasts in wounds

During the repair process in a wound the fibroblasts lived in three distinct zones. (a) the growing edge, (b) the 'synthetic' zone and (c) the crosslinking area. At the growing edge fibroblasts tended to be rounded cells associated with the endothelial buds on the one hand and with the macrophage vanguard on the other. These cells lived in an environment in which the oxygen tension was low and did not appear to be able to divide unless they were close to a perfused capillary bud (Silver, 1969). When a new bud was formed and penetrated through a collection of these rounded cells and a new circulatory path was established, the cells in the immediate vicinity of the new capillary started to divide. It was not clear whether the dividing cells were brought in as satellites of the endothelium or whether they represented cells which had changed from a quiescent to an active state *in situ*. Fibroblasts in this zone although apparently inactive and living in a hostile environment yet seemed to be capable of synthesis of tropocollagen since flooding of the area with fluorescent anticollagen solution rapidly led to the demonstration of a haze of very fine fibrils of fluorescent material surrounding each cell and filling the intercellular space. This material

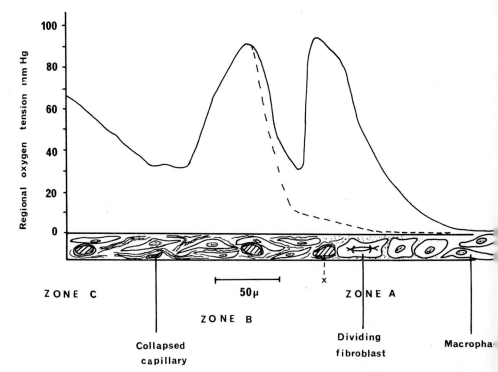

Fig. 1. Diagrammatic cross section of cells in a rabbit ear chamber near the healing wound edge. Above the section is the oxygen tension profile. The dotted line shows the change in profile that occurred when capillary X closed. Dotted area in zone A indicates region of immunologically demonstrable collagenous material. Zone B, zone of thin fiber synthesis; Zone C, zone of well established fibrosis.

appeared to condense on the surface of the capillary buds to form a network around the endothelial cells. The environment at the wound edge varied in oxygen tension from almost zero in the wound deadspace to between 60 and 90 mm Hg at the first perfused capillary (Fig. 1). Most of the dividing cells were found in the zone between 60 and 30 mm Hg. Cells lying in the zone of 30–10 mm were able to synthesise antigenically identifiable collagenous material while cells below the

10 mm level did not appear to produce any extracellular substance. The CO_2 levels in normal well perfused wounds varied from about 60 mm Hg in early wounds in the dead-space to about 40 mm in the tissue. In later, healing wounds the CO_2 levels dropped even in the dead-space. Fibroblasts in the area of the growing edge were in a fragile condition as could be demonstrated by fixation with normal ultrastructural fixatives such as glutaraldehyde. Fixation of cells at the wound edge often showed a considerable number of mitochondria which were swollen or disrupted and great care was necessary to prepare material from this zone.

Measurements from cells in this region show that the intracellular pH was in the order of 6·8–6·7 and although the membrane potential was normally maintained at this pH a very slight increase in hydrogen ion concentration beyond this level was usually associated with failure of the sodium pump.

In zone B the synthetic area, the fibroblasts could be more readily identified as being the typical elongated cells with characteristic morphology. These cells were surrounded by fibers which could be stained at the light microscope level as collagen and which showed typical cross banding at E.M. level. The oxygen tension in this zone showed much more gentle gradients than at the wound edge (Fig. 1), the average level being 45–50 mm Hg with a minimum around 25 mm Hg. Levels of PCO_2 were usually low (about 35–45 mm Hg) and the pH in the cells themselves was about 6·9–6·95. This zone was usually well perfused by arteriolar capillaries and has been described by Remensnyder and Majno (1968) as the 'arterialised zone'.

In zone C the fiber crosslinking was obvious, the cells were more sparse and the capillary network less profuse. The general conditions were those which have been described for normal fibrous tissue above.

Effects of hypovolemia

Reduction in blood volume is followed by immediate homeostatic mechanisms within the body which affect all organs to some extent. The changes which occur are directed towards the maintenance of systemic blood pressure and tissues which are less than vital to bodily survival tend to fare badly in terms of blood flow. In hemorrhagic shock fibrous tissue, muscle, gut and healing wound tissue all suffer from under-perfusion and in extreme conditions the blood flow may cease entirely. This leads to dramatic changes in tissue environment. Fibroblasts in the resting stage appear to be remarkably resistant to severe changes in environment but those which are in a proliferative phase are extremely sensitive to such changes.

Measurements in a rabbit ear chamber in which the wound deadspace had been almost eliminated by new tissue and which therefore contained growing and proliferating cells at the center, cells in the synthetic phase in the middle zone and quiescent cells in a well-formed fiber layer at the periphery showed that many vessels throughout the area were closed down when blood was withdrawn from the animal. This occurred well before there was any fall in systemic blood pressure; if blood volume was lowered until there was a fall in systemic blood pressure, perfusion of the area ceased almost entirely for long periods at a time. This led to a drastic fall in PO_2 which was most marked at the wound edge where conditions approximated to zero (Fig. 1). In the synthetic zone and in the fiber zone there was also a drastic fall in tension but a basic level of PO_2 was usually maintained at between 0·5 and 3·5 mm Hg. Nevertheless, for periods of several minutes at a time there were phases of complete anoxia in both of these other areas. The main difference which could be seen between the effects of this environment on the cells in the different regions was that whereas in the growing zone prolonged hypoxia i.e. hypoxia lasting for more than a few minutes led to rapid changes in pH and ultimately to sodium leakage, in the synthetic zone and in the fiber zone although the pH changed the alteration was much less dramatic and there was little sign of sodium leakage. Electron microscopy revealed that in those areas where sodium leakage was a feature, mitochondrial disruption became obvious very shortly after the hypoxic episode began whereas in those areas where there was little or no sodium leakage, mitochondria retained their normal morphology.

If true hemorrhagic shock was allowed to develop to a non-lethal stage and was then reversed, growth of the wound edge ceased for several days after the insult. Nevertheless fiber formation in the 'synthetic' zone continued so that a modified arrangement of tissue occurred in the chamber with the fibrous zone extending almost to the wound edge five to six days after the hemorrhagic episode. Renewed growth at the wound edge did not take place until about six or seven days after the insult. The changes in PO_2, PCO_2 and pH which occurred in the three zones of the chamber in relation to fibroblasts in different states are set out in Table I.

Correction of hypovolemia with fluids of low colloid osmotic pressure did not result in a return to the conditions which existed prior to bleeding. Although perfusion of the tissue frequently appeared to return to normal as measured by hydrogen clearance rates, the local environment remained hypoxic. The reason for this was difficult to ascertain but it is possible that micro-edema developed preferentially

Table I

Normal values of PO_2, PCO_2 and pH in fibroblasts in three zones of healing tissue in a rabbit ear chamber, and changes during hypovolemia

	Extracellular PO_2 mm Hg		Extracellular PCO_2 mm Hg		Intracellular pH	
	mean	range	mean	range	mean	range
Normal values						
Zone A (wound edge)	23	90–1·5	48	68–40	6·88	7·1–6·7
Zone B (synthetic fiber zone)	48	100–25	41	60–35	6·95	7·2–6·8
Zone C (established zone)	55	85–35	42	60–35	7·0	7·15–6·8
Values after 3 h at B.P. 50 mm Hg						
Zone A	1·2	4·0–0	87	150–60	6·6	6·8–6·4
Zone B	1·5	4·0–0	88	150–60	6·8	7·1–6·6
Zone C	2·7	8·0–0	73	140–58	6·8	7·1–6·6

in the permeable vessels of the new tissue areas and that this significantly increased the oxygen diffusion distance even though adequate perfusion was apparently maintained. Levels of CO_2 on the other hand were greatly reduced when the re-perfusion of the tissues started, in spite of the low PO_2. The features of hypovolemia followed by blood volume expanders is shown in Table II.

Table II

Changes in PO_2, PCO_2 and pH following correction of hypovolemia by injection of physiologically isotonic electrolyte solution of low colloid osmotic pressure

	Extracellular PO_2 mm Hg (mean)	Extracellular PCO_2 mm Hg (mean)	Intracellular pH (mean)
During hypovolemia			
Zone A	1·2	87	6·6
Zone B	1·5	88	6·8
Zone C	2·7	73	6·8
After restoration of blood vol.			
Zone A	1·8	62	6·6
Zone B	2·9	48	6·85
Zone C	68	37	7·1

Effects of anticollagen antibodies

Anticollagen antibodies raised against soluble collagen were applied directly to the tissue in the chamber after removal of the cover-slip. Ultraviolet fluorescence microscopy showed that most of the antibody became fixed around cells near the growing edge of the tissue and to a lesser extent in the synthetic zone. In the fibrous zone only thin fibers became fluorescent whereas the main body of the collagen remained unstained. Small fluorescent fibers could be seen forming networks around individual fibroblasts in the tissue. The effect on fibroblast environment of the application of anticollagen was to produce an Arthus type reaction with inflammation and emigration of leukocytes (Gaunt, 1972). There was thrombosis of the young vessels at the growing edge of the tissue and perivascular white cell cuffing of the capillaries further back from the edge. There did not appear to be any effect on the well-established vessels in the fibrous zone. The effect on the environment of the application of the anticollagen was similar to that seen in developing inflammation from any cause. A difference from the general inflammatory response was the very dramatic fall in PO_2 which occurred in the synthetic zone presumably due to increase in diffusion distance caused by the perivascular packing of white cells, together with the uptake of oxygen by these same cells. Following application of anticollagen, growth ceased for 7–10 days in our chambers and fibroblast disruption was a common feature. Before growth restarted the thrombosed vessels at the edge of the area and the cuffed vessels were repaired and a circulation was re-established.

Ultraviolet spectrofluorimetry

The intensity of fluorescence at 460 nm from mitochondrial and cytosolic NAD/NADH is an indication of the redox state of a cell. To excite this fluorescence spots of 366 nm uv light of 5 μ diameter were shone on to single fibroblasts in different zones of the healing area. The critical external oxygen tension of fibroblasts in respect of NAD reduction at any stage of their activity was found to be between 1·0 and 1·5 mm Hg. Ultraviolet light at 460 nm excites fluorescence at 585 nm from flavoprotein in the mitochondrial compartment of the cell. Measurements at this wavelength indicated a critical external oxygen tension of about 7·5 mm for flavoprotein reduction. From a combination of oxygen electrode and uv fluorimetry measurements with ultramicroscopy it would appear that a reduction of the extracellular environment to below 0·5 mm Hg (40% reduction of NAD) is incompatible with long-term survival of the cell if it is in an actively synthetic or growing state, but is compatible with survival when the cell is in a quiescent phase.

Measurements in monolayer cultures

Fibroblasts in unstirred non-confluent monolayer cultures are surrounded by a microclimate which extends, in the case of oxygen gradient, for about 100 μ from the cell surface. Gradients of CO_2 were impossible to detect with the probes used in this work.

The cell membrane potential was approximately 60 mV and was maintained under conditions of hypoxia for longer than the cell would tolerate the presence of a recording electrode. Prolonged hypoxia (3–4 h) led to a small proportion of cells showing lowered membrane potentials and evidence of sodium leakage into the cell, but the condition was almost always reversible.

Changes of intracellular pH never exceed 0·2 pH units in this series of observations even after 4 h of hypoxia.

DISCUSSION

Local cellular environment is a product of the interaction between the individual cell on the one hand and the activity of other cells in its neighbourhood as opposed to the supply of nutrients on the other. The environment will therefore change either if the cell activity changes without change in nutrient supply or if the nutrient supply changes with or without changes in cellular activity. It is clear that these two factors interact with each other but it is not clear to what extent they control each other. Tissue homeostatic mechanisms tend to maintain a *status quo* by negative feed-back so that if there is increased cellular activity there will be an automatic increase in the nutrient supply and vice versa. We do not know what is the 'set point' of this system and to what extent this set point can vary within what may be called physiological limits. If the set point strays beyond the physiological limit then conditions become pathological. Cellular adaptation to a hostile change in environment is at first reversible, but will eventually become irreversible if environmental conditions are not returned to normal.

In the healing situation it seems that fibroblast growth takes place within a rather narrow range of oxygen tension while synthesis occurs in a slightly different environment, whereas quiescent cells may exist over a whole range of conditions. It is also clear that growing cells cannot survive in a hostile environment which may be only very slightly different from the one in which they are normally found. This may account for the delicacy in balance in the healing process between inactivity, overactivity and normal activity.

Healing tissue in spite of its biological worth is clearly regarded by the body homeostatic mechanisms as dispensable in times of crisis. This does not seem to affect either the synthetic zones or the fiber zone on a long-term basis but it may have a drastic effect on the growing area if the blood supply to this delicate region is reduced to the level at which the local dividing fibroblast population is irreversibly damaged. The exact mechanism of damage is not clear but the present preliminary indications are that a reduction in oxygen supply and carbon dioxide removal is accompanied by an increase in anaerobic respiration with the production and intracellular accumulation of lactic acid, a fall in pH and finally a failure of the sodium pump. When the sodium pump fails the osmotic balance and the ionic balance of the cell cease to be maintained against the external environment there is an increase in intracellular sodium concentration, a decrease in potassium concentration and an increase in intracellular water. This is followed or accompanied by swelling of the mitochondria and disruption of their internal structure and finally disruption of the external mitochondrial membranes. If this damage develops widely within a cell it eventually becomes irreversible. This process, which can be seen in some fibroblasts at the edge of normal healing tissue suggests that many cells living near the dead-space are existing in an environment that is at the limit of their tolerance as well as being on the edge of the tissue. Oxygen supply is a critical factor in this region. It has been shown by Hunt et al. (1967) and by Silver (1969) that fibroblasts in the growing zone will utilise extra oxygen if it is supplied to them artificially. Any small changes which affect the environment under natural conditions are usually deleterious and will push the cells from being precariously in balance to being out of balance. They must then be replaced from elsewhere if the healing process is to survive. It is interesting that fibroblasts in the edge of a wound appear to be very much more susceptible to mitochondrial damage than are the pericytes around the capillaries. The pericytes can be seen at the E.M. level to retain their normal morphology even when they are surrounded by disrupted fibroblasts.

The measurements of intracellular redox state merely reinforce what has been emphasised previously (Chance, 1965) that cells are capable of functioning efficiently at very low oxygen tensions because of the great affinity for oxygen of the enzymes of the respiratory chain. A point of some interest is that in this study fibroblasts showed a critical oxygen tension not only similar to that reported by Hunt et al. (1972) but also in the same range as that found for brain cells or the liver cells (Chance and Silver, unpublished) whereas macrophages, presumably a final product of a similar stem cell to the fibroblast, are able to maintain oxidation of their pyridine nucleotides and structural integrity in very low external oxygen tension where fibroblasts already show signs of disruption. This may of course merely be an indication that the pyridine nucleotides of the macrophages have no substrate and therefore are oxidised because they have nothing with which to be reduced. Macrophages are however remarkably resistant to changes of pH both intracellularly and extracellularly, an adaptation which is clearly of survival benefit in the conditions in which they usually find themselves.

There is something of a paradox in the apparent resistance of fibroblasts to adverse conditions since they can often be cultured from material taken from animals which have been dead for some time and even (perhaps apocryphally) from sausage meat, yet growing fibroblasts in tissues are very susceptible to short-term hypoxia. In the current controversy over the origin of wound fibroblasts this would tend to add weight to the theory that new fibroblasts arise from pre-existing cells in the local tissue rather than that they come from a distant source such as the bone marrow (Büchner et al., 1970; Ross et al., 1970), it also suggests that they must come from cells which are quiescent and which are therefore resistant to insult by hypoxia.

The normal habitat of the fibroblast in healing tissue is an extremely variable one because of the great lability of the micro-circulation. All capillaries have the tendency to open and close in response to systemic demands but the capillaries in newly formed tissue are even more susceptible to these imposed changes than are those which are in well established tissues. For many minutes at a time normal wound tissue may be unperfused, it may then be perfused in one direction followed by a reverse perfusion. These changes cause considerable local alterations of the micro-environment and maybe an important feature in determining in which direction tissue will grow.

The measurements in monolayer culture confirm the observations of Werrlein and Glinos (1972) in respect of oxygen gradients but they are somewhat at variance with the measurements of pH and membrane potential in fibroblasts in healing tissue.

Fibroblasts in culture do not show the more dramatic changes in intracellular pH, and leakage of potassium from the cell, during periods of hypoxia, that appear to be characteristic of wound tissue hypoxia. They also fail to show mitochondrial damage following prolonged hypoxic episodes of several hours. One must therefore conclude that the environment of the fibroblast during the natural healing process is more hostile to the growing cells than are conditions in cultures. Alternatively, it could be argued that those cells which grow in artificial cultures are unrepresentative of wound populations and are 'selected' for survivability in an unnatural situation.

REFERENCES

Büchner, T., Junge-Hülsing, G., Wagner, H., Oberwittler, W. and Hauss, W. H. (1970). *Klin. Wochenschr.* **48,** 867–872.
Chance, B. (1965). *J. gen. Physiol.* **49,** 163–188.
Chance, B. and Legallais, V. (1959). *Rev. Sci. Instrum.* **30,** 732–735.
Gaunt, S. J. (1972). B.Sc. Thesis, University of Bristol.
Glinos, A. D. (1967). *In* "Control of Cellular Growth in the Adult Organism" (H. Teir and T. Rytömaa, eds.), pp. 41–53. Academic Press, London.
Hunt, T. K., Niinikoski, J. and Zederfeldt, B. (1972). *Acta Chir. Scand.* **138,** 109–110.
Hunt. T. K., Twomey, P., Zederfeldt, B. and Dunphy, J. E. (1967). *Amer. J. Surg.* **114,** 302–307.
Remensnyder, J. P. and Majno, G. (1968). *Amer. J. Path.* **52,** 301–323.
Ross, R. Everett, N. B. and Tyler, R. (1970). *J. Cell Biol.* **44,** 645–654.
Silver, I. A. (1966). *In* "Oxygen Measurements in Blood and Tissues" (J. P. Payne and D. W. Hill, eds.) pp. 135–153. Churchill Livingstone, London.
Silver, I. A. (1969). *Prog. Resp. Res.* **3,** 124–135.

Silver, I. A. (1972). *In* "Epidermal Wound Healing" (H. Maibach and D. Rovee, eds.) pp. 291–305. Medical Year Book Publ. Chicago Illinois.
Thomas, R. C. (1970). *J. Physiol. Lond.* **210,** 82P.
Walker, J. L. (1971). *Analyt. Chem.* **43,** 89A.
Werrlein, R. J. and Glinos, A. D. (1972). *In Vitro* **7,** 266.
Wood, S., Lewis, R., Mulholland, J. H. and Knack, J. (1966). *Bull. Johns Hopkins Hosp.* **119,** 1–5.

Probes for the Measurement of the Microenvironment

I. A. SILVER

Department of Pathology, University of Bristol
Bristol, England

Three types of probe were used in the experiments outlined in the previous chapter. These were: (1) The standard KCl-filled glass micropipette for measuring membrane potentials of single cells. (2) The 30% iridium-platinum in glass microelectrode for measuring oxygen tension and which if covered with palladium black can also be used for measuring hydrogen clearance and therefore give an indication of capillary perfusion of an area. (3) The 'Thomas' type glass membrane electrodes for measuring sodium, potassium or hydrogen ion concentration. The last of these if modified can be used as a micro-CO_2 electrode. A fourth type of electrode which is still in the development stage is the micro-glucose probe which utilises the action of glucose oxidase in the production of hydrogen peroxide (Clark and Sachs, 1968) which is subsequently reduced at a platinum anode (Chance, 1950). This probe is dependent on an adequate supply of oxygen for its efficient function.

CONSTRUCTION OF PROBES

Glass micropipettes

Electrodes for measuring membrane potentials were made from standard borosilicate glass, 2 mm outside diameter and shaped with a vertical electrode puller. They were used with tip sizes of 0·1–1·0 micron.

Oxygen microelectrodes

Electrodes for PO_2 were made either to the design of Silver (1965) or by electro-polishing 200 μ diameter 30% iridium in platinum wire in sodium cyanide and sodium hydroxide to a point of about 0·5 μ and then pushing the 'pencil point' so obtained through molten glass which was compatible in thermal expansion with platinum-iridium (Normalglas 16/111, Jena). The outside of the electrode was coated with a water repellant layer (Siliclad, Clay Adams, Inc.) in order to prolong its life and to reduce the possibility of micro-cracking. The tip of the electrode was subsequently covered with a Rhoplex membrane (Rohm-Haas, Philadelphia) to minimise 'poisoning'.

Glass membrane ion-selective electrodes

These were made by the method of Thomas (1970) in which an outer pyrex capillary is drawn on a standard electrode puller to a tip diameter appropriate to the investigation and a second capillary of ion selective glass is then drawn to a size slightly smaller than the pyrex capillary. The tip of the second capillary is sealed, the hollow cone thus formed is cut off and dropped inside the first capillary in such a way that the cone of sensitive glass lodges in the taper of the outer electrode so that its sealed end stops just short of the tip of the outer electrode. The outer electrode is then fused to the inner glass cone by a 25 μ micro-heating wire (20% rhodium-platinum). The body of the electrode is filled with electrolyte by the method of Zeuthen (1971) and a silver/silver chloride electrode is used to make contact. An electrode of this design which has a pH sensitive glass membrane can be transformed into a CO_2 electrode by filling the cavity at the tip with a gel containing 0·001 M sodium bicarbonate. Alternatively the tip can be filled with a solution of the same molarity bicarbonate and the end can be covered with a silicone rubber or Rhoplex membrane (Fig. 1).

Ion selective electrodes may also be made based on the pattern of Walker (1971) in which liquid ion exchangers are introduced into the tips of standard pyrex capillaries after treatment of the tip with Siliclad to make it water repellant. These probes are easy to make if one has a standard electrode puller but they cannot be converted into CO_2 electrodes and neither are they very effective as sodium monitors.

RECORDING

Most types of microelectrode have a rather high impedance and therefore require considerable care in shielding from both a.c. fields and magnetic interference. It is usually necessary to make measurements in a Faraday cage.

Amplification of the electrode signal from pNa and pK electrodes can be satisfactorily undertaken with a 10^{14} Ω input impedance amplifier such as an Analog Devices 311K. pH electrodes (and the derived PCO_2 electrode) can also be used with this type of amplifier, but, because of the high resistance of the pH probes it is better to use an amplifier of higher input impedance such as a vibrating capacitor pH meter.

PO_2 electrodes, of a size suitable for single cell investigations give a current of about 10^{-13} A/mm Hg when polarised by a suitable voltage source. Very great care is necessary when using equipment for measuring such low currents, that stray 'leak' currents and a.c. pick-up do not swamp the O_2 signal.

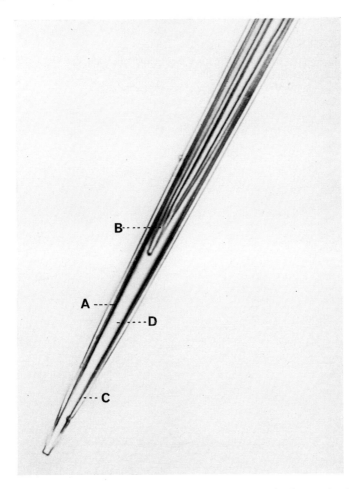

Fig. 1. Photomicrograph of a CO_2 sensitive microelectrode ($\times 1275$). A, pyrex capillary; B, pH glass cone; C, Rhoplex membrane, D, Space filled with 0·001 M $NaHCO_3$. This electrode was made with a large 'dead space' to demonstrate the arrangement of the various parts. For reasonable response times the tip of inner cone must be within 20 μ of the tip of the outer capillary.

CONCLUSION

All these probes are very fragile and must be used in a micromanipulator system. However, they are highly specific and they have very good resolution so that conditions obtaining at the surfaces of individual cells and even at different points on the surface of a cell may be measured. This may be of considerable interest in investigating the sites of actions of enzymes and so forth and also in the investigation of those factors which promote or inhibit fibroblast proliferation.

REFERENCES

Clark, L. C. and Sachs, G. (1968). *Ann. N.Y. Acad. Sci.* **148,** 133–153.
Chance, B. (1950). *Biochem. J.* **46,** 387–402.
Silver, I. A. (1965). *Med. Electron. Biol. Engng.* **3,** 377–387.
Thomas, R. C. (1970). *J. Physiol. Lond.* **210.** 82P.
Walker, J. L. (1971). *Analyt. Chem.* **43,** 89A.
Zeuthen, T. (1971). *Acta Physiol. Scand.* **81,** 141.

The Fibroblast and Inflammation

W. G. Spector

*Department of Pathology, St. Bartholomew's Hospital
Medical College, London, England*

One of the most characteristic features of pathological processes is their duality of effect so that they may act sometimes for the benefit of the host, sometimes for his detriment. Fibroblastic activity in the repair of wounds is a life-preserving mechanism which was essential for vertebrate evolution. Because of the duality referred to however, it is to be predicted that examples should exist in pathology of fibrosis as a harmful process. The following brief discussion will attempt to elaborate this theme and to suggest some general principles underlying inflammatory fibrosis.

Fibroblastic activity in inflammation can be said to occur in any of three situations:

1. acute inflammation,
2. chronic granulomas,
3. cryptogenic fibrosis.

In acute inflammation, fibroblasts proliferate and synthesise collagen and mucopolysaccharides only when there has been tissue destruction. If this is lacking, the inflammatory exudate disappears by simple resolution as in lobar pneumonia where all the destructive activity of bacteria and leucocytes is unleashed in spaces empty of everything except air. Destruction of tissue in acute inflammation occurs as a result of suppuration, tissue loss due to the injury itself or to ischaemia. In all cases however, the resultant organisation follows the same pattern as in wound healing. Here we can report one solid advance and one equally impressive block. The advance lies in the knowledge that the fibroblast is derived from local connective tissue cells and not as has been surmised from the mononuclear cells of the inflammatory exudate (Ross et al., 1970). The block is our lack of information about the mechanisms which activate and quieten fibroblasts in this situation.

In granulomas such as tuberculosis, leprosy, syphilis and rheumatoid arthritis, fibroblasts are taken for granted and generally assumed to represent attempts at healing, abortive or eventually successful. In pulmonary reactions to materials such as silica, it is not as widely appreciated as it should be that the massive nodules of collagen are associated with a diffuse granulomatous reaction. Silicosis because it has been so widely studied can serve as a model for this type of fibroblastic activity. The silica particles after inhalation are taken up by macrophages. Silicic acid is formed at the cell surface by interaction with water molecules. This substance is a powerful hydrogen bonding compound by virtue of being a hydrogen donor. Bonding occurs with the secondary amide groups of proteins and with phospholipids. There is damage to cell and lysosomal membranes and it is postulated that as a result of these interactions and of the hydrolases, cationic proteins etc. so released, collagen synthesis and possibly mitotic activity are stimulated in the adjacent fibroblasts (Allison, 1970).

Ultrastructural study of experimental granulomas produced by injection of BCG or carrageenan into skin, shows that in these lesions composed mainly of macrophages, fibroblasts are present from an early stage and that newly formed collagen fibres are to be found in their vicinity. These appearances are found long before the granuloma becomes in any sense fibrotic if indeed it ever does so. In the lung, BCG or carrageenan introduced into the bronchi is dealt with by an influx of macrophages from the circulation which ingest the irritants and transfer much of it out of the lungs via the trachea. The residue lies within macrophages in localised areas of the lung. Fibroblasts are found in these areas together with the collagen fibres and with the passage of time the proportion of fibroblasts to macrophages increases as does the proportion of collagen to fibroblasts (Velo and Spector, 1973).

Granulomas are of two sorts, of high or low turnover. High turnover lesions, e.g. provoked by BCG, B. pertussis vaccine, Freund adjuvant, have a high daily influx of monocytes and high rate of macrophage mitosis balanced by a high mortality rate amongst macrophages and associated with high toxicity of the irritant to macrophages. Low turnover granulomas due e.g. to carrageenan or carbon have a low daily influx of monocytes, a low mitotic rate and a low death rate amongst the macrophages with low toxicity of the irritant. The macrophages in effect form a long-lived population which is almost self-sufficient (Spector, 1969).

Fibroblastic activity appears earlier, is more obtrusive and proceeds to heavy collagen deposition faster in the high turnover than in the low turnover variety. In some low turnover granulomas, fibrosis may be absent months or years after initiation.

These findings are most simply interpreted by the hypothesis that fibroblastic activity in granulomas occurs when macrophages with phagocytosed irritant lie in proximity to locally derived fibroblasts. The macrophages release substances, probably of lysosomal origin which cause fibroblasts to proliferate and synthesise collagen and mucopolysaccharides. The more rapidly and extensively the macrophages are damaged, the faster this process occurs. Since the presence of irritants seems to hinder final fibrosis it seems likely that macrophages also release substances which interfere with collagen deposition, e.g. collagenase or lysosomal protease. This particular secretory activity could possibly be attributed to epithelioid cells which are macrophage derivatives specialised for secretion rather than phagocytosis and digestion (Papadimitriou and Spector, 1971). It is certainly true that epithelioid granulomata are slow to fibrose. Eventually as all the irritant is removed, digested or neutralised (e.g. by coating with endogenous material) collagen deposition and scarring can proceed uninterrupted to completion as in the healed tuberculous lesion or the non-functional silicotic lung.

Cryptogenic fibrosis includes a number of interesting conditions, some common, some rare. In the former category comes cirrhosis of the liver in which *inter alia* there is extensive fibrosis. The collagen probably originates from proliferation of pre-existing fibroblasts adjoining the sinusoids which are lined by mononuclear phagocytes as well as endothelium (McGee and Patrick, 1972). A preceding inflammation albeit slight, probably occurs before fibroblast activation. In chronic glomerulonephritis a similar situation of inflammation with mononuclear phagocytes and fibroblasts in close proximity exists. Both these diseases are consistent with the hypothesis of macrophage/fibroblast activation outlined above. In the less common condition of fibrosing alveolitis of the lung, collagen deposition by fibroblasts in the alveolar walls is frequently preceded by demonstrable evidence of macrophage accumulation in the same area, so this mysterious disease too may depend upon a similar mechanism. Needless to say in none of these three situations can one do more than speculate as to the cause of the macrophage damage which is postulated as an initiating event. Most interesting of all is diffuse systemic sclerosis where dense collagen deposits occur in the gastro-intestinal tract, skin, lungs and elsewhere. Here the fibrosis is accompanied by atrophy of the parent tissue, e.g. the glandular elements of skin. Early cases often show evidence of vasculitis and it may be that much of the fibrosis is a sequel to slow infarction of the tissue due to gradual arterial obliteration. The associations of the disease make it likely that if vasculitis exists it is of allergic nature. The involvement of so many systems certainly suggests a primary vascular origin of the disease.

This survey has been far from comprehensive but at present there seems little reason to abandon the simple view that pathological fibrosis results either from replacement of dead tissue or from specialised macrophage/fibroblast interaction. Indeed it may be that repair itself is triggered by macrophage involvement so that these two mechanisms could in fact be one.

ACKNOWLEDGEMENTS

Much of the personal research described in this article was financed by the Research Fund of St. Bartholomew's Hospital and by the U.S. Army Research and Development Group (Europe).

REFERENCES

Allison, A. C. (1970). *In* "Mononuclear Phagocytes" (R. van Furth, ed.) p. 422–440. Blackwell, Oxford.
McGee, J. O'D. and Patrick, R. S. (1972). *J. Pathol.*, **106**, p. vi.
Papadimitriou, J. M. and Spector, W. G. (1971). *J. Pathol.* **105**, 187–203.
Ross, R., Everett, N. B. and Tyler, R. (1970). *J. Cell Biol.* **44**, 645–660.
Spector, W. G. (1969). *Int. Rev. Exp. Pathol.* **8**, 1–51.
Velo, G-P. and Spector, W. G. (1973). *J. Pathol.* **109**, 7–19.

The Biological Response to Silica

A. G. HEPPLESTON

Department of Pathology, University of Newcastle upon Tyne
England

The *in vivo* response to the accumulation of crystalline silicon dioxide may be envisaged as having two main components. The initial step is phagocytosis of inhaled particles by alveolar macrophages and their subsequent dissolution. As a result it appears that a non-lipid factor is liberated and that this reacts with fibroblasts to stimulate the formation of hydroxyproline (HOP). At the same time silica evidently stimulates the production of lipids, especially phospholipids, by (a) the alveolar epithelium (type B or type II cells, the granular pneumocytes), and (b) the interaction with macrophages. Phospholipid may then induce the mononuclear phagocytic system to augment the central and local supply of pulmonary macrophages so as to replenish this population of cells as silica takes its toll. The hypothesis attempts to unify what appear to be the principal elements of the silicotic reaction and the evidence on which it rests must now be adduced.

THE MECHANISM OF THE FIBROBLASTIC RESPONSE

In order to separate the processes of phagocytosis and fibrosis resort was made to culture techniques (Heppleston and Styles, 1967a, b). Peritoneal macrophages were allowed to ingest quartz particles in culture and after being in contact for 24 h the cells were disintegrated. Having removed the particles and cell debris by centrifugation, the supernatant was applied to chick embryo fibroblasts grown as independent cultures. After further maintenance of the treated fibroblasts, their DNA and hydroxyproline contents were estimated chemically. Control procedures included the application to fibroblasts of extracts from macrophages incubated without quartz and from macrophages disintegrated before exposure to the dust. The extract from the interaction of quartz and living cells stimulated HOP production but DNA levels were unaffected, suggesting that cell proliferation was not a significant feature but that function was augmented. To produce an active extract, the macrophages had to be viable, since silica in contact with disintegrated macrophages failed to produce an active extract. Moreover, prior treatment of these cells with a suitable concentration of polyvinylpyridine-N-oxide (a compound known to inhibit silicotic fibrosis *in vivo*) abolished the stimulatory response. Titanium dioxide (anatase), which is virtually inert as a fibrogenic agent *in vivo*, was devoid of *in vitro* effect, as were quartz particles or 'soluble' silica applied directly to the fibroblasts. The evidence thus suggests that the quartz-macrophage reaction may be specific.

To attribute the formation of the macrophage factor simply to membrane damage is inadequate, since extracts of non-dusted, disrupted macrophages showed a much reduced or entirely negative effect *in vitro* in our cultures and in the *in vivo* experiments of Webster *et al.* (1967), where extracts of normal alveolar macrophages injected subcutaneously elicited no reaction or merely a minimal granulomatous response. On present evidence it may be suggested that silica in some way reacts with cell constituents, possibly lysosomal, to produce or activate a relatively

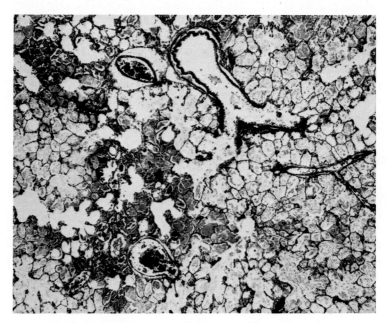

Fig. 1. Rat lung 21 weeks after inhalation exposure to a high concentration of quartz for 12 weeks. Alveoli throughout the lung are occupied by eosinophilic acellular material that varies in density. Alveolar walls are preserved and there is no fibrosis. (H and E. × 40)

soluble substance which is capable of stimulating collagen formation and which may act in low concentration.

Recent observations suggest that this substance is non-lipid in nature. A highly atypical pulmonary reaction was induced in specific pathogen-free (SPF) rats by inhalation of quartz (Heppleston, 1967; Heppleston et al., 1970) and the response was characterised by three main features:

(i) Wide areas of alveolar parenchyma were occupied by material (Fig. 1) which bore a striking histological and histochemical resemblance to that seen in human alveolar lipo-proteinosis. Neutral lipid, phospholipid and mucosubstances were prominent but all enzymatic activity was lacking. The resemblance extended to the ultrastructural features (Heppleston and Young, 1972). Under the electron microscope the alveolar material included numerous extracellular osmiophilic bodies (Fig. 2), whose origin was evidently the type B epithelial

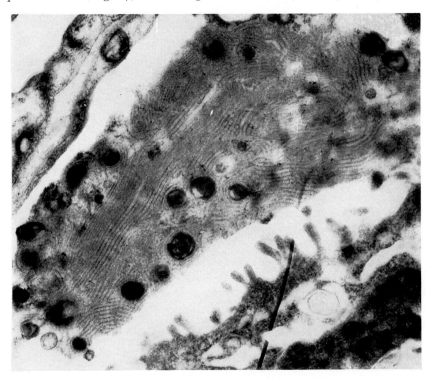

Fig. 2. Electron micrograph of experimental alveolar lipo-proteinosis. Between the attenuated epithelium of a type A cell (upper left) and the microvillous surface of a type B cell (lower right) lies a mass of phospholipid with a stacked lamellar structure. Incorporated into this deposit are many densely osmiophilic bodies, which correspond to the structures formed in type B epithelial cells. ($\times$ 30,000)

cells which were increased in number. In addition osmiophilic lamellae, consisting of phospholipid in a liquid-crystalline phase, commonly occupied the alveoli and assumed parallel or concentric layers, or took the form of a quadratic or hexagonal lattice. In this acellular, lamellar material quartz particles lay widely scattered (Fig. 3).

(*ii*) Chemically, the lungs possessed a high lipid content, but the total protein level was not elevated (Table I), although certain serum proteins were detected in the alveolar material by means of immuno-electrophoresis.

Table I
Experimental alveolar lipo-proteinosis
Chemical analysis of the lungs

	Dry wt. g	Ash g	Lipid g	Protein g	Glycogen mg
Control rats (8)	1·564	0·103	0·298	1·145	1·35
Experimental rats (13)	2·605	0·107	1·605	1·077	2·01

(*iii*) Mature silicotic nodules failed to develop over prolonged periods of observation. The isolation of quartz particles in the phospholipid matrix evidently prevented contact with macrophages and hence interfered with the inception of the fibrogenic mechanism. Moreover, Webster *et al.* (1967) extracted the lipid fraction from macrophages dusted *in vivo* and found it gave no reaction on subcutaneous injection, whereas the non-lipid residue produced a large granuloma.

It thus appears that, on the one hand, a raised lipid content of the lung is associated with inhibition of silicotic fibrogenesis and, on the other, that only the lipid-free material from macrophages that have ingested silica is capable of provoking a granulomatous response. Recent experiments have been designed to ascertain the factors which may determine the genesis of alveolar lipo-proteinosis. To this end rats were exposed to quartz inhalation under a variety of conditions to give an appreciable lung burden and the initial findings permit certain tentative conclusions.

(*i*) The SPF status of the rat is not an important factor, since conventional rats also develop the disease.

(*ii*) Reduction of the fat content of the diet does not affect the response, implying that the altered metabolic state is probably a local pulmonary phenomenon.

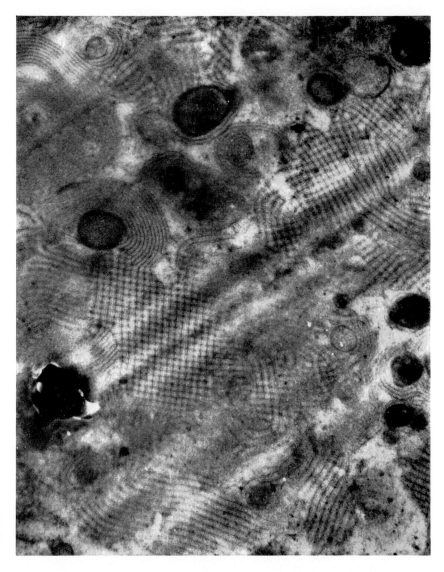

Fig. 3. Electron micrograph showing alveolar phospholipid in a liquid-crystalline phase with the formation of quadratic lattices (possessing four filaments to each square), and concentric and parallel lamellae. Denser amorphous discs may represent lipid in a gel phase. Embedded in the acellular material is a small group of quartz particles (lower left). ($\times$ 30,000)

(*iii*) Inhalation exposure to quartz in an urban as opposed to a rural atmosphere does not prevent the response.

(*iv*) The prime determining factor appears to be the rate at which quartz particles accumulate in the alveolar parenchyma. Whereas exposure to high atmospheric concentrations for most of the day over a period of 2–3 months led solely to lipo-proteinosis in all rats, exposure to low concentrations of quartz for a few hours per day over many months resulted in the development of silicotic nodules (Fig. 4). A lipo-

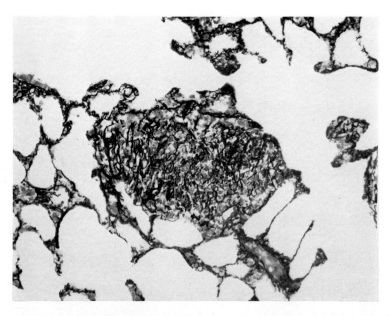

Fig. 4. An early silicotic nodule in a rat exposed to inhalation of quartz in low concentration for 25 weeks. The nodule is situated at the furcation of a respiratory air passage and shows connective tissue fibres (black) among the cellular elements. (Reticulin. × 150)

proteinaceous response also occurred but, initially at least, its severity was much reduced. It thus appears that, under these conditions, quartz and macrophages were able to interact at a sufficient degree of intensity to initiate fibrogenesis.

THE ACCUMULATION OF MACROPHAGES

A valid hypothesis of the mechanism of silicotic fibrogenesis must embrace a second major component, namely the continued production of macrophages. The weight of evidence now indicates that the bone marrow is the universal source of macrophages. However, it is possible that, before entering the alveoli, macrophages reside in the pulmonary interstitium (Bowden et al., 1969). Augmentation of macrophage production may thus be considered at systemic and local levels.

Systemic level

Silicotic lesions in the lung or peritoneum induced a prolonged stimulation of reticuloendothelial activity, as judged by the carbon clearance technique (Conning and Heppleston, 1966). This process may well be accompanied by a general stimulation and mobilisation of phagocytes, such as follows the administration of oestrogens, which also augment the phagocytic index. Sustained systemic stimulation of the bone marrow might thus provide the necessary number of macrophage-type cells or their precursors, whose emigration into silicotic lesions could maintain the local phagocytic population under the destructive action of silica as it is repeatedly released and reingested. Emigration will be facilitated if at the same time degenerate macrophages release a vascular permeability factor.

Pulmonary level

Local proliferation of marrow-derived mesenchymal cells in relation to aggregates of silica particles could also contribute to the prolonged supply of macrophages. In prolonged inhalation experiments, histologically recognisable lung burdens of coal or quartz were achieved in inbred mice and stathmokinetic techniques were applied. With neither dust was an increase of mitotic incidence detected in cells of alveolar walls contiguous with focal dust accumulations as compared with dust-free walls in the same animal (Heppleston and Brightwell, to be published). The early and temporary cellular proliferation in alveolar septa noted by Strecker (1965) with very small quartz burdens may be contributed by mesenchymal cells located in the interstitium at the time of the exposure, and it may be suggested that, after deposition and aggregation of silica particles, the demand is such that few macrophages pause to undergo mitosis in the pulmonary interstitium.

The stimulus to macrophage proliferation

Silicotic lungs are known to contain an excess of lipid (Marks and Marasas, 1960), a feature which was especially evident in lungs showing the lipo-proteinaceous reaction to inhaled silica. The lipids in the experimental condition have recently been characterised (Heppleston et al., 1972). The total lipid rose by a factor of 22—even greater than had earlier been found (Table I)—and to this increase phospholipid contributed much more than neutral lipid. Of the phospholipids, the phosphatidylcholines represented the major constituent, with dipalmityl lecithin as the principal variety, though the phosphatidylethanolamines and sphingomyelins were also elevated. Fatty acid analyses revealed a rise in the proportion of palmitate and an associated depression in the proportion of oleate, evident in both neutral lipid and the phospholipid fractions. Dipalmityl lecithin, believed to be a major component of lung surfactant, probably arises from type B epithelial cells, and its augmented production may represent a basic biochemical reaction of the lung to the irritant action of silica. In collaboration with Dr. K. Fletcher and Mr. I. Wyatt, the metabolic turnover of dipalmityl lecithin has been estimated in this disease, using tritiated palmitic acid (Table II). As compared with unexposed controls, quartz exposed rats

Table II

Incorporation of palmitic acid

	Control	Experimental
Total DPL palmitate (mg)	6·1	243
Total incorporation (% of dose)	0·53	1·70

Dose of tritiated palmitic acid 20 μC/rat; uptake calculated from 60 min values.

showed a threefold increase in the synthetic rate of dipalmityl lecithin but did not eliminate it, presumably because the eliminatory capacity of the lung, whether by bronchial evacuation or by degradation, was overtaxed. The silica-macrophage reaction may also contribute lipids to the lung. Munder et al. (1966) showed that lysolecithin was rapidly liberated from macrophages when they were incubated with tridymite, possibly through activation of a phospholipase A or B which attacked cell components, but titanium dioxide had no such effect. Simple lipids and material containing complex lipids are, moreover, known to be effective stimulants of reticulo-endothelial activity (see Conning and Heppleston, 1966). Despite the evidence against participation of lipid as a fibrogenic agent within silicotic nodules themselves, it nevertheless appears that lipid, when absorbed from or diffusing in the lesions, may possess a systemic and even a local role designed to maintain a supply of macrophages to replace those destroyed under the local action of silica.

This aspect now requires biochemical and stathmokinetic analysis both centrally and peripherally.

Whilst present evidence helps to explain some aspects of the biological response to silica, it is evident that the proximate mechanism of silicotic fibrogenesis has still to be elucidated.

REFERENCES

Bowden, D. H., Adamson, I. Y. R., Grantham, W. G. and Wyatt, J. P. (1969). *Arch. Pathol.* **88,** 540–546.
Conning, D. M. and Heppleston, A. G. (1966). *Brit. J. Exp. Pathol.* **47,** 388–400.
Heppleston, A. G. (1967). *Nature* **213,** 199.
Heppleston, A. G., Fletcher, K. and Wyatt, I. (1972). *Experientia* **28,** 938–939.
Heppleston, A. G. and Styles, J. A. (1967a). *Nature* **214,** 521–522.
Heppleston, A. G. and Styles, J. A. (1967b). *Fortschr. Staublungenforsch.* **2,** 123–128.
Heppleston, A. G., Wright, N. A. and Stewart, J. A. (1970). *J. Pathol.* **101,** 293–307.
Heppleston, A. G. and Young, A. E. (1972). *J. Pathol.* **107,** 107–117.
Marks, G. S. and Marasas, L. W. (1960). *Brit. J. Ind. Med.* **17,** 31–35.
Munder, P. G., Modolell, M., Ferber, E. and Fischer, H. (1966). *Biochem. Z.* **344,** 310–313.
Strecker, F. J. (1965). *Beitr. Silikoseforsch.* S-Bd. **6,** 437–464.
Webster, I., Henderson, C. I., Marasas, L. W. and Keegan, D. J. (1967) *In* "Inhaled Particles and Vapours II" (C. N. Davies, ed.) pp. 111–120. Pergamon Press, Oxford.

Experimental Fibrosis in Liver and Other Organs

J. Lindner and K. Grasedyck

Department of Pathology and I. Med. Clinic
University of Hamburg, Germany

DEVELOPMENT OF FIBROSIS

Fibroses are pathological increases of connective tissue in several tissues and organs, mainly induced by physical-mechanical, chemical, infectious-bacterial and other phlogistic stimuli. The development of fibroses depends on kind and duration of these phlogistic irritations, on their localization and the different reactions of the tissues and organs involved. Experimental fibroses are imitations of human fibroses for the purpose of investigating the basic processes of the disturbed connective tissue metabolism in the development of fibrosis and the therapeutical influences on it. These basic processes are similar but not identical in the fibroses of the different tissues and organs. Fibroses are mainly the consequences of chronic inflammation associated with progressive and more irreversible changes of structure and function of the affected tissues or organs.

EXPERIMENTAL FIBROSIS

One of the models mainly used for experimental fibrosis is disturbed wound healing, especially that of internal wound healing—the experimental granuloma induced by implanted foreign material demonstrated with histological examples of cotton pellet granuloma especially in the final stage of collagen-rich experimental fibrosis (Lindner and Gries, 1961; Gries and Lindner, 1963; Lindner, 1962, 1966, 1972).

Liver fibrosis

As an example of experimental liver fibrosis we show some of our results investigating the disturbed connective tissue metabolism during the development of fibrosis, and not only in the final stage of cirrhosis.

Fig. 1 shows that in this example the body weight of rats (in % of the controls) is markedly reduced within 4 months of thioacetamide (TAA) application (5 mg/100 g body weight/day). This effect is less distinct after treatment with prednisolone (0·2 mg/100 g body weight rat/day) but is pronounced by the combined treatment of these two substances.

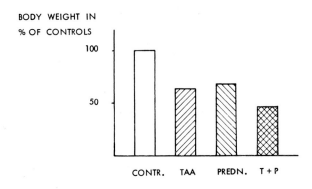

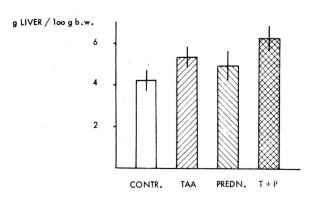

Fig. 1. Influence of treatment with thioacetamide and/or prednisolone on rat liver and body weight (body weight % of controls).

Similar results can be shown after 3, 5 and 6 months application, only partly dependent on the dose. Though the absolute liver weights are nearly constant, the relative liver weight (g liver/100 g body weight) is increased by TAA, prednisolone and especially by TAA plus prednisolone (lower part of Fig. 1), caused mainly by the increased amount of connective tissue and fat.

The summary of the histological results, regarding these materials, shows in Table I that the fat content, as an unspecific sign of the disturbed liver cell metabolism, is increased by prednisolone and/or

Table I
Fat content and increase of connective tissue in experimental liver cirrhosis

	% Fat	Connective tissue			
		±	+	++	+++
Controls	5	100%			
Prednisolone	15	95%	5%		
CCl$_4$ - cirrhosis	35		30%	40%	30%
CCl$_4$ + prednisolone	50		40%	40%	20%
TAA - cirrhosis	35		35%	40%	25%
TAA + prednisolone	50		20%	45%	35%

carbon tetrachloride and thioacetamide poisoning in this kind of experimental fibrosis. Prednisolone increases also the semiquantitatively evaluated histological connective tissue content of the liver as well as in combination with both poisons (Becker et al., 1964; Becker, 1967, 1969; Lindner et al., 1970). Steroids can produce experimental fibroses in several organs (liver, lung, arteries, skeleton muscle, stomach, bone and skin) depending on time and dose (Friberg, 1957; Kowalewski, 1961; Junge-Hülsing, 1963-1965; Asboe-Hansen, 1966; Lindner, 1966, 1969, 1971b, 1972; Hauss et al., 1968; Lindner and Breitenecker, 1968).

Collagen metabolism in liver fibrosis

Biochemical assays (together with Erl and Limbrock) show after collagen fractionation in NSC (0·45 M sodium chloride or neutral salt soluble collagen), ASC (0·5 M acetic acid or acid soluble collagen) and ISC (the insoluble residue) using methods described by Heikkinen (1968) an increase of insoluble collagen (ISC) in 4 and 6 months thioacetamide fibrosis. The soluble collagen fractions (NSC and ASC) and the insoluble fraction (ISC) as well as the total collagen content are diminished by prednisolone in the liver with and without developed TAA liver fibrosis. These results concern the liver only, as demonstrated in the upper part of Fig. 2. The total collagen content and the collagen fractions of other organs, in this case the skin, are mostly unchanged after 4 and 6 months TAA poisoning (further details about the primary co-reactions (= associated reactions) of other connective tissues in this and in other experimental fibroses below). However, after 4 and 6 months poisoning no correlating effect can be seen in the skin regarding the total contents of the collagen fractions. It must be distinguished, therefore, between connective tissue in general and the model of fibrosis in this case, where several mechanisms of regulation may have an

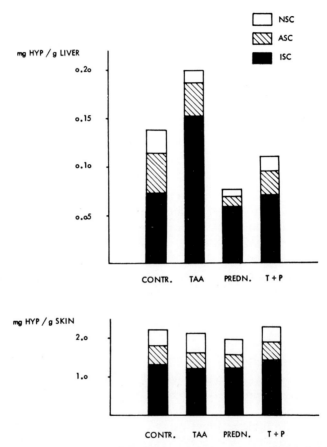

Fig. 2. Influence of thioacetamide and/or prednisolone on 3 collagen fractions of rat skin and liver. NSC, neutral salt soluble collagen; ASC, 0·5 M acetic acid soluble collagen; ISC, insoluble collagen.

influence. Liver cell damage is necessary for starting the connective tissue proliferation. It is stopped when the toxic substance is avoided in the first weeks of poisoning (Varga *et al.*, 1966; Bazin, 1972). Then it may be completely restituted, and this is the only but essential difference from naturally occurring liver fibrosis, where a mechanism of self-perpetuation continues the connective tissue proliferation in the liver even if there is no longer any toxic influence (but also after a given time of poisoning).

As mentioned above, the development of fibrosis with the involved connective tissue metabolic rates depends on several factors, including

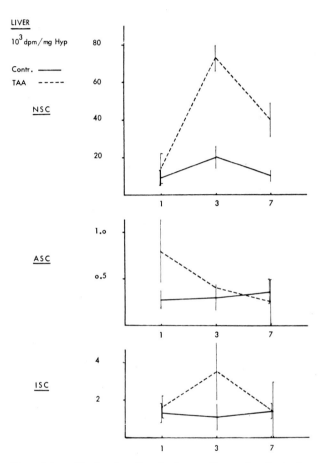

Fig. 3. Specific activity of hydroxyproline in the 3 collagen fractions of experimental liver fibrosis (1, 3 and 7 days after intraperitoneal injection of 0.5 µCi ^{3}H-proline, specific activity 1 mCi/mmole).

the duration of the influence of the fibrogenic factors. The protocollagen proline hydroxylase (PPH) activities increase more than the collagen content (Takeuchi and Prockop, 1969; see below).

Using the estimation of the specific activity of hydroxyproline (after application of labeled proline) as a parameter for collagen synthesis and turnover, we find an increase of collagen synthesis and turnover, especially in the neutral salt soluble collagen fraction of the liver (Fig. 3) (Prockop et al., 1961; Juva and Prockop, 1966). After 6 months we find a decrease of the NSC activity from the 1st to the 7th day after i.p. injection of labeled proline and a slower increase of the activity of isolated hydroxyproline in the ISC fraction. Prednisolone and estrogen decrease the collagen synthesis depending on dose, time and fibrosis stage. Exact investigations of the turnover times require much more material and a longer period of observation. Here we only wanted to state the increased collagen synthesis in our experiment of liver fibrosis, which is expressed by the higher specific hydroxyproline activity after TAA treatment.

By further investigation we can show that there is also an increased degradation. But the results of both processes, synthesis and degradation, are of importance and must be investigated separately. The higher collagen content after TAA application, as shown in Fig. 2, is to be attributed to the predominating synthesis.

Glycosaminoglycan metabolism in liver fibrosis

The same is true for the glycosaminoglycan synthesis, estimated by ^{35}S-sulfate-incorporation rate determination (as routine indicator method) as well as using the assay of the specific activity of the sulfated proteoglycans. These anabolic processes are enhanced in the development of liver fibrosis in animals and humans as shown by several authors (Junge-Hülsing, 1963/1965; Becker et al., 1964; Becker, 1967, 1969; Hauss et al., 1968; Lindner et al., 1970; Lindner, 1971b, 1972). However, in the final stage of liver fibrosis the in vivo ^{35}S-sulfate-incorporation rates are lower in the liver of rats treated by thioacetamide and/or prednisolone compared with untreated controls. Similar results are seen after incorporation in vitro which is especially suitable for investigation of human biopsy material.

Unspecific reactions in experimental fibrosis

The so-called unspecific mesenchymal co-reaction of other organs is not uniform, it is pronounced in the aorta of TAA-treated rats (Junge-Hülsing, 1963/1965; Hauss et al., 1968; Lindner et al., 1970; Lindner, 1971b, 1972). In the final stage of experimental liver fibrosis the proteoglycan pattern can be changed, compared with the normal situation with a relative and absolute increase of dermatan sulfate and heparan sulfate (Becker, 1967).

We have to distinguish between these anabolic and the catabolic processes in the disturbed connective tissue metabolism during the development of liver fibrosis and other experimental fibroses (see above). Together with

Table II

Collagen-peptidase activity in 5-month old rats after 18 weeks thioacetamide and/or 4 weeks penicillamine treatment

	Controls	TAA	TAA + Penicillamine	Penicillamine
Blood serum	1·24 ± 0·48 (6)	2·17 ± 0·36 (6)	1·95 ± 0·24 (6)	2·04 ± 0·20 (5)
Liver	15·14 ± 1·42 (6)	25·17 ± 4·84 (6)	19·32 ± 2·72 (6)	22·06 ± 3·62 (5)
Lungs	84·37 ± 9·79 (6)	116·93 ± 6·87 (5)	109·21 ± 8·49 (6)	104·26 ± 21·61 (5)

The means, standard deviations and number of determinations are indicated.

Prinz, Grade and Kölln we investigated the activity of the so-called collagen peptidases, using the substrate synthetized by Wünsch and Heidrich (1963) and methods developed by Strauch (1968). Table II shows an increase of this enzyme activity in serum, liver and lung of 5-months old rats after 18 weeks TAA and/or 4 weeks penicillamine treatment in comparison to the controls mainly in the supernatant of homogenized and centrifuged tissue. An increased hepatic collagen protease activity after carbon tetrachloride poisoning has been described by Bazin and Delaunay (1964, 1966) and Hirayama et al. (1970, 1971).

Similar to most anabolic and catabolic enzymes the activity of collagen peptidases varies depending on maturation and ageing in the various connective tissues, but not in the serum of animals and humans. However, we find an increased serum activity in humans with severe progressing hepatitis and cirrhosis similar to that in the development of experimental liver fibrosis (Grasedyck et al., 1971).

Regarding the unspecific mesenchymal co-reaction (= associated reaction) it is important to know (1) that the collagen peptidase activity shows a different pattern in the several organs demonstrated in Fig. 4, and (2) that

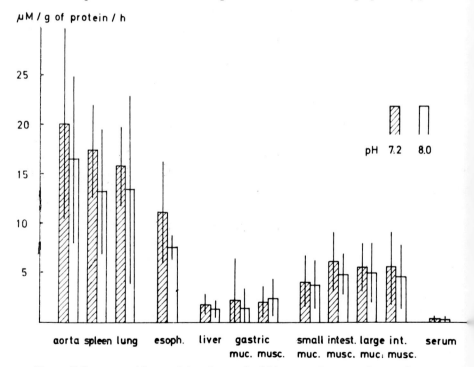

Fig. 4. Collagen peptidase activity of several rabbit organs in comparison to the serum activity: μmole split substrate/g prot./h (PZ-substrate after Wünsch and Heidrich, 1963). (pH-optimum 7·2 and 8·0 for PZ-substrate-cleaving enzyme activity.)

Table III

Protease and β-glucuronidase activity in liver and lungs of 5-month old rats after 18 weeks thioacetamide and/or 4 weeks penicillamine treatment

		Controls	TAA	TAA + Penicillamine	Penicillamine
Protease	liver	245·00 ± 63·82 (26)	619·45 ± 130·52 (12)	301·50 ± 150·18 (12)	182·00 ± 31·2 (14)
	lungs	309·80 ± 89·95 (18)	307·76 ± 178·25 (8)	347·83 ± 79·91 (12)	394·30 ± 54·8 (12)
β-Hlucuronidase	liver	2·11 ± 0·60 (22)	1·83 ± 0·54 (10)	2·44 ± 0·83 (12)	1·76 ± 0·51 (12)
	lungs	0·82 ± 0·34 (18)	2·73 ± 0·84 (8)	2·14 ± 0·62 (8)	2·28 ± 0·24 (10)

The means, standard deviations and number of analyses are indicated.

activity can increase in these co-reactions in the same order, especially in aorta, spleen and lung. The unspecific protease activity is enhanced only in the liver of TAA poisoned rats and is decreased after 4 weeks of penicillamine treatment alone (Table III).

As an example for enzymes involved in the proteoglycan breakdown we show that the β-glucuronidase activity is not significantly changed in the liver at the same time, but this enzyme activity is increased in the lung of rats treated with TAA and/or penicillamine (lower part of Table III). Parallel to this disturbance of the proteoglycan metabolism by penicillamine we find a decrease of ^{35}S-sulfate-incorporation in liver, skin and aorta as did Müller et al. (1971) in cotton granulomas. In contrast to these findings we observed an increase in the lung.

Effect of penicillamine in liver fibrosis

Penicillamine is known to inhibit collagen synthesis (Nimni and Bavetta, 1965; Chvapil et al., 1968; Ruiz-Torres, 1968; Uitto, 1969; Uitto et al., 1970) and favors its solubility by preventing crosslinks (Nimni, 1965; Nimni et al., 1967, 1969; Harris, 1966; Harris et al., 1966; Klein and Garg, 1967; Jaffe et al., 1968 etc.). Penicillamine (100 mg/kg/day) also reduced the relative liver weight, especially after 8 weeks treatment between the 14th and 22nd week of thioacetamide poisoning (Fig. 5). In this final stage of experimental liver fibrosis the 8 weeks penicillamine treatment is without influence on the 3

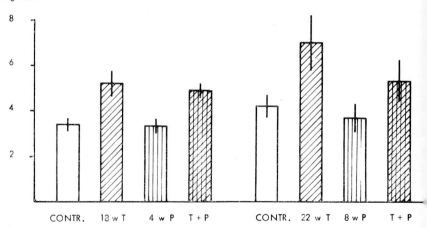

Fig. 5. Examples of various applications of thioacetamide and/or penicillamine on rat liver weight, in relation to body weight. T, thiocetamide; P, penicillamine, w, weeks.

collagen fractions. Like Fig. 2, Fig. 6 shows mg hydroxyproline/g liver dry weight in the three collagen fractions of rat liver. Here again the increase of ISC can be observed after TAA application. Penicillamine produces no alternations compared with untreated controls in this final stage of liver fibrosis. But markedly the 4 weeks penicillamine treatment between the 14th and 18th week of thioacetamide poisoning, and more evidently the penicillamine treatment in earlier stages of this

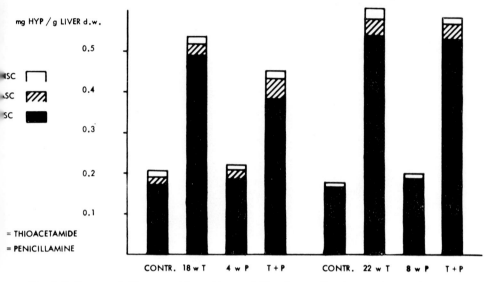

Fig. 6. Influence of thioacetamide and/or penicillamine application on the 3 collagen fractions (NSC, ASC and ISC) of rat liver (in mg hydroxyproline/g liver dry weight).

kind of experimental fibrosis, show the expected increase of the NSC (according to our previous results).

The results in Fig. 6 are interesting findings which we did not expect at all and should be investigated further. Until now we cannot say whether there is an adaptation of the penicillamine effect or whether there is a fatigue in collagen degrading mechanisms in the final stages of liver fibrosis before the development of liver cirrhosis. Further research with collagen turnover studies and enzyme determinations (especially regarding the penicillamine influence on the PPH activity) will give more detailed answers to these questions.

GENERAL ASPECTS OF FIBROSIS

In summary, we see in the development of experimental liver fibrosis, as in humans, by the long-term influence of fibrogenic factors, a progressive disturbance of the connective tissue metabolism with increased anabolic and catabolic processes and turnover rates of cells, proteoglycans and collagen, finally with a decrease of hematogenic and histiogenic cells, with pathologically changed relations between cells and intercellular substances, especially of collagen, which is mainly enhanced in the final stage of liver fibrosis as of other experimental fibroses (as in the first mentioned experimental granuloma).

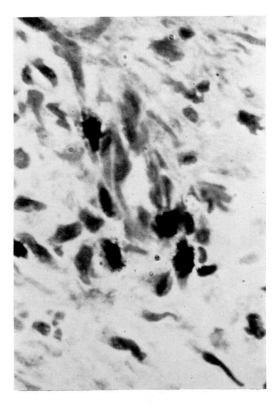

Fig. 7. ^{3}H-thymidine autoradiograph of labeled hematogenic and histiogenic cells in the DNA-synthesis-phase before cell reduplication.

The consequences, changes of structure and function, are shown by one example by Hauss et al. (1968): the collagen enrichment around liver cells with intact mitochondria, glycogen granules and dilated endoplasmic reticulum (unpublished results). Hauss et al. (1968) held the enlargement of the so-called 'transit way' between sinusoids and liver cells by the increasing amount of proteoglycans and, later, collagen responsible for further development of this fibrosis.

Finally, some main points in the development of experimental fibroses:

1. The increase of the number and turnover of connective tissue cells in progressive inflammation is developed by hematogenic and histiogenic cells, shown by ^{3}H-thymidine autoradiography (Fig. 7), with labeled hematogenic monocytes and proliferating pericytes of the vascular wall. Looking for simple routine methods for quantitative

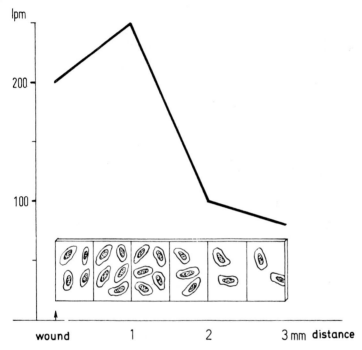

Fig. 8. Schematical demonstration of comparative quantitative autoradiographic and incorporation measurement of histological tissue sections. Decrease of ^{3}H-thymidine incorporation from incision to periphery (7-day old incision wound of rat skin). (For further details see text and Lindner, 1972.)

results, in an example of experimental post-traumatic inflammation and fibrosis (Fig. 8), there is shown the decrease of ^{3}H-thymidine incorporation from the wound center to the periphery measured in the order of magnitude of tissue sections. This corresponds to the results of the usually estimated thymidine-indices (which are comprised below) see also legend to Fig. 8).

2. Regarding the feature of fibroblasts in the final stages of experimental fibrosis, Fig. 9a, left part, shows an example of synovial fibrosis

with increased collagen content. In this final stage of experimental fibrosis, Lindner (1968) found, together with Überberg, cells with an increase of the intracellular filamentous fibril content combined with a reduction to the cytoplasmic organelles (Fig. 9b, right). The question arises whether these typical pictures of fibrosis are dependent on disturbances of the protocollagen proline hydroxylase (PPH), of the collagen extrusion, or on the synthesis of noncollagen protein, perhaps abnormal ones as in ageing (Lindner, 1968, 1969, 1971a and b, 1972).

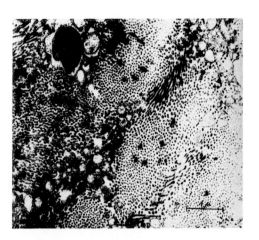
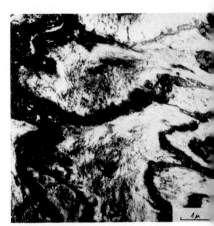

Fig. 9. (a) Increased collagen synthesis and content in later stages of hyperergic knee joint synovitis, arthritis and the fibrosis of synovial tissue. (b) Increase of intracytoplasmatic, filamentous material in synovial cells (with simultaneous decrease of cell organelles; so-called 'Verfaserung' of synovial cells with following cell death. (For further details see text and Lindner, 1971b.)

Furthermore the relationship between these cells and the ageing smooth muscle cells must be discussed in comparison with the results on the so-called fibromuscular cells of the Dupuytren contraction shown by Gabbiani (p.139) and demonstrated by Haust and More (1960), Wissler (1968), Ross (1970) and others on arterial connective tissue, especially regarding the possibility that in synovial fibrosis basal membranes and elastin-like materials are synthesized by these cells also (for a review see Lindner, 1969, 1971b, 1972).

3. We have to remember that in experimental fibrosis of one tissue, here of the synovialis of the classic hyperergic arthritis (Klinge, 1933), there exist unspecific mesenchymal co-reactions ($=$ associated reactions) of other connective tissues as shown by an example of the ^{35}S-sulfate-incorporation indicator method. The incorporation rates are enhanced

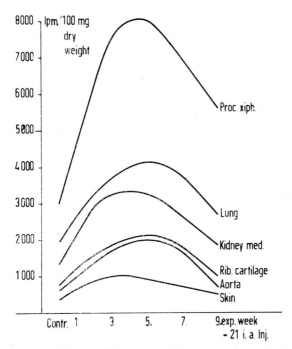

Fig. 10. Demonstration of the so-called non-specific, associated reaction in connective tissues of non-thymectomized rabbits with hyperergic arthritis (course of ^{35}S-sulfate-incorporation as indicator method).

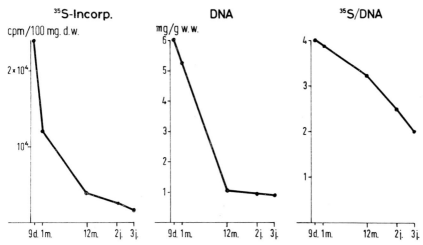

Fig. 11. Ageing-dependent decrease of ^{35}S-sulfate-incorporation (left) in relation to dry weight, of the DNA content in relation to wet weight (middle), and course of ^{35}S-sulfate-incorporation in relation to the DNA (cell) content in ageing vascular connective tissue (rat aorta). (For details see Lindner, 1971b.)

in several organs during this time (until the 9th week) mostly without decrease to normal values (See Fig. 11) (Junge-Hülsing, 1963/1965; Hauss et al., 1968; Lindner, 1966, 1968, 1972).

4. As mentioned above there exist some similarities between fibrosis and ageing especially in some muscle organs. Therefore ageing is mainly regarded as fibrosis and is also used as a model for it. Regarding the ageing dependency of one metabolic rate (the routine indicator method) we found a decreased ^{35}S-sulfate-incorporation rate of rat aorta during maturation and ageing (in terms of dry weight) (Fig. 11). The DNA content decreases, especially during maturation, and also the ^{35}S-sulfate-incorporation related to the DNA content. Thus in ageing as in fibrosis the decrease of cell number is evident (Lindner, 1969, 1971a, b, 1972).

Finally regarding the analogy of both processes, Fig. 12 recapitulates the findings of the literature concerning the several parameters of the disturbed arterial connective tissue metabolism in ageing: decrease of synthesis and breakdown of proteoglycans and collagen, decreased contents of the most proteoglycan fractions and the soluble collagen fractions, increase of the insoluble collagen content, decreased turnover rates and increased half-life times of the proteoglycan and the collagen fractions.

This increase of the total collagen content in the ageing vascular connective tissue is in favour of the insoluble collagen fraction, whereas the total amount of the two soluble collagen fractions decreases as early as during maturation and more intensely during ageing of the vascular wall as in fibrosis of other connective tissues. As to the degradation of collagen during ageing of vascular connective tissue the following facts emerged. Collagenolytic enzyme activities in the aorta decrease depending on age, most intensely during maturation (Lindner, 1971b). Since the enzyme activities have not yet been expressed in terms of the cell content, the question remains unsolved as to whether the total collagen increase of the vascular wall depending on ageing is achieved by augmented synthesis with constant degradation, or, as seems more probable, by an essentially constant synthesis and decreased degradation.

The decrease of the collagen degradation caused by ageing is essentially determined by the relative and absolute increase of insoluble collagen, the degradation capacity of which decreases with age also in the vascular wall because of an increased consolidation of collagen by crosslinkages. That is also true for any kind of experimental fibrosis.

Thus all these changes of metabolic rates in vascular ageing can also be observed in final stages of experimental fibrosis (after the previous results). But these parameters must be further investigated separately

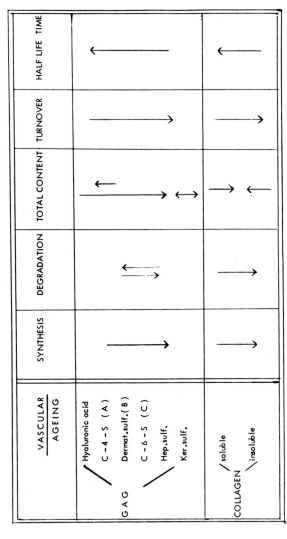

Fig. 12. Schematical summary of ageing-dependent changes of metabolic rates of glycosaminoglycan and collagen fractions with ageing of vascular connective tissue. (For further details see text and Lindner, 1971b.)

in the development of experimental fibroses for better understanding of the whole process and for a perhaps possible therapy which should be at least one purpose of experimental fibrosis investigations.

SUMMARY

Fibroses are characterized by pathologically augmented connective tissue caused by several facts. In an example for experimental fibroses, in thioacetamide (TAA) induced liver fibrosis a significant increase of collagen could be observed, especially in the insoluble fraction. This result is due to an accelerated collagen synthesis as shown by incorporation of labeled proline. An increased activity of collagen peptidases could also be stated. In parallel, the proteoglycan synthesis investigated by labeled sulfate incorporation (as indicator method) is activated by TAA. There is no alteration in the β-glucuronidase activity, whilst unspecific proteases are enhanced.

The pathological connective tissue formation may be inhibited to a certain degree by penicillamine. Generally the number and turnover of connective tissue cells is increased in experimental as in other types of fibrosis, accompanied by ultrastructural alterations in these cells. In addition to the local fibrotic effects unspecific reactions can be shown in other connective tissues of the same animals. Finally the similarities between fibrosis and ageing are pointed out.

REFERENCES

Asboe-Hansen, G. (1966). "Hormones and Connective Tissue". Munksgaard, Copenhagen.
Bazin, S. and Delaunay, A. (1964). *Ann. Inst. Pasteur (Paris)* **106**, 543–552.
Bazin, S. and Delaunay, A. (1966). *In* "Biochimie et Physiologie du Tissu Conjonctif" (P. Comte, ed.) pp. 569–578. Imprim. du Sud-Est, Lyon.
Bazin, S. (1972). (Personal communication.)
Becker, K. (1967). *Zschr. exp. Med.* **144**, 222–235.
Becker, K. (1969). *Zschr. exp. Med.* **151**, 10–17.
Becker, K., Szarvas, F. and Lindner, J. (1964). *Med. Welt* 1622–1625.
Chvapil, M., Hurych, J. and Ehrlichová, E. (1968). *Hoppe-Seyler's Z. Physiol. Chem.* **349**, 218–222.
Friberg, O. (1957). *Ann. Med. Exp. Biol. Fenn.* **35**, Suppl. **8**, 1–70.
Grasedyck, K., Ropohl, D., Szarvas, F. and Lindner, J. (1971). *Klin. Wschr.* **49**, 163–164.
Gries, G. and Lindner, J. (1963). *Arch. exper. Path. Pharmakol.* **245**, 87.
Harris, E. J. (1966). *J. Clin. Invest.* **45**, 1020.
Harris, E., Jaffe, I. and Sjoerdsma, A. (1966). *Arthritis Rheum.* **9**, 509.
Hauss, W. H., Junge-Hülsing, G. and Gerlach, U. (1968). "Die unspezifische Mesenchymreaktion. Zur Pathogenese der reaktiven Mesenchymerkrankungen". Thieme, Stuttgart.
Haust, M. D. and More, R. H. (1960). *Am. J. Path.* **37**, 377–389.

Heikkinen, E. (1968). *Acta Physiol. Scand. Suppl.* **317**, 1–69.
Hirayama, C., Morotomi, I. and Hiroshige, K. (1970). *Biochem. J.* **118**, 229–232.
Hirayama, C., Morotomi, I. and Hiroshige, K. (1971). *Experientia* **27**, 893–894.
Jaffe, I. A., Merriman, P. and Jacobus, D. (1968). *Science* **161**, 1016–1017.
Junge-Hülsing, G. (1965). "Untersuchungen zur Pathophysiologie des Bindegewebes". Habil. Schrift Münster 1963, Theoret. und Klin. Med. in Einzeldarstellungen Bd. 24. Hüthig, Heidelberg.
Juva, K. and Prockop, D. J. (1966). *Analyt. Biochem.* **15**, 77–83.
Klein, L. and Garg, B. D. (1967). *Arthritis Rheum.* **10**, 292.
Klinge, F. (1933). *Ergebn. allg. Path. Anat.* **27**, 1–336.
Kowalewski, K. (1961). *Acta Endocrinol.* **38**, 427–431.
Lindner, J. (1962). *Langenbecks Arch. klin. Chir.* **301**, 39–70.
Lindner, J. (1966). *Folia Histochem. Cytochem.* **4**, 21–46.
Lindner, J. (1968). *Verh. Dtsch. Ges. Inn. Med.* **74**, 1315–1349.
Lindner, J. (1969). *In* "Atherosclerosis" (F. G. Schettler and G. S. Boyd, eds.) pp. 73–140. Elsevier, Amsterdam.
Lindner, J. (1971a). *In* "Molekulare und zelluläre Aspekte des Alterns" (D. Platt and H. G. Lasch, eds.) pp. 51–62. Schattauer, Stuttgart-New York.
Lindner, J. (1971b). *In* "Handbuch der Allgemeinen Pathologie", Vol. VI/4, pp. 245–368. Springer, Berlin, Heidelberg and New York.
Lindner, J. (1972). *In* "Handbuch der plastischen Chirurgie" (E. Gohrbrandt, J. Gabka and A. Berndorfer, eds.), Vol. I. pp. 1–153. De Gruyter, Berlin.
Lindner, J. and Breitenecker, G. (1968). *Therapiewoche* **50**, 2239 and **52**, 2357–2362.
Lindner, J. and Gries, G. (1961). *10th Congr. Lega Internat. Rheum.* Vol. II, pp. 1362–1366.
Lindner, J., Grasedyck, K., Ropohl, D., Szarvas, F., Erl, D. and Limbrock, G. (1970). *Verh. Dtsch. Ges. Path.* **54**, 688.
Müller, U. S., Wagner, H., Wirth, W., Junge-Hülsing, G. and Hauss, W. H. (1971). *Arzneimittelforsch.* **21**, 679–683.
Nimni, M. E. (1965). *Biochim. Biophys. Acta* **111**, 576–579.
Nimni, M. E. and Bavetta, L. A. (1965). *Science* **150**, 905–907.
Nimni, M. E., Gerth, N. and Bavetta, L. A. (1967). *Nature* **213**, 921–922.
Nimni, M. E., Deshmukh, K., Gerth, N. and Bavetta, L. A. (1969). *Biochem. Pharmacol.* **18**, 707–714.
Prockop, D. J., Udenfriend, S. and Lindstedt, S. (1961). *J. Biol. Chem.* **236**, 1395–1398.
Ross, R. (1970). *J. Cell. Biol.* **47**, 175–191.
Ruiz-Torres, A. (1968). *Arzneimittelforsch.* **18**, 594–597.
Strauch, L. (1968). *Mitteilungen Max-Planck-Ges.* **1**, 40–56.
Takeuchi, T. and Prockop, D. J. (1969). *Gastroenterology* **56**, 744–750.
Uitto, J. (1969). *Biochim. Biophys. Acta* **194**, 498–503.
Uitto, J., Helin, P., Rasmussen, O. and Lorenzen, I. (1970). *Ann. Clin. Res.* **2**, 228–234.
Varga, F., Méhes, G. and Molnár, Z. (1966). *Acta Physiol. Acad. Sci. Hung.* **29**, 69–74.
Wissler, R. W. (1968). *J. Atheroscler. Res.* **8**, 201–213.
Wünsch, E. and Heidrich, H. -G. (1963). *Physiol. Chem.* **332**, 300–304.

Role of the Fibroblast in Controlling Rate and Extent of Repair in Wounds of Various Tissues

WALTON VAN WINKLE, JR., J. CHRISTOPHER HASTINGS, ELLEN BARKER and WALTRAUD WOHLAND

Division of Surgical Biology, Department of Surgery University of Arizona Medical Center, Tucson, Arizona 85724, USA

Some years ago Howes *et al.* (1939) suggested that all wounds were repaired at the same rate regardless of the type of tissue wounded. They believed that the strength of the healing wound at any time post-wounding would be the same regardless of location, but the extent of repair, expressed in terms of wound strength as a percentage of tissue strength, would depend solely on the absolute strength of the unwounded tissue. This was perhaps the first indication that there might exist a special 're-pair fibroblast'.

If there is a fibroblast concerned solely with repair, as distinct from growth and maintenance, then all wounds would be expected to heal at the same rate regardless of the characteristics of the tissue in which the wound is made. On the other hand, if the fibroblast concerned with repair is similar in activity to that concerned with growth and maintenance then wounds in different tissues might be expected to heal at different rates, perhaps at rates related to normal metabolic turnover of connective tissue in the wounded organ.

The fibroblast is the principal architect involved in construction of repair tissue. Fibroblastic activity in wounded tissue has most frequently been assayed by measuring physical properties of the collagen deposited by this cell. Thus, many studies use breaking strength to measure progress of repair. More recently, it has been realized that extracellular events such as crosslinking, remodelling, and collagenolytic activity affect the physical properties of wounds (Madden and Peacock, 1968, 1971). We believe that to investigate wound healing adequately, one must measure as many parameters of synthesis, crosslinking and degradation of wound tissue as possible.

Because scar tissue does not resemble normal connective tissue either structurally or functionally, we decided to test the hypothesis that the repair process proceeds at the same rate in all tissues. The following experiments were performed.

METHODS

Sixty-six adult dogs of both sexes were used. Under sterile operating room conditions, utilizing pentobarbital anesthesia and assisted respiration, a midline incision from xiphoid to pubis was made. A 12-cm incision was made along the anterior greater curvature of the stomach and then closed in two layers with continuous sutures. In the bladder wall an anterior and posterior incision was made extending from just below the dome to just above the trigone. These wounds were also closed in two layers with continuous sutures. The colon was isolated and a 12-cm longitudinal incision was made on the antimesenteric border which in most dogs extended from a point about 6 cm distal to the ileo-colic junction to a point about 6 cm proximal to the rectum. This wound was closed in the same manner as the stomach wound. The abdominal incision was closed with figure of eight sutures, and the skin closed with either interrupted or continuous subcuticular sutures. The animals were fed a soft diet for the first 5 days post-operatively and then returned to their normal ration of dog food. Six dogs were lost post-operatively.

Dogs were killed at 5, 14, 21, 28, 70 and 120 days post-operatively. At sacrifice, the dogs were anesthetized and the cardio-pulmonary system supported to maintain tissue perfusion during the time of wound sampling. Skin wounds were excised *in toto* leaving a 1·5-cm margin on both sides. A strip of normal skin parallel to the wound and 3 cm wide was excised on the left side. The abdomen was opened maintaining careful hemostasis and the stomach removed and opened along the lesser curvature. The bladder was removed and opened with an incision midway between the two wounds. The entire colon was removed from cecum to rectum and opened along the mesenteric border.

Samples for biochemical and histological analysis were taken as follows. Two samples, each weighing approximately 0·5 g, were cut from each wound for biochemical analysis and two adjacent samples placed in 10% formalin for histological examination. Samples of normal unwounded tissue were removed for biochemical analysis, exercising care to take these samples from the same anatomical location in each dog. The samples for biochemical analysis were minced with scissors and placed in cold incubation media as described by Uitto (1970). Five μCuries of ^{14}C-proline was added to each sample and the tissues were incubated for 4 h in a shaking water bath at 37°C. The separation of ^{14}C-hydroxyproline from ^{14}C-proline was performed by the method of Juva and Prockop (1966). The specific activity of ^{14}C-hydroxyproline was used to measure rate of collagen synthesis.

Samples for measuring physical properties of the wound were taken and treated as follows. The remaining wounded and normal tissues were cut with a double-bladed knife into strips 0·5 cm wide and breaking strengths measured on a Model TM Instron using a 'C-Cell' with ranges of 0–500 to 0–25,000 g. Approximately 10–14 determinations were made on each skin, stomach, and colon wound and 4–6 determinations on each bladder wound. The same number of breaking strength determinations were made on unwounded tissue. Strips of normal colon were cut proximally and distally to the wounded colon and recorded separately. Likewise, the strength of unwounded skin from upper and lower abdomen was measured separately.

The histologic specimens were cut at 5 μ and stained with hematoxyilin and eosin and Masson's trichrome stains.

RESULTS

All wounds showed approximately the same absolute gain in breaking strength during the first two weeks post-wounding (Fig. 1).

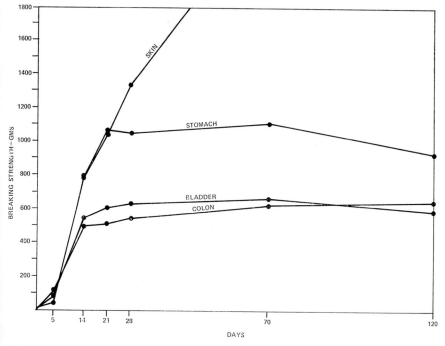

Fig. 1. Breaking strengths of 5-mm wound strips of skin, stomach, bladder and colon of dogs.

A sharp decrease in rate of strength gain was seen at 14–21 days in bladder and colon wounds. A similar sharp decrease in rate of strength gain was observed in stomach wounds after 21 days. Skin wounds, on the other hand, continued to gain strength at approximately the same rate throughout the 120 day observation period.

The breaking strength of unwounded colon was observed to increase sharply from proximal to distal ends. Since the wound extended into both the proximal and distal segments, the wound strengths were measured separately in these two segments. No significant difference in wound strength was observed between wounds in the distal and proximal colon (Fig. 2).

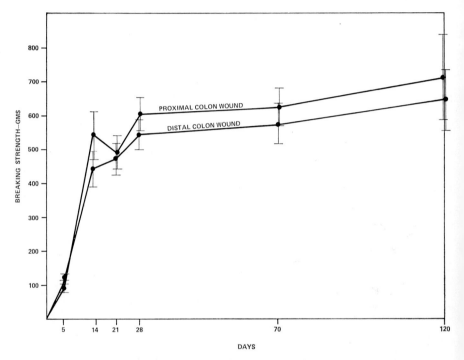

Fig. 2. Breaking strengths of 5-mm wound strips of the proximal and distal colon of dogs. The vertical lines at each point represent the standard error of the mean. The breaking strength of normal unwounded proximal colon is 845 ± 45 g and of the distal colon, 1609 ± 91 g.

Figs. 3–6 show rate of collagen synthesis expressed as specific activity of ^{14}C-hydroxyproline (per μg of total hydroxyproline) in wounds. The dotted lines show wound breaking strength. The dotted horizontal line shows breaking strength of unwounded tissue and the solid horizontal line represents rate of collagen synthesis in normal unwounded tissue.

All wounds showed a significantly increased rate of collagen synthesis, compared to unwounded tissue, by the 5th postoperative day. Although the shallow drop in the rate at 14 days may be artifactual, the obvious peak synthetic activity occurred at 21–28 days and dropped thereafter.

In skin wounds, 70–120 days old, rate of collagen synthesis was not significantly different from normal unwounded skin, yet the breaking strength of these wounds continued to increase (Fig. 3).

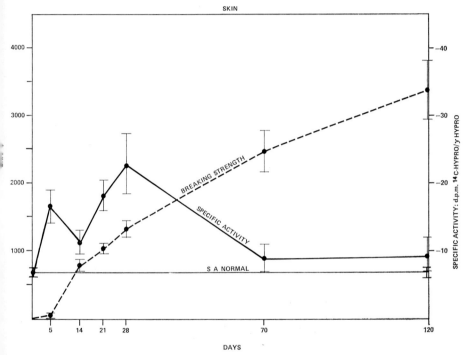

Fig. 3. Breaking strength and rate of collagen synthesis of skin wounds in dogs. The rate of collagen synthesis is expressed as specific activity of ^{14}C-hydroxyproline. The horizontal solid line represents the rate of collagen synthesis in normal, unwounded skin. The breaking strength of 5-mm strips of unwounded skin is 7391 ± 426 g for the upper abdomen and 6094 ± 487 g for the lower abdomen.

Bladder wounds had a rate of collagen synthesis significantly below normal unwounded bladder tissue at days 70 and 120 (Fig. 4). Histological preparations of these wounds revealed a mature scar with a population of fibroblasts considerably smaller than found in adjacent unwounded connective tissue. Notably, normal collagen synthesis of bladder tissue greatly exceeded other tissues studied.

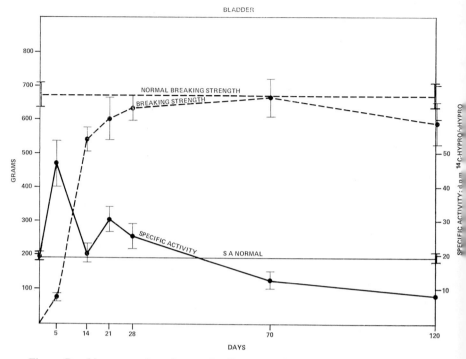

Fig. 4. Breaking strength and rate of collagen synthesis of bladder wounds in dogs. The horizontal solid line represents the rate of collagen synthesis in normal unwounded bladder and the dotted horizontal line represents the breaking strength of normal unwounded bladder. The control values represent the means of 60 dogs.

Stomach wounds had a peak rate of collagen synthesis at 5 days, rather than at days 21–28, as in other tissues. Furthermore, even after 120 days the rate of collagen synthesis in the wound was slightly, but significantly, elevated over unwounded stomach, while the wound gained no appreciable strength after 21 days (Fig. 5).

Colon wounds had a markedly increased rate of collagen synthesis even at 120 days, amounting to approximately five times that of normal unwounded colon. Yet, at this time the colon was only gaining strength slowly, if at all (Fig. 6).

To study the general extent of wound healing with time, the strength of each wound was calculated as a percentage of unwounded tissue strength. The strength gain curves shown in Fig. 7 have a slightly different shape from those shown in Figs. 3–6. The control values for breaking strength shown in these latter figures are based on a composite mean strength value for all dogs in the study. The calculations for Fig. 7

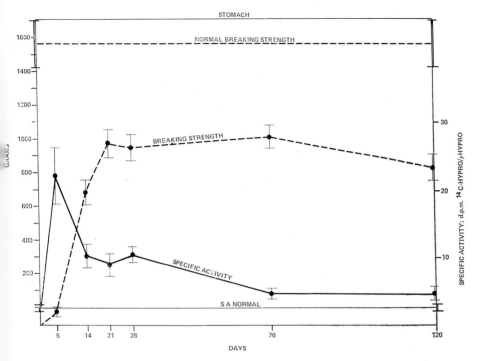

Fig. 5. Breaking strength and rate of collagen synthesis of stomach wounds in dogs.

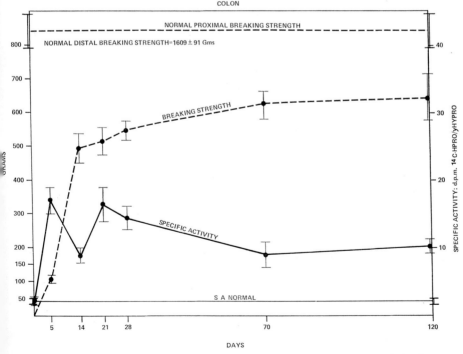

Fig. 6. Breaking strength and rate of collagen synthesis of colon wounds in dogs.

were made using each individual dog as its own control. The values derived from different methods of calculation are not sufficient to alter the conclusions drawn.

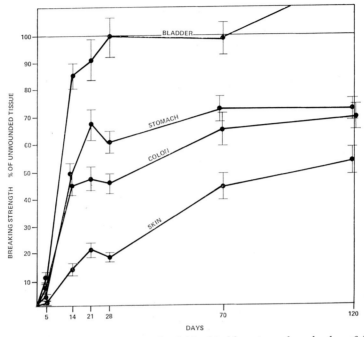

Fig. 7. The extent of healing of wounds of skin, bladder, stomach and colon of dogs in terms of breaking strength as a percentage of the strength of the respective normal unwounded tissue. Each dog serves as his own control.

Table I shows that extent of healing, as measured by wound strength, is inversely proportional to strength of unwounded tissue.

Table I
Relation of normal tissue strength to extent of wound healing at 120 days

Tissue	Control strength (g/5-mm strip)	Wound strength (% of unwounded strength)*
Skin	6742 ± 426	53 ± 5
Stomach	1662 ± 66	72 ± 5
Colon (proximal)	845 ± 45	75 ± 5
Bladder	671 ± 38	120 ± 10

* Mean ± standard error

DISCUSSION

The breaking strength of a wound represents not only fibroblastic activity of synthesizing and depositing collagen, but also a number of extracellular events occurring at various rates and extents during the entire period of repair. For example, if the enzyme lysyl oxidase were inhibited, preventing the first step in collagen crosslinking, the result would be a wound having little or no strength. Degradation of collagen by tissue collagenase is an important factor in normal remodelling of scar tissue, yet excessive degradation failing to maintain an equilibrium between synthesis and lysis would lessen wound strength.

Collagen synthesis rates directly measure the functional activity of fibroblasts. If fibroblasts always synthesize collagen at a constant rate, then differences in observed rates of synthesis in wounded or normal tissues would reflect the number of cells in the synthetic phase. There are no data indicating fibroblastic activity is an 'all or none' phenomenon. Simply counting fibroblast population density by light microscopy in normal or wounded tissue does not indicate how many cells are synthesizing collagen. In our present state of knowledge, we cannot correlate directly the histological picture with the biochemical data.

Because all wounds gain strength at approximately the same rate during the first 14–21 days and because the maximum rate of collagen synthesis occurs at the same time, it would suggest a period of maximum fibroblastic activity. Microscopic examination of these wounds show the density of fibroblasts is similar in all tissues. After 14–21 days, however, the rate of collagen synthesis declines and in visceral wounds, the rate of strength gain decreases markedly. One is tempted to seek a cause and effect relationship between collagen synthesis and strength gain, except that such a relationship does not hold for skin wounds. With skin there exists a paradoxical situation of continued strength gain even though the rate of collagen synthesis returns to normal. In contrast, the stomach and colon show little or no gain in strength after 21 days, but the rate of collagen synthesis remains elevated. The explanation for these differences among skin, colon and stomach wounds probably lies in extracellular events that influence remodelling and crosslinking of collagen in later phases of wound healing. These effects may be tissue specific.

The bladder wound is unusual. During the first 14–21 days, its rate of healing is similar to other wounds. However, since normal breaking strength of unwounded bladder tissue is the lowest of any tissue studied, the wound strength attains normal values in two weeks. The rate of collagen synthesis in the bladder wound decreases rapidly after the 21st day and by 70 days is actually less than normal bladder tissue. This coincides with the wound fibroblast population which decreases after 21 days, being practically acellular by 120 days. In this instance, histological information appears to explain biochemical data.

Neither rate of collagen synthesis in unwounded tissue, peak rate of synthesis in wounded tissue nor rate of synthesis in the 120-day old wound are correlated with breaking strength at 120 days (Table II).

Table II
Extent of wound healing related to normal, peak and 120-day rate of collagen synthesis

Tissue	Extent of healing % of Control strength*	Specific activity of ^{14}C-Hypro*		
		Normal tissue	Wound peak	120 days
Skin	53 ± 5	6.7 ± 0.7	22.9 ± 4.6	9.0 ± 2.8
Stomach	72 ± 5	2.3 ± 0.3	22.0 ± 4.1	4.4 ± 0.7
Colon	75 ± 5	1.9 ± 0.2	16.3 ± 2.5	10.2 ± 1.4
Bladder	120 ± 10	19.3 ± 1.7	47.1 ± 6.7	9.0 ± 1.3

* Mean ± standard error

This again suggests extracellular factors dominate the healing process after the first 14-21 days. These extracellular events may be determined by the nature of the collagen synthesized and deposited by the fibroblast during early stages of healing as suggested by recent works of Forrest et al. (1972) and Bailey and Robins (1972). These workers have shown that patterns of collagen crosslinking are different in various tissues and also differ with age of the tissue. Since not only the number but the type of crosslink varies, it is possible that during synthesis collagen is 'programmed' for particular types of crosslinks. The question arises—do fibroblasts program wound collagen to resemble the immediately surrounding tissue or is there a separate 'repair fibroblast' that produces a collagen programmed for particular types of crosslinks which differ from the tissue in which the wound occurs? The work of Forrest et al. (1972) would favor the concept of 'repair fibroblasts' since they found a different pattern of crosslinks in wounded and unwounded dermis.

Data based on ascorbic acid requirements of embryonic and young guinea pigs suggest that fibroblasts synthesizing collagen rapidly have an absolute requirement for ascorbic acid. Fibroblasts involved in normal growth and maintenance can utilize other reducing substances as a co-factor for proline hydroxylase (Woessner and Gould, 1957; Gould, 1958, 1960; Manner and Gould, 1961; Robertson, 1961). Peterkofsky and Udenfriend (1965) and Hutton *et al.* (1967) showed that ascorbic acid could be replaced by pteridine derivatives under certain circumstances. The finding by Gabbiani *et al.* (1972) of apparently specialized contractile 'myofibroblasts' in granulation tissue suggests that there may actually be a 'repair fibroblast'. Our observations that the rate of wound healing is the same during the period of maximum collagen synthesis regardless of the tissue wounded lends support to the concept of 'repair fibroblasts'. Furthermore, the observation that colon wounds attain the same strength regardless of location, whereas normal colon exhibits a 2-fold difference in strength from one end to the other, adds support to this view.

SUMMARY

Wounds in skin, stomach, colon and bladder heal at approximately the same rate during the first 14–21 days post-wounding, as measured by rate of breaking strength gain. All wounds show a maximal rate of collagen synthesis during this period. Thereafter, the visceral wounds either fail to gain further strength, or gain strength very slowly while skin wounds continue to gain strength rapidly throughout the 120 day observation period. After 14–21 days, the rate of collagen synthesis declines in all wounds, reaching normal levels of unwounded tissue in skin but remaining elevated in stomach and colon. The rate of collagen synthesis in bladder wounds becomes significantly lower than normal bladder tissue after 70 days and is correlated with the disappearance of fibroblasts in the scar tissue.

It is suggested that 'repair fibroblasts' control the rate of wound healing during the first 14–21 days of repair, but thereafter extracellular events such as crosslinking, collagen degradation, and remodelling determine the subsequent course of healing in a specific manner for each tissue.

The extent a wound heals within any given time period appears related to strength of normal unwounded tissue but not to the normal rate of collagen synthesis, to the peak rate of synthesis, nor to the duration of increased collagen synthesis.

ACKNOWLEDGEMENTS

Our research was supported in part by a grant from Ethicon, Inc., Somerville, N. J. and N.I.H. Grant AM14047.

We wish to acknowledge the technical assistance of Terry Werstlein, Erik Krutzsch and Narcisco Tellez in carrying out these studies. J. Christopher Hastings was supported by the University of Illinois Training Grant, 1930-03.

REFERENCES

Bailey, A. J. and Robins, S. P. (1972). *FEBS Letters* **21**, 330–334.
Forrest, L., Shuttleworth, A., Jackson, D. S. and Mechanic, G. L. (1972). *Biochem. Biophys. Res. Comm.* **46**, 1776–1781.
Gabbiani, G., Hirschel, B. J., Ryan, G. B., Statkov, P. R. and Majno, G. (1972). *J. Exp. Med.* **135**, 719–734.
Gould, B. S. (1958). *J. Biol. Chem.* **232**, 637–649.
Gould, B. S. (1961). *Ann. N. Y. Acad. Sci. USA* **92**, 168–174.
Gould, B. S. and Manner, G. (1960). *Ann. N. Y. Acad. Sci. USA* **85**, 385–396.
Howes, E. L., Harvey, S. C. and Hewitt, C. (1939). *Arch. Surgery* **38**, 934–945.
Hutton, J. J., Jr., Tappel, A. L. and Udenfriend, S. (1967). *Arch. Biochem. Biophys.* **118**, 231–240.
Juva, K. and Prockop, D. J. (1966). *Analyt. Biochem.* **15**, 77–83.
Madden, J. W. and Peacock, E. E., Jr. (1968). *Surgery* **64**, 288–294.
Madden, J. W. and Peacock, E. E., Jr. (1971). *Ann. Surg.* **174**, 511–520.
Peterkofsky, B. and Udenfriend, S. (1965). *Proc. Natl. Acad. Sci. USA* **53**, 335–342.
Robertson, W. Van B. (1961). *Ann. N. Y. Acad. Sci. USA* **92**, 159–167.
Uitto, J. (1970). *Biochim. Biophys. Acta* **201**, 438–445.
Woessner, Jr. J. F. and Gould, B. S. (1957). *J. Biophys. Biochem. Cytol.* **3**, 685–695.

Intercellular Matrix of Hypertrophic Scars and Keloids

Suzanne Bazin, Claude Nicoletis and Albert Delaunay

Institut Pasteur, 92380 Garches, France

In wound healing, the metabolic activity of fibroblasts is activated first in the wound margins, then in the granulation tissue formed between these margins. Large amounts of connective tissue matrix components are synthesized in the granuloma; later on the tissue progressively turns into a normal biochemical composition and a biophysical structure.

In abnormal, hypertrophic scars and in keloids, an overproduction of connective tissue has been histologically observed. Pathologists easily describe such abnormal scars as fibrotic tissue. Consequently morphologic alterations and tough consistence of the repaired tissue have been suspected as results of an overproduction of collagen fibres or of an augmented crosslinkage of the newly formed collagen.

However, up to the present time, very few biochemical data are available to support such a hypothesis. In a preliminary investigation, we tried to obtain quantitative data about the amounts of collagen and of glycosaminoglycans in human hypertrophic scars or keloids (50 samples), normal scars (e.g. completely healed wound tissue) (10 samples) and in normal skin (8 samples). These tissues had been taken from patients during operations of reparative or aesthetic surgery; as far as possible the samples of normal skin were taken from sites similar to those where scars had developed. Clinically, keloids can be distinguished from hypertrophic scars during the early phase of their development only. Histologically there is no apparant difference between the two types of scars, at least at the time they are surgically excised. Following several authors (Asboe-Hansen, 1960; Mancini and Quaife, 1962), we have gathered all the abnormal scars we examined under the name 'Hypertrophic scars'.

METHODS

The pooled minced tissues were divided into four aliquots.

1. The first aliquot was used to determine the water content.
2. The second aliquot was used to determine the collagen content by the method of Fitch et al., 1955.
3. The third aliquot was used to estimate the degree of collagen crosslinkage through the relative proportions of four fractions of increasingly crosslinked collagen, extracted by a step-wise procedure:

   (a) neutral salt soluble collagen, extracted by phosphate buffer $0.15\ M$, pH 7.8, at 4°C,

   (b) acid soluble collagen, extracted by citrate buffer $0.1\ M$, pH 3,5 at 4°C,

   (c) 60° labile collagen, extracted by phosphate buffer $0.04\ M$, in NaCl $0.15\ M$, pH 6.6, at 60°C,

   (d) insoluble collagen, gelatinized in 5% trichloracetic acid at 90°C.

The three first solubilized fractions were purified by twice repeated NaCl precipitation and acetic acid dissolution, then by dialysis.

The collagen content of each fraction was estimated through hydroxyproline determination (Stegemann, 1958) after hydrolysis in 6N HCl, at 105°C, for 24 h, and chromatographic separation of hydroxyproline (Partridge and Elsden, 1961).

4. The fourth aliquot was used for the determination of glycosaminoglycans. It was first defatted then homogenized in phosphate buffer $0.15\ M$, pH 7.8 and three times extracted with the same buffer, at 4°C. The supernatants and residues of extraction were digested exhaustively with papain at pH 5.5 and 65°C, for at least 2 or 3 days, then with trypsin at pH 8 and 37°C for 8 h.

Total glycosaminoglycans were precipitated by 5 volumes ethanol at pH 5 and their hexosamines determined according to Pearson (1963) after hydrolysis in 2N HCl at 105°C for 15 h. Sialoglycoproteins were estimated through sialic acid determination (Svennerholm's method, 1958).

Acid glycosaminoglycans were precipitated by cetylpyridinium chloride and fractionated by the method of Schiller et al., 1961, then estimated through uronic acid determination by the carbazol method (Dische, 1955).

RESULTS

The content of the examined tissues of water, glycosaminoglycans and collagen and the relative proportions of the separated four collagen fractions is given in Table I.

The content of sialoglycoproteins and acid glycosaminoglycans in fractions soluble and insoluble in phosphate buffer is given in Table II. In the same table are given the results of chromatographic separation of acid glycosaminoglycans: two main fractions were separated, one containing hyaluronate (eluted by $0.5\ M$ NaCl), the other containing

Table I
Collagen and total glycosaminoglycans content

	Water*	Collagen content†	Fractions of collagen‡				Glycosamino-glycans§
			Neutral salt soluble	Acid soluble	60°C labile	Insoluble	
Normal skins	65·2 (± 2·48)	72·8 (± 5·73)	0·73 (± 0·4)	4·43 (± 1·9)	27·9 (± 9·5)	66·9 (± 19)	2·07 (± 0·54)
Normal scars	70·6 (± 7·52)	80·5 (± 16)	0·29 (± 0·2)	10·27 (± 4·5)	40·2 (± 15·5)	48·9 (± 18·4)	2·32 (± 1·03)
Hypertrophic scars	72·1 (± 5·32)	75·3 (± 13·7)	2·02 (± 0·9)	6·89 (± 0·8)	38·9 (± 11)	52·2 (± 9·5)	5·25 (± 1·97)

* in % of wet tissue.
† in mg hydroxyproline/g dry tissue.
‡ in % of total collagen.
§ in mg hexosamines/g dry tissue.

In the parentheses the standard deviation.

Table II
Sialoglycoprotein and glycosaminoglycan content

	Sialoglycoproteins*		
	Total content	Soluble fraction in phosphate buffer pH 7·8	Insoluble fraction in phosphate buffer pH 7·8
Normal skins	0·94 (±0·68)	0·63 (±0·21)	0·31 (±0·09)
Normal scars	1·91 (±1·03)	0·84 (±0·29)	1·07 (±0·75)
Hypertrophic scars	2·71 (±1·84)	1·95 (±1·42)	0·76 (±0·45)

	Acid glycosaminoglycans†				
	Total content	Soluble fraction in phosphate buffer pH 7·8	Insoluble fractions in phosphate buffer pH 7·8	Hyaluronate fraction	Chondroitin-dermatan-sulphate fraction
Normal skins	1·14 (±0·45)	0·76 (±0·09)	0·38 (±0·07)	0·49 (±0·020)	0·40 (±0·13)
Normal scars	1·33 (±0·27)	1·01 (±0·21)	0·32 (±0·26)	0·78 (±0·10)	0·65 (±0·05)
Hypertrophic scars	2·92 (±1·54)	2·33 (±1·37)	0·59 (±0·38)	0·62 (±0·36)	1·68 (±0·72)

* in mg N-acetylneuraminic acid/g dry tissue.
† in mg glucuronic acid/g dry tissue.

chondroitin sulphate (6 and 4 isomers) and dermatan-sulphate (eluted by 1·5 M NaCl).

It can be seen from these tables that the results of collagen and glycosaminoglycan determinations are affected by rather large variations. This fact results from the random collection of samples which came from donors not similar in age, sex or race or from different localizations on the body. No correlation could be found in regard to localization nor to the duration of the scar development. Thus the results of all samples were gathered together to calculate mean values and standard deviations. Some statistically significant results could however been obtained:

1. The *water content* of scars (normal scars and hypertrophic scars) is slightly higher than the water content of normal skin (Table I).

2. The *collagen content* of scars (normal scars and hypertrophic scars) is not significantly different from that of normal skin (Table I).

3. The *degree of crosslinkage of collagen* as revealed by the relative proportions of the four successively extracted fractions, is less in normal and hypertrophic scars than in normal skins. It is not significantly different in normal scars and in hypertrophic scars (Table I).

4. The *total glycosaminoglycan content* is higher in normal and hypertrophic scars than in normal skin, the highest content being observed in hypertrophic scars (Table I).

5. The *high content of glycosaminoglycans* in scars is characterized:

(a) in normal scars by a content of sialoglyco-proteins higher than in normal skin (especially the fraction insoluble in phosphate buffer) and by a content of acid glycosaminoglycans not significantly different of that of normal skin (Table II).

(b) in hypertrophic scars by contents of sialoglycoproteins and of acid glycosaminoglycans higher than in normal skin (fractions soluble and insoluble in phosphate buffer) and higher than in normal scars (fraction of sialoglycoproteins soluble in phosphate buffer and fractions of acid glycosaminoglycans soluble and insoluble in phosphate buffer, Table II).

6. The *acid glycosaminoglycans* are similar in their nature and repartition into soluble and insoluble fractions in phosphate buffer in normal scars and in normal skin.

In hypertrophic scars, the acid glycosaminoglycan content is much higher than in normal scars and in normal skins. Such high content is due to a chondroitin-dermatan sulphate fraction much more abundant (approximately three times) than in normal scars and in normal skin, the hyaluronate fraction being not significantly different in the three types of studied tissues.

DISCUSSION

We have observed that the collagen and the degree of collagen crosslinkage are not significantly different in hypertrophic scars and in normal scars. In both types of scars, the collagen is however less crosslinked than in normal skin. Similar results had been previously obtained with regard to the collagen content by Ito et al. (1960), Sibeleva et al. (1965), Amante et al. (1963), Harris and Sjoerdsma (1966) in normal and hypertrophic scars, and, with regard to the collagen crosslinkage by Banfield and Brindley (1959) and by Verzar and Willnegger (1961) in normal scars, by Rasmussen et al. (1964), Harris and Sjoerdsma (1966) and by Samohyl and Pospišilova (1970) in hypertrophic scars. Thus it appears that hypertrophic scars are not characterized by a high content or a high crosslinkage of collagen, as it should be in a fibrotic tissue.

Biochemical differences between normal scars and hypertrophic scars are revealed by the glycosaminoglycans of these tissues:

(a) the glycosaminoglycan content is much higher in hypertrophic scars than in normal scars,

(b) a high proportion of easily soluble sialoglycoproteins have been found in hypertrophic scars, in contrast with a high proportion of insoluble sialoglycoproteins in normal scars,

(c) the acid glycosaminoglycan content is higher in hypertrophic scars than in normal scars.

These three observations show that hypertrophic scars offer characteristics similar to those of a not fully developed granulation tissue (Bazin and Delaunay, 1963, 1967, 1968).

The most important difference concerns the acid glycosaminoglycans. Their high proportion in hypertrophic scars had been described histochemically by Zhuravlyova and Orlovskaya (1965) and observed chemically by Sibeleva et al. (1965).

We have confirmed such results and moreover we have been able to get a new result: at least in these samples we have studied the chondroitin-dermatan-sulphate fractiononly appears increased in hypertrophic scars, the hyaluronate fraction being approximately the same as in normal scars.

Such increased acid glycosaminoglycan fractions should have been separated in its components: chondroitin-sulphate and dermatan-sulphate; unfortunately we had not enough material to carry out such separation in every sample; we were able to do it in a few samples only using the electrophoretic method; in this limited number of samples we found a much higher proportion of chondroitin-sulphate than of dermatan-sulphate (approximately five times more).

Thus our preliminary results show that hypertrophic scars appear characterized by an abnormally high proportion of the chondroitin-dermatan sulphate fraction of acid glycosaminoglycans and mainly chondroitin-sulphate. Chondroitin-sulphate is known to be present in tissues as the polysaccharide moiety of high molecular weight proteoglycan. Such macromolecules may induce abnormal physical interactions in the intercellular matrix. It may be that abnormal interactions might be correlated to the abnormal consistence of the scar tissue.

On the other hand, acid glycosaminoglycans have been found accumulated in a high proportion in granulation tissues during the first stage of their development; later on, the glycosaminoglycans in excess are removed from the tissue, first hyaluronate, then chondroitin-sulphate (Delaunay and Bazin, 1964). Our observation of an excessive accumulation of chondroitin-sulphate in hypertrophic scars supports the view that chondroitin-sulphate might not be normally removed from this granulation tissue. Thus in hypertrophic scars, the metabolic activity of fibroblasts appears to be lacking in the regulation normally effective in granulation tissues, as it is in normal scars. Is the disturbance of such regulation due to a lack of messenger molecules? Or to an alteration of receptors on the fibroblast membranes? Perhaps this last hypothesis could be supported by observations of some abnormal cytologic characters in fibroblasts of hypertrophic scars, as noted by Gillette and Conway (1959) and Zhuravlyova and Orlovskaya (1965) but these preliminary studies need to be extended.

ACKNOWLEDGEMENTS

We should like to express our thanks to Nicole Briquelet and Jean-Claude Allain for their technical assistance.

REFERENCES

Amante, L., Birati, A. and Pernis, B. (1963). *Boll. Soc. Ital. Biol. Sper.* **39**, 1224–1227.
Asboe-Hansen, G. (1960). *Dermatologica* **120**, 178–184.
Banfield, W. G. and Brindley, D. C. (1959). *Surg. Gynecol. Obstet.* **109**, 367–372.
Bazin, S. and Delaunay, A. (1963). *Ann. Inst. Pasteur (Paris)* **105**, 624–634.
Bazin, S. and Delaunay, A. (1967). *Ann. Inst. Pasteur (Paris)* **112**, 316–328.
Bazin, S. and Delaunay, A. (1968). *Ann. Inst. Pasteur (Paris)* **115**, 899–907.
Delaunay, A. and Bazin, S. (1964). *Int. Rev. Connect. Tissue Res.* **2**, 301–325.
Dische, Z. (1955). *Methods Biochem. Anal.* **2**, 325–332.
Fitch, S. M., Harkness, M. L. H. and Harkness, R. D. (1955). *Nature* **176**, 163.
Gillette, R. W. and Conway, H. (1959). *Exp. Cell. Res.* **18**, 313–317.

Harris, E. D. and Sjoerdsma, A. (1966). *Lancet* 707–711.
Ito, T., Miyazawa, I. and Sano, S. (1960). *Jap. J. Dermatol.* **70**, 139.
Mancini, R. E. and Quaife, J. V. (1962). *J. Invest. Dermatol.* **38**, 143–150.
Partridge, S. M. and Elsden, D. F. (1961). *Biochem. J.* **80**, 34P.
Pearson, C. H. (1963). *Biochem. J.* **88**, 540–545.
Rasmussen, M., Khalil, G. W. and Winkelmann, R. K. (1964). *J. Invest. Dermatol.* **43**, 349–355.
Samohyl, J. and Pospisilova, J. (1970). Proceedings of 3rd International Congress in Burns Research, Prague, pp. 140–141.
Schiller, S., Slover, G. A. and Dorfman, A. (1961). *J. Biol. Chem.* **236**, 983–987.
Sibeleva, K. F., Zenkevich, G. D. and Laufer, A. L. (1965). *Vopr. Med. Khim.* **11**, 55–60.
Stegemann, H. (1958). *Hoppe Seyler's Z. Physiol. Chem.* **311**, 41–45.
Svennerholm, L. (1958). *Acta Chem. Scand.* **12**, 547–554.
Verzar, F. and Willenegger, H. (1961). *Schweiz. Med. Wochenschr.* **91**, 1234–1236.
Zhuravlyova, M. V. and Orlovskaya, G. V. (1964). Symposium: Mechanisms Sclerotic Processes and Cicatrization, Novosibirsk, USSR, February 24–28, pp. 242–252.

Factors Influencing Wound Healing in Rheumatoid Arthritis

KAUKO VAINIO
*Rheumatism Foundation Hospital
Heinola, Finland*

One of the special features of the peripheral parts of the extremities in rheumatoid arthritis is the impaired circulation due to vascular changes. The tissues are in a hypoxaemic stage which according to Niinikoski (1969) reduces collagen synthesis and may cause delayed wound healing.

The main factors causing hypoxaemia are: (a) arteritis, (b) impaired venous outflow due to pathologic dilatation of the venules and congestion caused by hydrops and oedema of the tissues and (c) the general tendency of rheumatoid arthritis patients to have anaemia.

Arteritis

Arteritis of the small arteries is a common feature in rheumatoid arthritis.

Incidence. Virtama (1959) found in post mortem brachial angiographies in rheumatoid arthritis patients local obliterations and post stenotic dilatations of the digital arteries as well as hypervascularisation and dilations of the arterial lumens near bony erosions. Scott *et al.* (1960) found in arteriography of ten patients either vascular occlusion or distortion in all examined cases. In 100 consecutive biopsies of the tissues of the hand taken at our clinic, Ritama (unpublished) was able to detect vascular lesions in 80 cases.

The histological picture. This varies from a mild indeterminate arteritis to occlusion of the vessels. In some cases a very heavy perivascular infiltration resembling periarteritis nodosa may be detected (Fig. 1).

Clinical picture. It is well known that the hands and feet of rheumatoid arthritis patients are cold, sweating and cyanotic. In more advanced cases there may be necrotic areas in the skin, most often in the nail-fold, nail-edge or digital pulp due to small infarcts. Even gangrene

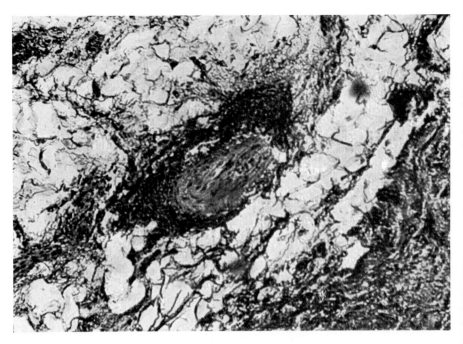

Fig. 1. Occlusive arteritis with perivascular infiltration. (H and E. ×100)

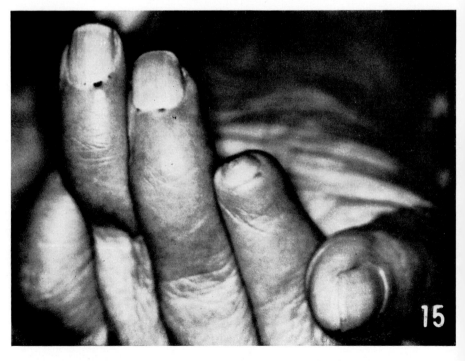

Fig. 2. Ischaemic infarcts in the nail-folds. The tip of the index finger was amputated because of gangrene.

of the finger tips can be seen (Fig. 2). In extreme cases the vasculitis may cause a gangrene of all four limbs (Bywaters, 1957).

Possibilities to prevent the sequels of postoperative hypoxaemia

Surgery should be avoided in the peripheral parts of the extremities with severely impaired circulation. Sometimes, if surgical interventions are indispensable, the condition of the tissues can be improved with sympathectomy. To avoid the deleterious effect of general anaemia to the peripheral circulation the operation should be delayed until haemoglobin values are improved at least to 100 g/l. Tourniquet time should be limited to the minimum. Personally I do not like to have the ischaemia for a longer time than $1\frac{1}{4}$h. Flatt (1971) considers 2 h as an absolute maximum.

Anaesthesia. Supraclavicular, brachial or sciatic nerve blocks cause a long-lasting improvement of the circulation of the peripheral parts of the limbs after release of the tourniquet. Because of this fact these blocks are widely used at our clinic. Incisions have to be planned so that no 'critical' flaps or corners are made. The sutures should never be tied tightly.

Amyloidosis

Amyloidosis is a fairly common complication. In severe amyloidosis the skin is oedemic and fragile. The sutures cut through easily and wound disruption may occur. Thus, surgery should be avoided in severe amyloidosis. In slight cases I have not seen disturbances in wound healing.

Effects of drugs upon wound healing

Corticosteroids inhibit the mitotic activity of cells and the proliferation of fibroblasts (Ruhman and Berliner, 1965). This causes a delayed wound healing. Thus, hypercortisonism or simultaneous treatment with high doses of corticosteroids is a relative contra-indication for surgery.

Intradermally injected corticosteroids cause local thinning of the skin. Slim and Henderson (1972) showed that corticosteroids applied to the skin in an alcohol solution have a profound inhibitory effect to the collagen synthesis. In treating tenosynovitis with local corticosteroids injections into the tendon tissue should be carefully avoided because of the risk of tendon ruptures (Moberg, 1965). I have seen a cause of rupture of the Achilles tendon after repeated intratendinous injections (Fig. 3).

Cytostatic drugs, used in rheumatoid arthritis which cannot be controlled with other measures, diminish the tensile strength of skin wounds (Kari and Kulonen, 1969). Tetracycline, often used to control chronic infections, also diminishes the tensile strength of the wound (Aine and Kulonen, 1969). It even inhibits the synthesis of collagen in bone formation as well as the mineralization of bone (Kaitila et al., 1970). Finally, according to Boström (1969) acetylsalicylic acid inhibits

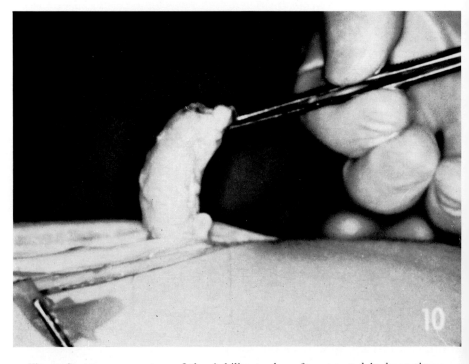

Fig. 3. Spontaneous rupture of the Achilles tendon after repeated hydrocortisone injections.

mucopolysaccharide synthesis in rats, but to a lesser degree than salicylic acid. However, I have not noticed any deleterious effect of therapeutic doses of acetylsalicylic acid on rheumatoid arthritic patients subjected for surgery.

REFERENCES

Aine, E. and Kulonen, E. (1969). Scand. J. Clin. Lab. Investigation, Suppl. **109**, 18.
Boström, H. (1969). Scand. J. clin. Lab. Investigation suppl. **108**, 5.
Bywaters, E. G. L. (1957). Ann. rheum. Dis. **16**, 84–103.
Flatt, A. (1969). Ann. rheum. Dis., Suppl. **28**, 80.

Kaitila, I., Wartiovaara, J., Laitinen, O. and Saxen, L. (1970). *J. Embryol. exp. Morph.* **23**, 185–211.
Kari, A. and Kulonen, E. (1969). *Scand. J. clin. Lab. Investigation, Suppl.* **109**, 21.
Moberg, E. (1965). *Am. J. Surg.* **109**, 353–355.
Niinikoski, J. (1969). *Acta physiol. Scand. Suppl.* **334**.
Ruhman, A. G. and Berliner, D. L. (1965). *Endocrinology* **76**, 916–927.
Scott. J. T., Hourihane, D. O., Doyle, F. H., Steiner, R. E., Laws, J. W., Dixon, A. St. J. and Bywaters, E. G. L. (1961). *Ann. rheum. Dis.* **20**, 224–234.
Slim, A. W. and Henderson, P. S. (1972). *Scand. J. clin. Lab. Investigation, Suppl.* **123**, 38.
Virtama, P. (1959). *Acta rheum. Scand.* **5**, 304–313.

Respiratory Patterns of Fibroblasts in Reparative Tissue

T. HUNT, R. ALLEN, D. KITTRIDGE, J. NIINIKOSKI
and P. EHRLICH

*Department of Surgery, University of California
San Francisco, California, 94122, USA*

Oxygen supply and its relation to fibroplasia has been mentioned several times during this symposium. Oxygen supply has been discussed in connection with hydroxylation of proline (D. J. Prockop, this volume p. 311), in relation to energy supply and maintenance of intracellular pH (I. A. Silver, this volume p. 507), and in reference to human wounds and the rate of repair (Niinikoski et al., 1972).

When oxygen tension was measured in experimental and human wounds it was found to be low (5 to 20 mm Hg) in the central dead space. Oxygen tension in human wounds was slightly higher.

The PO_2 of the dead space of a wound represents the low point of a PO_2 gradient which begins at the last intact capillary and falls as the wound space is approached. Silver described the anatomy of this gradient earlier in this book.

The low PO_2 suggests that the function of fibroblasts may be limited by oxygen supply since extracellular oxygen tensions in this range are 'critical' in some other systems.* Therefore, the dimensions of this gradient are important because they define the number and type of cells functioning at the various oxygen tensions.

Microelectrodes measure the gradient directly. However, numerically useful results are difficult to obtain since to define the mean PO_2 at the capillary requires an infinite number of microelectrode measurements. Therefore, we devised another technique based on antimycin A as a source of mathematically more useful data. Several assumptions have been made in this method, but the results are reproducible and

---

* The 'critical' oxygen tension is the oxygen tension below which the metabolic rate becomes dependent upon PO_2.

compare well with estimates derived from microelectrode studies. Antimycin A inhibits cellular respiration by blocking the movement of electrons through the mitochondrial transport system. Applying antimycin A to the surface of granulation tissue should eliminate most of the oxygen consumption of the tissue distal to the last functioning capillaries. Oxygen consumption probably continues deep to these capillaries. With the cytochrome system of the 'distal wound' blocked, the PO_2 in the dead space should rise to approximate the mean capillary PO_2 in the most distal capillary.

Four stainless steel wire mesh cylinders were implanted in each of 24 adult rabbits by a previously described method (Ehrlich et al., 1972). Ten to fifteen days later, each animal was placed in a restraining cage with its head in a transparent plastic head mask and the wire mesh cylinders on the back were exposed for sampling. Groups of animals breathed 10%, 20% (air), or 45% oxygen atmospheres for at least 2 h before sampling. After the baseline values were determined by aspirating wound fluid from one of the cylinders and arterial blood from an ear vein, 0·1 mg of antimycin A dissolved in 0·1 ml of polyethylene glycol was injected into another wound. Thirty minutes later, fluid was withdrawn from this wound and its respiratory gas tensions and pH were determined.

Results are shown in Fig. 1 which shows to the left of its midpoint the gradient from arterial PO_2 to 'mean capillary PO_2'. The right side of the graph represents the 'mean capillary' to wound space gradient. The midpoint of the mean capillary to wound space gradient is an estimate of the average PO_2 to which the cells in this area of the wound are exposed. The mean capillary PO_2 is comparable to but slightly lower than that estimated by Remensnyder and Majno (1968), Silver (1969), and Niinikoski et al. (1972) when microelectrodes were used.

The oxygen gradient in the wound obviously increases as arterial PO_2 rises. One must then conclude that the cells in the gradient area use more oxygen if more is made available. If oxygen consumption did not increase, the wound fluid PO_2 would rise as much as the mid-capillary PO_2. The increased gradient on breathing 45% oxygen has been confirmed by microelectrode measurement.

The estimates·of mean-tissue PO_2 are useful for at least two purposes: (1) to correlate collagen production with changes in oxygen environment; and (2) to characterize the environment so that correlations can be made with tissue and cell culture experiments. For example, the relationship of collagen production to arterial PO_2 has been determined in previous experiments performed essentially as described above. A curvilinear relationship is dictated since any possible linear

RESPIRATION OF FIBROBLASTS IN REPARATIVE TISSUE 587

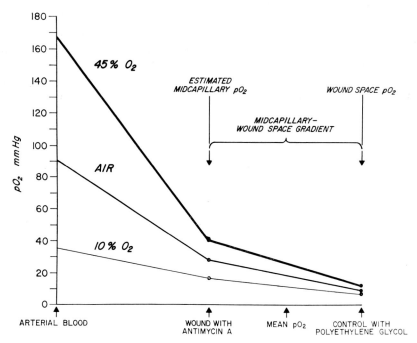

Fig. 1. Oxygen gradients from arterial blood, to 'mean capillary' and to wound space as determined by antimycin A modified wound space PO_2.

relationship among the determined points predicts collagen production at zero PO_2, a known impossibility. If these values are replotted, using Fig. 1 to determine the mean-tissue PO_2 which corresponds to the given arterial PO_2, one arrives at the results in Fig. 2. This figure shows that collagen synthesis is a linear function of mean-tissue PO_2 with the zero intercept at, or very close to, zero collagen production.

The mean-tissue PO_2 can be reproduced in cell culture and correlated with oxygen consumption. To do this, second or third generation fibroblasts obtained from wire mesh cylinder wounds in rabbits were trypsinized and washed, and the cells were suspended in tissue culture medium. The cells and medium were then placed in a sterile glass chamber stirring bath at 37°C (Yellow Springs Instrument Co. No. 5301). The oxygen electrode then measured the rate of fall of PO_2 as the dissolved oxygen supply was consumed by the fibroblasts. To prove that the decay curve was not influenced by accumulation of H^+ or CO_2, the PO_2 was allowed to fall to zero, and a small bubble of oxygen was placed in the chamber. The PO_2 rose immediately while the pH and PCO_2 remained the same. Oxygen consumption again resumed at the same rate as before and the second decay curves were identical to the first. The PO_2 fell linearly to an inflexion area in the region of 10–15 mm Hg, the same area as mean-tissue PO_2 in the wound. Below this point oxygen consumption rapidly diminished as PO_2 fell.

Obviously, the oxygen environment of granulation tissue is in a critical range in which minor changes in PO_2 cause great changes in oxygen consumption. If more oxygen was made available to cells existing in this range, more oxygen would be used. The cell culture

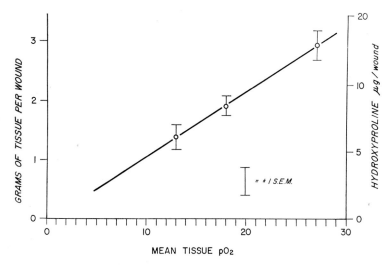

Fig. 2. Connective tissue synthesis bears a linear relation to mean-tissue PO_2. (Reproduced with permission from Ehrlich et al., 1972.)

experiments, therefore, prove the somewhat unusual hypothesis that oxygen consumption in healing tissue *normally* is a function of oxygen supply.

What are the limiting metabolic steps which fail as oxygen tension falls? The answer may be complex. The relatively high PO_2 of the critical area suggests that the cytochrome system is not limiting since cytochrome a_3 remains saturated until PO_2 at the mitochondrion falls below 1 mm Hg. In most tissues, this translates to an extracellular PO_2 of approximately 5 mm Hg. Perhaps the flavoprotein hydrogen transport system or some oxygenase reactions such as prolyl hydroxylase or lysyl oxidase are the first to be limited by oxygen deficiency.

Greater than normal oxygen supply induces a complex metabolic response over a period of days (Hunt and Pai, 1972). The effect of prolonged hyperoxia in wound healing is not explained simply by greater energy production.

SUMMARY

The oxygen environment of cells in granulation tissue was studied in wounds in rabbits. The environment was found to be hypoxic. The PO_2 in which fibroblasts of granulation tissue exist is in a critical range in terms of the rate of oxygen consumption. The rate of collagen production is also related directly to the average tissue oxygen tension.

ACKNOWLEDGEMENTS

This work was supported by U.S. Public Health Service Grant GM 12829.

REFERENCES

Ehrlich, H. P., Grislis, G. and Hunt, T. K. (1972). *Surgery* **72**, 578–583.
Hunt, T. K. and Pai, M. (1972). *Surg. Gynecol. Obstet.* **135**, 561–567.
Niinikoski, J., Hunt, T. K. and Dunphy, J. E. (1972). *Am. J. Surg.* **123**, 247–252.
Remensnyder, J. P. and Majno, G. (1968). *Am. J. Pathol.* **52**, 301–323.
Silver, I. A. (1969). *Prog. Resp. Res.* **3**, 124–135.

DISCUSSION

Dr. Winter:

Wound surfaces, it seems, are generally relatively anoxic and the speed of healing may be governed by the rate of supply of oxygen to the regenerating tissues. This is the conclusion drawn from observations on the speed of epidermal regeneration on shallow wounds in porcine skin under several different environmental conditions. Epidermal cell migration takes place twice as fast under an oxygenpermeable film of polyethylene than under a normal dry scab. Migration is slower if the wound is covered with a film which has low oxygen permeability, for example, polyester film (Winter, 1972). The normal wound surface has a PO_2 of only about 10 mm Hg beneath the scab, where epidermal migration takes place. On wounds, occluded by polytetrafluoroethylene or polythene film, the PO_2 in the vicinity of the migrating epidermis is some 6 or 7 times higher (Silver, 1972).

Evidently, when wounds are covered with polythene, or other oxygen-permeable films, oxygen from the air diffuses through the film, is dissolved in the exudate on the wound surface and can be utilised by the migrating epidermal cells.

Unlike fibroblasts, whose activities are apparently inhibited by too much oxygen, epidermal cells can use all the oxygen that can be supplied to them. When pigs were treated intermittently with 100% oxygen at 2 atmospheres absolute pressure, in a hyperbaric oxygen chamber, the epithelisation of standard shallow wounds was accelerated by about 30% (Winter and Perrins, 1970).

In the healing of skin wounds the prevailing oxygen concentration affects connective tissue regeneration directly in ways which are being investigated by Hunt and Niinikoski and indirectly because epithelisation influences the course of underlying connective tissue regeneration and the speed of epithelisation is very sensitive to the amount of oxygen in the immediate environment of the moving epidermal cells. It is suggested that the best results are to be expected from wound dressings that are freely permeable to oxygen.

REFERENCES TO DISCUSSION

Silver, I. A. (1972). In "Epidermal Wound Healing" (H. I. Maibach and D. T. Rovee, eds.) pp. 291–305. Year Book Medical Publishers, Chicago.

Winter, G. D. (1972). Ibid. pp. 71–122.

Winter, G. D. and Perrins, D. J. (1970). In "Proceedings of the 4th International Congress on Hyperbaric Medicine" (J. Wada and T. Iwa, eds.) pp. 363–368. Igakushoin, Tokyo.

Oxygen and Wound Healing:
A New Technique for Determining Respiratory Gas Tensions in Human Wounds

JUHA NIINIKOSKI and JAAKKO KIVISAARI

*Department of Surgery and the Department of Medical Chemistry
University of Turku, Turku, Finland*

Oxygen dynamics of healing wounds gained practical importance after it was shown that the rate of tissue repair varies directly with the supply of oxygen (Niinikoski, 1969; Stephens and Hunt, 1971). The main targets of oxygen in healing wounds appear to be the synthesis of collagen and differentiation of fibroblasts (Niinikoski, 1969). The oxygen supply of the cells is very much diffusion limited. Oxygen gradients are steep between the capillary and the healing tissue a few microns away and significant portions of any injured tissue exist in conditions of low oxygen tensions which are far from optimal (Silver, 1969; Niinikoski et al., 1972a). These findings on the tissue microclimate have been obtained by ultramicro electrode techniques under direct vision. However, clinically it is not feasible to locate the sampling site microscopically. Therefore a method which gives an average reading in the healing tissue is needed.

Oxygen and carbon dioxide tensions of human tissue can be measured by implanting a Silastic tube into the target organ. The tube is then perfused with anoxic saline which equilibrates to the average PO_2 and PCO_2 of the surrounding tissue medium because Silastic is highly permeable to respiratory gases (Niinikoski et al., 1972b). The method has been applied for determinations in the subcutaneous tissue (Niinikoski et al., 1972b) and healing bone (Niinikoski and Hunt, 1972a) and has been proved to be a useful tool for assessing the viability of skin flaps (Myers et al., 1972). The values obtained with this method do not represent those of an intact tissue. The tissue to be measured is injured during implantation and therefore, the sequence of healing is reflected in the determinations.

In the present investigation the previous method has been modified so that the external conducting apparatus used for perfusion has been omitted. The tonometer tube is filled with anoxic saline which equilibrates to the average PO_2 and PCO_2 of the surrounding tissue within a short period. The equilibrated fluid is collected in an Astrup-type glass capillary tube which is then emptied into a cuvette containing either an oxygen or carbon dioxide electrode. Both tissue gas tensions are obtained within a period of ten minutes which saves considerable time and increases the clinical applicability of the method.

MATERIALS AND METHODS

The tonometer was made from a sterile silicone elastomer tube (Atrial Catheter-J, A 190, Extracorporeal Medical Specialities Inc. Mount Laurel, New Jersey) 16 cm long, impregnated with silver (O.D. 1·35 mm; I.D. 1·10 mm). Five human male volunteers acted as experimental subjects. The tube was implanted under local anesthesia with 1% lidocain in the lateral side of the upper arm of each of the volunteers by introducing it through a wide-bore needle so that 14 cm of the tube remained under the skin. The puncture wounds were sealed with Nobecutan spray (Bofors, Nobel-Pharma, Sweden) and each end of the tube was fixed with a Steri-Strip (3M, Minnesota Mining and Manufacturing Co.).

Oxygen and carbon dioxide tensions were measured by filling the Silastic tube with anoxic saline for two minutes. During this period an equilibration of 90–95% was achieved in the gas tensions between saline and the tissue in contact with Silastic. After the 2-min period the equilibrated fluid was sampled into an Astrup-type glass capillary tube by filling the tonometer with another dose of anoxic saline from a 1-ml glass syringe (Fig. 1). The

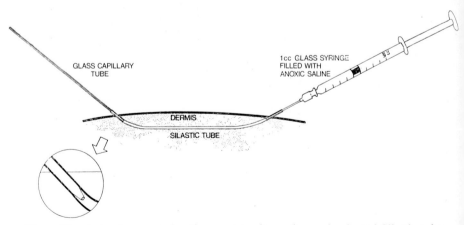

Fig. 1. The design for measuring tissue gas tensions using an implanted Silastic tube and the capillary sampling technique.

volumes of the glass capillary and Silastic tube were equal (90 μl). The Astrup capillary was then inserted into a micro sample injector (Radiometer, Copenhagen, Denmark) and the sample emptied into a thermostatted 70 μl cuvette containing either a Clark-type oxygen electrode or Severinghaus carbon dioxide electrode. The electrodes were connected with a Radiometer gas monitor PHM 71.

Zero adjustment of the O_2 electrode was obtained with gaseous nitrogen and calibration took place with aerated saline, PO_2 150 mm Hg, using capillary sampling technique. The CO_2 electrode was also calibrated by the capillary sampling technique using saline solutions of two known CO_2 tensions (26 and 57 mm Hg).

The relative efficiencies of various silicone elastomer tubes were tested *in vitro* by bathing the tubes in normal saline of known PO_2

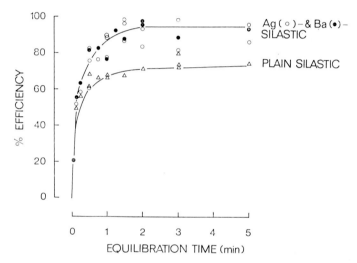

Fig. 2. The relative efficiences of various Silastic tubes used to measure fluid oxygen tension. One hundred per cent efficiency is a PO_2 level of 150 or 550 mm Hg. Temperature 37°C.

(150 and 550 mm Hg) and PCO_2 (57 mm Hg) at 37°C. Plain Silastic (Dow Corning Corp., Medical Products Division, Midland, Michigan) (O.D. 1·15 mm; I.D. 0·95 mm) and tubes impregnated with silver (O.D. 1·35 mm; I.D. 1·10 mm) or barium (Atrial Catheter-J., B190, Extracorporeal Medical Specialities Inc., Mount Laurel, New Jersey) (O.D. 1·50 mm; I.D. 1·20 mm) were tested. In both oxygen tensions a 90–95% efficiency was achieved in two minutes using Ag- and Ba-Silastic whereas plain Silastic of smaller wall thickness was less permeable to oxygen (Fig. 2). When the CO_2 permeability was tested

Ag-Silastic was most permeable showing a 85–90% efficiency at two minutes (Fig. 3). Ba-Silastic was less efficient probably because of the chemical affinity of CO_2 to barium and a greater wall thickness. Increase in the equilibration time added very little to the relative efficiencies and thus a two-minute equilibration period and Ag-Silastic were chosen for the experiments. During each measurement several capillaries for the assay of wound PO_2 and PCO_2 were filled. In some experiments response of tissue gas tensions to breathing of pure oxygen or a

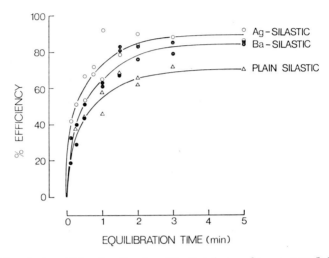

Fig. 3. The relative efficiencies of various Silastic tubes used to measure fluid carbon dioxide tension. One hundred per cent efficiency is a PCO_2 level of 57 mm Hg. Temperature 37°C.

mixture of 96% O_2 and 4% CO_2 was tested and the capillary blood PO_2 was assessed serially in samples taken from fingertips. Prior to the measurements the Silastic tube was thoroughly rinsed with anoxic saline to avoid contamination by air. If samples were not measured immediately the glass capillaries were sealed with wax and stored on crushed ice. O_2 and CO_2 permeabilities of the tonometers remained unchanged while being implanted in the tissue.

RESULTS

Immediately after implantation the oxygen tension varied between 60 and 80 mm Hg (Fig. 4). This was followed by a gradual decline in PO_2 over the following seven days until a minimum value of 40 to 45 mm Hg was achieved. Between days 7 and 10 the PO_2 remained

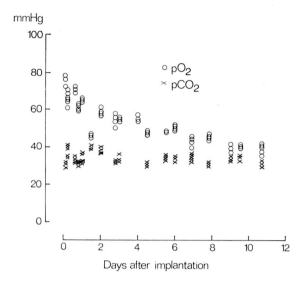

Fig. 4. Tissue gas tensions in the arms of five human volunteers measured with an implanted Silastic tube and capillary sampling technique. Each value represents one determination.

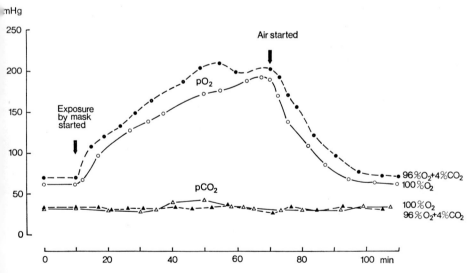

Fig. 5. Response of tissue gas tensions to breathing of pure oxygen or a mixture of 96% O_2 and 4% CO_2. Tonometry with Silastic tube *one* day after implantation. Each value represents one determination by the capillary sampling technique. Gas concentrations inside the mask were checked with a Radiometer gas monitor.

unchanged. Wound PCO_2 values remained between 30 and 40 mm Hg throughout the observation period.

The high initial PO_2 was probably due to fresh trauma since the highest responses of tissue PO_2 to breathing of pure oxygen were recorded during the first days after implantation (cf. Figs. 5 and 6). Oxygen breathing was usually continued for one hour before a maximum steady value was achieved. Adding of carbon dioxide into

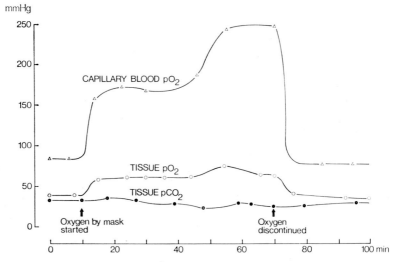

Fig. 6. Response of capillary blood PO_2 and tissue PO_2 and PCO_2 to breathing of pure oxygen. Tonometry with Silastic tube *nine* days after implantation. Each value represents one determination by the capillary sampling technique. Oxygen concentration inside the mask was checked with a Radiometer gas monitor.

pure oxygen in the breathing gas almost doubled the respiratory frequency. It had no effect on the maximal response of tissue PO_2, and the tissue PCO_2 showed no change during breathing of pure oxygen or increased CO_2 tensions (Fig. 5). Alteration in the capillary blood PO_2 were clearly reflected in the tissue PO_2 (Fig. 6).

DISCUSSION

Reported 'average' PO_2 values in healing tissues have varied depending on the techniques used. The results of this work are comparable to the earlier human studies in which tissue gas tensions were measured by continuous perfusion of the implanted Silastic tonometer (Niinikoski *et al.*, 1972b). The normal sequences of change of PO_2 (Fig. 4) are identical and the responses to change in ambient PO_2 are in general similar to the findings with the earlier method.

Niinikoski and Hunt (1972b) reported a mean of 20 to 25 mm Hg in the subcutaneous wounds of rabbits. In the skin flaps of pigs the average PO_2 has varied between 21 and 53 mm Hg and the PCO_2 between 41 and 53 mm Hg (Myers et al., 1972). All these results were obtained using Silastic tonometers. Waring and Pearce (1964) recorded a mean PO_2 of 53 mm Hg in subcutaneous gas pockets in infants. On the other hand, investigators using semimicro needle electrodes often do not give absolute values, but only the change in arbitrary units.

In the present study the tissue PO_2 remained unchanged after reaching the level of 40 to 45 mm Hg. This is consistent with findings with ultramicro oxygen electrodes in a fully established rabbit ear chamber (Niinikoski et al., 1972a). Seventy days after implantation, even the center of the chamber is invaded by the growing tissue. Vascularization in the center is very dense and the 'average' tissue PO_2 is about 45 mm Hg at a mean capillary PO_2 of 60 mm Hg.

The baseline PCO_2 recorded in the present study varied between 30 and 40 mm Hg, i.e. the range of arterial blood PCO_2 (Fig. 4). No changes were noted in the tissue PCO_2 during one-hour exposure to an increased CO_2 concentration or pure oxygen. This is consistent with findings of Myers and his group who also found that it is extremely difficult to affect the PCO_2 in the subcutaneous tissue and, that if any changes occur, they are very slow (personal communication). In other tissues, e.g. in the intestine, the tissue PCO_2 responds very rapidly to an increase in the ambient carbon dioxide concentration (Kivisaari and Niinikoski, unpublished data).

Niinikoski et al. (1972b) have shown that the tissue reaction around the Silastic tonometer consists of only two to four cell layers one week postimplantation. Blood vessels frequently approach to a distance of 40 to 50 μ from the tube. Studies conducted in our laboratory have shown that the Silastic tubing perfused with anoxic saline measures and consumes oxygen from a distance of at least 1·0 mm (Kivisaari and Niinikoski, unpublished data). This clearly exceeds the thickness of the reactive area around the tonometer and therefore, most of the oxygen in the sample derives probably from normal subcutaneous tissue after the initial trauma phase is over.

The capillary sampling technique has some advantages compared with the continuous perfusion of the tonometer:

1. Shorter duration of the measurement increases the clinical applicability of the method. Samples for the assay of tissue gas tensions can be

collected in four minutes and measured in five by the capillary sampling techniques whereas production of the same data using continuous perfusion takes 1·5 h.

2. The external conducting apparatus used for perfusion is exposed to a possibility of leakage and occlusion which is excluded from the present method.

3. Each laboratory possessing an Astrup-type machinery for measurements of acid-base equilibrium and blood gases can use this method with minimal additional investments.

If measurements of respiratory gas tensions in intact tissues are preferred even the new and costly mass spectrometer has the same disadvantage as the Silastic tubing: the oxygen sensor has to be implanted into the target organ which produces trauma interfering with the supply of oxygen.

SUMMARY

Oxygen and carbon dioxide tensions were measured serially in one-hour to 11-day old human wounds by implanting a 14 cm Silastic tube subcutaneously in the arm. The Silastic tonometer was filled with anoxic saline, which equilibrated to the average PO_2 and PCO_2 of the surrounding tissue within a few minutes because Silastic is highly permeable to O_2 and CO_2. The equilibrated fluid was collected in an Astrup-type glass capillary tube which was then emptied into a micro cuvette containing either an oxygen or carbon dioxide electrode. Due to the chemical inertness of Silastic, inflammatory reaction around the tonometer is minimal.

ACKNOWLEDGEMENTS

This work was supported in part by Contract No. DAJA37-72-C-1573 from the U.S. Army through its European Research Office.

We are grateful to Dr. Thomas K. Hunt, M.D., San Francisco, California, Dr. M. Bert Myers, M.D., New Orleans, Louisiana, and Professor Eino Kulonen, M.D., Turku, Finland, for their great help and constructive criticism during this work. Our thanks are also due to Mrs. Heidi Pakarinen for skilful technical assistance.

The publishers of the *American Journal of Surgery* kindly permitted the reproduction of the figures.

REFERENCES

Myers, M. B., Cherry, G. and Milton, S. (1972). *Surgery* **71**, 15–21.
Niinikoski, J. (1969). *Acta Physiol. Scand. Suppl.* **334**, 1–72.
Niinikoski, J. and Hunt, T. K. (1972a). *Surg. Gynecol. Obstet.* **134**, 746–750.
Niinikoski, J. and Hunt, T. K. (1972b). *Surgery* **71**, 22–26.

Niinikoski, J., Hunt, T. K. and Dunphy, J. E. (1972a). *Am. J. Surg.* **123,** 247–252.
Niinikoski, J., Heughan, C. and Hunt, T. K. (1972b). *J. Surg. Res.* **12,** 77–82.
Silver, I. A. (1969). *Prog. Resp. Res.* **3,** 124–135.
Stephens, F. O. and Hunt, T. K. (1971). *Ann. Surg.* **173,** 515–519.
Waring, W. W. and Pearce, M. B. (1964). *Biochem. Clin.* **4,** 31–38.

Injury and Repair in Vascular Connective Tissue

Ib Lorenzen

Medical Department C, Gentofte Hospital, Copenhagen, Denmark

CONNECTIVE TISSUE IN HUMAN VASCULAR DISEASES

Changes in the vascular connective tissues are characteristic features of some of the most common diseases in our society. The nature of these alterations as well as their importance to the development of the diseases in question varies. Unfortunately our interpretation of the changes in the vascular connective tissue observed in the different diseases is based mainly on hypotheses.

Considering our present knowledge of the biological function of the connective tissue components the importance of these components to the normal function of the blood vessels seem well documented. Apart from the cells this is true of several of the intercellular macromolecules.

As to the role of the vascular connective tissue in diseases involving the vascular wall several possibilities exist:

(*i*) The alterations may be rather non-specific and reflect tissue damage and secondary reparative processes. This is probably the most common type, observed in a number of the connective tissue diseases and in arteriosclerosis.

(*ii*) The vascular connective tissue may also be involved in the development of the diseases at a higher level. Examples are the metabolic changes observed in the hereditary connective tissue diseases, such as Hurler's and Marfan's syndrome in which genetically determined changes of intercellular macromolecules are responsible for the diseases. Similar hormonaly induced alterations of the vascular connective tissue seem to be of importance in some endocrinological diseases e.g. diabetes mellitus and myxoedema.

Probably a number of other metabolic changes of primary pathogenetic importance exist. Thus it is generally assumed that changes in the connective tissue play a role in ageing as well as in some of the vitamin deficiencies. Due to the frequency of the diseases including these types of alterations the processes of injury and repair of the vascular connective tissue have been paid particular attention.

INJURY AND REPAIR

In the consideration of injury and repair in the vascular wall the following questions arise:

(*i*) Which are the causes of vascular injury?

(*ii*) Which of the connective tissue components takes part in the reaction to injury and in the repair processes?

(*iii*) What is the normal reaction pattern of the vascular connective tissue to injury?

(*iv*) Are some factors able to interfere with the normal processes of repair?

(*v*) What determines the susceptibility of the vascular wall to injury?

Arteriosclerosis is the most important of the vascular diseases in which processes of injury and repair are supposed to play a role. Consequently most studies on these phenomena have been performed on arteries. Arterial injury may be induced by several factors such as hydraulic phenomena, lipids, infections, immunological reactions and other factors causing hypoxia of the arterial wall, e.g. tobacco smoking (Astrup, 1972).

The importance of different types of vascular injury in the pathogenesis of human arteriosclerosis is indicated by the following phenomena: (*i*) The localization of the arteriosclerotic lesions to the site subjected to greatest haemodynamic strain. (*ii*) The unquestionable role of arterial hypertension to the development of arteriosclerosis. (*iii*) The predisposing role of the syphilitic and rheumatic aortitis to the localization of human arteriosclerosis and finally (*iv*), the similarity of some of the features of human arteriosclerosis to healing processes of wounds. Also the similarity between different types of experimental arteriosclerosis induced by vascular injury and human arteriosclerosis supports this hypothesis. Below is presented a short survey on our own experimental studies on the reaction of the vascular connective tissue to different types of injury.

EXPERIMENTAL MODELS

In all our experimental studies we have used the aorta of adult male albino rabbits. The aorta offers the advantage of yielding sufficient amounts of tissue for simultaneous morphological and biochemical studies. Furthermore, the rabbit aorta is rather sensitive to damage and the reactions are of a satisfactory reproducibility. Our investigations have mainly been restricted to the descending thoracic aorta. This part of the aorta is of the elastic type like the human aorta and is rarely the site of spontaneous arteriosclerosis in contrast to the first part of the aorta, from the aortic valve to the first intercostal arteries (Garbarsch et al., 1970). Different procedures have been used in our experiments to induce arterial damage: (i) Injections of catecholamines and thyroid hormones (Lorenzen, 1963), (ii) exposure to systemic hypoxia (Helin and Lorenzen, 1969) and (iii) mechanical dilatation (Helin et al., 1971a). Even if the mechanism of the damaging action of the listed agents varies, the morphological and biochemical alterations following these types of injury are closely related. Among the four types of damaging agents the mechanical injury performed by a single short lasting dilatation with a balloon catheter offers the advantage that the exact time of the vascular damage is known. One of the major difficulties in the estimation and interpretation of morphological and biochemical alterations in human arterial tissue as well as in several types of experimental models is the fact that the exact time of the start and consequently the time of existence of the alterations observed is usually unknown. This fact furthermore makes a correct evaluation of the interrelationship between coexisting changes observed at a 'snap shot' extremely difficult.

On this background we considered the unphysiological procedure represented by the mechanical dilatation justified. The observations reported in the following are primarily from this model (Helin et al., 1971a, b; Lindy et al., 1972).

The injury has been induced by pulling an inflated balloon catheter distally from the first intercostal arteries to the diaphragm. The duration of this procedure is about 1 s. The animals have then been killed at different time intervals up to half a year after the dilatation. The analytical procedures used in our studies include morphological and biochemical analyses as well as isotope studies.

GROSS AND MICROSCOPIC ALTERATIONS IN VASCULAR INJURY AND REPAIR

Two weeks after the dilatation the dilated aortas show gross arteriosclerotic changes consisting of white plaques surrounding the intercostal arteries, bean-shaped, arteriosclerotic lesions and sometimes large aneurysms (Figs. 1–4). We have observed no differences in the appearance and the extent of the gross alterations in the later stages i.e. up to half a year after the injury. The type of the gross lesions are thus widely independent of the age of the lesions. The microscopic alterations in the early stages, three and six days after the injury are dominated by necroses and degenerative changes in the intima as well as in the media. The endothelial layer is absent in large areas (Helin et

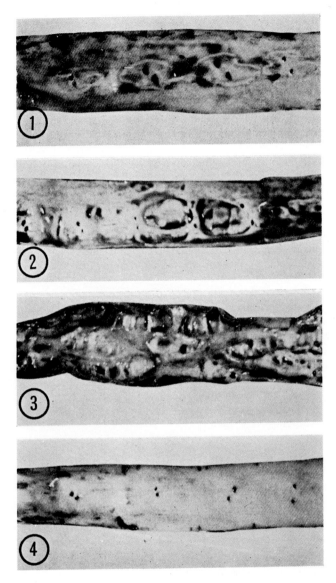

Figs. 1–4. Thoracic aorta of rabbits 2 weeks after mechanical dilatation (Figs. 1–3) and of a control rabbit (Fig. 4). (Helin et al., 1971a.)

al., 1971b). In the luminal half of the media, infiltration with granulocytes is observed. Two weeks after the injury the greater part of the inner surface area is re-endothelialized. The new endothelial cells estimated by ordinary light microscopy as well as by scanning electron microscopy (Christensen and Garbarsch, 1972) differ from the normal

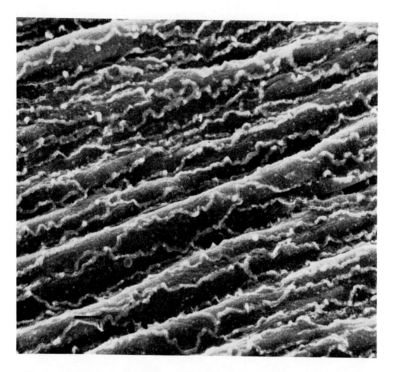

Fig. 5. Normal intima. Coarse longitudinal folds produced by the underlying elastic lamellae are covered by a rhombic pattern of endothelial cells. (bar 10 μm; × 800)

cells by a more irregular and polygonal appearance. (Figs. 5, 6 and 7, scanning electron micrographs of rabbit thoracic aorta with silver staining of the endothelial lining by B. C. Christensen M.D., Institute of Anatomy, University of Odense.) A subendothelial cellular thickening of the intima is a constant phenomenon. Furthermore new formation of collagen and elastic fibres are seen. The medial lesions include calcified and non-calcified foci surrounded by collagen and mononuclear cells as well as giant cells. Thirty, 60 and 180 days after the injury the changes are rather similar to those of the two-week stage, the only difference being a more pronounced intimal thickening and an increase in the amount of collagen fibres.

Histochemical investigation of the glycosaminoglycans (Helin *et al.*, 1971b) reveals metachromasia and alcianophilia of the intercellular

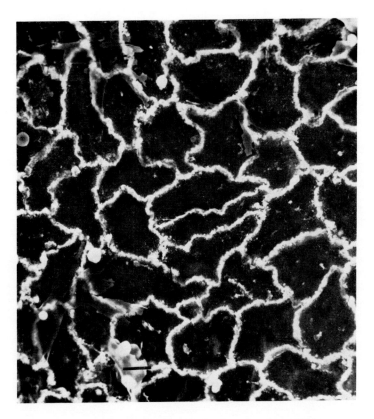

Fig. 6. Regenerated hexagonal endothelial cells. (bar 10 μm; × 800)

substance of the intima as well as of the media. The distribution depends somewhat upon the age of the lesions with a tendency of localization around the necrotic calcified foci, particularly in places with accumulation of collagen (Helin *et al.*, 1971b). Investigation of the metachromasia at different pH levels as well as after digestion with testicular hyaluronidase suggests the presence of hyaluronic acid as well as different sulphated glycosaminoglycans.

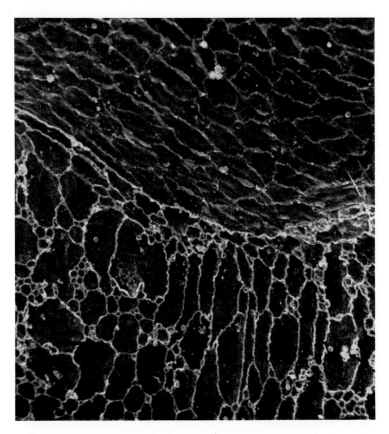

Fig. 7. Regenerated endothelial cells. Two typical patterns of silver lines are separated by a sharp boundary. Hexagonal cells above and cells of varying size and shape below. The smallest meshes are probably too small to represent individual cells. (bar 50 μm; × 240)

BIOCHEMICAL ALTERATIONS FOLLOWING VASCULAR DAMAGE

Biochemical analyses reveal an early increase in the aortic content of hyaluronic acid followed by an increase of chondroitin-4,6-sulphate and later heparan sulphate and dermatan sulphate (Fig. 8). Once increased

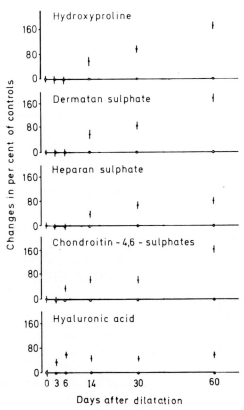

Fig. 8. Changes in aortic glycosaminoglycans and hydroxyproline following a single mechanical dilation. The bars represent means ± standard deviations of means (Helin et al., 1972).

the fractions remain elevated throughout the observation period. The *in vivo* uptake of ^{35}S-sulphate into chondroitin-4, 6-sulphate as well as in the hyaluronidase resistant fraction representing heparan sulphate and dermatan sulphate show an early increase followed by a decrease, compatible with a transitorily stimulated synthesis (Fig. 9). The aortic content of hydroxyproline, the total amount as well as the hydroxyproline content determined after extraction of collagen by autoclaving,

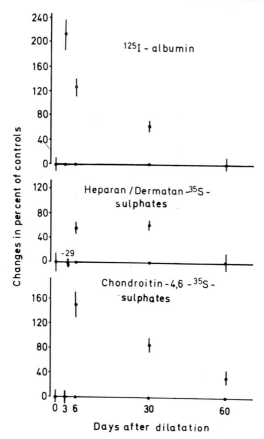

Fig. 9. The uptake of ^{35}S-sulphate and ^{125}I-albumin in aorta following a single mechanical dilatation (Helin et al., 1972).

shows a steady increase from the fourteenth day. This indicates an increase in the aortic collagen. According with this observation an increase in the protocollagen proline hydroxylase activity of aorta has been observed from the 6th day (Fig. 10) (Lindy et al., 1972).

The vascular permeability to ^{125}I-labelled albumin is increased immediately after the injury approaching normal values 60 days after the injury (Fig. 11). The total aortic content of DNA and RNA is elevated from the 3rd day (Figs. 11 and 12). This may reflect cellular proliferation as well as increased protein synthesis. Finally the LDH activity of the aorta increases from the 6th day (Fig. 13). The increase involves particularly the cathodically migrating isoenzymes which are composed of M subunits. This may reflect tissue hypoxia.

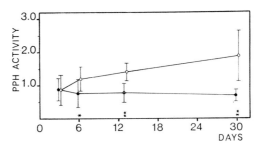

Fig. 10. Protocollagen proline hydroxylase (PPH) activity expressed as dpm $\times$ 10^{-4}-^{14}C-hydroxyproline synthesized in an aliquot of 50,000 dpm ^{14}C-labelled protocollagen substrate per g wet weight of tissue in injured (o) and control aortas (·). Mean $\pm$ SD of 5 to 10 samples. (*) $P \leqslant 0.05$. ($\ast\!\!\ast$) $P \leqslant 0.005$. For details see Lindy et al. (1972).

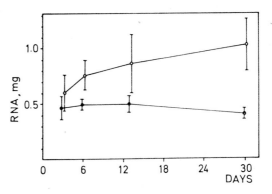

Fig. 11. Total content of RNA in injured (o) and control aortas (·). The bars indicate means $\pm$ SD of 5 to 10 samples. For details see Lindy et al. (1972).

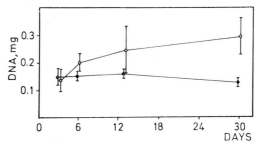

Fig. 12. Total content of DNA in injured (o) and control aortas (•). The bars indicate means $\pm$ SD of 5 to 10 samples. For details see Lindy et al. (1972).

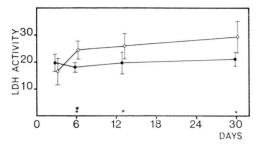

Fig. 13. LDH activity expressed as IU/g wet weight of tissue in injured (o) and control (•) aortas. Mean ± SD of 10 samples. For details see Lindy et al. (1972).

The reported time-dependent alterations in the glycosaminoglycans and collagen of the aortic wall following a single mechanical dilatation injury are related to the alterations observed in granulation tissue. In experimental granulomas Lehtonen (1968) observed an initial increase in the total amount of hyaluronic acid, which remained elevated for more than 120 days. A similar long-lasting increase was observed in the total amount of chondroitin sulphates. Niinikoski (1969) reported an increase in the total amount of collagen persisting for more than 15 days also in experimental granulomas. Finally oedema, leukocytic reaction, accumulation of amorphous metachromatic intercellular substance as well as fibroblast proliferation and collagen formation are characteristic features of the granulation tissue (Dunphy and Udupa, 1955; Taylor and Saunders, 1957). These observations support the concept that the alterations in the aortic wall following a single mechanical dilation injury are repair processes.

REACTION OF THE VASCULAR WALL TO DIFFERENT TYPES OF INJURY

The morphological and biochemical alterations in the aortic wall following a mechanical injury are closely related to the alterations which we have observed in previous studies on other types of vascular damage including vascular injury induced by exposure of the animals to systemic hypoxia and injection of catecholamines and thyroid hormones (Lorenzen 1963; Helin et al., 1971b). Similarly Ichida and Kalant (1968) observed an increase in the concentration, synthesis and turn-over of condroitin sulphates of aorta with cholesterol atheromatosis. Saxena and Nagchaudhuri (1969) found an increase of hyaluronic acid of aorta during the initial stages of atheroma formation in cholesterol-fed rabbits. The concentration decreased later on in the more advanced stages. Hilz and Utermann (1960) demonstrated an increased incorporation of

^{35}S-sulphate into chondroitin sulphate in rat aortas with arteriosclerotic changes induced by corticotrophin. Finally Hauss et al. (1960) observed an increased incorporation of radioactive sulphate into chondroitin sulphate in rabbit aorta damaged by bacterial infection and sensitization to foreign proteins. The similarities between the various forms of experimentally induced arterial lesions suggest that the reaction of the arterial wall to damage is rather uniform and non-specific, largely independent of the nature of the damaging agent, as already emphasized by Gillman and Hathorn (1957).

RELATION OF EXPERIMENTAL STUDIES TO HUMAN VASCULAR DISEASES

The changes in human arteriosclerosis have certain features in common with experimentally induced processes of injury and repair of the arterial wall (Lorenzen, 1966). The earliest changes of human arteriosclerosis are observed in the intercellular substance as intimal accumulation of amorphous metachromatic ground substance. The underlying changes in the glycosaminoglycans may be responsible for the deposition of lipids and calcium salts. The subsequent changes: cellular proliferation and formation of collagenous fibrils followed by degenerative changes show the characteristics of repair in connective tissue.

A direct comparison of the biochemical changes in the glycosaminoglycans and collagen in experimental vascular lesions with the observations reported from human vascular diseases is rendered difficult by two facts:

(i) The analytical results from human vascular connective tissue are usually only given as *concentrations* and only rarely as *total contents*. This may provide an erroneous picture of the metabolic alterations.

(ii) The start of the pathological processes and their time of existence in human vascular connective tissue is usually unknown.

In spite of these difficulties the comparison shows a number of similarities which seem to support the hypothesis that non-specific processes of injury and repair are of importance also in human vascular diseases, in particular in human arteriosclerosis and in several connective tissue diseases. Kumar et al. (1967) found an increase in the concentration of chondroitin-4,6-sulphate and dermatan sulphate in the early stages of human arteriosclerosis (fatty streaks). In the more advanced stages (fibrous plaques) a decrease was observed. Klynstra et al. (1967) observed an early increase in the concentration of heparan sulphate and chondroitin-4,6-sulphate followed by a later decrease. Buddecke (1962) reported an increase in the concentration of chondroitin-6-sulphate, heparan sulphate and dermatan sulphate and a decrease in the concentration of hyaluronic acid.

From the working hypothesis on the importance of injury and repair processes in human vascular diseases it seems relevant in future studies to test suspected agents of their vascular damaging properties, to elucidate how the normal repair mechanism of the vascular wall can be disturbed and to investigate factors of importance to the susceptibility of the vascular connective tissue to injury.

SUMMARY

Non-specific processes of injury and repair are supposed to play a role in different types of vascular diseases, among these arteriosclerosis and connective tissue diseases. A survey is given on the gross, microscopic and biochemical alterations in the aortic wall of rabbits following injury.

As a prototype of arterial injury the mechanical damage following a single short-lasting dilatation with a balloon catheter is selected.

Two weeks after the injury gross arteriosclerotic changes are visible. The appearance of the gross alterations are widely independent of the age of the lesions. The microscopic changes in the early stages consist of necrosis and degenerative alterations followed in the later stages by cellular intimal thickening, and new formation of cells, amorphous metachromatic intercellular substance as well as collagen and elastic fibres in the media.

Biochemical analyses reveal an early increase in the aortic content of hyaluronic acid followed by an increase in chondroitin-4,6-sulphate, heparan sulphate and dermatan sulphate. Once increased the fractions remain elevated throughout an observation period of 6 months. The *in vivo* uptake of ^{35}S-sulphate into the sulphated glycosaminoglycans shows an early increase followed by a decrease. The aortic content of hydroxyproline and the activity of the enzyme protocollagen proline hydroxylase is increased after the injury. Similarly there is an increase in the content of DNA, RNA and lactate dehydrogenase activity. Finally the vascular permeability to ^{125}I-labelled albumin is increased immediately after the injury approaching normal values 60 days after the injury.

The alterations show great similarity to those following quite different types of vascular damage. This indicates a uniform reaction pattern of the arterial wall to injury. Furthermore the alterations are similar to those observed in granulation tissue and are therefore in all probability to be considered as processes of repair. The biochemical and morphological alterations following experimental injury have a number of features in common with the alterations in human arteriosclerosis. The experimental models may therefore be of relevance to the study of this disease.

ACKNOWLEDGEMENTS

I should like to express my thanks to The American Heart Association, Inc. for permission to reproduce Figs. 10, 11, 12 and 13.

REFERENCES

Astrup, P. (1972). *Brit. Med. J.* **4,** 447–449.
Buddecke, E. (1962). *J. Atheroscler. Res.* **2,** 32–36.
Christensen, C. B. and Garbarsch, C. (1973). (In Press.)
Dunphy, J. E. and Udupa, K. N. (1955). *New Engl. J. Med.* **253,** 847–851.
Garbarsch, C., Mathiessen, M. E., Helin, P. and Lorenzen, I. (1970). *Atherosclerosis* **12,** 291–300.
Gillman, T. and Hathorn, M. (1957). *J. Mt. Sinai Hosp.* **24,** 857–868.
Hauss, W. H., Junge-Hülsing, G. and Strobel, W. (1960). *Z. Rheumaforsch.* **19,** 161–164.
Helin, P. and Lorenzen, I. (1969). *Angiology* **20,** 1–12.
Helin, P., Lorenzen, I., Garbarsch, C. and Matthiessen, M. E. (1971a). *Atherosclerosis* **13,** 319–331.
Helin, P., Lorenzen, I., Garbarsch, C. and Matthiessen, M. E. (1971b). *Circ. Res.* **29,** 542–554.
Helin, P., Lorenzen, I., Garbarsch, C. and Matthiessen, M. E. (1972). *Angiology* **23,** 183–187.
Hilz, H. and Utermann, D. (1960). *Biochem Z.* **332,** 376–379.
Ichida, T. and Kalant, N. (1968). *Canad. J. Biochem.* **46,** 249–260.
Klynstra, F. B., Böttcher, C. J. F., Van Melsen, J. A. and Van der Laan, E. J. (1967). *J. Atheroscler. Res.* **7,** 301–309.
Kumar, V., Berenson, G., Ruiz, E. R., Dalferes Jr., E. R. and Strong, J. P. (1967). *J. Atheroscler. Res.* **7,** 583–590.
Lehtonen, A. (1968). *Acta physiol. scand. Suppl.* **310,** 1–74.
Lindy, S., Turto, H., Uitto J., Helin, P. and Lorenzen, I. (1972). *Circulation Res.* **30,** 123–130.
Lorenzen, I. (1963). "Experimental Arteriosclerosis". Munksgaard, Copenhagen.
Lorenzen, I. (1966). *In* "Hormones and Connective Tissue" (G. Asboe-Hansen, ed.) pp. 157–166. Munksgaard, Copenhagen.
Niinikoski, J. (1969). *Acta physiol. scand. Suppl.* **334,** 1–72.
Saxena, I. D. and Nagchaudhuri, J. (1969). *Indian J. med. Res.* **57,** 96–102.
Taylor, H. E. and Saunders, A. M. (1957). *Amer. J. Path.* **33,** 525–528.

Connective Tissue, Smooth Muscle and Leukocytes in Repair Processes of the Arterial Wall

S. BJÖRKERUD

Departments of Histology and Internal Medicine I
University of Göteborg, Göteborg, Sweden

Together with the heart, arterial tissue is in the unique position of being subjected to continuous oscillating strain during the entire life-time of the individual. This and other facts indicate that a major quality of the arterial tissues is the capacity to withstand mechanical strain and injury. An analysis of this capacity and its relationship to the different tissues of the arterial wall will be the basis of the following discussion.

The arterial wall is not one tissue, but is normally composed of at least three different tissues. These tissues are (*i*) *endothelium* on the lumenal side and in the vasa vasora of the adventitia and, in larger arteries, of parts of the media, (*ii*) *smooth muscle tissue* of the media and of the subendothelial zone, if this is present, and (*iii*) the *connective tissue* of the adventitia. As these tissues have a number of distinctive features, it can be assumed that they differ with regard to their response to injury. This would result in an extremely complex pattern of changes in the arterial wall when all of them are subjected to instantaneous injury (Björkerud, 1972a). In certain conditions the arterial wall also contains components of other tissues as reflected in the presence of thrombocytes, neutrophil leukocytes, monocytes and foreign body giant cells. This adds considerably to the complexity of the changes occurring in the arterial wall.

We have aimed to devise a model system for experimental arterial injury with high selectivity with regard to the type of arterial tissue and the extension of the injury. The availability of such a system could enable the analysis of the contribution of each different arterial tissue in the response of the arterial wall to different types of injury and provide a means for the study of basal repair and regeneration mechanisms *in vivo*.

The discussion will be focused on the morphological patterns of the remodelling of the tissue after different types of injury. The methods used enable the study of whole arterial segments without sectioning to the resolution limit of the light microscope (Björkerud, 1972b). The integrity of the arterial endothelium was determined with dye exclusion tests carried out as supravital perfusions *in situ* (Björkerud and Bondjers, 1972).

RESPONSE PATTERNS AFTER SPECIFIC TYPES OF ARTERIAL INJURY

By means of a microsurgical instrument equipped with microscopic diamonds as cutting edges (Björkerud, 1969a; Fig. 1), graded mech-

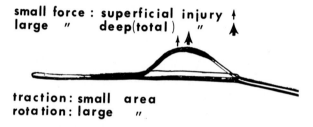

Fig. 1. The distal end of the microsurgical instrument used for the *in vivo* induction of defined mechanical injury in the aorta. An inner Perlon line can be bent out through a slit in an outer Nylon catheter by pushing the free proximal end of the line into the catheter. The part of the line which can be exposed through the slit is coated with microscopic diamonds which serve as cutting edges. Pushing the inner line with small force creates superficial, i.e. subtotal injury of the intima and media; large force leads to total injury. A small area is injured by traction of the whole instrument; a large area is injured by rotating the instrument. (Reproduced from Björkerud (1969a) after slight modification.)

anical injury could be induced to the aortic wall in the living experimental animal (rabbit and rat). The *location* in the aorta, the radial extension, i.e. the *depth*, and the circumferential extension, i.e. the *surface area*, could be adjusted at will. The response of the arterial wall to the following types of injuries was analyzed: (1) superficial, i.e. subtotal, injury with small area, (2) deep injury, i.e. total necrosis of the intima and media, with small area, and (3) superficial, i.e. subtotal, injury with large area.

Subtotal injury with small area

A subtotal injury with small area is characterized initially by an acute inflammatory response which is followed by a rapid formation of an intimal thickening by multiplication of smooth muscle cells encompassing the injured region and by migration of such cells through the internal elastic membrane to a subendothelial position (Fig. 2). After 2–3 weeks a considerable subendothelial thickening has formed.

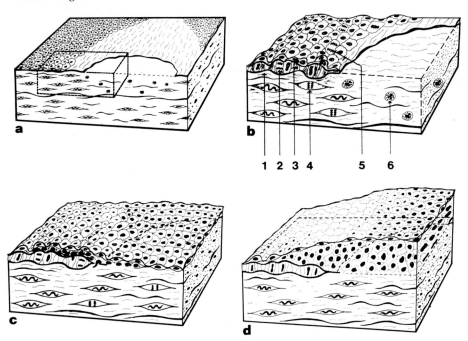

Fig. 2. Schematic outline of the changes in the arterial wall following a subtotal injury with small area. Initial changes (a) are an acute inflammatory reaction followed by (b) regeneration of the endothelium (5) and smooth muscle cells by cell divisions. A marked intimal thickening forms (c, d) by migration of cells from the media (3), by cell divisions, and by formation of extracellular components in the subendothelial space. After completion of the endothelial covering, the intimal thickening disappears and the gross structure of the segments is restored. (Reproduced from Björkerud, 1969a.)

The height of the thickening may exceed that of the media. Partly, the increasing thickness of the intima is due to the formation of extracellular components as collagen and elastic fibers. The smooth muscle tissue of the injured media regenerates during the formation of the intimal thickening and the defect of the endothelial lining is repaired by ingrowth from the adjacent endothelium. After the integrity of the endothelium has been restored the intimal thickening recedes, and the lesion is rarely detectable after 8 weeks except for the defects of the internal elastic membrane. Lipids do not accumulate in the lesions (Björkerud, 1969a, b; Bondjers and Björnheden, 1970).

This type of repair reaction has a number of properties in common with repair in other tissues, e.g. skin. However, the very hyperplastic nature of the reaction seems to be rather specific for the arterial wall. Both the properties of the thickening and the change in the geometry of the injured segment due to the presence of the thickening suggest that it may play a pathophysiological role by at least two mechanisms. Collagen and elastic fibers are abundant in the thickening. The elastic components are oriented in concentric planes to form bridges over the injured region.

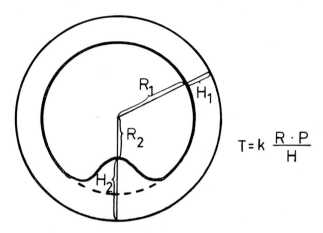

Fig. 3. Schematic illustration of the change in geometry in a tubular arterial segment due to the formation of an intimal thickening and the relationship of different geometric parameters to the tangential wall tension (T) as expressed by the equation of Frank-Laplace. R, radius of lumen; H, thickness of the wall; P, blood pressure; k, constant.

The edges of the elastic fibers are attached to the internal elastic membrane lateral to the damage (Björkerud, 1969a). It is likely that the fibrous components relieve the injured sector from tangential strain. In other words, the subendothelial thickening may act as an adhesive tape or suture. The change of the geometry of the arterial segment due to the presence of a thickening may also have the effect of reducing the tangential strain as is apparent from the relationship between the geometrical parameters of the segment and the tangential wall tension (Fig. 3). Both an increase in wall thickness and a decrease of the radius will reduce the tangential wall tension in the thickened sector, i.e. in the damaged area. Therefore, it is possible that the repair reaction following a subtotal injury with restricted area reflects a delicate repair mechanism, specific for the arterial wall, and which is elicited by injuries keeping within a physiological range. The appearance of this type of lesion is similar to the non-atherosclerotic process, progressive 'diffuse' intimal thickening.

Deep injury with small area

The reaction which follows the induction of a deep injury, i.e. a total necrosis of intima and media, with a small area, is initially characterized by an acute inflammatory reaction in the injured tissue. The necrotic tissue calcifies in a few days (Fig. 4). The number of smooth muscle cells

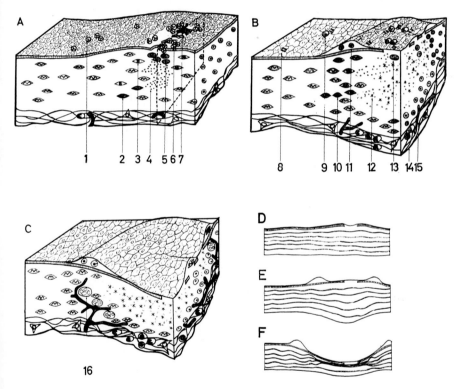

Fig. 4. Scheme illustrating the changes following the induction of a total injury of intima and media with small area. An initial (A) acute inflammatory reaction (1) is followed (B) by rapid reendothelialization, calcification of the necrotic region (12), and increased cellularity of the encompassing tissue to a capsule-like structure. Finally, (C,F), the region is dilatated in an aneurysm-like fashion. Foreign body giant cells (16) are present in the capsule and capillaries grow towards and into the capsule from the adventitia (15). (Reproduced from Björkerud and Bondjers, 1971a, Fig. 2.)

of the encompassing tissue increases markedly, and they form a dense capsule-like structure around the calcified necrotic region. Towards the necrosis foreign body giant cells are present in the capsule. Wide tortuous capillaries grow towards and into the capsule from the adventitia. The general features of this response pattern seem to be rather non-specific. Similar changes are well known from the pathology of other tissues after injury (cf. Letterer, 1959). The morphological properties of this type of arterial lesion are very similar to those of media sclerosis in man (Mönckeberg sclerosis) (Björkerud and Bondjers, 1971a).

Subtotal injury with large area

The response after a subtotal injury with large area can be divided in three phases: (a) an *early phase;* (b) a subsequent phase with *defective reendothelialization;* and (c) a final phase with *late reendothelialization* (Björkerud and Bondjers, 1973.)

(*a*) *The early phase* (Figs. 5A and B)

Most of the aortic endothelium is initially desquamated after the induction of the injury. Simultaneously with an acute inflammatory reaction an intimal thickening is formed by multiplication and migration of smooth muscle cells encompassing the injured region. Reendothelialization takes place by ingrowth of endothelium from the aortic branches. In regions close enough to branches to be reendothelialized within 2–3 weeks the structure of the wall is restored in a way similar to that following the induction of a sub-total injury with small area (see above).

(*b*) *Phase of defective reendothelialization* (Fig. 5C)

In regions more remote from the aortic branches the time required for reendothelialization is increased to a varying degree. In such regions the intimal thickening does not revert as long as the surface is devoid of endothelial lining but increases in height and lipids accumulate in the intima and the inner media (Björkerud, 1969b). Small thrombi and flaps of detached endothelium are frequent along the margin of the growing endothelium which may be reasons for the marked retardation of the reendothelialization (cf. Fig. 6). An interesting feature of the lesions in this phase of development is the virtual absence of oriented mature thick elastic fibers which were prominent in the reversible non-atherosclerotic lesions following subtotal injury with a small area (see above). In contrast the elastic component of the progressive type of experimental intimal thickening is composed of thin short fibrils with no specific orientation (Björkerud, 1969b).

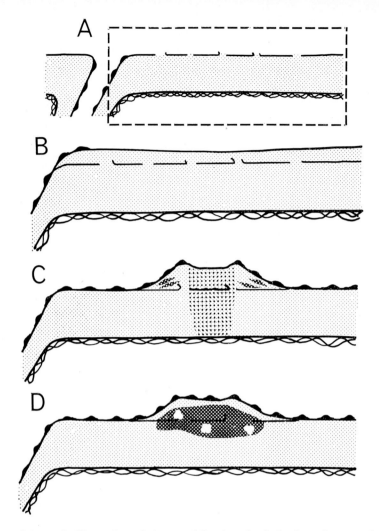

Fig. 5. Schematic illustration of changes following the induction of a superficial, i.e. subtotal injury with large area. A: Initially, a large proportion of the aortic surface is desquamated (subsequent changes in enframed sector are depicted in figures below). B: The *early phase* is characterized by an acute inflammatory reaction, the formation of an intimal thickening by multiplication and migration of smooth muscle cells, and reendothelialization of regions adjacent to branches by ingrowth of endothelium from the branches. In regions which are close enough to branches to be reendothelialized rapidly, the structure is restored (see left side of C). C: The thickness of the intima increases markedly, and lipids accumulate mainly extracellularly in the non-endothelialized regions of the plaques (dotted area). Smooth muscle form cells occur in the reendothelialized regions of the plaques. Small thrombi and flaps of desquamated endothelium are frequently found along the edge of the regenerating endothelium. Following *late reendothelialization* (D), the underlying tissue succumbs more or less completely. Masked lipids are liberated, and the necrotic tissue stains strongly for lipids (dark area). Monocytes and foreign body giant cells (white), containing lipid inclusions, are frequent in the necrotic tissue. The lipids disappear in a few days, and similar changes follow as those after the direct induction of a necrosis (cf. Fig. 4), i.e. the necrosis is calcified and surrounded by a dense capsule containing foreign body giant cells. Capillaries grow towards the necrotic region from the adventitia.

A decreased number of cells per unit tissue mass in the media underlying non-regressive intimal thickenings suggests additional injury instead of healing in the non-regressive type of lesions (Björkerud, 1970). This may be hypothetically explained by the increased distance for

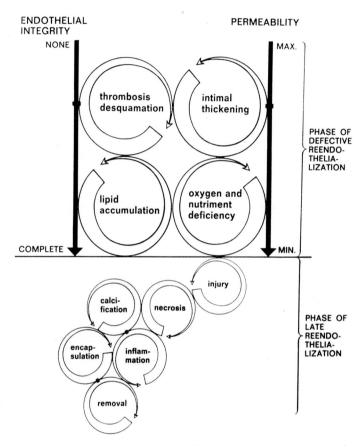

Fig. 6. Scheme illustrating possible relationships between some coacting and counteracting factors during the development of experimental aortic atherosclerotic lesions induced by subtotal injury with large area. Pairs of factors connected by bars indicate counteraction.

diffusion of nutriments and oxygen in this situation (cf. Fig. 6). It is probable that the thickened subendothelial zone in the regressive lesion represents a factor which promotes healing (see above) both for geometrical and mechanical reasons. However, if long-lasting and excessive the intimal thickening may have a quite opposite effect due to inferior mechanical and changed diffusion properties rendering repair impossible. During the phase of defective reendothelialization the lesions are very similar to early atheromata in man.

(c) Phase of late reendothelialization (Fig. 5D)

Following final reendothelialization of the lesion the tissue underlying the endothelium succumbs more or less completely. The tissue is changed to a necrotic mass which stains strongly for lipids, which gives the lesion an appearance very similar to the so-called fatty streaks in man. Monocytes and foreign body giant cells containing lipid inclusions are numerous. In a few days the lipids disappear, and the lesion is changed in a way similar to that following the direct experimental induction of a total necrosis, i.e. the region is calcified, encapsulated and capillarized. The final appearance is similar to that of media sclerosis in man.

DIFFERENT ROLES OF THE ENDOTHELIUM

As with the subendothelial thickening the presence of endothelium also seems to have opposite effects on the underlying tissue in different situations. In the regressive type of response, which leads to restitution, lipids do not accumulate even though the injured sector is devoid of endothelium for a period of about two weeks (Björkerud, 1969a). It has been shown (Björkerud and Bondjers, 1971b) that an excessive flow of serum occurs into the lesion during the denuded stage (cf. Fig. 6). This fact, taken together with the absence of accumulated lipids, strongly indicates a considerable capacity of the arterial tissues for the elimination of blood serum constituents, as e.g. cholesterol.

Instead of being a negative factor the absence of endothelium and the increased serum influx could constitute a growth-promoting factor, favoring repair in the underlying tissues in analogy with the growth-stimulating effect of serum for tissue culture *in vitro* (e.g. Clarke and Stoker, 1971). Many constituents of serum could exert a growth-stimulating effect. The fact that insulin stimulates the formation of endoplasmic reticulum membrane lipids *in vitro* in proliferating arterial tissue (Björkerud, 1973), taken together with the enhancement of growth response of cultured fibroblasts to serum by insulin (Clarke and Stoker, 1971), suggests that insulin may be one such serum factor (Björkerud, 1972c). Such a growth-regulating action of the endothelium, mediated by influx of serum, could be subjected to feedback by as simple a mechanism as changes of the mechanical stability of the underlying tissue.

On the other hand, upon prolonged exposure to an excessive serum inflow as in the phase of defective reendothelialization (Fig. 6) lipids do accumulate in the lesion (Björkerud, 1969b; Bondjers and Björkerud, 1973a). As outlined above, quite dramatic changes occur upon final reendothelialization of this type of lesion (Fig. 6). These data indicate different effects of the endothelium depending on the situation in the underlying tissue. The presence of an intact endothelium is according to available data to an advantage for the underlying tissue in certain situations. However, this may be the case only if the thickness and the diffusion characteristics of the tissue is compatible with the presence of the inflow-restricting factor represented by the endothelium (Björkerud and Bondjers, 1971b; Bondjers and Björkerud, 1973a, b). Otherwise, gross injury and necrosis may follow.

The results discussed above display an extreme complexity of the different response patterns after different types of arterial injury. The results clearly indicate interdependence of the different arterial tissues in several respects. However, such interdependence seems not to be simple, and it is probable that the elucidation of such relationships are needed for the understanding of the biological meaning of the presumably even more complex patterns of the biochemical parameters.

ACKNOWLEDGEMENTS

These studies were supported by grants from the Swedish Medical Research Council (No. B72-19X-2589-04C) and the Swedish National Association against Heart and Chest Diseases.

REFERENCES

Björkerud, S. (1969a). *Virchows Arch. (Pathol. Anat.)* **347**, 197–210.
Björkerud, S. (1969b). *J. Atheroscler. Res.* **9**, 209–213.
Björkerud, S. (1970). In "Atherosclerosis: Proceedings of the Second International Symposium" (R. J. Jones, ed.) pp. 126–129. Springer-Verlag, Berlin.
Björkerud, S. (1972a). *Angiology* (In Press.)
Björkerud, S. (1972b). *Atherosclerosis* **15**, 147–152.
Björkerud, S. (1972c). *Eur. J. Clin. Invest.* **2**, 273–274.
Björkerud, S. (1973). *Adv. Metab. Disord., Suppl.* **2**, 55–63, 66.
Björkerud, S. and Bondjers, G. (1971a). *Atherosclerosis* **14**, 259–276.
Björkerud, S. and Bondjers, G. (1971b). *Atherosclerosis* **13**, 355–363.
Björkerud, S. and Bondjers, G. (1972). *Atherosclerosis* **15**, 285–300.
Björkerud, S. and Bondjers, G. (1973). *Atherosclerosis* (In Press.)

Bondjers, G. and Björnheden, T. (1970). *J. Atheroscler. Res.* **12,** 301–306.
Bondjers, G. and Björkerud, S. (1973a). *Atherosclerosis* **17,** 85–97.
Bondjers, G. and Björkerud, S. (1973b). *Atherosclerosis* **17,** 71–83.
Clarke, G. D. and Stoker, M. G. P. (1971). *In* "Growth Control in Cell Cultures" (G. E. W. Wolstenholme and J. Knight, eds.) pp. 17–28. Churchill Livingstone, Edinburgh and London.
Letterer, E. (1959) "Allgemeine Pathologie" Thieme, Stuttgart.

The Smooth Muscle Cell in Connective Tissue Metabolism and Atherosclerosis

Russell Ross

Department of Pathology, School of Medicine
University of Washington, Seattle, Washington, USA

It is now widely accepted that the arterial smooth muscle cell is of critical significance in the genesis of the lesions of atherosclerosis (French, 1966; Haust and More, 1963; Jones et al., 1967). The lesions of atherosclerosis take several forms, including the so-called fibromusculoelastic lesion, said to be the antecedent of the fatty streak, the fibrous plaque and the more complicated lesions. All of these entities have several features in common. Principal among these is the proliferation and accumulation of smooth muscle cells in the innermost layer of the artery, the intima. As these lesions progress the smooth muscle cells may become deformed, contain accumulations of lipid and are often surrounded by extracellular connective tissue fibers. The arterial smooth muscle cell has been shown to be responsible during growth and development for the synthesis and secretion of collagen, elastic fiber proteins, and probably for the proteoglycans present within the media of the arterial wall (Ross and Klebanoff, 1971).

Smooth muscle cells derived from the media of muscular arteries of primates (*Maccaca nemestrina*) can be grown *in vitro* (Ross, 1971). This has permitted us to investigate some of the factors responsible for the migration, proliferation and synthesis of connective tissue proteins by these cells.

It has been possible, through the use of an intra-arterial balloon catheter, to induce a lesion in monkeys that is morphologically identical to the fibromusculoelastic thickening seen in human vessels that occur at branch points, or regions thought to be susceptible to increased intravascular pressure (Stemerman and Ross, 1972).

This chapter will summarize initial observations utilizing these two approaches in studying the role of primate arterial smooth muscle in atherosclerosis.

TISSUE CULTURE STUDIES

Recent studies have demonstrated that smooth muscle derived from the media of the aorta have characteristic growth patterns in culture that are different from those observed for other connective tissue forming cells (Ross, 1971 and 1972). More recently, it has been possible to grow primate smooth muscle cells in culture under similar circumstances and to use pooled serum derived from the same genus of primate (*Maccaca nemestrina*) as an additive to the culture medium to examine the effects of monkey serum and its substituents upon the growth of these cells.

Lipids, i.e. serum cholesterol and cholesterol ester together with various fractions of lipoproteins have been implicated as being responsible for the genesis of the lesion of atherosclerosis. The cell culture system permits a systematic analysis of each of these components in terms of the role each may potentially play in proliferation and migration of smooth muscle, as well as in connective tissue protein synthesis by these cells. Serum constituents have been shown to be responsible for several of these phenomena for other types of cells in culture. Holley and Kiernan (1971) and Todaro *et al.* (1967) and Temin (1971) have fractionated serum in several different ways and have suggested that there were particular molecular components in serum responsible for DNA synthesis and consequently for proliferation of fibroblasts in culture. More recently, Lipton *et al.* (1971) have been able to separate fractions from serum, one of which appears to be specifically responsible for migration of fibroblasts, whereas another plays a role in proliferation of fibroblasts in this system. Several studies have appeared in the literature to suggest that low density lipoproteins and cholesterol or cholesterol esters will also cause smooth muscle cells to grow out from explants and proliferate in culture. Thus a systematic analysis of the role played by serum components will be necessary if we are to understand their significance *in vivo*.

Smooth muscle cells derived from the media of the thoracic aorta of the primate *Maccaca nemestrina* have characteristic growth curves dependent upon the age of the donor. If the donor is less than four years of age the growth curves of the cells are similar (Fig. 1). When cells from donors of this age are grown in the presence of 5% or 10% pooled serum from the same genus of monkey, they grow logarithmically for 10–14 days before entering stationary phase. Cells from older donors appear to grow logarithmically for shorter time periods.

When seen by phase microscopy in culture the smooth muscle cells are large and contain extensive bundles of myofilaments (Ross, 1972).

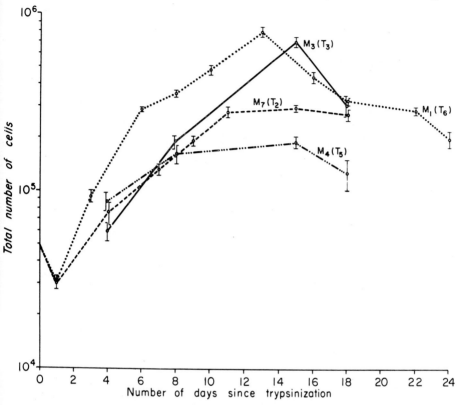

Fig. 1. This graph demonstrates typical growth curves of arterial smooth muscle cells derived from the media of the thoracic aorta from monkeys, *Maccaca nemestrina*, of three different ages. All the cells were grown in the Dulbecco-Vogt modification of Eagle's medium containing 10% pooled serum obtained from the same genus of primate. Animals M-1 and M-3 were 1–2 years old. Animal M-7 was $4\frac{1}{2}$ years old. Animal M-4 was 9 years old. Subsequent growth have reinforced the impression that cells derived from monkeys aged 1–2 years grow logarithmically (under these conditions) for 12–14 days whereas they grow logarithmically for shorter periods of time when derived from animals that are older.

Figs. 2–4 are electron micrographs of these cells. In cultures they retain their characteristic bundles of myofilaments together with interfilament dense bodies normally present *in vivo*. The cells also contain regions of cytoplasm having aggregates of ribosomes, and small deposits of glycogen and peripheral vesicles as can be seen in Figs. 3 and 4.

After the cells have grown *in vitro* for several weeks extracellular microfibrils (Ross, 1971; Ross and Bornstein, 1969) together with collagen begin to appear (Figs. 4 and 5). Previous studies (Ross, 1971)

Fig. 2. A low power electron micrograph of a medial smooth muscle cell derived from *Maccaca nemestrina* after 4 weeks of growth in culture. Most of the cytoplasm can be seen to consist of bundles of myofilaments with interfilament dense bodies. (×4500)

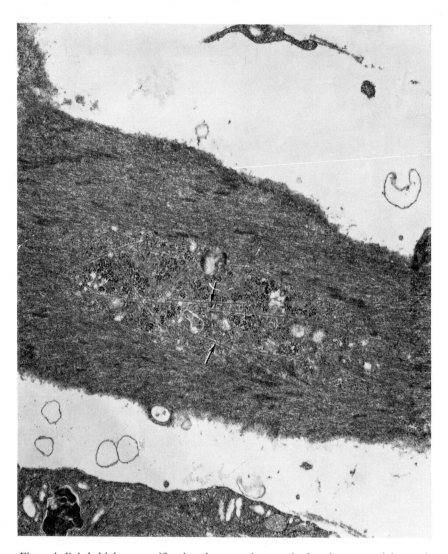

Fig. 3. A slightly higher magnification electron micrograph of a primate arterial smooth muscle cell grown *in vitro* and illustrates not only the large numbers of myofilaments present within the cytoplasm of these cells, but individual microtubules (arrow) and aggregates of ribosomes as well. ($\times 12000$)

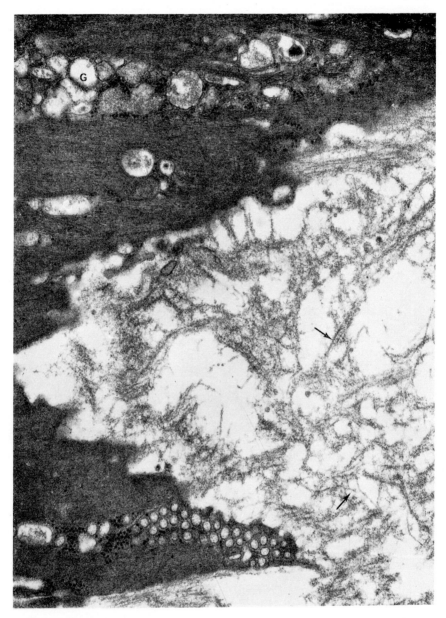

Fig. 4. An electron micrograph of portions of thoracic aorta smooth muscle cells from *Maccaca nemestrina* grown in culture for 6 weeks. Numerous peripheral vesicles characteristic of smooth muscle can be seen in a segment of the cell in the lower portion. Cell organelles including rough endoplasmic reticulum and Golgi constituents (G) can be seen in the upper part. Elastic fiber microfibrils are present in the extracellular space surrounding the cell, (arrow). ($\times$ 33,000)

Fig. 5. This electron micrograph demonstrates two components found in the extracellular space surrounding smooth muscle cells that have grown *in vitro* for 8 weeks. Collagen fibrils with their characteristic periodic banding can be seen in the right portion. Elastic fiber microfibrils with a beaded appearance that sometimes gives the appearance of a periodicity are noted on the left. (×52,000)

have demonstrated that these extracellular microfibrils have an amino acid composition identical with that previously described for microfibrils associated with elastic fibers (Ross and Bornstein, 1969).

Both collagen and elastic fiber microfibrils form during similar periods of time in culture as is clearly demonstrated in Fig. 5, where the morphological difference in these two extracellular connective tissue matrix proteins can be seen and where the periodic banding of the collagen fibrils is clearly demonstrated.

In vivo STUDIES

Stemerman and Ross (1972) have demonstrated that it is possible to produce a lesion *in vivo* identical in appearance to the fibromusculoelastic thickening, normally considered to be preatherosclerotic in humans, in the common iliac arteries of a series of monkeys. This lesion is induced by placing an intravascular balloon catheter in the external femoral artery. The balloon of the catheter is then inflated and the catheter is pushed into the abdominal aorta and pulled back rapidly into the femoral artery. This procedure selectively removes the lining endothelium of the arteries throughout the region where the expanded balloon has passed through the lumen of the vessels. Examination of such a treated artery between 10 min and 24 h after ballooning demonstrates that the endothelial cells are missing, that the rest of the vessel wall is intact, and that platelet microthrombi are adherent to the exposed elastic fibers of the internal elastic lamina and to remnant fragments of basement membrane.

Within 3–7 days after deendothelialization, regenerating endothelial cells can be seen, presumably derived from remnant patches of endothelium and/or from accessory vessels. During this same time, smooth muscle cells were observed migrating through fenestrae of the internal elastic lamina, which remains intact, from the media into the intima of the artery wall. As a result of this migration and proliferation, within one month after deendothelialization, the intima of the vessel that normally contains no cells, or only an occasional single smooth muscle cell, contains from 5 to 10 cell layers of smooth muscle cells between the internal elastic lamina and the regenerated endothelium. The lesion reaches its maximum size three months after deendothelialization, and contains up to 15 layers of smooth muscle cells surrounded by small immature elastic fibers, collagen fibrils, and proteoglycans. Six months after deendothelialization, the lesion is markedly reduced in size to 2–3 cell layers suggesting that in the presence of normal plasma constituents and in the absence of further injury to the vessel wall, the lesion is reversible.

The opportunity to selectively and reproducibly induce such a lesion permits an analysis of the effects of increased serum constituents upon the extent of the formation of this lesion, the potential reversibility of the lesion, and its duration.

Thus, an understanding of the metabolism of this cell and the factors responsible for its accumulation in atherosclerotic vessels is of paramount importance. The opportunity to study arterial smooth muscle cells *in vitro* and correlate the observations of the effects of serum constituents upon the various growth parameters *in vitro* together with their effects upon the genesis of a lesion induced by the removal of endothelium *in vivo* will permit us to obtain a much clearer understanding of how these cells participate in the formation of the lesion of atherosclerosis and to determine which of the factors are important in its etiology.

THE FIBROBLAST VERSUS THE SMOOTH MUSCLE CELL

A perplexing question remains—what is the relationship between the fibroblast and the smooth muscle cell? Both appear to be differentiated cells present in specialized connective tissues, capable in each of these tissues of synthesizing different amounts and kinds of connective tissue matrix components. Both appear to be drived during embryogenesis from undifferentiated mesenchyme. There is no evidence, to date, that either cell type can give rise to the other. The important question of such a possible interrelationship remains to be elucidated.

ACKNOWLEDGEMENTS

This work was supported in part by a grant from the USPHS No. HL 14823, and grant No. GM 13543.

The author would like to acknowledge the assistance of Miss Kathleen Sabo in these studies.

REFERENCES

French, J. E. (1966). *Int. Rev. Exp. Pathol.* **5,** 253–353.
Haust, M. D. and More, R. H. (1963). *In* "Evolution of the Atherosclerotic Plaque" (R. J. Jones, ed.) pp. 51–63. Chicago University Press, Chicago.
Holley, R. W. and Kiernan, J. A. (1971). *In* "Growth Control in Cell Cultures" (J. Knight and G. E. W. Wolstenholme, eds.) pp. 3–10. Churchill Livingstone, London.
Jones, R., Daoud, A. S., Zumbo, O., Coulston, F. and Thomas, W. A. (1967). *Exp. Mol. Pathol.* **7,** 34–57.
Lipton, Al, Klinger, I., Paul, D. and Holley, R. W. (1971). *Proc. Natl. Acad. Sci. USA* **68,** 2799–2801.
Ross, R. (1971). *J. Cell Biol.* **50,** 172–186.

Ross, R. (1972). *In* "The Pathogenesis of Atherosclerosis" (R. W. Wissler and J. C. Geer, eds.) pp. 147–163. The Williams and Wilkins Co. Baltimore.
Ross, R., and Bornstein, P. (1969). *J. Cell Biol.* **40,** 366–381.
Ross, R. and Klebanoff, S. J. (1971). *J. Cell Biol.* **50,** 159–171.
Stemerman, M. Ross, R. (1972). *J. Exp. Med.* **136,** 769–789.
Temin, H. M. (1971). *J. Cell Physiol.* **78,** 161–170.
Todaro, G., Matsuya, J., Bloom, S., Robbins, A. and Green, H. (1967). *In* "Wistar Institute Symp. Monogr." (V. Defendi and M. Stoker, eds.) pp. 87–97. Wistar Institute Press, Philadelphia.

In Vitro Incorporation of Labelled Precursors in the Macromolecules of the Polymeric Stroma of Normal and Pathological Arterial Wall

L. ROBERT, S. JUNQUA, A. M. ROBERT, M. MOCZAR
and B. ROBERT

*Laboratoire de Biochimie du Tissu Conjonctif, Faculté de Médecine
Université Paris-Val de Marne, 94000 Créteil, France*

The intercellular matrix of the arterial wall is composed of four major types of macromolecules: elastin, collagen, structural glycoproteins and proteoglycans (for a review see Robert, 1970). The ratio of these four types of macromolecules varies with the calibre and anatomical location of the artery, as the result of the adaptation to the local mechanical and rheological requirements of the functionally differentiated cellular elements of the arterial wall. The qualitative and quantitative composition of these macromolecules changes also during pathological processes such as arteriosclerosis. The rate of synthesis and degradation of these four classes of intercellular macromolecules can therefore be considered as a convenient tool for the study of the mechanism of differentiation of the arterial wall. It also can be used for the study of the pathological modifications of their rate of synthesis and degradation.

CONCEPTUAL BASIS AND EXPERIMENTAL APPROACH

In order to gain information on the biosynthetic processes concerning the macromolecules of the arterial wall a simplified procedure had to be worked out, suitable for serial determinations on relatively small samples, such as fragments of rat or rabbit aortas. Labelled molecules have been selected which are incorporated in all four classes of macromolecules (collagen, elastin, proteoglycans, glycoproteins). In the experiments described here ^{14}C-lysine was used because of its importance in crosslink formation in elastin and collagen as well as ^{14}C-galactose in order to assess specifically the relative rate of incorporation of this precursor in the glycans attached to these macromolecules (Moczar et al., 1972). In other experiments ^{3}H-proline with

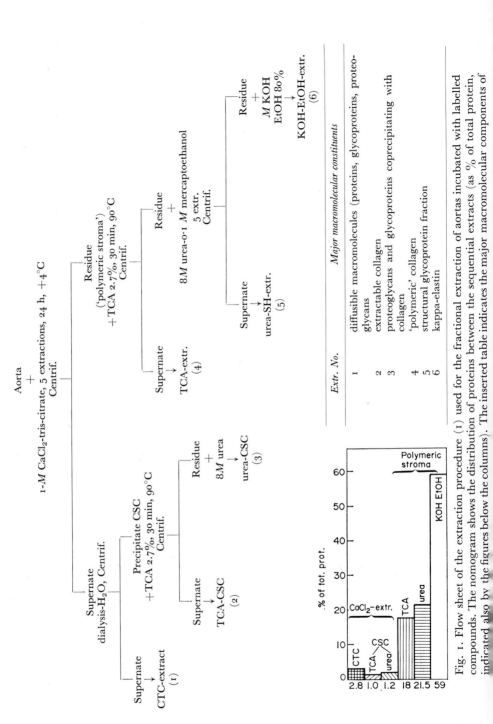

Fig. 1. Flow sheet of the extraction procedure (1) used for the fractional extraction of aortas incubated with labelled compounds. The nomogram shows the distribution of proteins between the sequential extracts (as % of total protein, indicated also by the figures below the columns). The inserted table indicates the major macromolecular components of

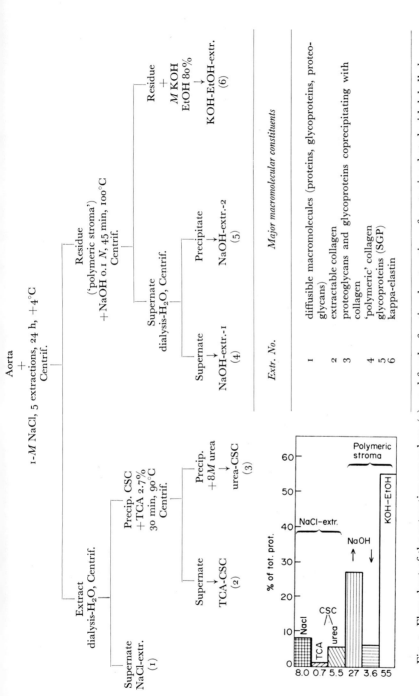

Fig. 2. Flow sheet of the extraction procedure (2) used for the fractional extraction of aortas incubated with labelled compounds. The nomogram shows the distribution of proteins between the sequential extracts (as % of total protein, indicated also by the figures below the columns). The inserted table indicates the major macromolecular components of the extracts.

^{14}C-glucosamine were used (in double labelling) to follow proteoglycan and fibrous protein synthesis (Junqua et al., 1972).

Earlier studies on other connective tissues such as cornea, sclera or tendon showed that these tracers were actively incorporated in tissue slices during 4–5 h of incubation (Robert and Parlebas, 1965a, b)

After incubation with labelled precursors in Krebs-Ringer-phosphate medium in the presence of ATP (5 μM) (in some experiments a complete amino acid mixture was also added) for 4 h at 37°C, the finely sliced tissues were thoroughly washed in an excess of 'cold' lysine (or proline + glucosamine or galactose). The washed tissues were then subjected to a sequential extraction procedure yielding a complete 'dissolution' of the tissue with as minimal hydrolytic breakdown as possible. Two different procedures were used:

1. M CaCl$_2$-tris-citrate (CTC) extraction followed by hot trichloracetic acid (TCA) extraction and again $8M$ urea-$0 \cdot 1 M$ mercaptoethanol extraction (see Fig. 1).

2. M NaCl extraction followed by hot $0 \cdot 1$N NaOH extraction (see Fig. 2).

The subfractionation of the salt-soluble extracts and of the 'polymeric stroma' was carried out as described (Moczar and Robert, 1970) and is shown in Figs. 1 and 2.

COMPOSITION OF THE EXTRACTS

The major macromolecular components of the extracts are shown in the tables on Figs. 1 and 2. For detailed description of these components see Moczar and Robert (1970). The distribution of these protein fractions is similar with both extraction methods, except for the more selective subfractions of the salt insoluble residue obtained with method (1), above.

PRODUCTION OF ARTERIAL LESIONS

Two different methods were used to produce arterial lesions: a mild cholesterol administration in the diet (1%) and immunisation with kappa-elastin. The immunisation with kappa-elastin was performed as described by A. M. Robert et al. (1971), using kappa-elastin from human aorta in Freund's complete adjuvant.

DISTRIBUTION OF RADIOACTIVITY IN THE EXTRACTS

Table I shows the specific radioactivity (dpm/mg prot.) of the extracts obtained from adult (4–5 kg) rabbit aortas with no macroscopic signs of arteriosclerosis. The table shows one experiment (I) carried out by the first extraction procedure (see Fig. 1) and two experiments (II) carried out with extraction procedure (2). All the fractions obtained with both methods are labelled. The distribution of radioactivity in the extracts is similar with both extraction methods used (see Figs. 1 and 2) with the exception of the TCA-extract of the CSC-fraction: its specific activity was higher in the experiment performed with the second extraction procedure using NaCl than with the first method using $CaCl_2$.

The distribution of specific activity in the three extracts obtained from the polymeric stroma is similar with both methods shown in Figs. 1 and 2. The lowest specific activity was found in the TCA and NaOH extracts of the stroma containing mainly polymeric collagen. The structural glycoprotein containing fractions (urea-extract in method 1 and the dialysis-precipitate of the NaOH extract in method (2) have the highest specific activity, of the same order or higher than that of the salt soluble (CTC or NaCl) extracts. A significant radioactivity is associated with the elastin fraction also (KOH-EtOH extracts).

NATURE OF THE LABEL INCORPORATED INTO THE ELASTIN FRACTION

The apparent incorporation of ^{14}C-lysine in polymeric, insoluble elastin observed by two different extraction procedures can be explained either by the presence of nonelastin components in this final residue (remaining even after 45 min boiling in 0·1 N NaOH) or by a slow but measurable incorporation of proelastin precursors into insoluble elastin during the incubation time.

It was shown by electron microscopy that purified elastin preparations contain two morphologically distinct elements: translucent, homogeneous lamellae, supposed to represent crosslinked proelastin molecules and 'microfibrillar' elements varying in density and localisation according to the age of the tissue and the method of preparation (for references see B. Robert et al., 1971; Ross and Bornstein, 1969; Kadar et al., 1973). We presented arguments in favour of the identity of these 'micro-fibrillar' elements with structural glycoproteins (B. Robert et al., 1971). It is currently assumed that in the full grown animal the elastic tissue does not show any measurable turnover (Banga, 1969; Kao et al., 1961; Walford et al., 1964). It has also been observed however that in pathological conditions such as in the early phases of arteriosclerosis (Kadar et al., 1969) or after blood vessel transplantation (Veress et al., 1970) or in some malignant tumours a limited neosynthesis of elastin may occur (Jackson and Orr, 1957; Williams et al., 1972). All of these observations remain however mainly if not entirely on morphological grounds.

In the light of these considerations we could expect an incorporation of labelled molecules either in the 'microfibrillar' glycoprotein fraction or in the elastin fraction or in both components, present in the final insoluble residue as obtained by both methods shown in Figs. 1 and 2. As this final residue was exhaustively extracted in 8 M urea.

Table I
Incorporation of ^{14}C-lysine in normal rabbit aorta
4 h incubation at 37°C, for details see text.

	Specific activity of extracts					
	Salt soluble fractions			Stroma		
Exp. No.	CTC* or NaCl	TCA	CSC urea	TCA† or NaOH-1 supern.	urea† or NaOH-2 precip.	KOH-EtOH
I	2100	320	1900	200	3000	680
II (a)	1770	1120	1740	360	1790	640
(b)	1450	2690	1270	620	3010	1000

* CTC extract for exp. (I) and NaCl for exp. (II).

† TCA and urea extract for exp. (I) and NaOH-1 and NaOH-2 for exp. (II).

In experiment (I) extraction method (1) was used (see Fig. 1), in experiments (II) extraction method (2) (see Fig. 2) was used. II(a) and II(b) represent two separate experiments. The designations of the extracts are the same as in Figs. 1 and 2. The specific activity of the extracts is given in dpm/mg prot. In experiments (I) the figures indicated represent the average of four parallel incorporation experiments, carried out with 10 aortas. Experiments II(a) and (b) were carried out with three aortas each, incubated together.

Mercaptoethanol in the first method (see Fig. 1) and with boiling 0·1 N NaOH for 45 min in the second method (see Fig. 2), its 'microfibrillar' glycoprotein content was reduced below 1% as was shown in several elastin samples purified by five different procedures (B. Robert et al., 1971). Therefore if only the glycoproteins associated with elastin were labelled in our experiments, the specific activity of the samples could not exceed one per cent of the specific activity of the urea-extracts.

The range of specific activities of these glycoproteins for normal aortas were from 1790 to 3010 dpm/mg prot. The specific activity of the final residues (elastin fraction) was found between 640 to 1000 dpm/mg prot. (see Table I). The calculated specific activity assuming even 5% of glycoprotein in the final residue would be 90 to 150 dpm/mg, significantly lower than the experimentally found values. Therefore it appears more probable that the results observed are due partly at least to the incorporation of labelled lysine into the elastin fraction during the incubation period. This contention was further enforced by the coacervation of the ^{14}C-lysine labelled rabbit aorta kappa-elastin with purified 'cold' porcine aorta kappa-elastin (Junqua et al., 1973).

Similar results were obtained when ^{3}H-proline was used as a tracer. ^{14}C-glucosamine added at the same time in a double labelling experiment was on the contrary, not incorporated appreciably into the insoluble elastin fraction, demonstrating again that the contribution of labelled glycoprotein fractions to the radioactivity of the elastin fraction cannot be the explanation of the reported results (Junqua et al., 1973).

INCORPORATION OF ^{14}C-GALACTOSE INTO THE GLYCANS OF MACROMOLECULES OF THE ARTERIAL WALL

The relatively high specific activity of the urea extract (method 1, see Table I) confirmed former findings obtained with other connective tissue preparations, showing that the structural glycoprotein fraction

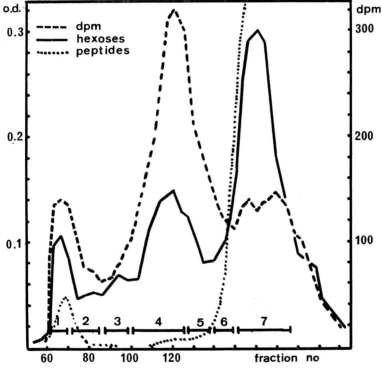

Fig. 3. Elution diagram of the collagenase and pronase hydrolysate of the insoluble stroma (extracted with M CaCl$_2$, see Fig. 1) of the media of pig aorta incubated with ^{14}C-galactose. Sephadex G-50 medium column (4.8 × 800 cm for 2.6 g hydrolysate). Elution with 0.1 M acetic acid at 5 ml/3 min; ———, orcinol (carbohydrates); - - - - -, radioactivity dpm in fraction; ·······, optical density at 280 nm.

of the insoluble stroma acquires a significantly higher specific activity during *in vitro* incubation with tracers than the other macromolecular components (Robert and Parlebas, 1965a, b). To obtain more information on this aspect, a different approach was chosen (Moczar *et al.*, 1972).

Pig aortas, stripped of adventitia were cut into fine slices and incubated with 1-^{14}C-galactose for 4 h in the same conditions as above. The washed aortas were then extracted exhaustively with M $CaCl_2$ and the insoluble stroma was hydrolysed with collagenase and pronase. The hydrolysate was fractionated on a Sephadex column and the glycopeptides localised by the orcinol method. Fig. 3 shows the results obtained.

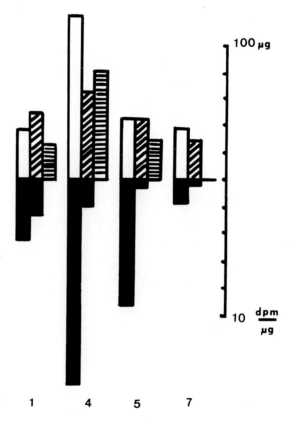

Fig. 4. Molecular ratio (upper part of graph) and specific activity (lower part of graph) of the hexoses isolated by paper chromatography from the hydrolysates of glycopeptides obtained from the stroma of pig aorta incubated with ^{14}C-galactose, as indicated in Fig. 3. □, glucose; ▨, galactose; ☰, mannose; ■, specific activity dpm/μg hexose.

Three main regions can be distinguished on the elution profile: the first major peak (fraction No. 1 on Fig. 3) contains uronic acid and ester sulphate containing glycosaminoglycans. The second region, with several peaks and shoulders (fractions Nos. 2, 3, 4 and 5 on Fig. 3) contains galactose, glucose, mannose, glucosamine, fucose and sialic acid and was shown to be derived from the structural glycoprotein fraction of aorta (Moczar et al., 1970). The last major peak (Nos. 6 and 7 on Fig. 3) contains nearly exclusively hydroxylysyl-galactoside and hydroxylysyl-galactoside-glucoside. The distribution of radioactivity is also shown on the elution profile of Fig. 3. The first glycosaminoglycan peak has a moderate radioactivity. The second glycopeptide region shows the strongest radioactivity and the third peak has again a moderate radioactivity.

Fig. 4 shows the ratio of hexoses in the individual glycopeptide fractions eluted from the Sephadex column (see Fig. 3) and their specific activity (Moczar et al., 1973). It can be seen that the specific activities of the hexoses reflect the above mentioned distribution of radioactivity found with lysine incorporation: the highest specific activity is observed in the intermediary peaks (Nos. 4 and 5 on Fig. 3) containing glycans derived from the structural glycoproteins. Labelled galactose and glucose were isolated from the glycopeptide hydrolysates; but mannose was not labelled. In similar experiments performed with calf corneas incubated with ^{14}C-galactose, labelled galactose, glucose and mannose could be isolated (Moczar et al., 1973).

The results obtained with this method using ^{14}C-galactose as well as those obtained using ^{14}C-lysine indicate a relatively high incorporation rate in the structural glycoprotein fraction of aorta. These experiments show also that surviving aorta can utilise externally added galactose for glycoprotein biosynthesis and convert it efficiently to glucose.

INCORPORATION OF ^{14}C-LYSINE IN ARTERIOSCLEROTIC AORTAS

Experiments identical to those reported in Table I were performed on aortas of young (1–1.5 kg) rabbits immunised with kappa-elastin isolated from human aorta or fed a cholesterol containing diet. Most of these aortas developed arteriosclerotic lesions (A. M. Robert et al., 1971). The specific activity of the extracts obtained with method (1) (see Fig. 1) from the insoluble stroma of these aortas incubated 4 h at 37°C with ^{14}C-lysine is shown in Table II.

The specific activity of the TCA (collagen) and KOH-EtOH-(elastin)-extracts is similar to those found in older (4–5 kg) rabbits

Table II
Incorporation of ^{14}C-lysine in the polymeric stroma of control and atherosclerotic rabbit aortas

Treatment of rabbit*	No. of animals	Specific activity of extract dpm/mg prot.		
		TCA	urea†	KOH-EtOH
Freund's adjuvant	4	393 ± 183	4020	631 ± 330
Cholesterol containing regime	3	393 ± 131	3000	591 ± 259
K-elastin in Freund's adjuvant	2	152 ± 34	1290	192 ± 130

* For the details see methods.
† Average specific activity of purified subfractions (see Moczar and Robert 1970).
Incubation and extraction by method 1 (see Fig. 1). The results given are average specific activities (dpm/mg prot.) ± standard error of the mean.

(see Table I). The specific activity of the urea-extract (structural glycoproteins) is however somewhat higher.

The radioactivity of all three fractions obtained from the aortas of the immunised rabbits (kappa-elastin) was found appreciably lower than that of rabbits treated with Freund's adjuvant only. The incorporation is lower also in the extracts of the cholesterol-fed groups, manifesting itself mainly in the urea-extract (about 25% inhibition). This weak inhibition may be due to the relatively moderate amount of cholesterol fed for a limited period to the animals. The type of the lesions produced by these two treatments is not the same: cholesterol feeding produces a predominantly lipidic lesion (Grosgogeat and Lenegre, 1966), and immunisation with kappa-elastin produces lesions with early calcification and little or no lipid deposition (A. M. Robert et al., 1971).

DISCUSSION

The experiments reported confirm the biosynthetic capacity in surviving aorta slices for at least 4 h. Incorporation of sulphate and some other labelled molecules has been already reported in normal and pathological arterial tissue by several laboratories (Kresse and Wessels, 1969; Kresse and Buddecke, 1970; Hauss et al., 1968). Henry (1968) found a decreased incorporation of ^{14}C-leucine in the ribosomal fraction of the aorta of rabbits submitted to an atherogenic diet. Sulphate incorporation on the other hand was found to increase during the early phases of atherogenesis (Junge-Hülsing and Wagner, 1969).

Our results clearly indicate that rabbit aorta slices incubated *in vitro* incorporate lysine, galactose or other tracers used in all macromolecular fractions. A significant portion of total label is incorporated in the macromolecules of the insoluble, polymeric stroma (remaining after exhaustive salt extractions). The highest specific activity was associated with the $8\,M$ urea-soluble structural glycoprotein fraction. Similar results were obtained using two different extraction procedures (Junqua et al., 1973). It was also confirmed by using ^{14}C-galactose as a tracer for the glycans of the aorta macromolecules and by the analysis of separated glycopeptides of a collagenase-pronase digest of the insoluble stroma (Moczar et al., 1973).

A small but significant radioactivity was associated both with the insoluble (polymeric) collagen and elastin fractions when lysine and proline were used as tracers. It appears probable that part at least of the label is incorporated in elastin itself. It is possible that in the *in vitro* incubation conditions there is a partial derepression of elastin biosynthesis.

Experiments recently carried out with a sensitive high voltage-electrophoretic separation of crosslink amino acids (Moczar et al., 1972) did confirm this conclusion (Ouzilou et al., 1973).

In aortas rendered sclerotic by immunisation with kappa-elastin there was a significant decrease in incorporation in all fractions of the insoluble stroma. Only a slight inhibition was found in aortas of rabbits on a diet moderately rich in cholesterol.

Our results suggest the possibility of an inhibition of the biosynthetic activity of arterial smooth muscle cells by circulating anti-elastin antibodies. As anti-elastin antibodies were shown to be present in human sera (Stein et al., 1965) such an inhibition might play a role in human atherogenesis.

SUMMARY

1. Rabbit aortas were shown to incorporate *in vitro* ^{14}C-lysine and ^{14}C-galactose in all the macromolecular fractions obtained by three different methods (two salt extraction procedures and by the separation of glycopeptides by gel filtration of proteolytic digests).

2. A significant fraction of total label is incorporated in the insoluble, polymeric stroma. The highest specific activity is associated with the urea extractable structural glycoprotein fraction. A low but significant radioactivity is found in the insoluble collagen and elastin fractions.

3. The ^{14}C-lysine incorporated into the insoluble elastin fractions may be present partly at least in elastin and only partly in the 'microfibrillar' glycoprotein components. This is suggested by the coacervation of labelled rabbit aorta kappa-elastin with 'cold' pig aorta kappaelastin.

4. ^{14}C-galactose is efficiently used by surviving aorta slices for glycoprotein synthesis. It is partly converted to ^{14}C-glucose which could be isolated from the glycopeptide fractions obtained from proteolytic digests. No radioactivity was found in isolated mannose.

5. In sclerotic aortas of rabbits immunised with human aorta kappa-elastin a significant decrease of incorporation was found in all the fractions isolated from the insoluble stroma. The possibility is discussed that this inhibition is due to circulating antielastin antibodies, shown to be present in human sera.

ACKNOWLEDGEMENTS

This research was supported by the C.N.R.S. (Equipe de Recherche No. 53), the D.G.R.S.T. (contract No. 71.7.2809) and I.N.S.E.R.M. (contract libre No. 714054).

The able technical assistance of F. Chavarot and M. T. Seguin is thankfully acknowledged.

REFERENCES

Banga, I. (1969). In "Biochemistry of the Vascular Wall" Frieberg 1968, Part II, pp. 18–92. S. Karger, Basel and New York.
Grosgogeat, Y. and Lenegre, J. (1966). *Arch. Mal. du Coeur, Suppl.* 2, **59**, 26–33.
Hauss, W. H., Junge-Hülsing, G. and Gerlach, U. (1968). In "Die unspezifische Mesenchymreaktion", Thieme-Verlag, Stuttgart.
Henry, J. C. (1968). *C.R. Acad. Sci. (Paris)* **266**, 1449–1450.
Jackson, D. G. and Orr, J. W. (1957) *J. Path. Bact.* **74**, 265.
Junge-Hülsing, G. and Wagner, H. (1969). In "Ageing of Connective and Skeletal Tissue" (A. Engle and T. Larsson, eds.) pp. 213–228. Nordiska Bokhandelns Förlag, Stockholm.
Junqua, S., Robert, A. M. and Robert, L. (1973). (Submitted for publication.)
Kadar, A., Veress, B. and Jellinek H. (1969). *Exp. Mol. Pathol.* **11**, 212–223.
Kadar, A., Robert, B. and Robert, L. (1973). *Pathol. Biol.* (In Press.)
Kao, K. T., Hilker, D. M. and McGavack, T. H. (1961). *Proc. Soc. Exp. Biol. Med.* **106**, 335–338.
Kresse, H., and Wessels, G. (1969). *Z. Physiol. Chem.* **350**, 1605–1610.
Kresse, H. and Buddecke, E. (1970). *Z. Physiol. Chem.* **351**, 151–156.
Moczar, M. and Robert, L. (1970). *Atherosclerosis* **11**, 7–25.

Moczar, M., Moczar, E. and Robert, L. (1970). *Atherosclerosis* **12,** 31–40.
Moczar, E., Robert, B. and Robert, L. (1972) *Analyt. Biochem.* **45,** 422–427.
Moczar, M., Moczar, E. and Robert, L. (1973). (Submitted for publication.)
Ouzilou, J., Courtois, Y., Moczar, M., Robert, A. M. and Robert, L. (1973). (In preparation).
Robert, A. M., Grosgogeat, Y., Reverdy, V., Robert, B., and Robert, L. (1971). *Atherosclerosis* **13,** 427–449.
Robert, B., Szigeti, M., Derouette, J. C., Robert, L., Bouissou, H. and Fabre, M. T. (1971). *Eur. J. Biochem.* **21,** 507–516.
Robert, L. and Parlebas, J. (1965a). *C.R. Acad. Sci.* **261,** 842–844.
Robert, L. and Parlebas, J. (1965b). *Bull. Soc. Chim. Biol.* **47,** 1853–1866.
Robert, L. (1970). *In* "Atherosclerosis" Proc. Second Int. Symp. (R. J. Jones, ed.) pp. 59–68, Springer-Verlag,
Ross, R., and Bornstein, P., (1969). *J. Cell. Biol.* **40,** 366–381.
Stein, F., Pezess, M. P., Robert, L., and Poullain, N., (1965). *Nature* **207,** 312–313.
Veress, B., Kadar, A., Bartos, G. and Jellinek, H., (1970). *Acta Morphologica Acad. Sci. Hung.* **18,** 63–72.
Walford, R. L., Carter, P. K. and Schneider, R. B. (1964). *AMA Arch. Path.* **78,** 43–45.
Williams, G., Green, F. H. Y. and Ahmed, A. (1972). *Scand. J. Clin. Lab. Invest.* **29,** Suppl. 123, 35.

VI
Summarizing Remarks

The Fibroblast, a Retrospective View of the Meeting

RUSSELL ROSS

University of Washington, School of Medicine
Seattle, Washington 98195, USA

This symposium began with an introduction by Professor Kulonen who pointed out the important and ubiquitous nature of the fibroblast in terms of the role it plays in granulation tissue, in tissue repair and in pathological alterations leading to clinical problems for the patient such as in rheumatoid disease, hypertrophic scars, sclerosis, fibrosis, some degenerative diseases and in hereditary morphological defects.

He pointed to the deficiencies in our knowledge in terms of understanding the factors affecting cell proliferation, the synthesis of the various connective tissue matrix components and the degradation of the extracellular matrix.

Professor Kulonen also pointed to studies showing the response of granulation tissue to pharmacological agents that act on smooth muscle, such as vasopressin, histamine and serotonin—effects which were to become of even greater interest as a result of the presentation by Dr. Gabbiani, who demonstrated a highly developed system of contractile filaments in fibroblasts in granulation tissue that appear to be actomyosin-like, raising the question of the relationship between the fibroblast and the smooth muscle cell—relationships to which I shall return later.

METABOLISM OF CARBOHYDRATES

The discussion of the first day then turned to carbohydrate metabolism. Drs. Holtzer and Dorfman discussed the differentiation of chondroblasts. They studied the effects of BrdU upon the ability of these cells to retain their phenotypes and to synthesize collagen and chondroitin-SO_4. Holtzer found that both intact cartilage and liberated chondrocytes produced typical $\alpha 1$-type collagen and chondroitin SO_4, whereas BrdU treated cells synthesized both $\alpha 1$ and $\alpha 2$ chains and hyaluronic acid, but no chondroitin SO_4. Dorfman found a decrease in chondroitin SO_4 synthesis by liberated BrdU-treated chondrocytes together with a decrease in hyaluronic acid. Both investigators found that the shape of the cells is related to their environment. Dorfman demonstrated that a high cell density was an equally important factor for the cells to retain the cartilage phenotype, an observation that was reinforced by the observations of Fitton-Jackson in her studies of bone-forming and cartilage-forming cells in culture.

Equally interesting were the observations of Dorfman that BrdU caused a decrease in sugar transferases, a decrease in SO_4 incorporation and a decrease in acetate incorporation by cartilage cells. He noted that the turnover of BrdU was greater than was originally anticipated, based upon studies of guanosine and thymidine turnover. These and other aspects of this interesting compound and its relation to cartilage differentiation remain to be explored.

In studying the relationship of matrix components to the synthesis of chondroitin SO_4, Helen Muir found that hyaluronic acid decreased the incorporation of $^{35}SO_4$ whereas chondroitin SO_4 had no effect—an interesting observation in light of the data subsequently presented by Drs. Balazs and Wasteson, who showed that hyaluronic acid depressed the migration of fibroblasts *in vitro*, in addition to causing some fibroblasts to aggregate *in vitro*. Interestingly, this aggregation of cells appeared to be dependent upon the chain length of the polysaccharide. Dr. Muir also noted that other agents that may act at the cell surface, such as lysozyme and concanavalin both caused an increase in $^{35}SO_4$ incorporation. These observations point to the importance of obtaining a better understanding of the interactions that occur at the cell surface and the relationships between these effects and various cellular activities, such as protein and glycosaminoglycan synthesis.

The synthesis of two other proteoglycans, dermatan SO_4 and heparin both of which involve L-iduronic acid, was discussed by Drs. Fransson and Lindahl. Fransson's observations suggested that fibroblasts in culture secrete a factor that converts L-iduronic acid to D-glucuronic acid, resulting in a glucuronic acid-rich dermatan SO_4. In contrast, Lindahl suggested that L-iduronic acid synthesis occurs by the epimerization of D-glucuronic acid residues already present in the polysaccharide chain, and pointed to sulfation as a necessary part of the epimerization. Hence post-polymeric modification of the molecule may be an important phenomenon.

The intracellular localization of connective tissue polyanions, using basic dyes and reagents that will block the reactivity of DNA, such as copper phalocyanin, or by varying the salt concentration, permitted the localization of these polyanions in Hurler's fibroblasts in culture

(Dr. Scott). Improved intra and extracellular localization of these molecules at the ultrastructural level remains an elusive, but important problem.

A group discussion of a number of inborn errors of metabolism-possibly better explained as lysosomal diseases took place between Drs. Aula, Fransson and Dorfman.

Dorfman pointed to the absence of an L-iduronidase in Hurler's and Scheie's disease, an N-acetylglucosaminidase missing in Sanfilippo B disease and a β-glucuronidase deficiency similar to the lack of a β-hexosaminidase, in Tay-Sachs's disease.

The great complexity of the various pathways of degradation of these polysaccharides and in particular the ability of the cells to degrade them, has to be appreciated if we are going to understand these different hereditary disease entities.

METABOLISM OF COLLAGEN

The second half of the first day turned to a discussion of collagen metabolism. An excellent review of collagen synthesis, secretion and fibril aggregation was presented by Darwin Prockop that set the stage for the ensuing discussion. The discovery of procollagen, an intracellular precursor that contains cystine and is relatively poor in glycine, proline, hydroxyproline and hydroxylysine was described. The cystine containing segment of the procollagen molecule is present at the amino terminus of each $\alpha 1$ chain, and is apparently cleaved by a specific enzyme, procollagen peptidase, prior to fibril aggregation, analogous to the conversion of fibrinogen to fibrin. It is not clear at what stage the 3 α chains of collagen, or procollagen, form into a triple helix, nor is it entirely clear as to whether the phenomenon of cleavage of the extra piece from the procollagen chain to create α chains or intact collagen molecules occurs intracellularly or extracellularly, possibly at the cell surface.

The enzymes necessary for the hydroxylation of proline and lysine were also discussed and their necessary cofactors, ferrous iron, O_2, α-ketoglutarate and ascorbate were pointed out. It has been possible to elute and purify proline hydroxylase by the use of an affinity column consisting of agarose and procollagen. Experiments by Bjørn Olsen, describing the molecular configuration of the molecule by electron microscopic negative staining techniques were discussed, and the use of synthetic polypeptides containing either a random configuration or a triple helical configuration, demonstrated that the enzyme can act on either substrate. Hence peptidyl proline can be hydroxylated when it is in the triple helical form.

Prockop also discussed a system for interfering with proline and lysine hydroxylation using a metal chelator, α, α'-dipyridyl. During the use of this agent collagen secretion is depressed and protein accumulates intracellularly.

The nature of procollagen was further discussed in George Martin's presentation. He was able to separate procollagen from collagen, using chromatography on DEAE following by CM cellulose. Extraction of procollagen yielded dimers, due to disulfide bonds between the pro-α chains. Martin referred to the fact that procollagen chromatographs differently from dermatosparactic collagen, later discussed by Charles Lapière, suggesting that these two forms may have different lengths of polypeptides at their respective amino termini, possibly due to previous cleavage of the dermatosparactic molecule of procollagen.

A discussion of whether hydroxylation of proline and/or lysine is necessary for collagen secretion to occur, ensued as a result of Prockop's earlier data suggesting this to be the case. A discussion of this problem turned to published data concerning a newly discovered disease of hydroxylysine deficiency in culture. In fibroblasts, in both cases, either hydroxylysine-deficient or hydroxyproline-deficient collagen appeared to be secreted by the cells, suggesting that neither hydroxylation nor glycosylation of the collagen molecule may be necessary prior to secretion of the molecule. Further data relating to this was presented by Dr. Peter Müller, who was able to uncouple hydroxylation from triple helical formation of the α chains, by showing that in calvaria in culture the connective tissue forming cells secreted randomly coiled, underhydroxylated α chains into the medium in the presence of α, α'-dipyridyl. Müller suggested that perhaps a different enzyme is responsible for triple helix formation.

At this point Bornstein discussed the possible intracellular pathways that could be taken by collagen precursors, and their possible modes of secretion. A number of inhibitors, including vinblastin, colchicine, α, α-dipyridyl, cytochalasin B, D_2O, and high pressure were used in the studies he presented. In all but dipyridyl, hydroxylation was normal, procollagen peptidase activity was normal, and yet secretion was impaired. Collagen synthesis was slightly inhibited whereas non-collagen protein synthesis appeared to be at normal levels. His results suggested that microtubules might form pathways within the cell for organelles such as the rough endoplasmic reticulum and Golgi complex. With loss of microtubules, secretion may lack an intracellular directional component and thus be inhibited. The possibility of direct secretion from the rough endoplasmic reticulum of collagen precursors versus that of proteoglycans via the Golgi complex was discussed.

Clearly more information will be necessary before this problem can be resolved.

A new model for proline hydroxylase was presented by Dr. Hurych based upon calculations for physico-chemical interactions between model compounds in which bond energies were studied to determine the sequence and mode of interaction of these compounds.

Allen Bailey discussed the role of lysine and hydroxylysine in crosslink formation and pointed to the differences between these in skin collagen. He presented data to show that dihydroxylysine derived crosslinks decrease in number with increase in age in skin and tendon, apparently as a means of increasing intermolecular stability.

Finally collagen degradation was discussed by David Jackson who pointed to the possible role of proteolytic enzymes in the degradation of insoluble collagen by first preparing collagen fibrils through a series of degradative steps prior to collagen becoming susceptible to known collagenases, a subject dealt with further by John Dingle.

In his discussion Dr. Dingle pointed to the role of lysosomal enzymes in matrix degradation, particularly the role of invading pannus in rheumatoid arthritis in cartilage degradation. He stressed the ability of cathepsin B_1 to degrade intact collagen in contrast to cathepsin D that degrades proteoglycan, but not collagen. He demonstrated the extracellular localization, by using a specific immunofluorescent antibody directed against these enzymes, and stressed their extracellular role not only in disease, but in normal degradation during physiologic turnover of the matrix. A fascinating recent observation by his group relates to the fact, that *in vitro* in organ culture, cartilage degradation does not occur with vitamin A administration in the absence of perichondrial or associated fibroblasts. This interrelationship may prove to be an important one, and further studies of this problem will be of great interest.

EXTRACELLULAR SPACE AND THE CELL SURFACE

Discussions concerning the role of the extracellular space, in particular the effect of the environment upon the cells and the nature of the extracellular matrix in which they find themselves, occupied much of the second day. The effects of proteoglycans and other matrix constituents upon the diffusibility of solutes and upon the cell surface was stressed in the presentations by Drs. Wasteson and Balazs as they relate to cell migration and cell aggregation.

The discussion then turned to the importance of understanding the nature of the cell surface. Drs. Warren, Kent and Pasternak talked about the importance of surface constituents such as sialic acid and the role of specific transferases in regulating surface changes, via their constituent glycoproteins.

Dr. Pasternak presented interesting observations demonstrating his ability to separate cells in culture in different parts of the DNA synthetic cycle by separating them in Ficoll gradients according to size. This approach should enable a further examination of the presence of various cell activities during particular parts of the cycle—information critical to our understanding of how and when cells make their surface constituents, among other properties of interest.

CELL CULTURE

A discusssion of serum constituents, a very complex issue, and their importance in cell proliferation was presented by Dr. Glinos in his studies of suspension cultures. The important various factors relating to differentiation in culture, in bone and cartilage such as cell density was elaborated by Dr. Fitton-Jackson.

Charles Levene presented a series of *in vitro* observations concerning the role of ascorbate in collagen synthesis, and as mentioned earlier, was able to produce a collagen with underhydroxylated proline that was found extracellularly. Interestingly, both lysine hydroxylation and glycosylation was less affected in his system. The determination of the intracellular and extracellular molecular form of collagen in these studies will be most important if the role of hydroxylation and reducing agents is to be better understood in these phenomena.

Dr. Pontén discussed ageing of cells and pointed to different numbers of doubling abilities of fibroblasts in culture based upon the age and source of the donor. In an ingenious experimental system he was able to suggest that ageing is a manifestation of the number of times a cell has divided and not the chronologic age of the cells.

INJURY AND REPAIR OF CONNECTIVE TISSUE

The final day of the symposium turned to a series of discussions of factors responsible for connective tissue activation and repair. Dr. Castor presented evidence for a connective tissue-activating peptide of low molecular weight, that stimulated hyaluronic acid synthesis by synovial fibroblasts. He found elevated levels of this peptide in the sera of patients that had several different types of inflammatory joint disease.

The localization and presence of enzymes associated with inflammation and repair in wounds at varying times was presented by Dr. Raekallio, followed by a discussion of the role of inflammatory cells in wounds by Ross. The importance of the neutrophil and the monocyte in wound repair was discussed. The studies of Dr. David Simpson

(1972)* who examined wounds in guinea pigs made neutropenic by a specific antineutrophil serum was mentioned. Dr. Ross noted that in the absence of overt infection, wounds containing no neutrophils at any point in time, healed exactly as did control wounds containing large numbers of neutrophils.

The importance of the macrophage and the possibility of interactions between macrophages and fibroblasts was emphasized in the work of Spector and Heppleston where cell/cell interactions, as well as cellular responses possibly stimulated by secretions from macrophages that had been presented with substances such as silica, can occur.

Differences in wound repair in different visceral organs, in terms of tensile strength gain were reported by Van Winkle and the elaboration of increased proteoglycans in keloids was emphasized by Dr. Bazin.

The ability to measure tissue levels of O_2 for many of the reactions necessary to connective tissue formation were presented. The observations by Winter concerning the relationship between various factors in the environment and bone formation were also of interest. His studies suggested that in subcutaneously implanted polyhydroxyethylmethacrylate sponges, that the size of the holes in the sponges, was related to whether or not a calcifiable matrix would form.

VASCULAR WALL

The last session of the symposium dealt with the components of the vascular wall. Drs. Lorenzen, Ross, and Björkerud each presented studies in which the intima of vessels was injured by the use of different types of intravascular catheters. The ensuing lesion that formed in these studies appeared to be related to the extent of the injury i.e. whether or not the media of the arterial wall was injured in addition to the intima. In addition Dr. Robert discussed studies in which elastin peptides were used as antigens and injury to the arterial wall as a result of the immune response was investigated.

All of these studies pointed to the importance of the smooth muscle cell in the arterial media as not only the principal contractile cell of the vessel wall, but as the principal connective tissue synthetic cell during development as well as in the response to injury of the intima. It is the smooth muscle cell that participates in the proliferative response after intimal damage occurs, and thus the factors controlling the migration of these cells, their proliferation, synthesis of connective tissue matrix constituents and ability to metabolize lipids remain important areas of future investigations. The investigations by Ross and his colleagues of

* Simpson, D. M. and Ross, R. (1972). *J. Clin. Invest.* **51**, 2009–2023.

arterial smooth muscle cells in culture provide one important approach to further understanding of these cells.

This symposium was varied and contained a great deal of information of interest and importance. One of the principal reasons for its success was the emphasis placed upon the primary cell of the connective tissue—the fibroblast—and its various relatives. Thus the thematic handling of the information presented provided a perspective that would have been otherwise difficult to obtain.

Postscript and Acknowledgements

J. Pikkarainen and E. Kulonen

When finishing our work for the symposium and for the proceedings we are aware that we were unable to achieve a comprehensive review of fibroblasts. This is due to not only our own limitations but also the uneven development of the knowledge. It is understandable that the specific synthetic functions of the fibroblasts have been in the focus of attention. However, during the preparation of the programme we realized how meagre was our knowledge of the pharmacology, toxicology and endocrinology of the fibroblasts, so that we have to rely on analogies to other cells. It is a timely problem to reveal the receptor functions and energy metabolism of the fibroblasts and thus differentiate them from other related mesenchymal cells like macrophages, blood cells, fat cells, chondrocytes or bone cells. In other words, the factors which stimulate and regulate the proliferation should be investigated systematically.

To understand the differentiation of fibroblasts we should pay attention to their evolution and to the changes which appear in the translational system during the various phases of development. When this knowledge is available in the future, the cells of the diseased connective tissue can be characterized. Today there are simply too few defined characteristics of the connective tissue cells to be used as references in the studies on the cells from rheumatoid or degenerated synovial tissue or from fibrotic tissues of liver, kidney, lung, vascular wall or the gut. Methods are gradually becoming available for the analysis of the cells from the solid granulomatous tissues, both experimental and human.

Although, in the first place, the cells are sick and the changes in the intercellular matrix only secondary, there may be feed-back influences on the cell surface which is now realized to be important in regulating the cell functions. According to the concept of the free movement of proteins in the lipid bilayer of plasma membranes, the organization of active groups on the cell wall may be determined by adjacent molecules of the matrix. This may be the source of the polarity of the cells and have a bearing on the morphogenesis.

We were interested to include as much information on the diseases of fibroblast proliferation as possible, although some advisers were quite pessimistic. Giving all credit to the distinguished contributors, we have to state our belief that a wide field is here open before us. Not so much genius but dedication and skill are required in analysing the specific features of the cells from diseased human tissues.

Besides the dissemination of the frontier knowledge with the emphasis in the critical review, our main purpose was to facilitate a new combination of ideas. We were aware that the meeting was in danger of remaining a copy of other meetings or just a social occasion. We avoided giving too much guidance to the participants on their subject, which would annoy the speakers and prevent us from hearing new things. Furthermore, we wished to work against the threatening specialization of minds, asking the members to be patient with their fellows with problems and methods eventually sounding strange and irrelevant.

The practice of distributing scientific information has changed. People do not read as much any longer: oral communication and automated data-watching have somewhat replaced the literature. The meetings are not ideal either. The information actually released will usually be available in print in a few months. Nobody wishes to advertise his frustrations, which actually would help other people, or to speak on his ideas which may be either futile or successful. Much cannot be done against this except to create conditions for improved personal relationships and to develop organizational means to avoid unnecessary competition.

Nevertheless, reading still remains the most important form of the continued education. This is the reason why the edited proceedings seem to us worthy of publication, to extend the meeting to those who for various reasons could not be present.

In the editorial work we have not attempted too much and cannot aspire to the standards of the prestige journals. Those participants who are used to the dictum 'publish or perish' have been the most obliging, as a rule. The ideal contribution to a volume like this would be of essay-type with references to technical details. The proceedings should not be a collection of progress reports but provide vision and inspiration with sources to further information. We admit that this volume is not consistent in style but rather reflects the diversity of human personalities. Only the written discussion has been forwarded to the publishers because we believe that a prepared statement is more valuable than a casual remark.

It would be of great benefit to small countries if the means of scientific information and contacts would develop so that we could better integrate our fundamental research to the universal search and dissemination of knowledge. The main emphasis would be to avoid duplication. Some international organization, like UNESCO, should develop a research administration maintained by individual governments on the basis of the gross national products. The cost of fundamental research should be included in those of general humanitarian enterprises such as aid to underdeveloped countries. This would also moderate the local excesses of the universal cultural revolution, which has disturbed the research activities everywhere. We think that temporary occasions like meetings will break ground to more organized systems in that direction.

On behalf of the whole Organization Committee we state our gratitude to the Sigrid Jusélius Foundation for selecting this topic as the subject for their symposium for which public funds would not have been available in Finland. Especially we wish to thank the Managing Director of the Foundation, Mr. M. Nykopp, and the General Secretary. Professor N. Oker-Blom, for their generous attitude and goodwill.

Furthermore, we wish to acknowledge the willing contributions of the staff of the Sigrid Jusélius Foundation and of the Department of Medical Chemistry, University of Turku. Several participants have given their valuable advice and we have had the privilege to consult also Dr. G. E. W. Wolstenholme of the Ciba Foundation. Dr. Russell Ross deserves a special tribute for the summarizing remarks. We are especially indebted to the staff of Academic Press Inc. Ltd. who accepted the responsibility for the revision of style and grammar. For the preparation of this volume we have also used time actually belonging to our employers (J. Pikkarainen in the Central Public Health Laboratory) or sponsors (E. Kulonen on the tenure of a grant from the Research Council for Medical Sciences, Finland, and of a contract with the Association of the Finnish Life Assurance Companies), and their benevolent forbearance deserves recognition.

Last, but not least, we should thank the members of the meeting who have sacrificed their comfort, summer vacation and sometimes their own resources in order to contribute. Without mentioning the names, we think with warm feeling of those who have complied so promptly with the correspondence, prepared their papers so well and handed in manuscripts which followed the instructions so conscientiously. With all the organizers we hope that everybody has gained some reward from the mutual contacts, creative associations and enriched ideas which either emerged in their fertile and prepared minds during the stay in Turku or may come out of the perusal of these proceedings.

Author Index

A

Aalto, M., 6, 7
Abbott, J., 61, 62, 63, 64, 65, 67, 68, 73, 74, 76, 77, 78, 82, 88, 91
Abelev, G. I., 268, 272
Abercrombie, M., 43, 58, 151, 152, 153, 273, 286
Ackers, J. P., 291, 298, 299, 301
Adair, G. S., 26, 39
Adamson, I. Y. R., 535, 537
Ahmed, A., 641, 649
Ahonen, J., 5, 7, 169, 170, 171, 172, 173, 175, 178, 180, 181, 182
Aine, E., 582, 582
Akpata, M., 287, 301
Algvere, P., 251, 252
Aleo, J. J., 399, 408, 409
Ali, S. Y., 123, 125
Allgöwer, M., 127, 137
Allison, A. C., 307, 308, 526, 528
Alvarez, T. R., 6, 7
Ambrose, E. J., 93, 96, 101, 205, 208
Amsterdam, A., 323, 336
Anderson, H., 63, 65, 77
Anderson, H. C., 88, 71, 121, 123, 125
Anderson, J. S., 229
Anderson, S. O., 29, 39
Andreae, U., 457, 462
Anfinsen, C. B., 363, 363
Ansay, M., 315, 316, 317, 319, 333, 337, 339, 343, 344, 345, 347, 379, 380, 384, 401, 410
Anseth, A., 441, 443, 448, 464, 471
Antonopoulos, C. A., 232, 236, 441, 443, 446, 448, 464, 471
Antonowicz, I., 401, 409
Anwar, R. A., 37, 38
Armstrong, R. L., 172, 182
Aro, H., 313, 319, 366, 372
Aronson, R. B., 314, 319, 365, 372
Arstila, A. U., 473, 478, 478, 479

Asboe-Hansen, G., 541, 556
Ash, J. F., 151, 153, 154, 332, 338
Asherson, G. L., 232, 236
Astrup, P., 614
Atkinson, P. H., 300
Augl, C., 269, 272
Auranen, A., 180, 181, 182
Austin, J. H., 457, 462
Auruch, J., 265
Autio, S., 473, 478, 478, 479
Axelsson, O., 431, 435, 436, 438

B

Babad, H., 323, 336
Bach, G. S., 455, 460, 461
Bäck, O., 205, 208
Backström, G., 432, 434, 435, 438, 439, 448
Bader, J., 270, 271, 272
Badonnel, M. C., 153, 154
Bailey, A. J., 15, 38, 382, 384, 386, 387, 389, 390, 392, 393, 394, 395, 403, 406, 408, 409, 410, 568, 570
Bajusz, E., 121, 125
Baker, J. B., 270, 271, 272
Balazs, E. A., 101, 237, 243, 249, 251, 252
Balo, J., 38
Banga, I., 38, 641, 648
Bard, J. B. L., 43, 45, 58, 166, 167, 209, 210, 211
Barker, S. B., 374, 377, 486, 499
Barnes, D. W. H., 136, 137
Barnes, M. J., 32, 38, 389, 393, 394
Barnett, J. E. G., 299, 300
Barrett, A. J., 203, 203
Bartlett, G. R., 262, 265
Barton, R. W., 450, 461, 471, 471
Bartos, G., 641, 649
Baserga, R., 197, 198, 198
Basset, C. A. L., 121, 124, 125

Basu, M., 284, *286*
Bates, C. J., 325, *336*, 389, *394*, 401, 403, 404, 406, 408, *410*
Bauer, G. E., 323, *336*
Baum, H., 476, *478*
Bavetta, L. A., 548, *557*
Baxter, R. M., 299, *301*
Bazin, S., 94, *101*, 203, *203*, 542, 546, *556*
Bdolah, A., 323, *336*
Bearn, A. G., 424, *429*, 450, *461*, 474, *478*
Becker, K., 287, *300*, 541, 544, *556*
Becker, R. M., *301*
Bekesi, E., 287, *300*
Bekesi, J. G., 287, 288, *300*, *301*
Bell, E., 90, *91*
Bellamy, G., 327, 333, *336*, 339, 340, *346*, 354, *363*, 379, 380, *384*
Benditt, E. P., 140, 152, *154*, 322, *338*, 503, *505*
Bensch, K. G., 329, *338*
Bentley, J. P., 340, *347*
Ben-Zvi, R., 323, *336*
Berenson, G., 612, *614*
Berg, R. A., 313, 314, *318*, *319*
Berg, R. B., 135, *137*
Bergeron, J. J. M., 261, *265*
Berglund, R. K., 463, *471*, 478, *479*
Berliner, D. L., 581, *583*
Bernbi, L., 387, *395*
Bernfield, M. R., 175, *182*
Betchaku, T., 43, *58*
Betz, E. H., 382, *384*
Beutler, E., 458, *462*
Bhatnagar, R. S., 324, *336*, 349, 359, *363*
Bischoff, R., 67, 75, 77
Bitter, T., 466, *471*, 486, *499*
Björk, I., 223, 226, 227, *229*
Björkerud, S., 615, 616, 617, 618, 619, 620, 622, 623, 624, *624*, *625*
Björnheden, T., 618, *625*
Black, P. H., 273, 285, *286*, 295, *301*
Bladen, H. A., 418, *420*
Bloom, G., 323, *338*
Bloom, S., 628, *636*
Bloom, W., 139, 140, 141, *153*
Blough, H. A., 303, 304, 308, *308*
Blumberg, J. B., 121, *125*
Blume, A., 164, 165, *167*, *168*
Blumenkrantz, N., 6, *7*
Boezi, J. A., 172, *182*

Bolund, L., 193, 194, *194*
Bond, V. P., 136, *137*
Bondjers, G., 616, 618, 619, 620, 623, 624, *624*, *625*
Bookchin, R. M., 393, *394*
Borek, E., 177, *182*
Bornstein, P., 30, 31 36 *39*, 61, 77, 315, 316, 317, *318*, 326, 327, 328, 329, 330, 331, 332, 333, *336*, *337*, 339, 340, 343, 344, 345, *346*, *347*, 354, 361, *363*, 379, 380, *384*, 629, 634, *636*, 641, *649*
Borrero, L. M., 216, *229*
Bosman, H. F., 287, *300*
Bosmann, H. B., 231, *236*, 314, *318*, 385, *394*
Boström, H., 582, *582*
Böttcher, C. J. F., 612, *614*
Boucek, R. J., 6, 7, 199, *203*
Bouissou, H., 640, 641, 642, *649*
Bowden, D. H., 535, *537*
Bradley, M. O., 151, 153, *154*, 332, *338*
Bradley, R. M., 289, 291, 295, 296, *301*
Bradshaw, W. S., 89, *91*
Brady, R. O., 284, *286*, 289, 295, 296, *300*, *301*
Brante, G., 449, *461*
Breitenecker, G., 541, *557*
Bridges, C. H., *347*, 382, *384*
Bridges, J. B., 121, 124, *125*
Brighton, W. D., 151, *154*
Brown, A. H., 464, *471*, 443, *448*
Brown, D. M., 34, *39*
Brown, R. S., 418, *420*
Brunk, U., 183, *188*
Bryan, J., 66, *77*
Büchner, T., 518, *518*
Buck, C. A., 274, 275, 280, 281, 282, 283, 284, 285, *286*, 295, *300*
Buckingham, R. B., 483, 486, 490, 491, 492, *499*
Buddecke, E., 220, *229*, 612, *614*, 647, *648*
Burger, M. M., 166, *167*, 267, 270, 271, 272, 273, *286*
Burgeson, K. E., 315, *318*, 346
Büring, K., 121, *125*
Burk, R. R., 166, *167*
Burwell, R. G., 121, *125*
Bush, V., 36, *38*, *39*
Butchard, C. G., 292, 299, *300*

AUTHOR INDEX

Butler, J., 251, *252*
Butler, W. T., 385, 386, *394*
Bye, I., 398, *410*
Byers, P. H., 344, 345, *347*, *364*
Bywaters, E. G. L., 581, *582*, *583*

C

Cahn, R. D., 63, *77*
Campbell, G. L., 76, *78*
Cancilla, P. A., 141, *153*
R-Candela, J. L., 323, *337*
Cantero, A., 287, *301*
Cantz, M., 450, 455, 460, *461*, 463, *471*
Cardinale, G. J., 365, 371, *371*
Carlsson, S. A., 193, 194, *194*
Carnes, W. H., 21, 34, *39*
Carter, P. K., 641, *649*
Castle, J. D., 323, 331, *336*
Castor, C. W., 483, 484, 485, 486, 487, 488, 490, 491, 492, 493, 494, *499*
Castor, L. N., 155, 166, *167*
Cebra, J., 203, *203*
Chacko, S., 63, 65, 74, 77, 78, 85, 88, *91*
Chance, B., 509, 517, *518*, 521, *524*
Cherry, G., 591, 597, *598*
Choi, Y. S., 324, *336*
Chrambach, A., 450, *461*, 471, *471*
Christensen, C. B., 605, *614*
Christner, P. J., 325, *336*
Church, R. L., 333, *336*, 339, 344, *346*
Chvapil, M., 136, *137*, *138*, 349, 353, *363*, 399, *410*, 548, *556*
Cifonelli, J. A., 431, 435, 437, 447, *448*, 451, 454, 455, *461*
Clark, L. C., 521, *524*
Clark, W. R., 89, *91*
Clarke, G. O., 623, *625*
Cliff, W. J., *154*
Cohen, S. S., 197, *198*
Cohn, R. M., 181, *182*
Collipp, P. J., 262, *265*
Colowick, S. P., 376, 377, *377*
Coman, D. R., 273, *286*
Comper, W. D., 217, *229*
Comstock, J. P., 325, *337*, 359, *363*, 373, 374, *377*, 408, *410*
Conning, D. M., 535, 536, *537*
Constable, B., 389, 393, *394*
Coon, H. G., 63, 65, 74, *77*
Cooper, G. W., 312, *319*, 324, *336*, *337*
Cooper, M. R., 329, *336*

Coulombre, A. J., *338*
Coulston, F., 627, *635*
Coulter, P. R., 121, *125*
Coutois, Y., 647, *649*
Cox, R. W., *38*
Craddock, V. M., 177, *182*
Critchley, D., 274, *286*, 295, *301*
Cronkite, E. P., 136, *137*
Crowley, G. M., *301*
Cuatrecasas, P., 268, 270, *272*
Cumar, F. A., 284, *286*, *301*
Cunningham, D. D., 155, 166, *167*, *168*
Cunningham, L. W., 385, 386, *394*
Cuthbert, J., 406, *410*

D

Dabbs, G. H., 121, *125*
Dahmen, G., 150, *153*
Dalfares, Jr., E. R., 612, *614*
Danes, B. S., 424, *429*, 450, *461*, 474, *478*
Daniel, P. F., 299, *301*
Daniels, J. R., 418, *420*
Daoud, A. S., 627, *635*
Darzynkiewicz, Z., *101*, 243, *252*
Datta, A., 289, *301*
David, J. D., 90, *91*
Davidson, E. A., 63, *78*, 85, 88, *91*, 431, *438*, 439, *448*, 451, *462*
Davies, L. M., 163, *168*, 399, *410*
Davies, M., 220, *230*
Davis, H. F., 26, *39*
Davis, M., 127, *138*
Davis, N. R., 15, 37, *38*, 387, *394*
Davison, P. F., 418, *420*
Dawson, G., 455, *461*
De Chatelet, C. R., 329, *336*
Defendi, V., 253, 256, *260*, 285, *286*
De Grouchy, J., 463, *471*
Dehm, P., 313, 314, 315, 316, 317, *318*, *319*, *320*, 327, 329, 331, 333, *337*, 339, 344, 345, *346*, 361, 362, *363*, 379, 380, *384*
De la Haba, G., 67, *77*
Delaunay, A., 203, *203*, 546, *556*
Den, H., 284, *286*
Derouette, J. C., 640, 641, 642, *649*
De Rydt, J., 73, *77*
Deshmukh, K., 548, *557*
Detwiler, S. R., 72, 76, *77*

667

De Tyssonsk, E. R., 232, *236*, 441, 448, *448*
Devis, R., 151, 152, *154*
Devrim, S., 334, *338*
Diegelmann, R. F., 329, 333, *337*, 361, *363*, 402, *410*
Di Ferrante, N., 463, *471*, 478, *479*
Dijong, I., 289, 295, *301*
Dillon, R. T., *395*
Dingle, J. T., 199, *203*, 419, *420*
Dion, A. S., 197, *198*
Dische, Z., 443, *448*, 464, 468, *471*
Dixit, P. K., 323, *336*
Dixon, A. St. J., 579, *583*
Dlouhá, M., 136, *138*
Dodgson, K. S., 476, *478*
Donnelly, P. V., 463, *471*, 478, *479*
Dorfman, A., 16, *39*, 61, 67, 73, 74, 75, 77, *78*, 80, 86, 89, *91*, 94, 96, *101*, 431, *437*, *438*, 439, 447, *448*, 449, 450, 451, 454, 455, 458, 460, *461*, *462*, 463, *471*
Dorling, J., 421, 423, *429*
Dorstewitz, E. L., *483*, *499*
Doyle, F. H., 579, *583*
Drake, M. P., 418, *420*
Dubuc, F. L., 121, *125*
Dulbecco, R., 155, 166, *167*
Duncan, H. M., 262, *265*
Dunphy, J. E., 517, *518*, 585, *589*, 591, 597, *599*, 611, *614*
Dwek, R. A., 291, 292, 299, *300*, *301*

E

Eagle, H., 129, *137*, 485, *499*
Easty, G. C., 93, 96, *101*
Edelman, I. S., 218, *230*
Ege, T., 193, 194, *194*
Ehrlich, H. P., 315, 316, 317, *318*, 327, 329, 330, 331, 332, 333, *336*, *337*, 343, 344, 345, *346*, *347*, 361, *363*, 586, 588, *589*
Ehrlichová, E., 548, *556*
Ehrmann, R. L., 209, *211*
Eidam, C. R., 157, *167*
Eigner, E. A., 340, *347*, 350, 353, 354, *364*
Eisenstein, R., 122, *125*
Eisenthal, R., 292, *301*
Ellis, S., 503, *505*

Ellison, M. L., 72, 77, 93, 96, *101*
Elsdale, T., 41, 43, 45, 54, *58*, 166, *167*, 209, 210, *211*
Enzinger, F. M., 150, 151, *154*
Epstein, E. H. Jr., 61, *78*, 350, 356, 357, *363*, 393, *394*
Ericsson, J. L. E., 183, *188*
Erl, D., 541, 544, *557*
Essner, E., 183, *188*
Estensen, R. D., 332, *337*
Etherington, D. J., 419, *420*
Everett, N. B., 152, *154*, 518, *518*, 525, *528*
Eylar, E. H., 314, *318*, 326, *337*, 385, *394*
Eyring, H., 217, *229*

F

Fabish, P., *301*
Fabre, M. T., 640, 641, 642, *649*
Fairweather, R., 387, *395*
Farber, J., 164, 165, *167*, *168*
Faris, B., 19, *38*, *39*
Farquhar, M., 324, 331, *337*
Farr, A. L., 262, *265*, 291, *301*, 476, *479*
Farrow, L. J., 151, *154*
Fawcett, D. W., 139, 140, *153*
Feeney, L., 141, *154*
Ferber, E., 536, *537*
Ferguson, S. J., 292, 299, *300*
Fessler, J. H., 216, 223, *229*, 315, *318*, 345, *346*
Field, J. B., 323, *338*
Fink, C. J., 323, *337*
Finnegan, C. V., 94, *101*
Fischer, H., 536, *537*
Fisher, H. W., 166, *168*
Fitch, S. M., 129, *137*
Fitton Jackson, S., 231, *236*
Flatt, A., 581, *582*
Fletcher, K., 537, *537*
Fliedner, T. M., 136, *137*
Folch, J., 262, *265*
Foley, R., 41, 45, *58*
Forrest, L., 568, *570*
Fowler, B. J., 93, *101*
Fowler, I., 93, *101*
Fowler, L. J., 15, *38*, 386, *394*
Fox, T. O., 271, *272*
Franco-Browder S., 73, 77
Franssila, K., 7

AUTHOR INDEX

Fransson, L.-A., 431, 432, 434, *437, 438*, 439, 441, 443, 444, *448*, 451, *461*, 464, 465, 470, *471*
Franzblau, C., 19, *38, 39*, 387, *395*
Fratantoni, J. C., 450, *461*, 478, *479*
Frederickson, R. G., 94, *101*
Freeman, M. I., 251, *252*
French, J. E., 627, *635*
Friberg, O., 541, *556*
Friedman, R., 455, *461*
Froese, G., 166, *167*
Fuhrer, J. P., 274, 281, 282, 283, 284, 285, *286*
Fukushima, K., 203, *203*
Fullmer, H. M., 418, *420*
Furth, J., 432, *437*
Furthmayr, H., 345, *347*, 380, 381, 382, *384*

G

Gabbay, K. H., 332, *338*
Gabbiani, G., 7, 139, 140, 141, 142, 143, 144, 145, 148, 149, 150, 151, 152, 153, *154*, 569, *570*
Gallagher, J. T., 299, *301*
Gallop, P. M., 15, *39, 327*, 392, 393, *394, 395*
Garbarsch, C., 603, 604, 605, 606, 608, 609, 611, *614*
Gardell, S., 232, *236*, 441, 443, 446, *448*, 464, *471*
Gardner, D. L., 36, *38, 39*
Garg, B. D., 548, *557*
Gaunt, S. J., 514, *518*
Gelotte, B. J., 22, 24, *38*
Georgiev, G. P., 170, *182*
Gerlach, U., 541, 544, 550, 554, *556*, 647, *648*
Gerth, N., 548, *557*
Gey, G. O., 209, *211*
Ghadially, F. N., 152, *154*
Ghidoni, J. J., 136, *137*
Gibson, Q. H., 486, *499*
Gilbert, F., 164, 165, *167*
Gilbert, I. G. F., 216, *229*
Gilden, R. V., 269, *272*
Gillman, T., 612, *614*
Glaid A. J. 376 *377*
Glasser R. M. 288, *301*
Glick, M. C., 274, 275, 280, 281, 284, 285, *286*, 295, *300*

Glimcher, M. J., 326, *388*, 386, *395*, 401, *410*
Glinos, A. D., 156, 159, 160, 161, 163, *167*, 507, 518, *518*, *519*
Glock, K., 419, *420*
Glynn, L. E., 145, *154*
Gnoh, G. H., 203, *203*
Godale, F., 152, *154*
Goldberg, B., 136, *137*, 151, 152, *154*, 163, 164, 166, *167*, *168*, 322, *337*
Goldberger, R. F., 363, *363*
Goldman, D., 151, 152, *154*
Gomez-Acebo, J., 323, *337*
Gontcharoff, M., 89, *91*
Good, R. A., 449, *462*
Goodwin, B. C., 164, *167*
Gotte, L., 27, *39*
Gottlieb, A. A., 324, *337*, 349, *363*
Gottschalk, A., 36, *38*
Gould, B. S., 569, *570*
Gouterman, M., 366, *372*
Graham, J. M., 261, 262, *265*
Grant, M. E., 311, 312, 313, 314, *318*, 324, *337*
Grantham, W. G., 535, *537*
Grasedyck, K., 541, 544, 546, *556*, *557*
Gray, W. R., 34, *39*
Green, F. H. Y., 641, *649*
Green, H., 136, *137*, 151, 152, *154*, 160, 163, 164, 166, *167*, *168*, 315, 317, *319*, 322, *337*, 340, 344, 345, *347*, 399, *410*, 628, *636*
Greenlee, T. K. Jr., 32, *38*, 140, *154*
Greider, M. H., 323, *337*
Gribble, T. J., 325, *337*, 359, *363*, 373, 374, *377*, 408, *410*
Gries, G., 539, *556*, *557*
Griffith, P. C., 121, *125*
Griffiths, J. B., 155, 166, *168*
Grimes, W. J., 284, *286*
Grislis, G., 586, 588, *589*
Grobéty, J., 144, 148, 149, *154*
Grobstein, C. I., 93, *101*
Grosgogeat, Y., 645, 646, *648*, *649*
Gross, J., 61, *78*, 79, *91*, 321, *337*, 340, *347*, 401, *410*, 417, 419, *420*
Guenther, H. L., 234, *236*

H

Habuchi, O., 63, *78*, 435, *438*
Hagen, P., 432, *437*

Hakala, M. T., 376, *377*
Häkkinen, H.-M., 6, *7*
Hakomori, S., 270, 271, *272*
Hall, C. W. 450, *461*, 478, *479*
Halme, J., 313, *318*
Halpern, M., 163, *168*
Haltia, M., 473, *478*
Hamburger, R. N., 135, *138*
Hamburger, V., 80, *91*
Hamerman, D. A., 61, 67, *78*, 215, *230*
Hamilton, H. C., 80, *91*
Hanafusa, H., 270, *272*
Hanley, W. B., 449, *461*
Hanset, R., 381, 382, *384*
Hardingham, T. E., 231, 234, 236, *236*
Hargrove, D. D., 156, *167*
Harkness, M. L. R., 129, *137*
Harkness, R. D., 129, *137*
Harper, E., 419, 420, *420*
Harris, E. J., 548, *556*
Harris, H., 135, *137*, 193, *194*
Harris, R. C., 449, *461*
Harrison, R., 292, *301*
Hartley, B. S., 14, *39*
Hartman, J. L., 32, *38*
Hartmann, J. F., 285, *286*
Harvey, S. C., 559, *570*
Hass, G. M., 122, *125*
Hatanaka, M., 269, *272*
Hathorn, M., 612, *614*
Haug, A., 432, *437*, *438*, 439, *448*
Hausmann, E., 314, *319*
Hauss, W. H., 6, 7, 518, *518*, 541, 544, 550, 554, *556*, *557*, 612, *614*, 647, *648*
Haust, M. D., 424, *429*, 552, *556*, 627, *635*
Hay, E. D., 321, *338*
Hay, R. J., 184, *188*
Hayden, G. A., *301*
Hayflick, L., 135, *137*, 183, *188*
Heaysman, J. E. M., 151, 152, *153*, 273, *286*
Heidrich, H.-G., 546, *557*
Heikkinen, E., 541, *557*
Heinen, J. H., 121, *125*
Helin, P., 548, *557*, 603, 604, 605, 606, 608, 609, 610, 611, *614*
Helle, O., *319*
Helting, T., 435, *437*
Henderson, C. I., 532, *537*, *537*

Henderson, P. S., 581, *583*
Henry, J. C., 647, *648*
Hentel, J., 135, *137*
Heppleston, A. G., 529, 531, 535, 536, *537*
Herman, R. H., 181, *182*
Hers, H. G., 449, 450, *462*, 477, *479*
Hess, B., 503, *505*
Heughan, C., 591, 597, *599*
Hewitt, C., 559, *570*
Highberger, J. H., 321, *337*
Hilfer, S. R., 96, *101*
Hilker, D. M., 641, *648*
Hiller, A., *395*
Hilz, H., 611, *614*
Hinrichsen, 451, *462*
Hirayama, C., 546, *557*
Hiroshige, K., 546, *557*
Hirsch, C., 218, *230*
Hirsch, E. I., 432, *437*
Hirschel, B. J., 7, 139, 140, 141, 142, 144, 145, 148, 149, 151, 152, *154*, 569, *570*
Hirschhorn, K., 135, *137*
Hirst, G. K., 307, *308*
Hobza, P., 366, 367, 368, *372*
Hoffman, R., 366, *371*
Hoffman, R. C., 289, *301*
Hogan, M. J., 141, *154*
Holborrow, E. J., 145, 151, *154*
Holley, H. L., 223, *229*
Holley, R. W., 155, *168*, 628, *635*
Holliday, R., 187, *188*
Holman, G. D., 299, *300*
Holthausen, H. S., 85, 88, *91*
Holtzer, H., 61, 62, 63, 64, 65, 66, 67, 68, 72, 73, 74, 75, *76*, 77, *78*, 82, 85, 88, *91*, 93, 96, *101*
Holtzer, S., 62, 63, 65, 69, 74, 77
Höök, M., 432, *437*
Horn, J. R., 488, *499*
Hors-Cayla, M. C., 463, *471*
Horwitz, A. L., 80, 86, *91*
Houghton, K. T., 222, *230*
Hou-Jensen, K., 503, *505*
Hourihane, D. O., 579, *583*
Housley, T., 387, *395*
Hovi, T., *272*
Howell, R. R., 477, *479*
Howell, S. L., 323, 331, *337*
Howes, E. L., 559, *570*

AUTHOR INDEX

Hrgovcic, R., 463, *471*, 478, *479*
Huennekens, F. M., 262, *265*
Huggins, C. B., 121, *125*
Hulliger, L., 127, *137*
Humphreys, T., 270, 271, *272*
Hunt, T. K., 517, *518*, 585, 586, 588, *589*, 591, 596, 597, *598*, *599*
Hunter, J. A. A., 418, *420*
Hurych, J., 136, *137*, *138*, 314, *319*, 366, 367, 368, *372*, 399, *410*, 548, *556*
Hussels, I. E., 478, *479*
Hutton, J. J. Jr., 324, *337*, 350, 353, 360, *363*, *365*, *372*, 373, 374, *377*, 569, *570*

I

Ichida, T., 611, *614*
Igarashi, S., 61, *78*, 79, *91*
Illiano, G., 270, *272*
Imai, H., 152, *154*
Inbar, M., 273, *286*
Infante, A. A., 311, *319*, 349, *363*
Irlé, C., *154*
Ishikawa, H., 73, 76, 77
Isselbacher, K. J., 269, *272*

J

Jackson, D. S., 340, *347*, 418, *420*, 568, *570*
Jackson, J. G., 641, *648*
Jacobs, H. G., 386, *394*
Jacobson, B., 237, 249, 252, 431, *438*, 439, *448*
Jacobus, D., 548, *557*
Jacoby, F., 136, *137*
Jaffe, I. A., 548, *556*, *557*
Jahnke, A., 150, *154*
Jahnz, M., 135, *137*
James, A. T., 307, *308*
James, D. W., 151, 152, *154*
Jamieson, G. A., *301*
Jamieson, J. D., 323, 324, 331, *336*, *337*, *338*
Jänne, J., 195, *198*
Janners, M. Y., 80, 89, *91*
Jansson, L., 435, *438*
Jatzkewitz, H., 457, *462*
Jellinek H., 641, *648*, *649*
Jenner, F. A., 473, 474, 478, *479*

Jiminez, S. A., 313, 314, 315, 316, 317, *318*, *319*, *320*, 327, 333, *337*, 339, 344, 345, *346*, 362, *363*, 379, 380, *384*
Johnson, G. D., 145, *154*
Johnson, G. S., 270, 271, *272*
Johnston, J. P., 223, *229*
Jones, R., 152, *154*, 627, 635
Jung, C., 264, *265*
Junge-Hülsing, G., 6, 7, 518, *518*, 541, 544, 550, 554, *556*, 557, 612, *614*, 647, *648*
Junqua, S., 640, 643, 647, *648*
Juva, K., 312, 315, *319*, 324, *337*, *338*, 349, 360, *363*, *364*, 544, *557*, 560, *570*

K

Kadar, A., 30, 36, *38*, *39*, 641, *648*, 649
Kadlecova, V., 136, *137*
Kaitila, I., 582, *583*
Kalant, N., 611, *614*
Kaluš, M., 136, *137*
Kaneko, T., 323, *338*
Kanfer, J. N., 291, *301*
Kang, A. H., 61, *78*, 79, *91*, 390, *394*
Kano-Sueoka, T., 174, *182*
Kantor, T. G., 447, *448*
Kao, K. T., 641, *648*
Kaplan, A., 324, *337*
Kaplan, D., 449, *461*
Kari, A., 582, *583*
Kates, M., 307, *308*
Kawai, S., 270, *272*
Kawakami, T. G., 253, *260*
Keech, M. K., 209, *211*
Keegan, D. J., 530, 532, *537*
Keen, L. N., 160, *168*
Kefalides, N. A., 61, 77
Keith, A., 121, *125*
Kemp, J. D., 89, *91*
Kent, P. W., 288, 291, 292, 298, 299, 300, *301*
Kenzora, J. E., 326, *338*, 386, *395*, 401, *410*
Kerwar, S. S., 380, *384*
Kiernan, J. A., 155, *168*, 628, *635*
King, J., 435, *437*
Kirby, K. S., 169, *182*
Kirśchner, N., 331, *337*

Kishida, Y., 313, 314, *319*
Kivimäki, T., 473, 478, *479*
Kivirikko, K. I., 129, *137*, 313, 314, *318*, *319*, 324, 327, *336*, *337*, 353, 360, *363*, 365, 366, 367, *372*, 408, *410*, 417, *420*
Kivisaari, J., 178, *182*
Klebanoff, S. J., 627, *636*
Klein, L., 504, *505*, 548, *557*
Kletzien, R. F., 332, *337*
Klinge, F., 552, *557*
Klinger, I., 628, *635*
Klöti, R., 251, *252*
Klynstra, F. B., 612, *614*
Knack, J., 508, *519*
Knecht, J., 451, *461*
Knopf, P. M., 324, *336*
Knowles, J. R., 25, *39*
Kodicek, E., 389, *394*, 401, *409*, *410*
Kohn, L. D., *319*, 327, 333, *337*, 340, *347*, 361, *363*, 379, 380, 382, *384*
Kohn, P., 289, *301*
Kojima, T., 475, *479*
Kolodny, E. G., *301*
Kolodny, E. H., 284, *286*
Konigsberg, J., 93, *101*
Kostianovsky, M., 323, *337*
Kotásek, A., 138, *138*
Kowalewski, K., 541, *557*
Koyama, H., 67, 77
Kraemer, P., 234, *236*
Krane, S. M., 326, *338*, 386, *395*, 401, *410*
Kregar, I., 203, *203*
Kresse, H., 450, 455, *461*, 463, *471*, 647, *648*
Krolikowski, L. S., 43, *58*
Kroz, W., 220, *229*
Kruse, P. F. Jr., 160, *168*
Kuettner, K. E., 234, *236*
Kühn, K., 315, *319*, 361, *364*, 379, 380, 381, 382, *384*, 387, *395*
Kvist, T. N., 94, *101*
Kulonen, E., 5, 6, 7, 33, *39*, 169, 170, 178, *182*, 205, *208*, 582, *582*, *583*
Kumar, V., 612, *614*

L

La Croix P., 76, 77
Lacy, P. E., 323, 331, *337*

Lagunoff, D., 435, *438*
Laitinen, O., 129, *137*, 582, *583*
Lampiaho, K., 5, 7
Lampiois, R., *38*
Landing, B. H., 424, *429*
Lane, B. P., 141, *154*
Lane, J. M., 311, *319*, 350, 356, 357, *364*, 386 *394*
Langer, L. O. Jr., 449, *462*
Langness, U., 313, *319*, 373, *377*
Lanks, K. W., 176, *182*
Lapière, C. M., 315, 316, 317, *319*, 327, 333, *337*, 339, 340, 343, 344, 345, *347*, 361, *363*, *364*, 379, 380, 381, 382, 383, *384*, 401, *410*, 417, *420*
Larsen, B., 432, *437*, *438*, 439, *448*
Larsen, R. D., 150, *154*
Lash, J. W., 62, 65, 72, 74, 77, *78*, 93, 96, *101*, 312, *319*, 324, *337*
Lattes, R., 150, 151, *154*
Lauffer, M. A., 216, *229*
Laurent, T. C., 22, *39*, 216, 223, 224, 225, 226, 227, 228, *229*, 230
Laws, J. W., 579, *583*
Lawson, D. E. M., 303, *308*
Layman, D. L., 315, 316, *319*, 327, 333, *337*, 339, 344, 345, *347*, 379, 380, *384*, 406, *410*
Lazar, G. K., 160, *168*
Lazarides, E., 311, 312, *319*, 349, *363*
Lazarow, A., 323, *336*
Lazarus, G. S., 418, *420*
Lazarus, N. R., 334, *338*
Lebez, D., 203, *203*
Lees, M., 262, *265*
Legallais, V., 509, *518*
Lehtonen, A., 611, *614*
Leibovich, S. J., 418, *420*
Leloir, L. F., 475, *479*
Lemire, V., 121, *125*
Lenaers, A., 315, 316, 317, *319*, *320*, 327, 333, *337*, 339, 340, 343, 344, 345, *347*, 361, *363*, *364*, 379, 380, 382, 383, *384*, 401, *410*
Lenegre, J., 646, *648*
Lennox, E. S., 324, *336*
Lent, R. W., 19, *39*
Lerman, L. S., 422, *429*
Lerman, M. I., *182*
Lester, G., 323, *336*
Letterer, E., 620, *625*

AUTHOR INDEX

Levene, C. I., 325, *336*, 389, *394*, 398, 401, 403, 404, 406, 408, *409*, *410*
Leventhal, M., 141, *154*
Levine, S., 90, *91*
Levitt, D., 74, *78*, 89, *91*
Lewis, M. S., 340, *347*, 350, 353, 354, *364*
Lewis, R., 508, *519*
Ley, K. D., 157, *168*
Lieb, W. R., 216, 228, *229*
Lillywhite, J. W., 127, 136, *138*
Lim, D., 103, *125*
Limbrook, G., 541, 544, *557*
Lin, S.-Y., 76, *78*
Lindahl, A. W. Jr., 323, *336*
Lindahl, U., 431, 432, 434, 435, 436, *437*, *438*, 439, *448*
Lindahl-Kiessling, K., 205, *208*
Lindblad, B., *372*
Lindner, J., 287, *300*, *301*, 503, *505*, 539, 541, 544, 546, 551, 552, 553, 554, 555, *556*, *557*
Lindstedt, G., 370, *372*, 544, *557*
Lindstedt, S., 370, *372*
Lindy, S., 603, 609, 610, 611, *614*
Lipton, A. L., 628, *635*
Lithner, F., 183, *188*
Littlefield, J. W., 157, *168*
Lloyd, W. J., 292, *301*
Long, F. A., 217, *230*
Lorenzen, I., 548, *557*, 603, 604, 605, 606, 608, 609, 610, 611, 612, *614*
Lorincz, A. E., 449, *461*
Loutit, J. T., 136, *137*
Low, F. N., 94, *101*, 411, *416*
Lowe, I. P., 323, *338*
Lowry, Q. H., 262, *265*, 291, *301*, 476, *479*
Lozaityte, I., 215, *229*
Lucia, S. P., 127, *138*
Luck, J. V., 150, *154*
Luduena, M. A., 151, 153, *154*, 332, *338*
Lukens, L. N., 311, 312, *319*, 325, *338*, 349, 353, *363*, *364*

M

McAlister, W. H., *461*
Macbeth, R. A., *301*
McCabe, M., 220, *229*
McCall, C. E., 329, *336*
McClain, P. E., 61, *78*
McDonald, J. K., 503, *505*

MacFadyan, D., *395*
McFarland, V. W., 284, *286*, 289, 291, 295, *301*
McGavack, T. H., 641, *648*
McGee, J. O'D., 313, *319*, 373, *377*, 527, *528*
McGoodwin, E. B., 315, 316, *319*, 327, 333, *337*, 339, 340, 344, 345, *347*, 349, 350, 354, 356, 358, *364*, 379, 380, *384*
McHale, J. S., 184, *188*
McHale, J. T., 184, *188*
Macintyre, E., 183, *188*
McKusick, V. A., 449, *461*, 478, *479*
McMinn, R. M. H., 501, *505*
Macpherson, I., *286*, 295, *301*
Macek, M., 135, 136, *137*, *138*, 399, *410*
Mackie, J. S., 216, *229*
Mackler, B., 262, *265*
Madden, J. W., 559, *570*
Mäenpää, P. H., 175, *182*
Majno, G., 6, 7, 139, 140, 141, 142, 143, 144, 145, 148, 149, 150, 151, 152, 153, *154*, 511, *518*, 569, *570*, 586, *589*
Mäkinen, P. L., 503, 504, *505*
Makino, M., 475, *479*
Malaisse, W. J., 323, 332, *337*, *338*
Malaisse-Lagae, F., 323, *337*
Malamud, D., 197, 198, *198*
Malawista, S. E., 329, *338*
Malmström, A., 432, 434, *437*, *438*, 439, 444, *448*, 465, 470, *471*
Mammi, M., 27, *39*
Mandel, L. R., 177, *182*
Manner, G., 163, *168*, 569, *570*
Manning, G. S., 222, 223, *229*
Marasas, L. W., 530, 532, 535, *537*
Margolis, R. L., 325, *338*
Marks, G. S., 535, *537*
Maroteaux, P., 463, *471*
Maroudas, A., 222, *229*
Martin, G. R., 315, 316, *319*, 327, 333, *337*, 339, 340, 344, 345, *347*, 349, 350, 354, 356, 358, *364*, 379, 380, *384*, 406, *410*
Martin, G. S., *286*
Marx, J. L., 6, 7
Marzullo, G., 74, *78*
Massey, V., *499*
Masson, H. A., 121, *125*

Matalon, R., 67, 75, 78, 447, 448, 449, 450, 451, 454, 455, 460, 461
Matheson, D. W., 72, 73, 74, 75, 76, 77, 96, 101
Mathews, M. B., 215, 229
Mathiessen, M. E., 603, 604, 605, 606, 608, 609, 611, 614
Matsuya, J., 628, 636
Matthes, J. K., 6, 7
Mattsson, K., 473, 479
Matukas, V. J., 61, 69, 78, 79, 91
Maumenee, A. E., 449, 461
Maximow, A. A., 136, 138
Mayne, R., 62, 63, 64, 67, 68, 69, 72, 73, 74, 75, 76, 77, 78, 88, 91
Mazia, D., 89, 91
Meares, P., 216, 229
Mechanic, G., 15, 39, 568, 570
Medoff, J., 74, 78
Meezan, E., 273, 286, 295, 301
Meezan, I., 273, 285, 286
Méhes, G., 542, 557
Mehta, P. N., 152, 154
Menzies, R. A., 184, 188
Merchant, D. J., 157, 167
Merendino, K. A., 121, 125
Merlie, J. P., 303, 304, 308, 308
Merriman, P., 548, 557
Meyer, F. A., 217, 229, 230
Meyer, K., 61, 78
Meyer-Schwickerath, G., 251, 252
Michl, J., 129, 135, 137, 138
Miller, E. J., 61, 69, 78, 79, 91, 350, 356, 357, 363, 364, 386, 390, 383, 394
Miller, R. L., 314, 319, 327, 338
Mills, S. E., 135, 138
Milsom, D. W., 418, 420
Milton, S., 591, 597, 598
Mirow, S. M., 96, 101
Mitchison, J. M., 264, 265
Miura, Y., 89, 91
Moberg, E., 581, 583
Mochan, B., 67, 72, 73, 75, 77
Moczar, E., 645, 649
Moczar, M., 637, 640, 644, 645, 646, 647, 648, 649
Modolell, M., 536, 537
Mohos, S. C., 419, 420
Molnar, Z., 288, 300, 301, 542, 557
Monson, J. M., 315, 316, 317, 318, 327, 336, 346

Montandon, D., 154
Montreuil, J., 458, 460, 462
Monty, K. J., 22, 24, 39
Moore, S., 413, 416
Moorhead, P. G., 135, 137
Mora, P. T., 284, 286, 289, 291, 295, 296, 300, 301
More, R. H., 552, 556, 627, 635
Morgan, H. P., 184, 188
Morgan, P. H., 386, 394
Morganroth, J., 289, 301
Morotomi, I., 546, 557
Morris, S. C., 61, 78
Morrison, E., 152, 154
Morton, L. F., 393, 394
Moscona, A., 81, 91
Moss, N. S., 152, 154
Mouton, M. L., 184, 188
Muir, H. M., 231, 234, 236, 236, 466, 471, 486, 499
Muirden, K. D., 483, 499
Mukherjee, D. P., 29, 39
Mulholland, J. H., 508, 519
Müller, P. K., 339, 340, 344, 347, 349, 350, 354, 356, 358, 364
Müller, U. S., 557
Mulveny, T., 314, 319
Munday, K. A., 299, 300
Munder, P. G., 536, 537
Munn, R. J., 253, 260
Murison, G. L., 65, 78
Mussini, E., 373, 377
Myers, M. B., 591, 597, 598

N

Nagchaudhuri, J., 611, 614
Nameroff, M., 63, 65, 66, 78
Narayanan, A. S., 406, 410
Neri, G., 463, 471, 478, 479
Ness, O., 319
Neufeld, E. F., 450, 455, 460, 461, 462, 463, 471, 478, 479
Nevo, Z., 80, 91
Newsom, B. G., 216, 229
Nichols, B. L., 463, 471, 478, 479
Niinikoski, J., 517, 518, 579, 583, 585, 586, 589, 591, 596, 597, 598, 599, 611, 614
Nimni, M. E., 61, 78, 548, 557
Nirenberg, M., 164, 165, 167, 168
Noble, N. L., 199, 203

AUTHOR INDEX

Noonan, K. D., 166, *167*
Nordling, S., *272*
Nordwig, A., 314, *319*
Novikoff, A. B., 183, *188*
Novoa, W. B., 376, *377*
Nusgens, B. V., 315, 316, 317, *319*, 333, 337, 339, 343, 344, 345, *347*, 379, 380, *384*, 401, *410*

O

Oberwittler, W., 518, *518*
O'Brien, J. S., 456, 457, *461*
O'Brien, R. D., 292, *301*
Öbrink, B., 226, 228, *229*, *230*
O'Connell, J. J., 411, *416*
O'Dell, B. L., *38*
Odland, G. F., 152, *154*
Ohad, I., 323, *336*
Ogston, A. G., 215, 220, 223, 224, *229*, *230*
O'Hara, P. J., *347*, 382, *384*
O'Hare, M. J., 93, 96, *101*
Okada, S., 457, 460, *461*
Olsen, B. R., 313, 314, 315, *318*, *319*, 327, *337*, 339, 344, 345, *346*, 379, 380, *384*
Olsson, I., 441, *448*
O'Neal, R. M., 136, *137*
Ono, T., 67, 77
Onodera, K., 273, *286*, *301*
Orci, L., 332, *338*
Orekhovich, V. N., 316, *319*
Orgel, L. E., 187, *188*
Orr, J. W., 641, *648*
O'Shea, J. D., 152, *154*
Österlin, S., 249, *252*
Otten, J., 270, 271, *272*
Ouzilov, J., 647, *649*
Oyama, V. I., 485, *499*
Ozer, H. L., *301*

P

Pai, M., 588, *589*
Palade, G. E., 323, 324, 331, *336*, 337, *338*
Palo, J., 473, 478, *478*, *479*
Pänkäläinen, M., 313, *319*, 365, 366, *372*
Papadimitriov, J. M., 527, *528*

Pappenheimer, J. R., 216, *229*
Parakkal, P. F., 418, 419, *420*
Pardee, A. B., 155, 166, *167*, 267, *272*
Parilla, R., 323, *337*
Pariser, R. J., 166, *168*
Parker, F., 152, *154*
Parker, G., 92, *101*
Parlebas, J., 640, 644, *649*
Partridge, S. M., 14, 15, 16, 17, 18, 19, 20, 21, 22, 23, 26, 27, 28, 32, 33, 36, *38*, *39*
Pastan, I., 270, 271, *272*, 323, *338*
Pasternak, C. A., 261, 262, *265*
Patel, J. C., 150, *154*
Patrick, R. S., 527, *528*
Paul, D., 628, *635*
Paul, J., 127, *138*
Paulton, S., 218, *230*
Pawelek, J. M., 65, *78*
Peach, C. M., 386, 387, 392, *394*
Peacock, E. E. Jr., 559, *570*
Pearce, M. B., 597, *599*
Pegrum, S. M., 151, 152, *153*
Pei-Lee Ho, 16, *39*
Perch, C. M., 15, *38*
Perdue, J. F., 332, *337*
Perez-Tamayo, R., 418, 419, *420*
Perlin, A. S., 431, *438*
Perrins, D. J., 590, *590*
Persson, H., 223, 226, 227, *229*
Pertoft, H., 205, *208*
Pessac, B., 253, 256, *260*
Peterkofsky, B., 324, 329, 333, *337*, *338*, 349, 361, *363*, *364*, 402, *410*, 569, *570*
Peters, J. H., 245, *252*
Peters, R. A., 292, *301*
Petrakis, N. L., 127, *138*
Pezess, M. P., 647, *649*
Pezzin, G., 27, *39*
Pfahl, D., 418, *420*
Pfefferkorn, E. R., 308, *308*
Pfeiffer, S. E., 333, *336*, 339, 344, *346*
Phelps, C. F., 288, *301*, 326, *338*, 386, *395*
Phillips, B., 261, *265*
Pictet, S., 90, *91*
Pierard, G., 383, *384*
Pietruszkiewicz, A., 223, 226, 227, *229*
Piez, K. A., 61, *78*, 312, *320*, 340, 344, 345, *347*, 350, 353, 354, 356, 357, 361, 362, *363*, *364*, 386, 393, *394*

Pigman, W., 223, *229*
Pikkarainen, J., 33, *39*
Pinnell, S. R., 326, *338*, 386, *395*, 401, *410*
Pious, D. A., 135, *138*
Plagemann, P. G. W., 157, *168*, 332, *337*
Platt, D., 223, *229*
Podosin, R., 449, *462*
Pohjanpelto, P., 195, 196, 197, *198*
Pollitt, R. J., 473, 474, 478, *479*
Pons, M., 307, *308*
Pontén, J., 183, *188*
Popenoe, E. A., 314, *319*, 365, *372*
Porter, K. R., 321, *338*
Porter, R. R., 203, *203*
Poullain, N., 25, *39*, 647, *649*
Prager, S., 217, *230*
Press, E. M., 203, *203*
Preston, B. N., 217, 220, 222, 226, *229*, *230*
Priest, R. E., 163, *168*, 399, *410*
Prince, R. K., 483, *499*
Prockop, D. J., 129, *137*, 311, 312, 313, 314, 315, 316, 317, *318*, *319*, *320*, 324, 325, 327, 329, 331, 333, *336*, *337*, 339, 344, 345, *346*, 349, 353, 359, 360, 361, 362, *363*, *364*, 367, *372*, 379, 380, *384*, 408, *410*, 543, 544, *557*, 560, *570*
Prodi, G., 424, *429*
Prodi, M. P., 424, *429*
Prynne, C. J., 325, *336*, 408, *410*

Q

Quastel, J. H., 287, *301*
Quinton, B. A., *461*

R

Raekallio, J., 501, 502, 503, 504, *505*
Raina, A., 195, *198*
Ralph, A., 299, *300*
Ramaley, P. B., 326, *338*, 339, 344, *347*
Randall, R. J., 262, *265*, 291, *301*, 476, *479*
Rao, N. A., 262, *265*
Rasmussen, H., 323, 329, 331, *338*
Rasmussen, O., 548, *557*
Rauterberg, J., 387, *395*
Ray, R. D., 234, *236*

Read, W. K., *347* 382, *384*
Recant, L., 334, *338*
Regelson, W., 237, *252*
Regnault, F., 251, *252*
Reilly, T. J., 503, *505*
Rein, A., 155, 166, *168*
Reissig, J. L., 475, *479*
Remensnyder, J. P., 511, *518*, 586, *589*
Renkin, E. M., 216, *229*, *230*
Revel, J.-P., 321, *338*
Reverdy, V., 645, 646, *649*
Rezáčová, D., 138, *138*
Rhoads, R. E., 313, *319*, 365, 366, 370, 371, *371*, *372*
Richards, F. M., 25, *39*
Richmond, J. E., 288, 289, *301*
Riekkinen, P., 473, 478, *478*, *479*
Riggs, A. D., 76, *78*
Rimoin, D. L., *461*
Ringertz, N. R., 193, 194, *194*
Robbin, A., 628, *636*
Robbins, P. W., 273, 274, 285, *286*, 295, *301*
Robert, A. M., 640, 641, 643, 645, 646, *648*, *649*
Robert, B., 640, 641, 642, 645, 646, 647, *648*, *649*
Robert, L., 25, *39*, 637, 640, 641, 642, 643, 644, 645, 646, 647, *648*, *649*
Robertson, W. van B., 399, 403, *410*, 569, *570*
Robins, S. P., 387, 392, 393, *394*, *395*, 568, *570*
Robinson, C. V., 218, *230*
Robinson, G. B., *301*
Robinson, H. C., 67, *78*, 458, *462*
Rodén, L., 431, *438*, 439, 444, *448*, 464, *471*
Romane, W. M., *347*, 382, *384*
Ronzio, R. A., 89, *91*
Ropohl, D., 541, 544, 546, *556*, *557*
Roseborough, W. J., 476, *479*
Rosebrough, N. J., 262, *265*, 291, *301*
Roseman, S., 284, *286*
Rosenberg, J. C., 374, *377*
Rosenberg, J. M., 121, *125*
Rosenberg, R. N., 164, 165, *167*, *168*
Rosenbloom, J., 311, 312, *319*, 324, 325, 326, *336*, *338*, 339, 344, *347*, 349, 359, *363*, *364*
Rosenthal, M. S., 135, *137*

AUTHOR INDEX

Ross, R., 30, 31, 32, 36, *38*, *39*, 127, 136, *138*, 139, 140, 152, *154*, 155, *168*, 321, 322, *338*, *503*, *505*, 518, *518*, 525, *528*, *552*, 557, 627, 628, 629, 634, *635*, *636*, 641, *649*
Rothstein, A., 264, *265*
Rubenstein, A. H., 334, *338*
Rubin, A. L., 418, *420*
Rubin, H., 5, 7, 155, 163, 166, *168*, 267, 269, 270, 271, *272*
Ruhman, A. G., 581, *583*
Ruiz, E. R., 612, *614*
Ruiz-Torres, A., 548, *557*
Ruoslahti, E., 268, *272*
Rush, B. F., 374, *377*
Rush, J. D., 253, *260*
Russell, D. H., 195, *198*
Rutter, W. J., 89, 90, *91*
Ryan, G. B., 7, 139, 140, 141, 142, 144, 145, 148, 149, 151, 152, 153, *154*, 569, *570*
Rydell, N. W., 251, *252*

S

Sachs, G., 521, *524*
Sachs, L., 273, *286*
Saddington, K., 222, *229*
Saffiotti, U., 398, *410*
Sage, H. J., 331, *337*
Saito, H., 62, 63, *78*, 270, 271, *272*, 435, *438*, 443, *448*, 465, *471*
Sajdera, S. W., 123, *125*
Sakakibara, S., 313, 314, *319*, 324, *337*
Salcado, L. L., 19, *39*
Salpeter, M. M., 322, *338*
Samarina, O. P., *182*
Sandberg, L. B., 34, *39*
Sanders, T. G., 89, *91*
Sanderson, G. R., 431, *438*
Sandhoff, K., 457, *462*
Sanfilippo, S. J., 449, *462*
Sanger, J. W., 67, 68, 73, 75, 76, 77, *78*,
Santiago, R., 455, *462*
Sapolsky, A. J., 203, *203*
Sato, H., 329, *338*
Saunders, A. M., 611, *614*
Savolainen, H., 473, *479*
Saxen, L., 582, *583*
Saxena, I. D., *614*
Schafer, I. A., 399, 403, *410*
Schenkein, I., 324, *338*

Schiltz, J., 68, 69, *78*
Schmidt, A. J., 504, *505*
Schmitt, F. O., 321, *337*, 418, *420*
Schneider, R. B., 641, *649*
Schofield, J. G., 332, *338*
Schork, P., 136, *137*
Schramm, M., 323, *336*
Schrier, B. K., 164, *168*
Schubert, M., 61, *78*, 215, *230*, 447, *448*
Schulte-Holthausen, H., 63, *78*
Schultz, A. M., 284, *286*
Schumacher, G. F. B., 234, *236*
Schwert, G. W., 376, *377*
Scott, J. E., 421, 422, 423, 424, *429*
Scott, J. T., 579, *583*
Scott, R. D., 350, 356, 357, *363*
Scott, R. F., 152, *154*
Searls, R. L., 74, *78*, 80, 89, *91*, 96, *101*
Sefton, B., 267, 269, 270, 271, *272*
Segrest, J. P., 386, *394*
Selye, H., 121, *125*, 144, 148, *154*
Seppälä, M., 268, *272*
Serick, M. E., 449, *461*
Severtzov, A. N., *182*
Shea, S. M., 141, *154*
Sheinin, R., 273, *286*, *301*
Sheppard, J. R., 270, 271, *272*
Sherman, T. F., 220, 223, *229*
Shimokomaki, M., 390, 392, *394*, *395*
Shinano, S., 203, *203*
Shoshan, S., 389, *394*, 403, 404, *410*
Shotton, D. M., 14, *39*
Shpikiter, V. O., 316, *318*
Shudo, K., *319*, 324, *337*
Shulman, H. J., 61, *78*
Shuttleworth, A., 568, *570*
Siegel, L. M., 22, 24, *39*
Silbert, J. E., 432, 433, *438*
Silver, I. A., 507, 508, 509, 517, *518*, *519*, 521, *524*, 586, *589*, *589*, *590*, 591, *599*
Silverman, B. F., 121, *125*
Silverman, L., 399, 403, *410*
Simar, L. J., 382, *384*
Simons, K., 313, *318*, *319*, 366, *372*
Simpson, B. J., 103, *125*
Sinex, M., *38*
Sisson, J. C., 486, *499*
Sizer, I. W., 164, *167*
Sjöberg, I., 451, *461*
Sjoerdsma, A., 548, *556*

Slavkin, H. C., *101*
Slim, A. W., 581, *583*
Sloane-Stanley, G. H., 262, *265*
Sly, W. S., *461*
Smetana, K., 129, *138*
Smirnov, M. N., *182*
Smith, B. D., 19, *39*, 344, 345, *347, 364*
Smith, C., 67, *78*
Smith, D. W., 21, 34, *39*
Smith, R. W., 289, *301*
Smith, W. J., 331, *337*
Snellman, B., 218, *230*
Snowden, J. McK., 217, 220, 222, *230*
Snyder, S. H., 195, *198*
Søndergaard, J., 6, *7*
Spach, M. L., 444, *448*, 464, *471*
Spackman, D., 413, *416*
Speakman, P. T., 37, *39*, 339, *347*, 362, *364*, 418, *420*
Spector, W. G., 526, 527, *528*
Spencer, B., 476, *478*
Spiro, M. J., 314, *320*, 327, *338*
Spiro, R. G., 314, *320*, 326, 327, *338*, 385, 386, *395*
Spooner, B. S., 151, 153, *154*, 332, *338*
Spranger, J., 449, *462*
Springer, A., 332, *337*
Srinivasan, P. R., 177, *182*
Srivastava, S. K., 458, *462*
Stalder, K., 413, *416*
Starcher, B. C., 17, 18, 19, 27, 28, *39*
Stark, M., 315, *320*, 361, *364*, 379, 381, *384*, 387, *395*
Statkov, P. R., 7, 139, 140, 141, 142, 144, 149, 151, 152, *154*, 569, *570*
Stearns, M. L., 503, *505*
Stegemann, H., 413, *416*
Stehbens, W. E., 141, *154*
Stein, F., 647, *649*
Stein, W. D., 216, 228, *229*
Stein, W. H., 413, *416*
Steiner, D. F., 334, *338*
Steiner, R. E., 579, *583*
Stemerman, M., 627, 634, *636*
Stephens, F. O., 591, *599*
Steven, F. S., 418, *420*
Stevenson, R. E., 478, *479*
Stewart, J. A., 537, *537*
Stoker, M. G. P., 623, *625*
Stoolmiller, A. C., 86, *91*, 431, *438*
Strahs, K., 73, 76, 77

Strates, B. S., 121, *125*
Strauch, L., 420, *420*, 546, *557*
Strawich, E., 61, *78*
Strecker, F. J., 535, *537*
Strecker, G., 458, 460, *462*
Strehler, B. L., 184, *188*
Strobel, W., 612, *614*
Strominger, J. L., 475, *479*
Strong, J. P., 612, *614*
Strudel, G., 72, *78*, 93, 94, 96, *101*, 411, 413, *416*
Stumpf, D., 457, *462*
Styles, J. A., 529, *537*
Sudderth, S. B., 449, *461*
Sueoka, N., 174, *182*
Suga, K., 319, *319*
Sullivan, J. C., 399, 403, *410*
Summers, D. F., *300*
Summerson, W. H., 374, 377, *486*, *499*
Suzuki, S., 62, 63, *78*, 435, *438*, 443, *448*, 465, *471*
Sweeney, D. B., 251, *252*
Swoboda, B. E. P., *499*
Sylven, B., 218, *230*
Szarvas, F., 541, 544, 546, *556*, *557*
Szigeti, M., 640, 641, 642, *649*
Szirmai, J. A., 232, *236*, 441, 446, *448*

T

Takemoto, K. K., *301*
Takeuchi, T., 324, 325, *338*, 543, *557*
Tanese, R. L., 334, *338*
Tanzer, M. L., 15, *39*, 333, *336*, 339, 344, *346*, 387, 392, *394*, *395*
Tappel, A. L., 324, *337*, 350, 353, 360, *363*, 365, *372*, 374, 377, 569, *570*
Tarlo, L. B., 124, *125*
Tarrant, G. M., 187, *188*
Tartof, D., *499*
Taylor, B., 161, 163, *167*
Taylor, E. L. A., 151, 153, *154*, 332, *338*
Taylor, H. E., 611, *614*
Taylor, J. F., 151, *154*
Taylor, N. F., 292, *301*
Telser, A., 16, *39*, 67, *78*, 458, *462*
Theilen, G. H., 253, *260*
Temin, H. M., 628, *636*
Thomas, J., *420*
Thomas, J., 15, 16, *39*
Thomas, R. C., 507, *518*, 522, *524*
Thomas, W. A., 152, *154*, 627, *635*

Thompson, E. J., 164, *168*
Thompson, J., 455, *461*
Thorp, F. K., 61, *78*, 94, 96, *101*
Thunell, S., 441, *448*
Tiffany, J. M., 303, *308*
Timpl, R., 345, *347*, 380, 381, 382, *384*
Ting, R. C. Y., *301*
Tittor, W., 220, *229*
Tobey, R. A., 157, *168*
Todaro, G., 160, *168*, 399, *410*, 628, *636*
Todd, P., 288, *301*
Tofft, M., *372*
Toole, B. P., 61, *78*
Torloni, H., 150, 151, *154*
Toth, S. E., 183, *188*
Trelstad, R. L., 61, *78*, 79, *91*, 322, *338*
Trinkhaus, J. P., 43, *58*
Trueheart, R. E., 122, *125*
Tsai, R. L., 315, 317, *319*, 340, 344, 345, *347*
Tudball, N., 451, *462*
Turk, V., 203, *203*
Turto, H., 603, 609, 610, 611, *614*
Twomey, P., 517, *518*
Tyler, R., 152, *154*, 518, *518*, 525, *528*
Tyrrell, D. A. J., 307, *308*

U

Udenfriend, S., 313, *319*, 324, 325, *337*, *338*, 349, 350, 353, 359, 360, *363*, *364*, 365, 366, 370, 371, *371*, *372*, 373, 374, 377, 408, *410*, 544, 557, 569, *570*
Udupa, K. N., 611, *614*
Uhr, J. W., 324, *338*
Uitto, J., 311, 315, 316, 317, *319*, *320*, 548, 557, 560, 570, 603, 609, 610, 611, *614*
Urist, M. R., 121, *125*
Urry, D. W., 27, 28, *39*
Utermann, D., 611, *614*

V

Vaes, G., 419, *420*
Vaheri, A., 267, 268, 269, 271, *272*
Vail, T. M., 161, 163, *167*
Välimäki, M., 178, *182*
Van der Laan, E. J., 612, *614*
Van Hoof, F., 449, 450, *462*, 476, *479*
Van Melsen, J. A., 612, *614*
Van Slyke, D. D., 365, *372*, *395*

Varga, F., 542, *557*
Vastamäki, M., 178, *182*
Velo, G. P., 526, *528*
Vencelj, H., 420, *420*
Veress, B., 641, *648*, *649*
Verzar, F., *395*
Viljanto, J., 169, *182*, 205, *208*
Virijonandha, J., 299, *301*
Virtama, P., 579, *583*
Vogt, P. K., 270, 271, *272*
Von der Mark, K., 315, 316, 317, *318*, 327, 328, *336*, *346*
Voss, H., 287, *300*, *301*
Vuust, J., 312, *320*, 340, 344, 345, *347*

W

Wagner, B. M., 418, *420*
Wagner, H., 518, *518*, 557, 647, *648*
Waite, M. R. F., 308, *308*
Walford, R. L., 641, *649*
Walker, J. L., 507, *519*, 522, *524*
Wallach, D. F. H., *265*
Walther, B. T., 90, *91*
Wang, J. H., 216, 218, *230*
Ward, G. A., 157, *168*
Waring, W. W., 597, *599*
Warmsley, A. M. H., 261, 262, *265*
Warren, G., 435, *438*
Warren, L., 274, 275, 280, 281, 282, 283, 284, 285, *286*, 295, *300*, *301*
Wartiovaara, J., 582, *583*
Watson, H. C., 14, *39*
Watterson, R. L., 93, *101*
Webster, I., 530, 532, *537*
Weinstein, E., 314, *319*
Weinstein, I. B., 176, *182*
Weintraub, H., 63, 64, 67, 72, 73, 75, 76, 77, *78*
Weis-Fogh, T., 13, 29, *39*
Weiss, J., 418, *420*
Weiss, P. H., 504, *505*
Weissbach, H., 380, *384*
Weissman, N., 21, 34, *39*
Weissmann, B., 451, 455, *461*, *462*
Wiesmann, U., 450, *462*
Werrlein, R. J., 156, 159, 160, *167*, 507, 518, *519*
Wessels, G., 647, *648*
Wessels, N. K., 151, 153, *154*, 332, *338*
Westermark, B., 183, *188*

White, F. H., 292, *301*
White, R. J., 291, 298, 299, *301*
Whittle, W. L., 160, *168*
Wick, G., 380, 381, 382, *384*
Williams, G., 641, *649*
Williams, J. A., 323, 324, 329, 332, *338*
Williams, R. J. P., 292, 299, *300*
Wilson, S. H., 164, 165, *167*, *168*
Wilt, F. H., 89, *91*
Wichterle, O., 103, *125*
Winer, A. D., 376, *377*
Winter, G. D., 103, *125*, 589, 590, *590*
Winterburn, P. J., 288, *301*, 326, *338*, 386, *395*
Winzler, R. J., 5, 7, 287, 288, 289, *300*, *301*
Wirth, W., 6, *7*, *557*
Wise, D., 449, *461*
Wissler, R. W., 552, *557*
Woessner, Jr., J. F., 199, 203, *203*, 418, *420*, 569, *570*
Wolff, J., 323, 324, 329, 332, *338*
Wollman, S. H., 323, *338*
Wood, G. C., 209, *211*
Wood, S., 508, *519*
Wrenn, J. T., 151, 153, *154*, 332, *338*
Wright, D., 483, 486, *499*
Wright, J., 299, *301*
Wright, N. A., *537*
Wu, H. C., 273, 285, *286*, 295, *301*
Wünsch, E., 546, *557*
Wyatt, J. P., 535, *537*

Wyke, A. W., 315, 316, 317, *318*, 327, 328, 331, *336*, *346*, 361, *363*
Wynne-Roberts, C. R., 490, *499*

X

Xavier, A., 292, 299, *300*
Xavier, A. V., 291, 292, 299, *300*, *301*

Y

Yamada, K. M., 151, 153, *154*, 332, *338*
Yamada, T., 205, *208*
Yamagata, T., 62, 63, *78*, 435, *438*, 443, *448*, 465, *471*
Yamashina, I., 475, *479*
Yaron, M., 484, *499*
Yeh, J., 166, *168*
Young, A. E., 531, *537*
Young, D. A., 323, *337*

Z

Zagury, D., 324, *338*
Zahim, D., 181, *182*
Zahradník, R., 336, 367, 368, *372*
Zederfeldt, B., 517, *518*
Zembala, M. 232, *236*
Zerner, M., 366, *372*
Zeuthen, T., 522, *524*
Zor, N., 323, *338*
Zumbo, O., 627, *635*
Zwilling, E., 74, *78*

Subject Index

A

Acetylcholinesterase in fibroblasts, 164
N-Acetylhexosaminidases, 440, 450, 451
Acid mucopolysaccharides, 424
  in Hunter syndrome, 463 et seq.
  intracellular localization, 421
Actinomycin D, 190, 197
Activation
  of connective tissue, 4, 483 et seq.
  of enzymes at wound healing, 503
  of oxygen in hydroxylation of proline, 366
  of protocollagen proline hydroxylase, 373
  of synovial cells, 486
Activity of water, 11
Actomyosin in granulation tissue, 144
Adenosine triphosphatase, 261
  in liver, effect of granuloma, 180
Ageing, 658
  cellular, 183
  and crosslinks, 370
  and fibrosis, 552
Aggregation of cells and glycosaminoglycans, 253 et seq.
  inhibition of, 259
Alcian blue, 422
Aldimine bonds and crosslinks, 387
Allografts, hyaluronic acid effect on, 247
Alveolar epithelial cells, 530
Alveolitis, fibrosing, 527
Amino acid acceptors in granuloma, 175
Aminopeptidases and wound healing, 501
Amino sugar metabolism, control of, 288
  intermediates, 288
Amino sugars, toxicity, 287
  transport, 287
Amyloidosis and wound healing, 581
Angiotensin, 145
Antibody, against collagen, 508, 514

Antigenic determinants, carbohydrate, 287
Anti-inflammatory drugs, 493
Antimycin A in wound respiration, 586
Anti-serotonin agents, 145
Aorta, macromolecular structure, 637
  smooth muscle cells, 627
Arterial (see also Vascular)
Arterial injury, 602
Arteriosclerosis, 6, 602, 627
  immune, 645
Arteritis and wound healing, 579
Ascorbic acid, 321, 326, 389, 397 et seq., 568
  and protocollagen hydroxylation, 313, 366, 399
N-Aspartyl-β-glycosaminidase, 473
Aspartylglycosaminuria, 473 et seq.
Autoantibodies against fibroblasts, 189
  against smooth muscle, 145

B

Binding sites, 299
Biosynthesis, see the specific substances
Bladder, wound healing, 560
Blastoid modulation, suppression by hyaluronic acid, 243
Blood group substances, 287
Bone, heterotopic, 109 et seq.
  induction by sponge, 103 et seq.
Bradykinin, 6, 144
Bromodeoxyuridine, 62 et seq., 82 et seq.

C

Calcification of sponge implant, 104
Calcium in induction of bone, 104
Cartilage, differentiation, 61, 79, 93
  laryngeal, 231
Cathepsins, 657
  in collagen degradation, 418
Cathepsin D in granuloma, 199 et seq.

SUBJECT INDEX

Cell culture, over agar, 80
  synthesis of collagen, 399
  density-dependent enzymes, 164
  of foetal blood cells, 127
  at low density, 80
  of mesenchymal cells, 80
  of chondroblasts, 232
  of smooth muscle cells, 628
  of synovial cells, 484
  of transformed fibroblasts, 273
Cell-cycle age, 184
Cell cycle, effect of putrescine, 197
Cell cycle of fibroblast, 261
Cell density and protocollagen proline hydroxylase, 373
Cell, in extracellular space, 9
  motility, 47
  movement, 139
Cell surface, 2
  antigens, 287
  of cultured cells, 267, 273
  glycoproteins of, 248, 249
  and proliferation, 267
Chick embryo fibroblast, cell surface of, 267
Chondroblast, 61 *et seq.*, 69 *et seq.*, 653
  effects of bromodeoxyuridine, 61, 79
Chondroblasts, glycosaminoglycan synthesis, 61, 79
  collagen synthesis, 61
  transformation into fibroblasts, 62, 79
Chondrocyte, adult, 231
Chondrogenesis, 72, 79
Chondroitin sulphates, 79, 83, 451, 653
  and cell aggregation, 258
  in hypertrophic scars, 575
  synthesis of sulphated glycosaminoglycans, 234
  in vascular damage, 608
  in cell aggregation, 258
  and diffusion, 215 *et seq.*
Chromosome, antigens in, 189
Cirrhosis of liver, 5
Colchicine, 323, 347
Collagen, 655
  ageing *in vitro*, 136
  in aorta, 601
  biosynthesis, 61, 68, 311 *et seq.*, 339, 373, 379, 397
  in cartilage, 61, 62, 68, 79
  in chondrogenesis, 94

Collagen (*contd.*)
  CNBr peptides from, 356
  and connective tissue activation, 486
  degradation, 417
  embryonic, 393
  fibres, formation, 133, 380
    stabilization, 385
    for cell culture, 209
    in periaxial organs of chick embryo, 413
  fractions and fibrosis, 541
  and inflammation, 526
  intracellular forms, 349
  intracellular translocation, 321
  in keloids, 571
  lattices for cell culture, 209
  in liver fibrosis, 541
  macromolecular assembly, 362
  proline hydroxylase (*see* Protocollagen proline hydroxylase)
  secretion, 321
  synthesis, by foetal blood cells, 127
    regulated by cell density, 160
    in smooth muscle cell culture, 634
    and the supply of oxygen, 585
    rate of, 559, 562
  underhydroxylated, 349, 358
  in vascular repair, 618
Collagenase, 402
  effect on chondrogenesis, 98
  in collagen degradation, 417, 546
  inhibition of multilayering, 45
  in isolation of cells, 206
  and periaxial microfilaments, 414
Colloidal silica, in isolation of cells, 205
Colon, normal healing, 560
Concanavalin A and synthesis of glycosaminoglycans, 234
Connective tissue, activating peptide, 483 *et seq.*, 658
  isolation of cells, 205
  enzymes and wound healing, 501
Contact inhibition, 4, 43, 57, 183
Contractures of connective tissue, 146, 150
Critical electrolyte concentration, 421
Critical PO_2, 585
Crosslinks, ageing, 390
  aldimine bonds, 387
  and ascorbic acid, 403
  biosynthesis of, 14

SUBJECT INDEX

Crosslinks (contd.)
  borohydride reduction of, 18
  in collagen, 386
  in elastin, 15
  inhibition in copper deficiency, 21
  in lathyrism, 37
Culture, dense in suspension, 156
  long-term maintenance, 126
  on collagen, 209
  suspension vs. attached, 155
C vitamin, see Ascorbic acid
Cyclic AMP, in collagen synthesis, 323, 331
  and activation of connective tissue, 494
  in cultured fibroblasts, 270
  in plasma membrane, 267
Cycloheximide, 67, 197
Cystic fibrosis, 456
Cytochalasin B, 145
  in secretion of collagen, 323, 331
Cytochrome oxidase and wound healing, 501
Cytotoxicity, inhibition by hyaluronic acid, 247

D

Deamination of heparin, 434
Degradation of collagen, 417 et seq.
Degradation of glycosaminoglycans, 440, 449, 454
  in Hunter syndrome, 463
Density gradient, with colloidal silica, 205
  with Ficoll, 232, 261
Dermatan sulphate, 451
  biosynthesis, 85, 439
  in cell aggregation, 258
  degradation, 440, 470
  in Hunter syndrome, 463
  in hypertrophic scars, 575
  secreted vs. extracellular, 443
  in vascular damage, 608
Dermatosparaxis in cattle, 316, 333, 341, 380
  in sheep, 318
Desmosine, 15 (see also Crosslinks, Elastin)
Deuterium oxide, 327
Dextrans and synthesis of glycosaminoglycans, 234

Differentiation, 653
  and ageing, 183
  of chondroblasts, 62, 79, 93
  of foetal blood cells, 127
  in synthesis of collagen, 408
Diffusion in extracellular systems, 215 et seq.
Disulphide bonds in pro-α chains, 345
DNA, nucleotide turnover in, 86, 87
  synthetic cycle, 658
  synthesis, effect of putrescine, 197
    regulated by cell density, 157
  in vascular damage, 609
DNAase, in isolation of cells, 206
Drugs and connective tissue activation, 494
  and wound healing, 581
Dupuytren's disease, 150

E

Ear chamber, 507
Ehlers-Danlos' syndrome, 326
Elastase, 14
Elastic fibre, amorphous component, 32
  glycoprotein component, 32
  in smooth muscle cell culture, 634
  in vascular repair, 618
Elasticity, rubber-like, 13
α-Elastin, 25, 26, 27
Elastin, absorption of solutes, 22
  biosynthesis, 13 et seq.
  corpuscular structures, 24
  electron microscopy, 30
  macromolecular organization, 13 et seq.
  microfibrils, 38
  partial hydrolysis of, 25
  soluble precursors, 34
Elastogenesis, 33
Elastomers, oil-droplet, 29
Endoglycosidases, 440
Endoplasmic reticulum, in collagen synthesis, 321
  enzyme markers of, 261
Endothelium of aorta, 606, 617, 634
Enzymes (see the specific names)
Environment, cellular, 9
  assays on, 522
Epimerizations in glycosaminoglycans, 431, 439, 440, 470
Epinephrine, effect on myofibroblast, 145

Epitheloid cells from foetal blood, 131
Esterases and wound healing, 501
Ester sulphates, formation of iduronic acid, 434, 437
Exoglycosidases, 440
Experimental connective tissue, 199, 205 (*see also* Granulation tissue)
Extracellular matrix, 657
and diffusion, 215 *et seq.*
Extracellular microfilaments, 411
Extracellular space, 9

F
Fabry's disease, 456
Ferrous ions in the hydroxylation of proline, 313, 368
α-Fetoprotein, 268
desialylated, 284
Fibroblast, 653 (*see also* Cell)
adhesions to collagen, 210
aggregation of, 255
antigens, 189
arrays in culture, 43
various forms, 42
vs. foetal blood cells, 127
vs. chondroblast, 61
as a contractile cell, 139
cultures, from aspartylglycosaminuria, 471
on collagen matrix, 209
from mucopolysaccharidoses, 424, 450
dermatan sulphate biosynthesis, 439
collagen biosynthesis, 399
and macrophage extract, silicosis, 530
effects of putrescine, 195
in healing wounds, 509
and inflammation, 525 *et seq.*
localization of collagen, 321
lipid metabolism of, 303
microenvironment, 507
morphology in culture, 210
motility in culture, 47
plasma membranes, 261
proliferation, 267, 507
redox state, 517
repair, 559 *et seq.*
respiratory pattern, 585
secretion of collagen, 321
vs. smooth muscle cells, 635

Fibrosis, 5, 397, 527, 539
and silica, 530
Fixation of polyanions, 424
of counter ions, 222
N-Fluoroacetylglucosamine, 291, 299
Fluorosugars, toxicity, 292
L-Fucose, incorporation of, 274

G
Gangliosidosis, 456
Genetic defects, 449 (*see also* specific diseases)
Giant cells, 617
Glia cells, 183, 255
Glucose-6-phosphatase in liver, effect of granulation tissue, 180
Glucuronic acid in glycosaminoglycans, 431, 439, 465
β-Glucuronidase, 440, 452, 470, 548
and wound healing, 501
Glycolipid biosynthesis, 295
Glycoproteins, of cell surface, 273
of plasma membrane, 287
in chondrogenesis, 94
Glycosaminoglycans, and ascorbic acid, 399
biosynthesis, 231, 439
in chondrogenesis, 94
degradation, 449, 548
and diffusion, 215 *et seq.*
in hypertrophic scars, 571
intracellular localization, 421
sulphated, biosynthesis, 61, 65, 231 *et seq.*, 431, 439, 544
in vascular damage, 608
Glycosylation of hydroxylysine, 314, 326, 406
Glycosyltransferases, 86
Golgi complex in collagen synthesis, 321
Graft *vs.* host reaction, hyaluronic acid effect on, 245
Granulation tissue, 169, 653, 659
amino acid acceptors, 175
hyaluronic acid effect on, 250
nucleic acid metabolism, 169 *et seq.*
proteolytic enzymes in, 200
in vascular damage, 601
Granuloma, experimental, 169, 418
Granuloma, chronic, 525
Griseofulvin, conversion of procollagen, 329

SUBJECT INDEX

Ground substance and diffusion, 215 *et seq.*
Growth, density-dependent inhibition of, 164, 267
Growth-promoting factors, in arterial tissue, 623
  putrescine, 195
Growth-regulation, density dependent, 155 *et seq.*
  transport associated, 166

H

Haemorrhagic shock and connective tissue, 511
Healing of wounds (*see* Wound healing)
Heparan sulphate, ingenetic defects, 451, 463
  in cell aggregation, 258
  in vascular damage, 608
Heparin, 452
  biosynthesis, 431
  in cell aggregation, 258
  degradation, 434, 435
Heritable disease (*see* specific diseases)
Heterotopic bone, 109
Hexobarbital metabolism, 180
Histochemistry of wound enzymes, 501
Histamine, 6, 144
Hunter's syndrome, 449, 463
  excretory products, 463
Hurler's syndrome, 424, 449
Hyaluronic acid, 452, 653
  in cell aggregation, 253, 258
  and chondroblast, 61, 65
  connective tissue activation, 486
  and diffusion, 215 *et seq.*
  cell surface glycoproteins, 248
  effect on fibroblasts, 249
  effect on graft *vs.* host reaction, 245
  effect on granulation tissue, 250
  effect on lymphocyte cytotoxicity, 247
  effect on lymphocyte stimulation, 243
  effect on macrophage migration, 238
  effect on wound healing, 249
  metabolism in synovial cells, 483
  receptor for, 260
  skin allografts, 247
  and synthesis of sulphated glycosaminoglycans, 233, 234
  synthesis, 83
  therapeutic uses, 251
  in vascular damage, 608

Hyaluronidase, 451, 470
  effect on chondrogenesis, 99
  in isolation of cells, 206
Hybridization of RNA with DNA, 172
"Hydron", 104
Hydrophobic bonding of elastin, 27
Hydroxyapatite, 104
Hydroxylation, 655 (*see* Protocollagen proline and lysine hydroxylases)
  in biosynthesis of collagen, 324, 361
  effect of ascorbic acid, 403, 406
  of protocollagen lysine, 312, 349
Hydroxylation of protocollagen proline, 312, 349
Hydroxylysine, carbohydrate attachment, 311, 385
  in collagen stabilization, 311, 385
  formation of crosslinks, 311, 387
Hydroxyproline, 399
  formation and role of, 311
Hypertrophic scars, 5
Hypovolemia and wound healing, 511, 514
Hypoxemia and wound healing, 579
Hypoxia and vascular damage, 609

I

"I"-Cell disease, 450, 455
α-L-Iduronidase, 440, 454
L-Iduronic acid, in biosynthesis of heparin, 431
  in dermatan sulphate, 439, 470
  sulphatase, 460
Inborn errors of metabolism, 655 (*see also* specific diseases)
Incorporation of acetate, 86
  of fucose, 274
  of galactose, 643
  of glucuronic acid, 432
  of hexosamine, 63, 64, 290, 298
  of leucine, 83
  of lysine, 645
  of proline, 69 *et seq.*, 161, 298, 328, 340, 353 *et seq.*, 414, 560
  of sulphate, 86, 208, 233, 444, 544
  of threonine, 290
  of thymidine, 83, 184, 197, 262, 269, 299, 551
  of uracil or uridine, 83, 299
Induction in chondrogenesis, 72

Inflammation, 483
  and fibroblast, 525 et seq.
Inflammatory cells, 658
Inflammatory response, vascular repair, 616
Influenza virus, effect on lipid metabolism, 303
Inherently precise process of cell array, 54
Injury, experimental, of aorta, 602, 603
Insulin, effect on cell cultures, 267
Intimal thickening of arteries, 605, 617
Intravascular lesions, 603, 616, 634
N-Iodoacetylglucosamine, 291
Ion movement in extracellular matrix, 220 et seq.
Isodesmosine, 15 (see also Elastin and Crosslinks)

K

Keratan sulphates, 452
α-Ketoglutarate in hydroxylation of proline, 313, 368
Keloids, intercellular matrix, 571

L

L-cell, acetylcholinesterase regulation, 164
  preferential synthesis of collagen, 156
Lactate, effect on proline hydroxylase, 373
Lathyrism, 69, 354, 398
Limb bud mesenchyme, 80
Limb chondrogenesis, 96, 98
Limiting nutrients, microenvironmental, 166
Lipids and connective tissue, 6
Lipid metabolism, effects of virus, 303, 305
Lipids in silicosis, 530
Liver fibrosis, experimental, 539
Liver enzymes, effect of granuloma, 180
Liver phospholipid, 181
Liver protein, 181
Liver RNA, 181
Lung, silicosis in, 530, 586
Lymphocyte stimulation, effect of hyaluronic acid, 243
Lysine in crosslinks, 387
  hydroxylation, 406

Lysinonorleucine, 18, 407
Lysosomal diseases, 473 (see specific diseases)
Lysosomal enzymes, 473, 657 (see specific enzymes)
Lysosomes, secondary, 187
  ultrastructure in aspartylglycosaminuria, 474
Lysozyme, 292
Lysozyme and synthesis of glycosaminoglycans, 234
Lysyl oxidase, 403 (see also Crosslinks)
Lysyl protocollagen hydroxylase, 314 (see also Protocollagen lysine hydroxylase)

M

Macromolecular assembly of collagen, 362
Macrophage, 659
  and inflammation, 527
  and silicosis, 526, 530
Mastocytoma cells and glycosaminoglycans, 432
Matrix in hypertrophic scars, 571
Medical applications of connective tissue reactions, 5
Membrane potential and sodium leakage, 516
Merodesmosine, 18 (see also Elastin, Crosslinks)
Metachromasia in aorta, 606
Methylation of nucleic acids, 177
Microelectrodes (O_2, CO_2), 522
Microtubules, 656
  secretion of collagen, 324, 327
Migration of cells, effect of hyaluronic acid, 238
Mitochondria, damage, 516
  of fibroblast, 261, 262
Mitosis, formation of antigens, 189
  regulation by cell density, 157
  stimulated by α-fetoprotein, 269
Molecular exclusion by elastic fibres, 22
Monoamine oxidase and wound healing, 501
Messenger RNA in granulation tissue, 175
Multilayering of cells in culture, 45
Myo-fibroblast, 139 et seq.
  electron microscopy, 141

SUBJECT INDEX

N

NADH-oxidase, 262
Neuraminic acid, 4
Neuraminidase, effect on cell cultures, 267, 281
    effect on chondrogenesis, 99
Nitrous acid in degradation of heparin, 435
Notochord, 95
Nuclear magnetic resonance of fluoro-sugars, 292, 299
Nucleic acids during the development of granulation tissue, 169 *et seq.*
Nucleo-cytoplasmic protein migration, 189
Nucleolar segregation, 190
Nucleolus-specific antigens, 189
5'-Nucleotidase, 261

O

Oligosaccharides in glycoproteins and glycolipids, 287
OOH particle in hydroxylation of proline, 367
Osmotic pressure, 11
Osteogenesis, 103 *et seq.*
Oxamic acid and proline hydroxylase, 375
Oxygen, consumption by wounds, 586
    gradient in wounds, 585
    intoxication, 5
    measurements, 592
    supply, and collagen synthesis, 588
    in wounds, 585
    tension in fibroblast environment, 509
    therapy, 595, 596

P

Papain, 268
Papaverine, 144
Patchwork of cells in culture, 44, 45, 47, 48
Penicillamine and fibrosis, 548
Peptide, connective tissue activating, 484 *et seq.*
Periaxial extracellular material in embryos, 411
Perineural metachromatic material, 93
Permeability of tissue, 216 *et seq.*
Phagocytes, and hyaluronic acid, 249
Phagocytosis of silica particles, 530

Phases of granulation tissue, 169
3'-Phosphoadenylylsulphate, 431
Phospholipids, in plasma membrane, 261
    in silicosis, 530
Phthalocyanins, 421
Physicochemical properties of elastin, 21
Phytohaemagglutinin and synthesis of glycosaminoglycans, 234
Pigs, formation of bone, 104
Plasma infusion in Hunter syndrome, 463
Plasma membrane of fibroblasts, 4, 261, 267, 273
Pluripotential cells, 136
Polyelectrolytes and diffusion, 216 *et seq.*
Polyhydroxyethylmethacrylate (poly-HEMA), formation of bone, 104
    methacrylic acid co-polymer, formation of bone, 115
Polymerization of collagen, 380
Polysaccharide network and diffusion, 215, 216 *et seq.*
Polyurethane, reactions in sponge, 116
Polyvinyl alcohol, reactions in sponge, 116
Pore size of sponge, bone formation, 116
Precursors of collagen, 315, 328, 339, 381
Prednisolone and fibrosis, 540
Pro-α chains, 317, 327, 339, 349
Procollagen, 313, 316, 327, 339, 380, 655
    conversion, 360
    peptidase, 327, 340, 380, 655
Proline, hydroxylation of, 313 *et seq.*, 365, 403, 543
Pronase, in study of cell surface, 275
Propanolol, and connective tissue activation, 492
Prostaglandins, 6, 145
Prostheses, 124
Protamine, 235
Proteolytic enzymes, and collagen, 418
    in experimental granuloma, 199
Protocollagen, 312, 344, 349
Protocollagen proline hydroxylase, 365, 373, 408
    in aorta, 609
    and fibrosis, 543
    active complex, 366
    electron absorption spectra, 369
Pseudomonas fluorescens, 292
Putrescine, 195

Q

Quantum chemical calculations on hydroxylation of proline, 367

R

Redox state in fibroblasts, 517
Reparation, 5, 559, 617, 658
  in aorta, 603
Reparative tissue, respiratory pattern, 585
Resilin, 13
Rheumatoid arthritis, and inflammation, 5, 483
  wound healing, 579
RNA, and collagen, 311 (see also specific forms)
  effect of putrescine, 197
  extraction, 169
  in granulation tissue, 169 et seq.
  high molecular weight, 170, 173

S

Sanfilippo's syndrome, 450
Sandhoff disease, 459
Scars, matrix of, 571
Scheie's syndrome, 450
Sclerosis, diffuse systemic, 527
Secretion of collagen, 321, 655, 656
Senescence, 184 (see also Ageing)
Serotonin, effect on myo-fibroblast, 6, 144
Sialic acid, 282 (see also Cell surface, Glycoproteins, Neuraminic acid)
Sialoglycoproteins, in scars, 574
Sialyl transferases, 282
Silica, colloidal in gradients, 205
  biological response, 529 et seq.
Silicosis, 5, 397, 526
Skin, dermatosparactic, 380
  normal healing, 560
Slippage of cells in culture, 56
Smooth muscle auto-antibody, 145
Smooth muscle cell, 627, 630, 653
  proliferation, 617
  vascular repair, 617
Somites and chondrogenesis, 72
Spatial factors in tissue differentiation, 123
S-period, effect of putrescine, 197
Sponge, effect on formation of bone, 103
  viscose cellulose, 169
Stomach, wound healing, 560
Strength (breaking) of wounds, 559
Succinate dehydrogenase and wound healing, 501
Sugar uptake of cultured fibroblasts, 269
Sulfaminogroup, relation to formation of iduronic acid, 432
Sulfhydryl group in hydroxylation of proline, 365
Syndesine, 407 (see also Crosslinks)
Synovial cells, metabolism, 483
  normal and rheumatoid, 483
Synovial fluid, effect on cell migration, 240

T

Tay-Sachs disease, 459
Tendon collagen and cathepsin, 418
Tension of O_2 and CO_2, 513
Thioacetamide, 540
Tissue culture, 127, 511
  collagen synthesis, 373
Tissue PO_2 and PCO_2, 591 et seq.
Transcellular movement of collagen, 321, 329
Transformation, absence of, 183
  chondroblasts into fibroblasts, 62
  viral, 273
Transport form of collagen, 316
Transport in proteoglycan solutions, 216 et seq.
Trifluoroacetic acid in degradation of heparin, 434
Tropocollagen from collagen, 419 (see also Collagen)
Tropoelastin, 34
Trypsin, 268
  in isolation of cells, 206
  in study of cell surface, 274

U

UDP-glucuronic acid, and dermatan sulphate, 439
  and glycosaminoglycans, 431
UDP-glycosyl transferases, 314
UDP-iduronic acid in heparin biosynthesis, 431

Underhydroxylated collagen, 312, 324
Urinary polysaccharides in Hunter syndrome, 463
Uronic acids in glycosaminoglycans, 431

V

Vascular wall, 659
  ageing and fibrosis, 552
  endothelium, 617
  injury, 617
  lesions, scheme of development, 622
Vasopressin, 145
Vertebral chondrogenesis, 94, 96
Vinblastine, conversion of procollagen, 329
Viral effect on lipid synthesis, 303, 305
Viral envelope, lipid composition of, 303
Vitreous humour, effect on cell migration, 240

W

Water, 11
Wound, 658
  in cartilage, effect of hyaluronic acid, 249
  contraction, 146
  healing, 149, 501, 559 *et seq.*
    and arteritis, 579
    and fibroblast proliferation, 597
    in rheumatoid arthritis, 579
  effect of hyaluronic acid, 249
  strength, 559 *et seq.*

Z

Zonal centrifugation of subcellular particles, 262